Number Theoretic Methods

Developments in Mathematics

VOLUME 8

Series Editor:

Krishnaswami Alladi, *University of Florida, U.S.A.*

Aims and Scope

Developments in Mathematics is a book series publishing

(i) Proceedings of Conferences dealing with the latest research advances,

(ii) Research Monographs, and

(iii) Contributed Volumes focussing on certain areas of special interest.

Editors of conference proceedings are urged to include a few survey papers for wider appeal. Research monographs which could be used as texts or references for graduate level courses would also be suitable for the series. Contributed volumes are those where various authors either write papers or chapters in an organized volume devoted to a topic of special/current interest or importance. A contributed volume could deal with a classical topic which is once again in the limelight owing to new developments.

Number Theoretic Methods

Future Trends

Edited by

Shigeru Kanemitsu

*Graduate School of Advanced Technology,
University of Kinki, Iizuka, Japan*

and

Chaohua Jia

*Institute of Mathematics,
Academia Sinica, Beijing, China*

KLUWER ACADEMIC PUBLISHERS
DORDRECHT / BOSTON / LONDON

A C.I.P. Catalogue record for this book is available from the Library of Congress.

ISBN 978-1-4419-5239-4

Published by Kluwer Academic Publishers,
P.O. Box 17, 3300 AA Dordrecht, The Netherlands.

Sold and distributed in North, Central and South America
by Kluwer Academic Publishers,
101 Philip Drive, Norwell, MA 02061, U.S.A.

In all other countries, sold and distributed
by Kluwer Academic Publishers,
P.O. Box 322, 3300 AH Dordrecht, The Netherlands.

Printed on acid-free paper

Contents

vi

Preface

This volume contains the proceedings of the very successful second China-Japan Seminar held in Iizuka, Fukuoka, Japan, during March 12-16, 2001 under the support of the Japan Society for the Promotion of Science (JSPS) and the National Science Foundation of China (NSFC), and some invited papers of eminent number-theorists who visited Japan during 1999–2001 at the occasion of the Conference at the Research Institute of Mathematical Sciences (RIMS), Kyoto University.

The proceedings of the 1st China-Japan Seminar held in September 1999 in Beijing has been published recently (2002) by Kluwer as DEVM 6 which also contains some invited papers. The topics of that volume are, however, restricted to analytic number theory and many papers in this field are assembled.

In this volume, we return to the lines of the previous one "Number Theory and its Applications", published as DEVM 2 by Kluwer in 1999 and uphold the spirit of presenting various topics in number theory and related areas with possible applications, in a unified manner, and this time in nearly a book form with a well-prepared index. We accomplish this task by collecting highly informative and readable survey papers (including half-survey type papers), giving overlooking surveys of the hitherto obtained results in up-to-the-hour form with insight into the new developments, which are then analytically continued to a collection of high standard research papers which are concerned with rather diversed areas and will give good insight into new researches in the new century.

Survey papers range over Transcendence Theory (Nagata, on G-functions), Probabilistic Number Theory (Elliott, on the product representation by rationals, Li and Pomerance, on "primitive roots" modulo n), History of Number Theory (Miyake, on the history of interaction between algebraic number theory and analytic number theory and Pan, on the Goldbach-Vinogradov theorem in short intervals), Ramanujan's Work and Automorphic Forms (Berndt and Yee ,on Ramanujan's contributions of Eisenstein series, and Kanemitsu-Tanigawa-Yoshimoto, on Ramanujan's formulas and modular forms, both are half-survey papers), Analytic Number Theory (Jia, on the distribution of α_p, p ranging primes) and Diophantine Equations (Darmon and Levesque) while research papers range over Arithmetic Algebraic Geometry (Deninger, on explicit formulas in analytic number theory and Wen, on non-abelian zeta-functions), Transcendence Theory (Amou and Matala-aho, on the arithmetic nature of solutions of q-difference equations, Bundschuh, on applications of Galochikin's result, and Noguchi, on applications of Nevanlinna theory), Analytic Number Theory (Cai, on partitions modulo prime powers, Matsumoto and Weng, on the zeta-function of polynomials, Nakai, on Weyl sums as finite theta series, and Zhang, on hybrid mean values of L-functions), Partition Theory (Alladi and Berkovich), Automorphic Froms (Kohnen, on Waldspurger's formulas and central critical values of L-functions, Berndt and Yee, on Ramanujan's contributions of Eisenstein series, Kanemitsu-Tanigawa-Yoshimoto, on Ramanujan's formulas and modular forms) and two papers of authors who have physical background (Srinivasa Rao, on quantum angular momentum, and Shcherbina and Tirozzi, on neural network)

Now some passages about the seminar itself are in order. The length of time spent for preparation notwithstanding, the seminar itself elapsed away in an instant as with all other pleasant events. Our present feeling is best expressed in the following Haiku: "Omoshiroute yagate kanashiki hanabikana", which in English would read "Fireworks, momentary flowering in the darkness, followed by utter vacuousness".

We received several (e-mail) messages to the effect that it was a successful and memorable one as well as oral communications expressing their high-spiritedness in attending the seminar. We hope everyone felt in the same way and that this was a pleasant occasion and you got new ideas or expansion of your research fields.

Hereby we wish to express our hearty thanks to the great help given to us by Dean of University of Kinki, Kyushu School of Engineering, Professor K. Kikukawa whose presidential address given half in Chinese was very impressive. Although the second editor expressed his thanks to his ex-students for their devoted help before and during the seminar, we feel we are to thank them again for their unusually efficient help: Drs. H. Kumagai and M. Yoshimoto, Ms. K. Mashimo and Ms. K. Wada. Without their utmost painstaking effort, we could not have made it. We wish to thank Dr. Yoshimoto also for his devoted help in editing many files of papers in this volume.

The last but not the least thanks are due to Ms. Chiaki Yokoyama of the JSPS for her timely and most appropriate advices, and above all her patience before, during and after the seminar. Also we thank all the participants for their active contribution to the seminar and the authors of papers in this volume for contributing their important works before the deadline. Finally we wish to thank all anonymous referees for their thorough scrutinizing of the papers and for their important comments which improved some of the papers greatly. As usual we should express our thanks to Kluwer Academic Publishers for publishing this volume, and to the editor John Martindale and his secretary Angela Quilici for their devoted help.

Let's hope to meet again in another conference in 5 years (which is the length of time needed before the first editor will get energetized again) in Japan. Many of us will meet at the next China-Japan Seminar to be held in Xi'an in September, 2003.

July 2002

Shigeru Kanemitsu and Chaohua Jia (Editors)

List of participants

Professor Shigeki Akiyama (Niigata University)

Professor Krishnaswami Alladi (University of Florida)

Professor Masaaki Amou (Gunma University)

Professor Tsuneo Arakawa (Rikkyo University)

Professor Tianxin Cai (Zhejiang University)

Professor Dr. Christopher Deninger (Universität Münster)

Professor Shigeki Egami (Toyama University)

Dr. Jun Furuya (Nagoya University)

Professor Masayoshi Hata (Kyoto University)

Professor Mikihito Hirabayashi (Kanazawa Institute of Technology)

Mr. Park Kyung Ho (Saga University)

Dr. Taro Horie (Suzuka Inst. of Techn.)

Dr. Yumiko Ichihara (Nagoya University, now Waseda University)

Professor Chaohua Jia (Academia Sinica)

Professor Masanobu Kaneko (Kyushu University)

Professor Shigeru Kanemitsu (University of Kinki)

Professor Masanori Katsurada (Keio University)

Professor Kiyoshi Kikukawa (University of Kinki)

Professor Isao Kiuchi (Yamaguchi University)

Professor Dr. Winfried Kohnen (Universität Heidelberg)

Dr. Hiroshi Kumagai (Kagoshima Inst. of Techn.)

Professor Minggao Lu (Shanghai University)

Ms. Keiko Mashimo (University of Kinki)

Professor Tapani Matala-Aho (University of Oulu)

Professor Kohji Matsumoto (Nagoya University)

Mr. Hidehiko Misho (Nagoya University)

Professor Katsuya Miyake (Tokyo Metropolitan University)

Professor Yasuo Motoda (Yatsushiro National College of Technology)

Professor Leo Murata (Meiji Gakuin University)

Professor Kenji Nagasaka (Hosei University)

Professor Makoto Nagata (RIMS-Kyoto University)

Mr. Minoru Nakashima (Saga University)

Dr. Hirofumi Nagoshi (Keio University, now RIMS-Kyoto University)

Professor Toru Nakahara (Saga University)

Professor Yoshinobu Nakai (Yamanashi University)

Professor Shoichi Nakajima (Gakushuin University)

Professor Yoshimasa Nakano (University of Kinki)

Professor Junjiro Noguchi (University of Tokyo)

Professor Ryotaro Okazaki (Doshisha University)

Professor Chengbiao Pan (The China Agricultural University)

Professor Ryuji Sasaki (Nihon University)

Professor Ken'ichi Sato (Nihon University)

Professor Iekata Shiokawa (Keio University)

Professor Shigeshi Shirasaka (Kagoshima Inst. of Techn.)

Professor Masaki Sudo (Seikei University)

Mr. Masatoshi Suzuki (Nagoya University)

Professor Yoshio Tanigawa (Nagoya University)

Professor Brunello Tirozzi (University of Rome)

Professor Haruo Tsukada (University of Kinki)

Mr. Hiroyuki Tsutsumi (Shimane University)

Ms. Kaoru Wada (University of Kinki)

Professor Isao Wakabayashi (Seikei University)

Professor Lin Weng (Nagoya University, now Kyushu University)

Mr. Mao Xiang Wu (Nagoya University)

Dr. Masami Yoshimoto (RIMS Kyoto University, now Nagoya University)

Professor Wenpeng Zhang (Northwest University)

Who's who in the picture 1. M. Nagata 2. H. Tsutumi 3. T. Horie 4. M. Nakashima 5. M. Hata 6. Y. Motoda 7. Y. Nakai 8. P.-K. Ho 9. T. Matala-aho 10. T. Arakawa 11. C.-B. Pan 12. L. Wen 13. M. Sudo 14. S. Egami 15. S. Akiyama 16. C.-H. Jia 17. T.-X. Cai 18. Y. Tanigawa 19. C. Deninger 20. R. Okazaki 21. M. Kaneko 22. S. Kanemitsu (multi-national Kanemitsu Corporation, cf. p. p. 118) 23. K. Miyake 24. K. Matsumoto 25. W.-P. Zhang 26. K. Nagasaka 27. M. Amou 28. W. Kohnen 29. M. Yoshimoto 30. B. Tirozzi 31. K. Alladi 32. M. Hirabayashi 33. I. Kiuchi 34. I. Shiokawa 35. I. Wakabayashi 36. M. Suzuki 37. M. Katsurada 38. H. Nagoshi 39. M.-X. Wu 40. H. Misho 41 J. Furuya 42. Y. Ichikawa 43. K. Wada 44. K. Mashimo 45. H. Kumagai 46. K. Kikukawa

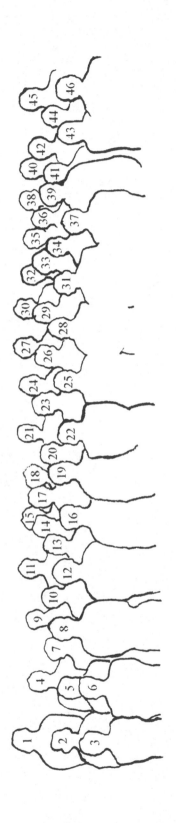

A LIMITING FORM OF THE q-DIXON $_4\varphi_3$ SUMMATION AND RELATED PARTITION IDENTITIES

Krishnaswami Alladi* and Alexander Berkovich*

Department of Mathematics, University of Florida, Gainesville, FL 32611, USA

alladi@math.ufl.edu   alexb@math.ufl.edu

Abstract   By considering a limiting form of the q-Dixon $_4\varphi_3$ summation, we prove a weighted partition theorem involving odd parts differing by ≥ 4. A two parameter refinement of this theorem is then deduced from a quartic reformulation of Göllnitz's (Big) theorem due to Alladi, and this leads to a two parameter extension of Jacobi's triple product identity for theta functions. Finally, refinements of certain modular identities of Alladi connected to the Göllnitz-Gordon series are shown to follow from a limiting form of the q-Dixon $_4\varphi_3$ summation.

Keywords: q-Dixon sum, q-Dougall Sum, weighted partition identities, Göllnitz's (Big) theorem, quartic reformulation, Jacobi's triple product, Göllnitz-Gordon identities, modular relations

1.   Introduction

The q-hypergeometric function $_{r+1}\varphi_r$ in $r + 1$ numerator parameters $a_1, a_2, \ldots, a_{r+1}$, and r denominator parameters $b_1, b_2, \ldots, b_r$, with base q and variable t, is defined as

$$_{r+1}\varphi_r \left(\begin{array}{c} a_1, a_2, a_3, \ldots, a_{r+1} \\ b_1, b_2, \ldots, b_r \end{array} ; q, t \right) = \sum_{k=0}^{\infty} \frac{(a_1; q)_k \ldots (a_{r+1}; q)_k t^k}{(b_1; q)_k (b_2; q)_k \ldots (b_r; q)_k (q; q)_k}.$$

$$(1.1)$$

When $r = 3$, for certain special choices of parameters a_i, b_j, and variable t, it is possible to evaluate the sum on the right in (1.1) to be a product.

*Research supported in part by National Science Foundation Grant DMS-0088975

More precisely, the $_4\varphi_3$ q-Dixon summation ([4], (II. 13), p. 237) is

$$_4\varphi_3\left(\begin{array}{c} a, -q\sqrt{a}, b, c \\ -\sqrt{a}, \frac{aq}{b}, \frac{aq}{c} \end{array}; q, \frac{q\sqrt{a}}{bc}\right) = \frac{(aq;q)_\infty(\frac{q\sqrt{a}}{b};q)_\infty(\frac{q\sqrt{a}}{c};q)_\infty(\frac{aq}{bc};q)_\infty}{(\frac{aq}{b};q)_\infty(\frac{aq}{c};q)_\infty(q\sqrt{a};q)_\infty(\frac{q\sqrt{a}}{bc};q)_\infty}.$$

$$(1.2)$$

Here and in what follows we have made use of the standard notation

$$(a;q)_n = (a)_n = \left\{\begin{array}{ll} \prod_{j=0}^{n-1}(1 - aq^j), & \text{if} \quad n > 0, \\ 1, & \text{if} \quad n = 0, \end{array}\right.$$

$$(1.3)$$

for any complex number a and a non-negative integer n, and

$$(a)_\infty = (a;q)_\infty = \lim_{n\to\infty}(a;q)_n = \prod_{j=0}^{\infty}(1 - aq^j), \text{ for } |q| < 1. \qquad (1.4)$$

Sometimes, as in (1.3) and (1.4), when the base is q, we might suppress it, but when the base is anything other than q, it will be made explicit.

Our first goal is to prove Theorem 1 is §2 which is a weighted identity connecting partitions into odd parts differing by ≥ 4 and partitions into distinct parts $\not\equiv 2\pmod{4}$. We achieve this by showing that the analytic representation of Theorem 1 is

$$\sum_{k=0}^{\infty} \frac{z^k q^{4T_k-1} q^{3k}(z^2 q^2; q^2)_k(1 + zq^{2k+1})}{(q^2; q^2)_k} = (-zq; q^2)_\infty(z^2 q^4; q^4)_\infty \quad (1.5)$$

and establish (1.5) by utilizing a limiting form of the q-Dixon summation formula (1.2). In (1.5), $T_k = k(k+1)/2$ is the k-th triangular number.

It is possible to obtain a two parameter refinement of Theorem 1 by splitting the odd integers into residue classes 1 and 3 (mod 4) and keeping track of the number of parts in each of these residue classes. This result, which is stated as Theorem 2 in §3, is a special case of a weighted reformulation of Göllnitz's (Big) theorem due to Alladi (Theorem 6 of [2]). In §3 we also state an analytic identity (see (3.3)) in two free parameters a and b that is equivalent to Theorem 2, and note that (1.5) follows from this as the special case $a = b = z$. Identity (3.3) can be viewed as a two parameter generalization of Jacobi's celebrated triple product identity for theta functions (see (3.5) in §3).

Identity (3.3) is itself a special case of *key identity* in three free parameters a, b, and c, due to Alladi and Andrews ([3], eqn. 3.14), for Göllnitz's (Big) Theorem. The proof of this key identity of Alladi and Andrews in [3] utilizes Jackson's q-analog of Dougall's summation for $_6\varphi_5$. Note that the left hand side of (3.3) is a double summation. On the other hand, the left side of (1.5) is just a single summation, and its proof requires only a limiting form of the q-Dixon summation for $_4\varphi_3$. Owing to the choice $a = b = z$, the double sum in (3.3) reduces to a single summation in (4.6) resembling (1.5), and this process is described in §4. Finally, certain modular identities for Göllnitz-Gordon functions due to Alladi [2] are refined in §5 using a limiting case of the q-Dixon summation (1.2).

We conclude this section by mentioning some notation pertaining to partitions. For a partition π we let

$$\sigma(\pi) = \text{the sum of all parts of } \pi,$$
$$\nu(\pi) = \text{the number of parts of } \pi,$$
$$\nu(\pi; r, m) = \text{the number of parts of } \pi \text{ which are } \equiv r(\text{mod } m),$$
$$\nu_d(m) = \text{the number of different parts of } \pi,$$
$$\nu_{d,\ell}(m) = \text{the number of different parts of } \pi \text{ which are } \geq \ell, \text{ and}$$
$$\lambda(\pi) = \text{the least part of } \pi.$$

2. Combinatorial interpretation and proof of (1.5)

Let $\mathcal{O}_4$ denote the set of partitions into odd parts differing by ≥ 4. Given $\tilde{\pi} \in \mathcal{O}_4$, a chain χ in $\tilde{\pi}$ is defined to be a maximal string of consecutive parts differing by exactly 4. Let $N_\lambda(\tilde{\pi})$ denote the number of chains in $\tilde{\pi}$ with least part $\geq \lambda$.

Next let $\mathcal{D}_{2,4}$ denote the set of partitions into distinct parts $\neq 2(\text{mod } 4)$. We then have

Theorem 1. *For all integers $n \geq 0$*

$$\sum_{\substack{\tilde{\pi} \in \mathcal{O}_4 \\ \sigma(\tilde{\pi}) = n}} z^{\nu(\tilde{\pi})}(1 - z^2)^{N_5(\tilde{\pi})} = \sum_{\substack{\pi \in \mathcal{D}_{2,4} \\ \sigma(\pi) = n}} z^{\nu(\pi;1,2)}(-z^2)^{\nu(\pi;0,2)}$$

So, for partitions $\tilde{\pi} \in \mathcal{O}_4$, we attach the weight z to each part, and the weight $(1 - z^2)$ to each chain having least part ≥ 5. The weight of $\tilde{\pi}$ is then defined multiplicatively. Similarly, for partitions $\pi \in \mathcal{D}_{2,4}$, each odd part is assigned weight z, and each even part is assigned the weight $-z^2$, where all these even parts are actually multiples of 4. For example, when $n = 10$, the partitions in $\mathcal{O}_4$ are $9 + 1$ and $7 + 3$, with weights $z^2(1 - z^2)$ and z^2 respectively. These weights add up to yield $2z^2 - z^4$. The partitions of 10 in $\mathcal{D}_{2,4}$ are $9 + 1, 7 + 3$, and $5 + 4 + 1$, with weights z^2, z^2, and $z^2(-z^2)$ respectively. These weights also add up to $2z^2 - z^4$, verifying Theorem 1 for $n = 10$.

We will now show that Theorem 1 is the combinational interpretation of (1.5).

It is clear that the product

$$\prod_{m=1}^{\infty}(1 + zq^{2m-1})(1 - z^2q^{4m}) \tag{2.1}$$

on the right in (1.5) is the generating function of partitions $\pi \in \mathcal{D}_{2,4}$, with weights as in Theorem 1. So we need to show that the series on the left in (1.5) is the generating function of partitions $\tilde{\pi} \in \mathcal{O}_4$ with weights as specified in Theorem 1. For this we consider two cases.

Case 1: $\lambda(\tilde{\pi}) \neq 1$.

If $\tilde{\pi}$ is non-empty, then $\lambda(\tilde{\pi}) \geq 3$.

Since the parts of $\tilde{\pi}$ differ by ≥ 4, we may subtract 0 from the smallest part, 4 from the second smallest part, 8 from the third smallest, ..., $4k-4$ from the largest part of $\tilde{\pi}$, assuming $\nu(\tilde{\pi}) = k$. We call this procedure the *Euler subtraction*. After the Euler subtraction is performed on $\tilde{\pi}$, we are left with a partition π' into k odd parts such that the number of different parts of π' is precisely the number of chains in $\tilde{\pi}$. If we denote by $G_{3,k}(q, z)$ the generating function of partitions $\tilde{\pi} \in \mathcal{O}_4$ with $\lambda(\tilde{\pi}) \neq 1, \nu(\tilde{\pi}) = k$, and counted with weight $z^{\nu(\tilde{\pi})}(1-z^2)^{N_5(\tilde{\pi})}$, then the Euler subtraction process yields

$$G_{3,k}(q, z) = z^k q^{4T_{k-1}} g_{3,k}(q, z), \tag{2.2}$$

where $g_{3,k}(q, z)$ is the generating function of partitions π' into k odd parts each ≥ 3 and counted with weight $(1 - z^2)^{\nu_{d,5}(\pi')}$.

At this stage we make the observation that if a set of positive integers J is given, then

$$\prod_{j \in J}\left(\frac{1 - twq^j}{1 - tq^j}\right) = \prod_{j \in J}(1 + (1 - w)\{tq^j + t^2q^{2j} + t^3q^{3j} + \dots\}) \tag{2.3}$$

is the generating function of partitions π^* into parts belonging to J and counted with weight $t^{\nu(\pi^*)}(1-w)^{\nu_d(\pi^*)}$. So from the principles underlying (2.3) it follows that

$$\sum_{k=0}^{\infty} g_{3,k}(q, z)t^k = \frac{1}{(1 - tq^3)}\prod_{j=0}^{\infty}\left(\frac{1 - tz^2q^{2j+5}}{1 - tq^{2j+5}}\right) = \frac{(tz^2q^5; q^2)_{\infty}}{(tq^3; q^2)_{\infty}}. \tag{2.4}$$

Using Cauchy's identity

$$\frac{(at)_{\infty}}{(t)_{\infty}} = \sum_{k=0}^{\infty}\frac{(a)_k t^k}{(q)_k}, \tag{2.5}$$

we can expand the product on the right in (2.4) as

$$\frac{(tz^2q^5;q^2)_\infty}{(tq^3;q^2)_\infty} = \sum_{k=0}^\infty \frac{t^k q^{3k}(z^2q^2;q^2)_k}{(q^2;q^2)_k}. \tag{2.6}$$

So by comparing the coefficients of t^k in (2.4) and (2.6) we get

$$g_{3,k}(q;z) = \frac{q^{3k}(z^2q^2;q^2)_k}{(q^2;q^2)_k}. \tag{2.7}$$

Thus (2.7) and (2.2) yield

$$G_{3,k}(q;z) = \frac{z^k q^{3k} q^{4T_{k-1}}(z^2q^2;q^2)_k}{(q^2;q^2)_k}, \text{ for } k \geq 0. \tag{2.8}$$

Case 2: $\lambda(\tilde{\pi}) = 1$.

Here for $k > 0$ we denote by $G^*_{1,k}(q,z)$ the generating function of partitions $\tilde{\pi} \in \mathcal{O}_4$ having $\lambda(\tilde{\pi}) = 1$, $\nu(\tilde{\pi}) = k$, and counted with weight $z^{\nu(\tilde{\pi})}(1-z^2)^{N_5(\tilde{\pi})}$. The Euler subtraction process yields

$$G^*_{1,k}(q,z) = z^k q^{4T_{k-1}} g^*_{1,k}(q,z), \tag{2.9}$$

where $g^*_{1,k}(q,z)$ is the generating function of partitions π' into k odd parts counted with weight $(1-z^2)^{\nu_{d,3}(\pi')}$. The principles underlying (2.3) show that

$$\sum_{j=1}^\infty g_1^*(q,z)t^k = \frac{tq}{1-tq}\prod_{j=1}^\infty \left(\frac{1-tz^2q^{2j+1}}{1-tq^{2j+1}}\right) = tq\frac{(tz^2q^3;q^2)_\infty}{(tq;q^2)_\infty}$$

$$= \sum_{k=0}^\infty t^{k+1} q^{k+1} \frac{(z^2q^2;q^2)_k}{(q^2;q^2)_k} \tag{2.10}$$

by Cauchy's identity (2.5). Thus by comparing the coefficients of t^k at the extreme ends of (2.10), we get

$$g^*_{1,k}(q,z) = q^k \frac{(z^2q^2;q^2)_{k-1}}{(q^2;q^2)_{k-1}}. \tag{2.11}$$

This when combined with (2.9) yields

$$G^*_{1,k}(q,z) = \frac{z^k q^k q^{4T_{k-1}}(z^2q^2 : q^2)_{k-1}}{(q^2;q^2)_{k-1}}. \tag{2.12}$$

Finally, it is clear that

$$\sum_{k=0}^\infty G_{3,k}(q,z) + \sum_{k=1}^\infty G^*_{1,k}(q,z) \tag{2.13}$$

is the generating function of partitions $\tilde{\pi} \in \mathcal{O}_4$ counted with weight as specified in Theorem 1. From (2.8) and (2.12), the sum in (2.13) can be seen to be

$$\sum_{k=0}^{\infty} z^k q^{3k} q^{4T_{k-1}} \frac{(z^2 q^2; q^2)_k}{(q^2; q^2)_k} + \sum_{k=1}^{\infty} z^k q^k q^{4T_{k-1}} \frac{(z^2 q^2; q^2)_{k-1}}{(q^2; q^2)_{k-1}}$$

$$= \sum_{k=0}^{\infty} z^k q^{3k} q^{4T_{k-1}} \frac{(z^2 q^2; q^2)_k}{(q^2; q^2)_k} + \sum_{k=0}^{\infty} z^{k+1} q^{k+1} q^{4T_k} \frac{(z^2 q^2; q^2)_k}{(q^2; q^2)_k}$$

$$= \sum_{k=0}^{\infty} \frac{z^k q^{3k} q^{4T_{k-1}} (z^2 q^2; q^2)_k (1 + z q^{2k+1})}{(q^2; q^2)_k} \tag{2.14}$$

which is the series on the left in (1.5).

From (2.14) and (2.1) it follows that Theorem 1 is the combinatorial interpretation of (1.5). Thus to prove Theorem 1 it suffices to establish (1.5) and this what we do next.

From the definition of the q-hypergeometric function in (1.1) we see that

$$(1 + zq)_4\varphi_3 \left(\begin{array}{c} z^2 q^2, -zq^3, \rho, \rho \\ -zq, \frac{z^2 q^4}{\rho}, \frac{z^2 q^4}{\rho} \end{array} ; q^2, \frac{zq^3}{\rho^2} \right)$$

$$= (1 + zq) \sum_{k=0}^{\infty} \frac{(z^2 q^2; q^2)_k (-zq^3; q^2)_k (\rho; q^2)_k (\rho; q^2)_k}{(q^2; q^2)_k (-zq; q^2)_k (\frac{z^2 q^4}{\rho}; q^2)_k (\frac{z^2 q^4}{\rho}; q^2)_k} \frac{z^k q^{3k}}{\rho^{2k}}$$

$$= \sum_{k=0}^{\infty} \frac{z^k q^{3k} (z^2 q^2; q^2)_k (1 + zq^{2k+1})}{(q^2; q^2)_k} \frac{(\rho; q^2)_k (\rho; q^2)_k}{(\frac{z^2 q^4}{\rho}; q^2)_k (\frac{z^2 q^4}{\rho}; q^2)_k} \frac{1}{\rho^{2k}}. \tag{2.15}$$

Next observe that

$$\lim_{\rho \to \infty} \frac{(\rho; q^2)_k}{\rho^k} = \lim_{\rho \to \infty} \frac{(1 - \rho)(1 - \rho q^2) \dots (1 - \rho q^{2k-2})}{\rho^k} = (-1)^k q^{2T_{k-1}} \tag{2.16}$$

and

$$\lim_{\rho \to \infty} \left(\frac{z^2 q^4}{\rho}; q^2 \right)_k = 1. \tag{2.17}$$

Thus (2.15), (2.16), and (2.17) imply that

$$(1 + zq) \lim_{\rho \to \infty} {}_4\varphi_3 \left(\begin{array}{c} z^2 q^2, -zq^3, \rho, \rho \\ -zq, \frac{z^2 q^4}{\rho}, \frac{z^2 q^4}{\rho} \end{array} ; q^2, \frac{zq^3}{\rho^2} \right)$$

$$= \sum_{k=0}^{\infty} \frac{z^k q^{3k} q^{4T_{k-1}} (z^2 q^2; q^2)_k (1 + zq^{2k+1})}{(q^2; q^2)_k} \tag{2.18}$$

which is the series on the left in (1.5).

At this stage we observe that the hypergeometric function $_4\varphi_3$ on the left in (2.18) is precisely the one in the q-Dixon summation (1.2) with the replacements

$$q \mapsto q^2, a \mapsto z^2q^2, b \mapsto \rho, c \mapsto \rho. \tag{2.19}$$

Thus with substitutions (2.19) in (1.2) we deduce that

$$(1 + zq) \lim_{\rho \to \infty} {}_4\varphi_3 \left(\begin{array}{c} z^2q^2, -zq^3, \rho, \rho \\ -zq, \frac{z^2q^4}{\rho}, \frac{z^2q^4}{\rho} \end{array} ; q^2, \frac{zq^3}{\rho^2} \right)$$

$$= (1 + zq) \lim_{\rho \to \infty} \frac{(z^2q^4; q^2)_\infty (\frac{zq^3}{\rho}; q^2)_\infty (\frac{zq^3}{\rho}; q^2)_\infty (\frac{z^2q^4}{\rho^2}; q^2)_\infty}{(\frac{z^2q^4}{\rho}; q^2)_\infty (\frac{z^2q^4}{\rho}; q^2)_\infty (zq^3; q^2)_\infty (\frac{zq^3}{\rho^2}; q^2)_\infty}$$

$$= (1 + zq) \frac{(z^2q^4; q^2)_\infty}{(zq^3; q^2)_\infty} = \frac{(z^2q^2; q^2)_\infty}{(zq; q^2)_\infty}$$

$$= \frac{(z^2q^2; q^4)_\infty (z^2q^4; q^4)_\infty}{(zq; q^2)_\infty} = (-zq; q^2)_\infty (z^2q^4; q^4)_\infty. \tag{2.20}$$

Thus (1.5) follows from (2.18) and (2.20) and this completes the proof of Theorem 1.

3.   Two parameter refinement

When a partition $\tilde{\pi} \in \mathcal{O}_4$ is decomposed into chains, the parts in a given chain all belong to the same residue class mod 4. This suggests that there ought to be a two parameter refinement of Theorem 1 in which we can keep track of parts in residue classes 1 and 3 (mod 4) separately. Theorem 2 stated below is such a refinement. Actually Theorem 2 is a special case of a refinement and reformulation of a deep theorem of Göllnitz [5] in three parameters a, b, and c due to Alladi ([2], Theorem 6) by setting one of the parameters $c = -ab$.

Theorem 2. *For all integers $n \geq 0$ and complex numbers a and b we have*

$$\sum_{\substack{\tilde{\pi} \in \mathcal{O}_4 \\ \sigma(\tilde{\pi}) = n}} a^{\nu(\tilde{\pi};1,4)} b^{\nu(\tilde{\pi};3,4)} (1 - ab)^{N_5(\tilde{\pi})}$$

$$= \sum_{\substack{\pi \in \mathcal{D}_{2,4} \\ \sigma(\pi) = n}} a^{\nu(\pi;1,4)} b^{\nu(\pi;3,4)} (-ab)^{\nu(\pi,0,4)}$$

Since Theorem 2 is a two parameter refinement of Theorem 1 which has the analytic representation (1.5), it is natural to ask for an analytic identity in two free parameters that reduces to (1.5). We will now obtain such a two parameter identity, namely, (3.3) below. Instead of deriving (3.3) combinatorially from Theorem 2 by following the method in §2, we will now illustrate a different approach which involves a certain cubic reformulation of Göllnitz's (Big) theorem due to Alladi [1], and its *key identity* in three free parameters a, b, and c due to Alladi and Andrews [3]. More precisely, we will show that (3.3) is the analytic representation of Theorem 2 after a discussion of the following special case $c = -ab$ of the key identity (3.14) of [3]:

$$\sum_{i,j \geq 0} \frac{a^i q^{(3i^2-i)/2}(ab;q^3)_i}{(q^3;q^3)_i} \cdot \frac{b^j q^{3ij} q^{(3j^2+j)/2}(ab;q^3)_j}{(q^3;q^3)_j} \cdot \left(\frac{1 - abq^{3(i+j)}}{1 - ab} \right)$$

$$= (-aq;q^3)_\infty (-bq^2;q^3)_\infty (abq^3;q^3)_\infty. \tag{3.1}$$

The cubic reformulation of Göllnitz's theorem in [1] in three free parameters a, b, and c, was in the form of an identity connecting partitions into distinct parts with a weighted count of partitions into parts differing by ≥ 3. When we set $c = -ab$ in Theorem 2 of [1], it turns out that within the set of partitions into parts differing by ≥ 3, we need only consider those partitions not having any multiples of 3 as parts; this is because (see [1], Theorem 2) the choice $c = -ab$ makes the weights equal to 0 if the partition has a multiple of 3 in it. Thus from the analysis in [3] and the specialization $c = -ab$ in Theorem 2 of [1], it follows that the combinatorial interpretation of (3.1) is

Theorem 3. *Let $\mathcal{D}$ denote the set of partitions into distinct parts. Let $\mathcal{D}_3$ denote the set of partitions into parts differing by ≥ 3, and containing no multiples of 3. Given $\tilde{\pi} \in \mathcal{D}_3$, decompose it into chains, where a chain here is a maximal string of parts differing by exactly 3. Then we have*

$$\sum_{\substack{\tilde{\pi} \in \mathcal{D}_3 \\ \sigma(\tilde{\pi}) = n}} a^{\nu(\tilde{\pi};1,3)} b^{\nu(\tilde{\pi};2,3)}(1 - ab)^{N_3(\tilde{\pi})}$$

$$= \sum_{\substack{\pi \in \mathcal{D} \\ \sigma(\pi) = n}} a^{\nu(\pi;1,3)} b^{\nu(\pi;2,3)}(-ab)^{\nu(\pi;0,3)}.$$

As an example, when $n = 9$, the partitions in $\mathcal{D}_3$ are 8+1 and 7+2 both having weights $ab(1 - ab)$. So these weights will add up to $2ab(1 -$

ab). The partitions of 9 in $\mathcal{D}$ with their corresponding weights are listed below:

Partitions :	$9, 8+1, 7+2,$	$6+3, 6+2+1, 5+4,$	$5+3+1, 4+3+2$
Weights :	$-ab, \quad ab, \quad ab, (-ab)^2, ab(-ab),$	$ab, ab(-ab), ab(-ab)$	

The above weights when added also yield $2ab(1-ab)$ thereby verifying Theorem 3 for $n = 9$.

Theorems 3 and 2 are really the same because the role of the modulus 3 in Theorem 3 is replaced by the modulus 4 in Theorem 2. More precisely, we may view $\mathcal{D}_3$ and $\mathcal{O}_4$ as sets of partitions into parts differing by $\geq k$ and containing only parts in the residue classes $\pm 1 (mod \quad k)$, for $k = 3, 4$. Similarly, we may think of $\mathcal{D}$ and $\mathcal{D}_{2,4}$ as sets of partitions into parts $\equiv 0, \pm 1 (mod \quad k)$, for $k = 3, 4$. Note that the functions $\nu(\pi; r, 3)$ and $\nu(\tilde{\pi}; r, 3)$ in Theorem 3 are replaced $\nu(\pi; r, 4)$ and $\nu(\tilde{\pi}; r, 4)$ in Theorem 2. Pursuing this line of correspondence, $N_3(\tilde{\pi})$ in Theorem 3 should be replaced by $N_4(\tilde{\pi})$ in Theorem 2, but this is the same as having $N_5(\tilde{\pi})$ in Theorem 2 because $\tilde{\pi} \in \mathcal{O}_4$.

Having observed the correspondence between Theorems 2 and 3, we deduce that the analytic representation of Theorem 2 (in the form of a two parameter q-hypergeometric identity) is obtained from (3.1) by the substitutions

$$q \mapsto q^{4/3}, a \mapsto aq^{-1/3}, b \mapsto bq^{1/3}, ab \mapsto ab, \tag{3.2}$$

which yields

$$\sum_{i,j \geq 0} \frac{a^i q^{2i^2-i}(ab; q^4)_i}{(q^4; q^4)_i} \cdot \frac{b^j q^{4ij} q^{2j^2+j}(ab; q^4)_j}{(q^4; q^4)_j} \cdot \left(\frac{1 - abq^{4(i+j)}}{1 - ab} \right)$$

$$= (-aq; q^4)_\infty (-bq^3; q^4)_\infty (ab; q^4; q^4)_\infty. \tag{3.3}$$

Identities (3.1) and (3.3) are interesting for another reason. They can be considered as two parameter extensions of Jacobi's celebrated triple product identity for theta functions. More precisely, if we put $ab = 1$ in (3.1) and (3.3), then on the left hand side of each of these identities, only the terms having either $i = 0$ or $j = 0$ survive, and so the identities reduce to

$$\sum_{i=-\infty}^{\infty} a^i q^{(3i^2-i)/2} = (-aq; q^3)_\infty (-a^{-1}q^2; q^3)_\infty (q^3; q^3)_\infty \tag{3.4}$$

and

$$\sum_{i=-\infty}^{\infty} a^i q^{2i^2-i} = (-aq; q^4)_\infty (-a^{-1}q^3; q^4)_\infty (q^4; q^4)_\infty \tag{3.5}$$

which are equivalent to Jacobi's triple product identity.

Identities (3.4) and (3.5) can be deduced combinatorially from Theorems 3 and 2 respectively, by setting $ab = 1$. This is because $ab = 1$ forces $N_3(\tilde{\pi}) = 0$ (resp. $N_5(\tilde{\pi}) = 0$) in Theorem 3 (resp. Theorem 2) and this brings about a drastic reduction in the type of partitions to be enumerated in $\mathcal{D}_3$ (resp. $\mathcal{O}_4$). We refer the reader to Alladi [1], [2] for these combinatorial arguments.

4.   Reduction to a single summation

The left hand side of identity (3.3) is a double summation. It turns out that if we set

$$a = b = z,$$

then the left side of (3.3) reduces to a single infinite sum. It is quite instructive to see how this happens, and so we describe it now.

First observe that

$$2i^2 - i + 2j^2 + j + 4ij = 2(i+j)^2 - (i+j) + 2j. \tag{4.1}$$

Thus if we set $a = b = z$ and reassemble the terms in (3.3) with $k = i+j$, then (4.1) shows that (3.3) becomes

$$\sum_{k=0}^{\infty} z^k q^{2k^2-k} \frac{(1-z^2 q^{4k})}{(1-z^2)} \left\{ \sum_{i+j=k} \frac{q^{2j}(z^2;q^4)_i(z^2;q^4)_j}{(q^4;q^4)_i(q^4;q^4)_j} \right\}$$
$$= (-zq;q^2)_\infty (z^2 q^4;q^4)_\infty. \tag{4.2}$$

At this point we take the product in Cauchy's identity (2.5) and decompose it as

$$\frac{(at)_\infty}{(t)_\infty} = \frac{(at;q^2)_\infty}{(t;q^2)_\infty} \cdot \frac{(atq;q^2)_\infty}{(tq;q^2)_\infty}. \tag{4.3}$$

If we now substitute the expansion in (2.5) for each of the products in (4.3), we get

$$\sum_{k=0}^{\infty} \frac{(a)_k t^k}{(q)_k} = \left(\sum_{i=0}^{\infty} \frac{(a;q^2)_i t^i}{(q^2;q^2)_i} \right) \left(\sum_{j=0}^{\infty} \frac{(a;q^2)_j t^j q^j}{(q^2;q^2)_j} \right). \tag{4.4}$$

By comparing the coefficients of t^k on both sides of (4.4) we obtain

$$\frac{(a)_k}{(q)_k} = \sum_{i+j=k} \frac{(a;q^2)_i(a;q^2)_j q^j}{(q^2;q^2)_i(q^2;q^2)_j}. \tag{4.5}$$

If we replace $q \mapsto q^2$ and $a \mapsto z^2$, we see that the sum in (4.5) becomes the expression within the parenthesis (namely the inner sum) on the left in (4.2). Thus with these replacements (4.5) implies that (4.2) can be written as the single summation identity

$$1 + \sum_{k=1}^{\infty} \frac{z^k q^{2k^2-k}(z^2q^2;q^2)_{k-1}(1-z^2q^{4k})}{(q^2;q^2)_k} = (-zq;q^2)_\infty(z^2q^4;q^4)_\infty.$$

(4.6)

Identity (4.6) is an analytic representation of Theorem 1. Note however that the series in (4.6) is different from the series in (1.5). The explanation of this is as follows.

If we add $G_{3,k}(q,z)$ and $G_{1,k}^*(q,z)$ for each $k \geq 1$, we get from (2.8) and (2.12)

$$G_{3,k}(q,z)+G_{1,k}^*(q,z) = \frac{z^k q^k q^{4T_{k-1}}(z^2q^2;q^2)_{k-1}}{(q^2;q^2)_k}\left\{1 - q^{2k} + q^{2k}(1 - z^2q^{2k})\right\}$$

$$= \frac{z^k q^{2k^2-k}(z^2q^2;q^2)_{k-1}(1 - z^2q^{4k})}{(q^2;q^2)_k}$$

which is the k-th summand in (4.6). The starting term 1 in (4.6) is to be interpreted as $G_{3,0}(q,z)$. On the other hand in (2.14) we are considering

$$G_{3,k}(q,z) + G_{1,k+1}^*(q,z), \text{ for } k \geq 0,$$

and this leads to the series in (1.5) which is different from (4.6).

The reason we preferred (1.5) to (4.6) is because (1.5) could be proved using only a limiting form of the q-Dixon summation of $_4\varphi_3$, whereas (4.6) would have required a limiting form of Jackson's $_6\varphi_5$ summation ([4], (II.21), p. 238).

5. Modular identities for the refined Göllnitz-Gordon functions.

The well known Göllnitz-Gordon identities are

$$G(q) = \sum_{n=0}^{\infty} \frac{q^{n^2}(-q;q^2)_n}{(q^2;q^2)_n} = \frac{1}{(q;q^8)_\infty(q^4;q^8)_\infty(q^7;q^8)_\infty}$$

(5.1)

and

$$H(q) = \sum_{n=0}^{\infty} \frac{q^{n^2+2n}(-q;q^2)_n}{(q^2;q^2)_n} = \frac{1}{(q^3;q^8)_\infty(q^4;q^8)_\infty(q^5;q^8)_\infty}.$$

(5.2)

Identities (5.1) and (5.2) are actually (36) and (34) on Slater's list [7], but it was Göllnitz [5] and Gordon [6] who realized their partition significance and their relationship with a continued fraction. More precisely, the Göllnitz-Gordon partition theorem is: *For i=1,2, the number of partitions of an integer n into parts differing by ≥ 2, with strict inequality if a part is even, and least part $\geq 2i-1$, equals the number of partitions of n into parts $\equiv 4, \pm(2i-1)$ (mod 8).*

In view of the form of the series-product identities (5.1) and (5.2), their partition interpretation given above, and their relationship with a certain continued fraction, the Göllnitz-Gordon identities are considered as the perfect analogues for the modulus 8 for what the celebrated Rogers-Ramanujan identities are for the modulus 5.

In [2] Alladi established four reformulations of Göllnitz's (Big) partition theorem using four quartic transformations. One of the reformulations yielded Theorem 2 of §3. Using one of the other reformulations, Alladi [2] deduced the modular identity

$$G(-q^2) + qH(-q^2) = (-q; q^4)_\infty (q^2; q^4)_\infty (-q^3; q^4)_\infty \qquad (5.3)$$

combinatorially. Alladi [2] then defined the *twisted* Göllnitz-Gordon functions

$$G_t(q) = \sum_{n=0}^{\infty} \frac{q^{n^2}(q; q^2)_n}{(q^2; q^2)_n}, \qquad (5.4)$$

and

$$H_t(q) = \sum_{n=0}^{\infty} \frac{q^{n^2+2n}(q; q^2)_n}{(q^2; q^2)_n}, \qquad (5.5)$$

and deduced the modular identity

$$G_t(q^2) + qH_t(q^2) = (-q; q^4)_\infty (-q^2; q^4)_\infty (q^3; q^4)_\infty \qquad (5.6)$$

combinatorially from the same reformulation of Göllnitz's (Big) theorem.

The twisted Göllnitz-Gordon functions do not have the product representations of the type $G(q)$ and $H(q)$ possess. But the modular identity (5.6) implies that

$$G_t(q^2) = \frac{(-q; q^4)_\infty (-q^2; q^4)_\infty (q^3; q^4)_\infty + (q; q^4)_\infty (-q^2; q^4)_\infty (-q^3; q^4)_\infty}{2}$$

$$(5.7)$$

and

$$H_t(q^2) = \frac{(-q; q^4)_\infty (-q^2; q^4)_\infty (q^3; q^4)_\infty - (q; q^4)_\infty (-q^2; q^4)_\infty (q^3; q^4)_\infty}{2q}.$$

$$(5.8)$$

In the absence of product reprentations, (5.7) and (5.8) show that $G_t(q^2)$ and $G_t(q^2)$ are arithmetic means of interesting products. From (5.3) it follows that $G(-q^2)$ and $H(-q^2)$ have representations similar to (5.7) and (5.8).

By utilizing a limiting form of the q-Dixon summation (1.2) for $_4\varphi_3$, we will now establish a more general modular identity (see (5.13) below) that contains both (5.3) and (5.6). To this end let $c = \delta\sqrt{aq}$ with $\delta = \pm 1$ in (1.2), and multiply both sides by $1 + \sqrt{a}$. This way we find

$$\sum_{k=0}^{\infty} \frac{\delta^k b^{-k} q^{k/2}(a)_k(b)_k(1 + \sqrt{a}q^k)}{(q)_k\left(\frac{aq}{b}\right)_k} = \frac{(a)_\infty\left(\frac{q\sqrt{a}}{b}\right)_\infty(\delta\sqrt{q})_\infty\left(\frac{\delta\sqrt{aq}}{b}\right)_\infty}{\left(\frac{aq}{b}\right)_\infty(\delta\sqrt{aq})_\infty(\sqrt{a})_\infty\left(\frac{\delta\sqrt{q}}{b}\right)_\infty}.$$

$$(5.9)$$

Analogous to (2.16) we now have

$$\lim_{b\to\infty} \frac{(b)_k}{b^k} = (-1)^k q^{Tk-1}.$$

$$(5.10)$$

Thus (5.9) and (5.10) imply that by going to the limit $b \to \infty$ we get

$$\sum_{k=0}^{\infty} \frac{(-\delta)^k q^{k^2/2}(a)_k(1 + \sqrt{a}q^k)}{(q)_k} = \frac{(a)_\infty(\delta\sqrt{q})_\infty}{(\sqrt{a})_\infty(\delta\sqrt{aq})_\infty}, \quad \text{for } \delta = \pm 1. \quad (5.11)$$

In (5.11) if we replace $q \mapsto q^4$ and $a \mapsto a^2 q^2$, we obtain

$$\sum_{k=0}^{\infty}(-\delta)^k q^{2k^2} \frac{(a^2 q^2; q^4)_k}{(q^4; q^4)_k}(1 + aq^{4k+1}) = \frac{(a^2 q^2; q^4)_\infty(\delta q^2; q^4)_\infty}{(aq; q^4)_\infty(\delta aq^3; q^4)_\infty}$$

$$= \frac{(a^2 q^2; q^8)_\infty(a^2 q^6; q^8)_\infty(\delta q^2; q^4)_\infty}{(aq; q^4)_\infty(\delta aq^3; q^4)_\infty} = (-aq; q^4)_\infty(\delta q^2; q^4)_\infty(-\delta aq^3; q^4)_\infty.$$

$$(5.12)$$

Note that the special case $\delta = 1, a = 1$ in (5.12) yields (5.3), whereas $\delta = -1, a = 1$ in (5.12) is (5.6). We may write (5.12) in the form of the modular identity

$$G_{a^2,\delta}(-q^2) + aqH_{a^2,\delta}(-q^2)$$
$$= (-aq; q^4)_\infty(\delta q^2; q^4)_\infty(-\delta aq^3; q^4)_\infty, \quad \text{for } \delta = \pm 1 \quad (5.13)$$

for the refined Göllnitz-Gordon functions which we define as

$$G_{a,\delta}(q) = \sum_{k=0}^{\infty} \frac{\delta^k q^{k^2}(-aq; q^2)_k}{(q^2; q^2)_k} \quad (5.14)$$

and

$$H_{a,\delta}(q) = \sum_{k=0}^{\infty} \frac{\delta^k q^{k^2+2k}(-aq; q^2)_k}{(q^2; q^2)_k}. \quad (5.15)$$

Here $G_{a,\delta}(q)$ is the generating function of partitions into parts differing by ≥ 2, with strict inequality if a part is even, where each odd part is assigned weight δ, and each even part given weight δa. The weight of the partition under consideration is the product of the weights of its parts. The function $H_{a,\delta}(q)$ has a similar interpretation, except for the added restriction that the least part is ≥ 3 for the partitions enumerated.

We note that the combinatorial arguments in [2] which yielded (5.3) and (5.6) could be used to derive the more general modular relation (5.13).

References

[1] K. Alladi, A combinatorial correspondence related to Göllnitz's (Big) partition theorem and applications, *Trans. Amer. Math Soc.* **349** (1997), 2721–2735.

[2] K. Alladi, On a partition theorem of Göllnitz and quartic transformations (with an appendix by B. Gordon), *J. Num. Th.* **69** (1998), 153–180.

[3] K. Alladi and G. E. Andrews, A new key identity for Göllnitz's (Big) partition theorem, *Contemp. Math.* **210** (1998), 229–241.

[4] G. Gasper and M. Rahman, Basic hyper-geometric series, *Encyclopedia of Mathematics and its Applications*, Vol.10, Cambridge (1990).

[5] H. Göllnitz, Partitionen mit Differenzenbedingungen, *J. Reine Angew Math.* **225** (1967), 154–190.

[6] B. Gordon, Some continued fractions of the Rogers-Ramanujan type, *Duke Math. J.* **32** (1965), 741–748.

[7] L. J. Slater, Further identities of Rogers-Ramanujan type, *Proc. London Math. Soc.* (2) **54** (1952), 147–167.

ARITHMETICAL PROPERTIES OF SOLUTIONS OF LINEAR SECOND ORDER q-DIFFERENCE EQUATIONS

Masaaki Amou*
Department of Mathematics, Gunma University, Tenjin-cho 1-5-1, Kiryu 376-8515, Japan
amou@math.sci.gunma-u.ac.jp

Tapani Matala-aho
Department of Mathematics, University of Oulu, PL 3000, 90014 University of Oulu, Finland
tma@cc.oulu.fi

Abstract   A linear independence measure is obtained for the values $F(t)$ and $F(qt)$ of solutions $F(z)$ of linear second order q-difference equations. From the result an irrationality measure for certain continued fractions is derived.

Keywords: linear independence measure, irrationality measure, q-difference equation, continued fraction

Dedicated to Professor Iekata Shiokawa on his 60th birthday

1.   Introduction

Throughout this paper, let K be an algebraic number field, κ its degree over the rational number field $\mathbf{Q}$, and $\mathcal{M}$ the set of all places of K. Let $q \in K$ and $v \in \mathcal{M}$ such that $|q|_v < 1$. The completion of K with respect to v is denoted by K_v. Let s be a positive integer, and $P(z), Q(z)$, and $M(z)$ polynomials in $K[z]$ with degrees k, ℓ, and m, respectively, satisfying $P(0)Q(0)M(0) \neq 0$. We then consider the linear

*The first named author was supported in part by Grant-in-Aid for Scientific Research (No. 13640007), the Ministry of Education, Science, Sports and Culture of Japan.

second order q-difference equation

$$N(z)F(q^2z) = P(z)F(qz) + Q(z)F(z), \quad N(z) = z^s M(z). \qquad (1.1)$$

In [8] the second named author treated solutions $F(z)$ on K_v of this kind of q-difference equations having general order $n \geq 2$, and proved at least two of the numbers

$$F(t), F(qt), ..., F(q^{n-1}t) \quad (t \in K)$$

are linearly independent over K under certain conditions on F, q, t, and the degrees of polynomials appearing in the functional equation. The result also contains certain quantitative assertion. Concerning about the present case, we need a condition

$$s > \max(2k, \ell + m) \qquad (1.2)$$

to guarantee the result in [8]. In particular, this condition is fulfilled by an interesting special case

$$qzF(q^2z) = -F(qz) + F(z),$$

which has a solution $F(z)$ such that $F(z)/F(qz)$ represents the Rogers-Ramanujan continued fraction. There are several works on arithmetical nature of the Rogers-Ramanujan continued fraction. We refer the readers to [3], [5], [6], [7], [11], and [12].

The purpose of this paper is to consider arithmetical nature of a solution $F(z)$ on K_v of (1.1), which is holomorphic in a neighbourhood of the origin, under the condition

$$s \geq \max(2k, \ell + m) \qquad (1.3)$$

admitting the equality in (1.2). We here explain briefly about the difference between the method used in [7], [8] and that used in this paper. In the former we start from (1.1), and replace z with $q^n t$ $(n \in \mathbf{N})$ to getting a sequence of good approximation forms for $F(t)$ and $F(qt)$. It works successfully under the condition (1.2), but fails under the condition (1.3) with the equality. On the other hand, in the following argument, we start from auxiliary functions $R(z) = A(z)F(qz) + B(z)F(z)$ with polynomials $A, B \in K[z]$ having large degrees, which are constructed by using Thue-Siegel lemma in the form given by Bombieri [2]. This kind of method has been applied to study Mahler functions (see e.g. [10]), and was introduced by Duverney [4] to study arithmetical nature of functions satisfying linear first order q-difference equations in qualitative nature and by the first named author, Katsurada, and Väänänen [1] in quantitative nature.

The plan of this paper is as follows. In the next section we introduce our object, that is, a meromorphic solution of (1.1) on K_v. After giving fundamental notation in Section 3, we state our main results in Section 4. The proofs of the main results are carried out in Sections 5, 6, and 7.

2. Solution of the functional equation

In this section we first show that the functional equation (1.1) has the unique solution

$$F(z) = \sum_{n=0}^{\infty} f_n z^n \quad (f_0 = 1) \tag{2.1}$$

in $K[[z]]$, under the condition

$$\alpha_0 := P(0) = -Q(0), \tag{2.2}$$

which converges on a certain disk $\mathbf{D} \subset K_v$ with center at the origin. Let us write the polynomials in (1.1) as

$$P(z) = \sum_{i=0}^{k} \alpha_i z^i, \quad Q(z) = \sum_{i=0}^{\ell} \beta_i z^i, \quad M(z) = \sum_{i=0}^{m} \gamma_i z^i.$$

Then the power series (2.1) satisfies (1.1) if and only if

$$\sum_{i \geq 0} \gamma_i q^{2(n-s-i)} f_{n-s-i} = \sum_{i \geq 0} \alpha_i q^{n-i} f_{n-i} + \sum_{i \geq 0} \beta_i f_{n-i}$$

for all n, where we use the convention that f_i, α_i, β_i, and γ_i are zero for all i in which they are not defined. It is fulfilled uniquely by f_n $(n = 1, 2, ...)$ satisfying the recursion formula

$$\alpha_0(1 - q^n) f_n = \sum_{i \geq 1} (\alpha_i q^{n-i} + \beta_i) f_{n-i} - \sum_{i \geq 0} \gamma_i q^{2(n-s-i)} f_{n-s-i}. \tag{2.3}$$

Hence (1.1) has a unique solution (2.1) in $K[[z]]$. It is easily seen that $|f_n|_v$ grows at most geometrically as n tends to infinity, that is,

$$|f_n|_v \leq c_0^n \quad (n = 0, 1, 2, ...) \tag{2.4}$$

with a positive constant c_0 depending only on q and the coefficients of P, Q, and M. This shows that the solution (2.1) of (1.1) converges on a disk

$$\mathbf{D} = \{z \in K_v \mid |z|_v < c_0^{-1}\}. \tag{2.5}$$

Then, by using (1.1) in the modified form

$$F(z) = Q(z)^{-1}(N(z)F(q^2 z) - P(z)F(qz))$$

repeatedly, the series (2.1) is continued meromorphically to the whole K_v and $F(t) \in K_v$ for all t satisfying $Q(q^n t) \neq 0$ ($n = 0, 1, 2, ...$). This function $F(z)$ on K_v is our object.

We next introduce a continued fraction attached to $F(z)$. Let us set

$$a_n(z) = z^s q^{s(n-1)} M(q^{n-1} z) Q(q^n z), \quad b_n(z) = -P(q^n z).$$

As was shown in [9, Theorem A], under an additional condition $|\alpha_0|_v \leq 1$,

$$Q(t) \frac{F(t)}{F(qt)} = \mathop{\mathbf{K}}_{n=1}^{\infty} \left(b_0(t) \,\middle|\, \begin{array}{c} a_n(t) \\ b_n(t) \end{array} \right) \tag{2.6}$$

holds for any $t \in K_v$ satisfying $F(qt) \neq 0$ and

$$M(q^n t) Q(q^n t) \neq 0 \quad (n = 0, 1, 2, ...), \tag{2.7}$$

where we use the notation

$$\mathop{\mathbf{K}}_{n=1}^{\infty} \left(b_0 \,\middle|\, \begin{array}{c} a_n \\ b_n \end{array} \right) = b_0 + \frac{a_1}{b_1 +} \frac{a_2}{b_2 +} \frac{a_3}{b_3 +} \cdots .$$

In the particular case where $M(z) \equiv q$, $P(z) \equiv -c$, and $Q(z) = a + bz$ with $ac \neq 0$, we have

$$(a + bt) \frac{F(t)}{F(qt)} = \mathop{\mathbf{K}}_{n=1}^{\infty} \left(c \,\middle|\, \begin{array}{c} aq^n t + bq^{2n} t^2 \\ c \end{array} \right).$$

This includes the Rogers-Ramanujan continued fraction for which $a = c = 1, b = 0$.

3. Notation

Let $K, \mathcal{M}, q$, and v be as above. For any $w \in \mathcal{M}$, we normalize the absolute value $|\ |_w$ of K so that $|p|_w = p^{-1}$ for a finite place w lying over the prime number p and $|x|_w = |x|$ ($x \in \mathbf{Q}$) for an infinite place w, where $|\ |$ denotes the ordinary absolute value. Let us set $\|\ \|_w = |\ |_w^{\kappa_w/\kappa}$, where $\kappa_w = [K_w : \mathbf{Q}_w]$. We also define the function $\delta(w)$ on $\mathcal{M}$ to be $\delta(w) = 0$ or 1 according as w is a finite or an infinite place.

The product formula

$$\prod_w \|\alpha\|_w = 1$$

holds for all nonzero $\alpha \in K$. The absolute height $H(\alpha)$ of $\alpha \in K$ is defined by

$$H(\alpha) = \prod_w \|\alpha\|_w^*, \quad \|\alpha\|_w^* = \max(1, \|\alpha\|_w).$$

For any vector $\underline{\alpha} = (\alpha_1, \alpha_2) \in K^2$, the absolute height $H(\underline{\alpha})$ of $\underline{\alpha}$ is defined by

$$H(\underline{\alpha}) = \prod_w \|\underline{\alpha}\|_w^*, \quad \|\underline{\alpha}\|_w^* = \max(1, \|\alpha_1\|_w, \|\alpha_2\|_w),$$

and two variants $\mathcal{K}_v(\underline{\alpha})$ and $\mathcal{L}_v(\underline{\alpha})$ of $H(\underline{\alpha})$ are defined by

$$\mathcal{K}_v(\underline{\alpha}) = \prod_{w \neq v} \max(\|\alpha_1\|_w, \|\alpha_2\|_w), \quad \mathcal{L}_v(\underline{\alpha}) = \min(\|\alpha_1\|_v, \|\alpha_2\|_v)\mathcal{K}_v(\underline{\alpha}).$$

We introduce the quantity $\lambda = \lambda_q$ defined by

$$\lambda = \frac{\log H(q)}{\log \|q\|_v}. \tag{3.1}$$

Note that $\lambda \leq -1$, and that $\lambda = -1$ if and only if $|q|_w \geq 1$ for all $w \neq v$

Finally, we recall the notion of linear independence measure and that of irrationality measure from [8]. Let θ_1 and θ_1 be elements of K_v which are linearly independent over K. Let ω be a positive number. If

$$|\alpha_1\theta_1 + \alpha_2\theta_2|_v > \frac{1}{(\mathcal{K}\mathcal{L}^\omega)^{\kappa/\kappa_v}} \tag{3.2}$$

holds for all $\underline{\alpha} = (\alpha_1, \alpha_2) \in K^2 \backslash \{\underline{0}\}$ with $\mathcal{K} = \mathcal{K}_v(\underline{\alpha})$ and $\mathcal{L} = \max(\mathcal{L}_v(\underline{\alpha}), \mathcal{L}_0)$, where $\mathcal{L}_0$ is a positive constant depending only on θ_1, θ_2, and ω, then we call ω a *linear independence measure relative to v of* θ_1 and θ_2. In the special case where $\theta_1 = \theta$ and $\theta_2 = 1$, if (3.2) holds for all $\underline{\alpha} = (\alpha_1, \alpha_2) = (1, -\alpha) \in K^2$ with $\mathcal{K}$ and $\mathcal{L}$ as above, then we call $\mu = \omega + 1$ an *irrationality measure relative to v of* θ.

Remark. Under the classical case where $K = \mathbf{Q}$ and $v = \infty$, we have $\mathcal{K}_v(\underline{\alpha}) \leq 1$ for all $\underline{\alpha} = (\alpha_1, \alpha_2) \in \mathbf{Z}^2$. Hence, if real numbers θ_1 and θ_2 are linearly independent over the rationals having $\omega > 0$ as a linear independence measure in the above sense, then

$$|\alpha_1\theta_1 + \alpha_2\theta_2|_v > \frac{1}{\max(\min(|\alpha_1|, |\alpha_2|), \mathcal{L}_0)^\omega}$$

holds for all $\underline{\alpha} = (\alpha_1, \alpha_2) \in \mathbf{Z}^2 \backslash \{\underline{0}\}$. This shows that the above notion for linear independence measure refines the ordinary notion for it in which we use $\max(|\alpha_1|, |\alpha_2|)$ instead of $\min(|\alpha_1|, |\alpha_2|)$.

4.   Main results

Let $F(z)$ be the solution of the functional equation (1.1) given in Section 2. For convenience, we write $F_0(z)$ instead of $F(z)$ if (1.1) is restricted to the case where $M(z) \equiv 1$ (and hence $m = 0$). We first state our result for general $F(z)$.

Theorem 1. *Assume that (1.1) is of the lowest degree, and that*

$$\frac{s}{2} + \lambda\tau = 0, \quad \tau = \max\left(k, \frac{s + \ell + m}{4}\right) \tag{4.1}$$

with $\lambda > -2$, where λ is the quantity given by (3.1). Then, for any nonzero $t \in K$ satisfying (2.7), $F(t)$ and $F(qt)$ are linearly independent over K. Moreover, an arbitrary positive number ω satisfying

$$\omega > \inf \frac{-\lambda(a + \tau a^2 + \tilde{\eta})}{\rho(\eta + s\rho/2) + \lambda(a + \tau a^2 + \tilde{\eta})} \tag{4.2}$$

with

$$a = 2 - \eta + \rho, \quad \tilde{\eta} = \frac{(s + 1)\eta^3}{6s(2 - \eta)}$$

is a linear independence measure relative to v of $F(t)$ and $F(qt)$, where the infimum in (4.2) is taken over all positive numbers η $(1 < \eta < 2)$ and ρ for which the quantities inside the infimum are positive.

Remarks. (a) The condition (4.1) implies the condition (1.3), and (4.1) is equivalent to the equality $s = \max(2k, \ell + m)$ if and only if $\lambda = -1$.

(b) Let us denote by $\omega(\eta, \rho)$ the quantity inside the infimum in (4.2). Under the condition (4.1), the denominator of $\omega(\eta, \rho)$ is expressed as

$$\rho((s + 1)\eta - (2s - \lambda)) + \lambda(2 - \eta + \tau(2 - \eta)^2 + \tilde{\eta}).$$

Hence $\omega(\eta, \rho)$ can take positive values in the range $1 < \eta < 2, \rho > 0$ if and only if $\lambda > -2$. As an explicit upper bound for $\omega(\eta, \rho)$, we can show for example

$$\omega(\eta, \rho) < (64s^2 + 200s + 21)/3$$

in the case where $\lambda = -1$.

(c) Under the condition

$$\frac{s}{2} + \lambda\tau > 0, \tag{4.3}$$

a result similar to Theorem 1 holds with a refined inequality

$$\omega > \frac{-\lambda\tau}{s/2 + \lambda\tau}$$

(see [8, Theorem 4.1]). This is the reason why we restrict ourselves to the case (4.1).

In Theorem 1 we have to assume generally that (1.1) is of the lowest degree, namely, $F(z)$ and $F(qz)$ are linearly independent over $K(z)$. In fact, for example,

$$zF_0(q^2z) = -F_0(qz) + (q^2z + 1)F_0(z), \quad F_0(z) = (1 + q)z + 1.$$

We will see in Section 7, Lemma 5 that if the condition (1.3) holds, then $F_0(z)/F_0(qz) \in K(z)$ implies $F_0(z) \in K[z]$. Hence, on noting the remark (a), we have the following result.

Theorem 2. *Assume that (4.1) with $\lambda > -2$ and $m = 0$ holds, and that $F_0(z)$ is not a polynomial. Then, for any nonzero $t \in K$ satisfying*

$$Q(q^n t) \neq 0 \quad (n = 0, 1, 2, ...), \tag{4.4}$$

$F_0(t)$ and $F_0(qt)$ are linearly independent over K. Moreover, an arbitrary positive number ω satisfying (4.2) is a linear independence measure relative to v of $F_0(t)$ and $F_0(qt)$.

Remark. We see from (1.1) that: (i) if $s > \max(k, \ell)$, then $F_0(z)$ is not a polynomial; (ii) if $s = \ell > k$, then $F_0(z) \in K[z]$ implies $\beta_s = q^{2n}$ with some nonnegative integer n.

It follows from Lemma 5 announced in the above together with the assertion (i) that $F_0(z)$ and $F_0(qz)$ are linearly independent over $K(z)$ if (1.2) holds.

The following corollary to Theorem 2 is a consequence of the relation (2.6) and this remark.

Corollary. *Assume that α_0 given by (2.2) satisfies $|\alpha_0|_v \leq 1$, and that (4.1) with $\lambda > -2$ and $m = 0$ holds. In the case where $s = \ell$, assume in addition that $\beta_s \neq q^{2n}$ for all nonnegative integers n. Then, for any nonzero $t \in K$ satisfying (4.4), the continued fraction*

$$\mathop{\mathbf{K}}_{n=1}^{\infty} \left(b_0(t) \,\middle|\, \frac{a_n(t)}{b_n(t)} \right)$$

with

$$a_n(t) = t^s q^{s(n-1)} Q(q^n t), \quad b_n(t) = -P(q^n t)$$

does not belong to K. Moreover, $\mu = \omega + 1$ with an arbitrary positive number ω satisfying (4.2) is an irrationality measure relative to v of the continued fraction.

In particular, by taking $q = 1/d$ with an arbitrary integer $d \neq \pm 1$, the real continued fraction

$$\overset{\infty}{\underset{n=1}{\mathbf{K}}} \left(1 \ \middle| \ \frac{q^n + q^{2n}}{1} \right)$$

is an irrational number, and not a Liouville number.

5.   Construction of auxiliary functions

As we will remark in the next section it is enough to treat $F(z)$ in the form (2.1) for proving our theorems. In what follows until the end of this paper, taking the remark into account, we assume that $F(z)$ is the series (2.1) converging on a disk $\mathbf{D}$ given by (2.5). The constants in O-notation are independent of n, which may be taken as large as we want.

We need the following result on the coefficients of $F(z)$ to construct auxiliary functions stated in Lemma 2 below.

Lemma 1. *There exists a positive number c such that, for any $w \in \mathcal{M}$,*

$$\|(q)_n f_n\|_w^* \leq C^{\delta(w)n} C_w^{n+1} \|q\|_w^{* \frac{s+1}{2s} n^2 + cn} \tag{5.1}$$

holds for all nonnegative integers n, where $(q)_n = (1-q) \cdots (1-q^n)$ with $(q)_0 = 1$, $C = 2(k + \ell + m + 1)$, and

$$C_w = \|\alpha_0\|_w^{-1} \max(\|\alpha_0\|_w, ..., \|\alpha_k\|_w, \|\beta_1\|_w, ..., \|\beta_\ell\|_w, \|\gamma_0\|_w, ..., \|\gamma_m\|_w).$$

Proof. Denoting $F_n = (q)_n f_n$ in the recursion formula (2.3), we have

$$\alpha_0 F_n = - \sum_{i \geq 1} (\alpha_i + \beta_i q^{n-i}) \frac{(q)_{n-1}}{(q)_{n-i}} F_{n-i} + \sum_{i \geq 0} \gamma_i q^{2(n-s-i)} \frac{(q)_{n-1}}{(q)_{n-s-i}} F_{n-s-i}.$$

It is easily seen that (5.1) holds for $n = 0$. Let us assume that (5.1) holds replacing n with $0, 1, ..., n-1$. Then we wish to show (5.1). Since

$$\|q^{n-i}\|_w^* \|(q)_{n-1}(q)_{n-i}^{-1}\|_w^* \leq 2^{\delta(w)(i-1)} \|q\|_w^{* in - i(i+1)/2},$$

we deduce from the above expression for F_n that

$$\|F_n\|_w^* \leq C^{\delta(w)n} C_w^{n+1} \|q\|_w^{* \max(d_i, e_j)},$$

where, for $1 \leq i \leq \max(k, \ell), 0 \leq j \leq m$,

$$d_i = \frac{s+1}{2s} n^2 + \left(c - \frac{i}{s} \right) n - ci + O(1),$$

$$e_j = \frac{s+1}{2s} n^2 + \left(c - \frac{j}{s} \right) n - c(s+j) + O(1)$$

with $O(1)$ independent of c. Hence (5.1) holds under a suitable choice of c. This completes the proof by induction on n. $\qquad\square$

The following lemma is analogous to a combination of Lemmas 1 and 2 of [1].

Lemma 2. *Let v_0 be an element of $\mathcal{M}$ which is different from v. Let η be an arbitrarily given positive number with $1 < \eta < 2$. Then, for any positive integer n, there exist polynomials*

$$A(z) = \sum_{i=0}^{n} a_i z^i, \quad B(z) = \sum_{i=0}^{n} b_i z^i,$$

not both zero, with $\underline{x} = (a_0, ..., a_n, b_0, ..., b_n) \in K^{2n+2}$, such that the function

$$R(z) = A(z)F(qz) + B(z)F(z) \qquad (5.2)$$

satisfies the following four conditions:

$$[\eta n] \;\leq\; \sigma \leq 2n + \max(s+m, k, \ell), \quad \sigma := \operatorname{ord} R(z), \qquad (5.3)$$

$$H(\underline{x}) \;\leq\; 2^{O(n)} H(q)^{\tilde{m}^2}, \quad \tilde{\eta} = \frac{(s+1)\eta^3}{6s(2-\eta)}, \qquad (5.4)$$

$$\|\underline{x}\|_w \;\leq\; 1 \quad \text{for all } w \neq v_0, \qquad (5.5)$$

$$|R(t)|_v \;\leq\; (n+2)^{\delta(v)} c_0^{\sigma+1} |t|_v^{\sigma} \quad \text{for } |t|_v \leq 1/(2c_0), \qquad (5.6)$$

where $\operatorname{ord} R(z)$ denotes the order of the zeros of $R(z)$ at $z = 0$ and c_0 is a constant given by (2.4).

Proof. We expand $R(z)$ given by (5.2) into power series including a_i, b_i. This gives a system of $[\eta n]$ linear homogeneous equations with $2n + 2$ unknowns to satisfy a lower bound condition of $\operatorname{ord} R(z)$ in (5.3). Applying Thue-Siegel Lemma in the form given by Bombieri [2], we can solve this system in K satisfying (5.4), (5.5), and (5.6). The proof is completely similar to that of Lemma 1 in [1] with the aid of Lemma 1 above. Hence we omit the details. There remains to show the upper bound of $\operatorname{ord} R(z)$ in (5.3). To this aim we use an another function $N(z)R(qz)$.

By the functional equation (1.1),

$$\tilde{R}(z) := N(z)R(qz) = \tilde{A}(z)F(qz) + \tilde{B}(z)F(z) \qquad (5.7)$$

where

$$\tilde{A}(z) = P(z)A(qz) + N(z)B(qz), \quad \tilde{B}(z) = Q(z)A(qz). \qquad (5.8)$$

Hence,

$$\begin{pmatrix} \tilde{R}(z) \\ R(z) \end{pmatrix} = \mathcal{R}(z) \begin{pmatrix} F(qz) \\ F(z) \end{pmatrix}, \quad \mathcal{R}(z) = \begin{pmatrix} \tilde{A}(z) & \tilde{B}(z) \\ A(z) & B(z) \end{pmatrix}. \quad (5.9)$$

We show that $\Delta(z) := \det \mathcal{R}(z) \neq 0$. By an elementary operation on the determinant,

$$F(qz)\Delta(z) = \begin{vmatrix} \tilde{R}(z) & \tilde{B}(z) \\ R(z) & B(z) \end{vmatrix}.$$

Since the lower bound for $\operatorname{ord} R$ in (5.3) implies that $\operatorname{ord} A = \operatorname{ord} B$, denoting the order by d and using (5.7), (5.8), we have

$$\operatorname{ord} B\tilde{R} = d + s + \sigma, \quad \operatorname{ord} \tilde{B}R = d + \sigma.$$

Hence

$$\operatorname{ord} F(qz)\Delta(z) = \operatorname{ord} \Delta(z) = d + \sigma,$$

which implies $\Delta \neq 0$, as desired. Since $\operatorname{ord} R \leq \operatorname{ord} \Delta \leq \deg \Delta$ and

$$\deg \Delta \leq \max(\deg A\tilde{B}, \deg \tilde{A}B) \leq 2n + \max(s + m, k, \ell)$$

by (5.8), an upper bound for $\operatorname{ord} R$ given in (5.3) holds. $\qquad \square$

6. Iteration of the functional equation

We first quote a result from [8].

Lemma 3. *For any nonnegative integer ν, there exists a matrix*

$$\mathcal{P}_\nu(z) = \begin{pmatrix} \tilde{P}_\nu(z) & \tilde{Q}_\nu(z) \\ P_\nu(z) & Q_\nu(z) \end{pmatrix}$$

with elements of $K[z]$ such that

$$[N(z)]_\nu \begin{pmatrix} F(q^{\nu+1}z) \\ F(q^\nu z) \end{pmatrix} = \mathcal{P}_\nu(z) \begin{pmatrix} F(qz) \\ F(z) \end{pmatrix} \quad (6.1)$$

where

$$[N(z)]_\nu := \prod_{i=0}^{\nu-1} N(q^i z) \quad ([N(z)]_{-1} := 1),$$

and that

$$\max \|\mathcal{P}_\nu\|_w^* \leq D_w^\nu \|q\|_w^{*\tau\nu^2} \quad \text{for all } w, \quad (6.2)$$

where $\|\mathcal{P}_\nu\|_w^*$ denotes the maximum of the values $\|\ \|_w^*$ for the coefficients of the elements of $\mathcal{P}_\nu$, τ is the quantity given in (4.1), and D_w ($w \in \mathcal{M}$)

are positive constants such that the product of all D_w is $O(1)$. Moreover, $\det P_\nu(z)$ does not vanish at $z = t$ satisfying (2.7).

Remark. Let t be an element of K satisfying (2.7), and (α, β) nonzero vector of K^2. It follows from (6.1) that, for any positive integer ν, there exists a nonzero vector $(\tilde{\alpha}_\nu, \tilde{\beta}_\nu)$ such that

$$\alpha F(qt) + \beta F(t) = \tilde{\alpha}_\nu F(q^{\nu+1}t) + \tilde{\beta}_\nu F(q^\nu t).$$

Hence, replacing t with $q^\nu t$ satisfying $q^\nu t \in \mathbf{D}$, we may assume $t \in \mathbf{D}$ for proving our theorems.

Using Lemmas 2 and 3, we show the main lemma as follows.

Lemma 4. *Let t be an element of $\mathbf{D}$ satisfying (2.7), where $\mathbf{D}$ is the disk (2.5). Let η $(1 < \eta < 2)$ and ρ be positive numbers, and set $a = 2 - \eta + \rho$. Then, for any positive number n, there exist two linearly independent linear forms*

$$R_{i,n} = M_{i,n} F(qt) + N_{i,n} F(t), \quad M_{i,n}, N_{i,n} \in K \quad (i = 1, 2)$$

such that

$$\max(\|M_{i,n}\|_w^*, \|N_{i,n}\|_w^*) \leq E_w^n \|q\|_w^{*(a+\tau a^2)n^2} \tag{6.3}$$

for all $w \in \mathcal{M}$ with $w \neq v_0$, where v_0 is an element of $\mathcal{M}$ which is different from v,

$$\max(\|M_{i,n}\|_{v_0}^*, \|N_{i,n}\|_{v_0}^*) \leq E_{v_0}^n \|q\|_{v_0}^{*(a+\tau a^2)n^2} H(q)^{\tilde{m}^2}, \tag{6.4}$$

$$|R_{i,n}|_v \leq E^n |q|_v^{\rho(\eta+s\rho/2)n^2}, \tag{6.5}$$

where E and E_w $(w \in \mathcal{M})$ are positive constants such that the product of all E_w is $O(1)$.

Proof. Let $R(z)$ be the function (5.2) satisfying the assertion of Lemma 2. Starting from (5.9), for any nonnegative integer ν, we define

$$\begin{pmatrix} \tilde{R}_\nu(z) \\ R_\nu(z) \end{pmatrix} := [N(z)]_\nu \begin{pmatrix} \tilde{R}(q^\nu z) \\ R(q^\nu z) \end{pmatrix} = \mathcal{R}(q^\nu z) \cdot [N(z)]_\nu \begin{pmatrix} F(q^{\nu+1}z) \\ F(q^\nu z) \end{pmatrix}. \tag{6.6}$$

Then, by Lemma 3,

$$\begin{pmatrix} \tilde{R}_\nu(z) \\ R_\nu(z) \end{pmatrix} = \mathcal{R}(q^\nu z) P_\nu(z) \begin{pmatrix} F(qz) \\ F(z) \end{pmatrix}, \tag{6.7}$$

which gives a pair of linear forms of $F(z)$ and $F(qz)$ for each ν. Since $\operatorname{ord} \Delta \geq [\eta n]$ and $\deg \Delta \leq 2n + O(1)$, where $\Delta = \det \mathcal{R}$, there exists a positive integer ν in the range

$$\rho n \leq \nu \leq \rho n + (2 - \eta)n + O(1) \tag{6.8}$$

such that $\det \mathcal{R}(q^\nu t) \neq 0$. We take such a ν, and set

$$\begin{pmatrix} R_{1,n} \\ R_{2,n} \end{pmatrix} = \begin{pmatrix} \tilde{R}_\nu(t) \\ R_\nu(t) \end{pmatrix}, \quad \begin{pmatrix} M_{1,n} & N_{1,n} \\ M_{2,n} & N_{2,n} \end{pmatrix} = \mathcal{R}(q^\nu t)\mathcal{P}_\nu(t). \quad (6.9)$$

Since $\mathcal{P}_\nu(t) \neq 0$ by Lemma 3, this gives two linearly independent linear forms of $F(t)$ and $F(qt)$. We show that the quantities above satisfy the conditions (6.3), (6.4), and (6.5).

In view of (6.9), (6.3) follows from (5.5), (6.2), and (6.8). Since

$$\|x\|_{v_0}^* = H(x) \leq 2^{O(n)} H(q)^{\tilde{m} m^2}$$

by (5.4) and (5.5), for $w = v_0$, we need an extra factor $H(q)^{\tilde{m} m^2}$ giving (6.4). By (6.6) and (5.6),

$$|R_\nu(t)|_v = |[N(t)]_\nu R(q^\nu t)|_v \leq |[N(t)]_\nu|_v (n+1)^{\delta(v)} c_0^{\sigma+1} |q^\nu t|_v^\sigma,$$

which with (5.3) and (6.8) implies (6.5) for $i = 2$. Since $\tilde{R}_\nu(t) = [N(t)]_{\nu+1} R(q^{\nu+1} t)$, (6.5) for $i = 1$ also holds. Thus the lemma is proved. $\square$

7.   Completion of the proofs

We now complete the proof of Theorem 1. Let t be an element of $\mathbf{D}$. Let ω be a positive number satisfying (4.2), and take η $(1 < \eta < 2)$ and $\rho > 0$ so that

$$\omega > -\lambda \frac{\Phi(\eta, \rho)}{\Psi(\eta, \rho)} > 0,$$

where

$$\Phi(\eta, \rho) = a + \tau a^2 + \tilde{\eta}, \quad \Psi(\eta, \rho) = \rho(\eta + s\rho/2) + \lambda\Phi(\eta, \rho)$$

with $a = 2 - \eta + \rho$. Let $\underline{\alpha} = (\alpha, \beta)$ be an arbitrarily given nonzero vector in K^2, and set

$$R = \alpha F(qt) + \beta F(t).$$

Then we wish to show that

$$|R|_v > \frac{1}{(\mathcal{KL}^\omega)^{\kappa/\kappa_v}}, \quad (7.1)$$

where $\mathcal{K} = \mathcal{K}(\underline{\alpha})$ and $\mathcal{L} = \max(\mathcal{L}(\underline{\alpha}), \mathcal{L}_0)$ with a suitably chosen positive constant $\mathcal{L}_0$.

In the case where $\beta = 0$, we have $\mathcal{K} = \|\alpha\|_v^{-1}$ by the product formula for α. Hence $(\mathcal{KL}^\omega)^{-(\kappa/\kappa_v)} = |\alpha|_v \mathcal{L}^{-(\omega\kappa/\kappa_v)}$, which is smaller than

$|\alpha F(qt)|_v$ if $\mathcal{L}_0 > |F(qt)|_v^{\kappa_v/(\omega\kappa)}$. This proves (7.1). The same holds also in the case where $\alpha = 0$.

In what follows we assume that both α and β are nonzero. By Lemma 4 with η and ρ as above, for any positive integer n, there exist $R_n = R_{i,n}$, $M_n = M_{i,n}$, and $N_n = N_{i,n}$ with $i = 1$ or 2 such that (6.3), (6.4), and (6.5) are fulfilled, and in addition that

$$\Delta = \begin{vmatrix} \alpha & \beta \\ M_n & N_n \end{vmatrix} \neq 0.$$

Let us assume that $\|\alpha\|_v \leq \|\beta\|_v$. By an elementary operation on the determinant,

$$F(t)\Delta = \begin{vmatrix} \alpha & R \\ M_n & R_n \end{vmatrix} = \alpha R_n - M_n R,$$

which implies that

$$|R|_v \geq \frac{|F(t)|_v}{2|M_n|_v}|\Delta|_v \tag{7.2}$$

if $|F(t)\Delta|_v \geq 2|\alpha R_n|_v$, or equivalently,

$$\log\|\Delta\|_v - \log\|\alpha\|_v - \frac{\kappa_v}{\kappa}\log|R_n|_v \geq \frac{\kappa_v}{\kappa}\log(2/|F(t)|_v). \tag{7.3}$$

Let us find n as small as possible for which (7.3) holds. On noting

$$|\Delta|_w \leq 2^{\delta(w)}\max(|\alpha|_w, |\beta|_w)\max(|M_n|_w, |N_n|_w)$$

for any $w \in M$, the product formula for Δ together with (6.3) and (6.4) implies that

$$\log\|\Delta\|_v = -\sum_{w\neq v}\log\|\Delta\|_w \geq -\log\mathcal{K}_v - \Phi(\eta,\rho)n^2\log H(q) + O(n). \tag{7.4}$$

On the other hand, it follows from (6.5) that

$$\frac{\kappa_v}{\kappa}\log|R_n|_v \leq \rho(\eta + s\rho/2)n^2\log\|q\|_v + O(n).$$

Hence the left-hand side of (7.3) is not smaller than

$$-\log\mathcal{L}_v - \Psi(\eta,\rho)n^2\log\|q\|_v + O(n).$$

Then, by taking $\mathcal{L}_0$ large enough, the least n satisfying

$$n \geq \sqrt{-\frac{\log\mathcal{L}}{\Psi(\eta,\rho)\log\|q\|_v}} + \log\log\mathcal{L} \tag{7.5}$$

28

ensures (7.3). Thus (7.1) follows from (7.2), (7.4), (7.5), and (6.3) with $w = v$.

Since the same argument can be carried out also in the case where $\|\alpha\|_v > \|\beta\|_v$, this completes the proof of Theorem 1.

The assertion of Theorem 2 follows directly from that of Theorem 1, if we have shown that the linear dependence of $F_0(z)$ and $F_0(qz)$ over $K(z)$ implies $F_0(z) \in K[z]$ under the condition (4.1) with $m = 0$. In fact, we can prove the last assertion under a weaker condition as follows.

Lemma 5. *Assume that $s \geq \max(k, \ell)$. Then $F_0(z)/F_0(qz) \in K(z)$ implies $F_0(z) \in K[z]$.*

Proof. Assume that $F_0(z)/F_0(qz) = A(z)/B(z)$ with relatively prime polynomials $A, B \in K[z]$. Note that $A(0) = B(0) \neq 0$. Let us divide the both sides of (1.1) by $F_0(qz)$ to get

$$z^s \frac{B(qz)}{A(qz)} = P(z) + Q(z)\frac{A(z)}{B(z)}.$$

Comparing the both sides, we see that if α is a zero of $A(qz)$ with multiplicity i, then α is a zero of $B(z)$ with multiplicity at least i. This implies that $A(qz)$ divides $B(z)$. Since

$$z^s B(qz)B(z) = A(qz)(P(z)B(z) + Q(z)A(z)), \qquad (7.6)$$

by comparing the degrees of the both sides,

$$\deg A = \deg B \quad \text{and} \quad s = \max(k, \ell).$$

The latter with the above observation implies that $B(z) = A(qz)$. Then, dividing the both sides of (7.6) by $B(z) = A(qz)$, we obtain

$$z^s A(q^2 z) = P(z)A(qz) + Q(z)A(z), \quad A(0) = 1.$$

Since $F_0(z) \in K[[z]]$ ($F_0(0) = 1$) is determined uniquely by (1.1), $F_0(z) = A(z)$, namely, $F_0(z) \in K[z]$. This completes the proof of the lemma, and hence that of Theorem 2. $\square$

Acknowledgments

The second named author would like to express his gratitude to Professor Masanori Katsurada for inviting him to Keio University, and to Professor Shigeru Kanemitsu for giving him the opportunity to talk in this nice conference. The authors are also indebted to the referee for careful reading of the manuscript and valuable comments.

References

[1] M. Amou, M. Katsurada, and K. Väänänen, Arithmetical properties of the values of functions satisfying certain functional equations of Poincaré, Acta Arith. **99** (2001), 389–407.

[2] E. Bombieri, On G-functions, in: *Recent Progress in Analytic Number Theory*, vol. 2, Academic Press, 1981, 1–68.

[3] P. Bundschuh, Ein Satz über ganze Funktionen und Irrationalitätsaussagen, Invent. Math. **9** (1970), 175–184.

[4] D. Duverney, Propriétés arithmétiques des solutions de certaines équations fonctionnelles de Poincaré, J. Théor. Nombres Bordeaux **8** (1996), 443–447.

[5] D. Duverney, Keiji Nishioka, Kumiko Nishioka, and I. Shiokawa, Transcendence of Rogers-Ramanujan continued fraction and reciprocal sums of Fibonacci numbers, Proc. Japan Acad. Ser. A, **73A** (1997), 140–142.

[6] D. Duverney and I. Shiokawa, On some arithmetical properties of Rogers-Ramanujan continued fraction, Osaka J. Math. **37** (2000), 759–771.

[7] T. Matala-aho, On Diophantine Approximations of the Rogers-Ramanujan Continued Fraction, J. Number Theory **45** (1993), 215–227.

[8] T. Matala-aho, On Diophantine approximations of the solutions of q-functional equations, Proc. Roy. Soc. Edinburgh Sect. A, **132A** (2002), 639–659.

[9] T. Matala-aho, On the values of continued fractions: q-series, submitted for publication.

[10] K. Nishioka, *Mahler Functions and Transcendence*, Lecture Notes in Math. 1631, Springer, 1996.

[11] C. F. Osgood, On the Diophantine Approximation of Values of Functions Satisfying Certain Linear q-difference equations, J. Number Theory **3** (1971), 159–177.

[12] I. Shiokawa, Rational approximations to the Rogers-Ramanujan continued fraction, Acta Arith. **50** (1988), 23–30.

RAMANUJAN'S CONTRIBUTIONS TO EISENSTEIN SERIES, ESPECIALLY IN HIS LOST NOTEBOOK

Bruce C. Berndt*
Department of Mathematics, University of Illinois
1409 West Green Street, Urbana, IL 61801, USA

berndt@math.uiuc.edu

Ae Ja Yee†
Department of Mathematics, University of Illinois
1409 West Green Street, Urbana, IL 61801, USA

yee@math.uiuc.edu

Abstract   The primary Eisenstein series considered in this paper are, in Ramanujan's notation, $P(q), Q(q)$, and $R(q)$. In more standard notation, $Q(q) = E_4(\tau)$ and $R(q) = E_6(\tau)$, where $q = \exp(2\pi i\tau)$. This paper provides a survey of many of Ramanujan's discoveries about Eisenstein series; most of the theorems are found in his lost notebook and are due originally to Ramanujan. Some of the topics examined are formulas for the power series coefficients of certain quotients of Eisenstein series, the role of Eisenstein series in proving congruences for the partition function $p(n)$, representations of Eisenstein series as sums of quotients of Dedekind eta-functions, a family of infinite series represented by polynomials in P, Q, and R, and approximations and exact formulas for π.

Keywords: Eisenstein series, Dedekind eta-function, partition function, congruences, formulas for power series coefficients, pentagonal number theorem, approximations to π, infinite series representations for $1/\pi$.

---

*Research partially supported by grant MDA904-00-1-0015 from the National Security Agency.
†Research partially supported by a grant from the Number Theory Foundation.
[1]2000 Mathematics Subject Classification: Primary, 11F11; Secondary, 11P83, 11F20.

1.   Introduction

In contemporary notation, the Eisenstein series $G_{2j}(\tau)$ and $E_{2j}(\tau)$ of weight $2j$ on the full modular group $\Gamma(1)$, where j is a positive integer exceeding one, are defined for $\operatorname{Im}\tau > 0$ by

$$G_{2j}(\tau) := \sum_{\substack{m_1,m_2 \in \mathbf{Z} \\ m_1,m_2 \neq (0,0)}} (m_1\tau + m_2)^{-2j}$$

and

$$E_{2j}(\tau) := \frac{G_{2j}(\tau)}{2\zeta(2j)} = 1 - \frac{4j}{B_{2j}} \sum_{k=1}^{\infty} \frac{k^{2j-1}e^{2\pi i k\tau}}{1 - e^{2\pi i k\tau}}$$

$$= 1 - \frac{4j}{B_{2j}} \sum_{r=1}^{\infty} \sigma_{2j-1}(r)q^r, \qquad q = e^{2\pi i\tau}, \qquad (1)$$

where B_n, $n \geq 0$, denotes the nth Bernoulli number, and where $\sigma_\nu(n) = \Sigma_{d|n}d^\nu$. The latter two representations in (1) can be established by using the Lipschitz summation formula or Fourier analysis. For these and other basic properties of Eisenstein series, see, for example, R. A. Rankin's text [45, Chap. 6].

In Ramanujan's notation, the three most relevant Eisenstein series are defined for $|q| < 1$ by

$$P(q) := 1 - 24 \sum_{k=1}^{\infty} \frac{kq^k}{1 - q^k}, \qquad (2)$$

$$Q(q) := 1 + 240 \sum_{k=1}^{\infty} \frac{k^3 q^k}{1 - q^k}, \qquad (3)$$

and

$$R(q) := 1 - 504 \sum_{k=1}^{\infty} \frac{k^5 q^k}{1 - q^k}. \qquad (4)$$

Thus, for $q = \exp(2\pi i\tau)$, $E_4(\tau) = Q(q)$ and $E_6(\tau) = R(q)$, which have weights 4 and 6, respectively. The function $P(q)$ is not a modular form. However, if $q = \exp(2\pi i\tau)$, with $\operatorname{Im} \tau = y$, then $P(q) - 3/(\pi y)$ satisfies the functional equation required of a modular form of weight 2 [45, p. 195]. In his notebooks [43], Ramanujan used the notations $L, M,$ and N in place of $P(q), Q(q),$ and $R(q)$, respectively.

Ramanujan made many contributions to the theory and applications of Eisenstein series in his paper [39], [42, pp. 136–162]. Applications were made to the theory of the divisor functions σ_ν; the partition function $p(n)$, the number of ways the positive integer n can be represented as a sum of positive integers with the order of the summands irrelevant; and $r_k(n)$, the number of representations of the positive integer n as a sum of k squares. Since Ramanujan did not consider any further applications of Eisenstein series to $r_k(n)$ in his lost notebook, we shall not further discuss his work on sums of squares. However, very exciting research on formulas for $r_{2k}(n)$ has recently emerged, and Eisenstein series play a prominent role. First, Z.–G. Liu [30] has found a simplified approach to Ramanujan's work in [39]. Secondly, new infinite classes of formulas for $r_{2k}(n)$ have been discovered by H. H. Chan and K. S. Chua [20], S. Milne [31], and K. Ono [32].

Among the most useful results on Eisenstein series established by Ramanujan in [39] are his differential equations [39, eqs. (30)], [42, p. 142]

$$q\frac{dP}{dq} = \frac{P^2(q) - Q(q)}{12}, \tag{5}$$

$$q\frac{dQ}{dq} = \frac{P(q)Q(q) - R(q)}{3}, \tag{6}$$

and

$$q\frac{dR}{dq} = \frac{P(q)R(q) - Q^2(q)}{2}. \tag{7}$$

For example, the differential equations for Q and R are needed to prove the following beautiful integral formula found on page 51 in Ramanujan's lost notebook [44], [37], [9, Thm. 3.1, pp. 82–83]. If $Q(q)$ and $R(q)$ are defined by (3) and (4), respectively, then

$$\int_{e^{-2\pi}}^{q} \sqrt{Q(t)}\frac{dt}{t} = \log\left(\frac{Q^{3/2}(q) - R(q)}{Q^{3/2}(q) + R(q)}\right), \qquad |q| < 1.$$

Several further results in this paper depend upon the differential equations (5)–(7).

Ramanujan recorded numerous results on Eisenstein series in his notebooks as well. Many are dependent on the following basic theorem in the theory of elliptic functions [3, Entry 6, p. 101]. Let, for $0 < x < 1$,

$$q := \exp\left(-\pi\frac{{}_2F_1\left(\frac{1}{2}, \frac{1}{2}; 1; 1-x\right)}{{}_2F_1\left(\frac{1}{2}, \frac{1}{2}; 1; x\right)}\right),$$

where ${}_2F_1\left(\frac{1}{2}, \frac{1}{2}; 1; 1-x\right)$ denotes the ordinary or Gaussian hypergeometric functionfunction, Gaussian hypergeometric. Then

$$z := {}_2F_1\left(\frac{1}{2}, \frac{1}{2}; 1; x\right) = \varphi^2(q),$$

where $\varphi(q)$ denotes the classical theta function

$$\varphi(q) := \sum_{n=-\infty}^{\infty} q^{n^2}.$$

The Eisenstein series Q and R can be represented in terms of z and x. These representations lead to many further series "evaluations" in terms of z and x; see [3, pp. 126–139]. Chapter 21 in Ramanujan's second notebook [43], [3, pp. 454–488] is entirely devoted to Eisenstein series. Eisenstein series arise in Ramanujan's "cubic" theory of elliptic functions; see [5, pp. 105–108]. Chapter 15 in his second notebook contains some of the results from [39] and other results not recorded in [39]; see [2, pp. 326–333]. Our brief summary has been by no means exhaustive; for further results of Ramanujan on Eisenstein series, consult [2]–[5].

The present survey of results on Eisenstein series found mostly in Ramanujan's lost notebook [44] comprises entries of different natures. In Section 2, we discuss Ramanujan's formulas for the coefficients of quotients of Eisenstein series. These were communicated to G. H. Hardy from a nursing home in 1918 and greatly extend the content of their last joint paper [24], [42, pp. 310–321]. In Section 3, we examine the role of Eisenstein series in proving congruences for the partition function $p(n)$. This material is found in a manuscript of Ramanujan on the partition and tau functions first published in handwritten form in [44] and then in [12] with commentary. The results in these two sections bring us to the natural investigation of possible congruences for the coefficients of quotients of Eisenstein series upon which we briefly focus in Section 4. At various places in his lost notebook, Ramanujan gives formulas for Eisenstein series in terms of quotients of Dedekind eta-functions. Several examples are given in Section 5. In [39], Ramanujan expresses families of Eisenstein series and related series as polynomials in P, Q, and R. A page in the lost notebook [44, p. 188] is devoted to another family of series, which in this case is related to the pentagonal number theorem; see Section 6 for some of these series. On another page in the lost notebook [44, p. 211], Ramanujan cryptically relates some formulas for Eisenstein series, which yield approximations to π in a spirit not unlike that for approximations to π given in his famous paper on modular equations and approximations to π [38]. We also indicate in Section 7 how Ramanujan's ideas lead to new series representations for $1/\pi$. In the final section, we discuss integrals of Eisenstein series associated with Dirichlet characters.

Lastly, we introduce further notation that we use in the sequel. As usual, set

$$(a; q)_\infty := \prod_{n=0}^{\infty} (1 - aq^n), \qquad |q| < 1.$$

Define, after Ramanujan,

$$f(-q) := (q; q)_\infty =: e^{-2\pi i \tau/24} \eta(\tau), \quad q = e^{2\pi i \tau}, \quad \text{Im } \tau > 0, \qquad (8)$$

where η denotes the Dedekind eta-function.

2. Formulas for the Coefficients of Quotients of Eisenstein Series

In their famous paper [23], [42, pp. 276–309], Hardy and Ramanujan found an asymptotic formula for the partition function $p(n)$, which arises from the power series coefficients of the reciprocal of the Dedekind eta-function, a modular form of weight $\frac{1}{2}$. As they indicated near the end of their paper, their methods also apply to several analogues of the partition function generated by modular forms of negative weight that are analytic in the upper half-plane. In their last published paper [24], [42, pp. 310–321], they considered a similar problem for the coefficients of modular forms of negative weight having one simple pole in a fundamental region, and, in particular, they applied their theorem to find interesting series representations for the coefficients of the reciprocal of the Eisenstein series $E_6(\tau)$. Although there are some similarities in the methods of these papers, the principal ideas are quite different in [24] from those in [23]. The ideas in [24] have been greatly extended by H. Poincaré [36], H. Petersson [33], [34], [35], and J. Lehner [27], [28].

While confined to nursing homes and sanitariums during his last two years in England, Ramanujan wrote several letters to Hardy about the coefficients in the power series expansions of certain quotients of Eisenstein series. These letters are photocopied in [44, pp. 97–126] and set to print with commentary in the book by Berndt and Rankin [13, pp. 175–191]. The letters contain many formulas for the coefficients of quotients of Eisenstein series not examined by Hardy and Ramanujan in [24]. Many of Ramanujan's claims do not fall under the purview of the main theorem in [24]. Ramanujan obviously wanted another example to be included in their paper [24], for in his letter of 28 June 1918 [13, pp. 182–183], he wrote, "I am sending you the analogous results in case of g_2. Please mention them in the paper without proof. After all we have got only two neat examples to offer, viz. g_2 and g_3. So please don't omit the results." This letter was evidently written after galley proofs for [24] were printed, because Ramanujan's request went unheeded. The functions g_2 and g_3 are the familiar invariants in the theory of elliptic functions and are constant multiples of the Eisenstein series $E_4(\tau)$ and $E_6(\tau)$, respectively.

Because the example from Hardy and Ramanujan's paper [24] is necessary for us in our examination of the deepest result from Ramanujan's letters, we state it below.

Theorem 1 *Define the coefficients p_n by*

$$\frac{1}{R(q^2)} =: \sum_{n=0}^{\infty} p_n q^{2n},$$

where, say, $|q| < q_0 < 1$. Let

$$\mu = 2^a \prod_{j=1}^{r} p_j^{a_j}, \tag{9}$$

where $a = 0$ or 1, p_j is a prime of the form $4m + 1$, and a_j is a nonnegative integer, $1 \le j \le r$. Then, for $n \ge 0$,

$$p_n = \sum_{(\mu)} T_\mu(n),$$

where μ runs over all integers of the form (9), and where

$$T_1(n) = \frac{2}{Q^2(e^{-2\pi})} e^{2n\pi}, \tag{10}$$

$$T_2(n) = \frac{2}{Q^2(e^{-2\pi})} \frac{(-1)^n}{2^4} e^{n\pi}, \tag{11}$$

and, for $\mu > 2$,

$$T_\mu(n) = \frac{2}{Q^2(e^{-2\pi})} \frac{e^{2n\pi/\mu}}{\mu^4} \sum_{c,d} 2\cos\left((ac + bd)\frac{2\pi n}{\mu} + 8\tan^{-1}\frac{c}{d}\right), \tag{12}$$

where the sum is over all pairs (c, d), where (c, d) is a distinct solution to $\mu = c^2 + d^2$ and (a, b) is any solution to $ad - bc = 1$. Also, distinct solutions (c, d) to $\mu = c^2 + d^2$ give rise to distinct terms in the sum in (12).

All of Ramanujan's formulas for the coefficients of quotients of Eisenstein series were established in two papers by Berndt and Bialek [6] and by Berndt, Bialek, and Yee [7]. The results proved in the former paper require only a mild extension of Hardy and Ramanujan's principal theorem. However, those in the latter paper are more difficult to prove, and not only did the theorem of Hardy and Ramanujan [24] needed to be extended, but the theorems of Poincaré, Petersson, and Lehner needed to be extended to cover double poles.

As one can see from Theorem 1, the formulas for these coefficients have a completely different shape from those arising from modular forms analytic in the upper half-plane. Moreover, these series are very rapidly convergent, more so than those arising from modular forms analytic in the upper half-plane, so that truncating a series, even with a small number of terms, provides a remarkable approximation. This is really very surprising, for when one examines the formulas for these coefficients, one would never guess how such amazingly accurate approximations could be obtained from just two terms.

We now offer an extension of the main theorems of Hardy and Ramanujan [24] and the aforementioned authors. Our theorem can be extended to vastly more general groups and to functions having several simple and double poles, but we only need an application for modular forms with only one double pole on a fundamental region for $\Gamma_0(2)$, and so we confine our statement to only this special case.

Theorem 2 *Suppose that $f(q) = f(e^{\pi i \tau}) = \phi(\tau)$ is analytic for $q = 0$, is meromorphic in the unit circle, and satisfies the functional equation*

$$\phi(\tau) = \phi\left(\frac{a\tau + b}{c\tau + d}\right)(c\tau + d)^n,$$

where $a, b, c, d \in \mathbf{Z}$; $ad - bc = 1$; c is even; and $n \in \mathbf{Z}^+$. Assume that $\phi(\tau)$ has only one pole in a fundamental region for $\Gamma_0(2)$, a double pole at $\tau = \alpha$. Suppose that $f(q)$ and $\phi(\tau)$ have the Laurent expansions,

$$\phi(\tau) = \frac{r_2}{(\tau - \alpha)^2} + \frac{r_1}{\tau - \alpha} + \cdots = \frac{\ell_2}{(q - e^{\pi i \alpha})^2} + \frac{\ell_1}{q - e^{\pi i \alpha}} + \cdots = f(q).$$

Then

$$f(q) = 2\pi i \sum_{c,d} \left\{ \frac{cr_2(n+2)}{(c\alpha + d)^{n+3}} - \frac{r_1}{(c\alpha + d)^{n+2}} \right\} \frac{1}{1 - (q/\underline{q})^2}$$

$$- 4\pi^2 r_2 \sum_{c,\underline{d}} \frac{1}{(c\alpha + d)^{n+4}} \frac{(q/\underline{q})^2}{(1 - (q/\underline{q})^2)^2}, \qquad |q| < 1,$$

where

$$\underline{q} = \exp\left(\left(\frac{a\alpha + b}{c\alpha + d}\right)\pi i\right),$$

*and the summation runs over all pairs of coprime integers (c, d) (with
c even) which yield distinct values for the set $\{\underline{q}, -\underline{q}\}$, and a and b are
any integral solutions of*

$$ad - bc = 1.$$

Furthermore,

$$r_1 = -\frac{i\ell_1}{\pi e^{\pi i \alpha}} + \frac{i\ell_2}{\pi e^{2\pi i \alpha}} \quad \text{and} \quad r_2 = -\frac{\ell_2}{\pi^2 e^{2\pi i \alpha}}.$$

Now define

$$B(q) := 1 + 24 \sum_{k=1}^{\infty} \frac{(2k-1)q^{2k-1}}{1 - q^{2k-1}}, \qquad |q| < 1,$$

and define the coefficients b_n by

$$\frac{1}{B(q)} = \sum_{n=0}^{\infty} b_n q^n,$$

where $|q|$ is sufficiently small. A formula for b_n was claimed by Ramanu-
jan in one of his letters to Hardy and was proved by Berndt, Bialek,
and Yee in [7]. We focus in this survey on the coefficients of $1/B^2(q)$,
however.

It is not difficult to see that $B(q)$ is the unique modular form of
weight 2 with multiplier system identically equal to 1 on the modular
group $\Gamma_0(2)$. It is also easy to show that $B(q)$ is related to $P(q)$ by the
simple formula

$$B(q) = 2P(q^2) - P(q).$$

We now state our principal application of Theorem 2.

Theorem 3 *Define the coefficients b'_n by*

$$\frac{1}{B^2(q^2)} =: \sum_{n=0}^{\infty} b'_n q^{2n},$$

where, say, $|q| < q_0 < 1$. Then,

$$b'_n = 18 \sum_{(\mu_e)} \left(n + \frac{3\mu_e}{2\pi} \right) T_{\mu_e}(n), \tag{13}$$

*where the sum is over all even integers μ_e of the form (9), and where
$T_{\mu_e}(n)$ is defined by (10)–(12).*

Using *Mathematica*, we calculated $b'_n, 1 \leq n \leq 10$, and the first two terms in (13). The accuracy is remarkable.

n	b'_n	$18\left((n + \frac{3}{\pi})T_2(n) + (n + \frac{15}{\pi})T_{10}(n)\right)$
1	-48	-48.001187
2	$1,680$	$1,679.997897$
3	$-52,032$	$-52,031.997988$
4	$1,508,496$	$1,508,496.002778$
5	$-41,952,672$	$-41,952,671.998915$
6	$1,133,840,832$	$1,133,840,831.996875$
7	$-30,010,418,304$	$-30,010,418,304.008563$
8	$781,761,426,576$	$781,761,426,576.003783$
9	$-20,110,673,188,848$	$-20,110,673,188,847.986981$
10	$512,123,093,263,584$	$512,123,093,263,584.006307$

To further examine the rapidity of convergence of (13), we calculated the coefficients of $1/B^2(q)$ up to $n = 50$. For $n = 20, 30, 40$, and 50, the coefficients have, respectively, 29, 43, 57, and 70 digits, while two-term approximations give, respectively, 29, 42, 55, and 66 of these digits.

3. Eisenstein Series and Partitions

Ramanujan utilized Eisenstein series to establish congruences for the partition function $p(n)$. Among his most famous identities connected with partitions are the following four identities:

$$\sum_{n=1}^{\infty} \left(\frac{n}{5}\right) \frac{q^n}{(1 - q^n)^2} = q \frac{(q^5; q^5)_\infty^5}{(q; q)_\infty}, \tag{14}$$

$$1 - 5\sum_{n=1}^{\infty} \left(\frac{n}{5}\right) \frac{nq^n}{1 - q^n} = \frac{(q; q)_\infty^5}{(q^5; q^5)_\infty}, \tag{15}$$

$$\sum_{n=1}^{\infty} \left(\frac{n}{7}\right) q^n \frac{1 + q^n}{(1 - q^n)^3} = q(q; q)_\infty^3 (q^7; q^7)_\infty^3 + 8q^2 \frac{(q^7; q^7)_\infty^7}{(q; q)_\infty}, \tag{16}$$

and

$$8 - 7\sum_{n=1}^{\infty} \left(\frac{n}{7}\right) \frac{n^2 q^n}{1 - q^n} = 49q(q; q)_\infty^3 (q^7; q^7)_\infty^3 + 8\frac{(q; q)_\infty^7}{(q^7; q^7)_\infty}, \tag{17}$$

where $|q| < 1$ and $\left(\frac{n}{p}\right)$ denotes the Legendre symbol. Identity (14) can be utilized to prove Ramanujan's celebrated congruence $p(5n + 4) \equiv 0 \,(\mathrm{mod}\,5)$, while (16) can be employed to establish Ramanujan's equally famous congruence $p(7n + 5) \equiv 0 \,(\mathrm{mod}\,7)$. The identities in (15) and

(17) are Eisenstein series on the congruence subgroups $\Gamma_0(5)$ and $\Gamma_0(7)$, respectively. These four identities were first recorded in an unpublished manuscript of Ramanujan on the partition and tau functions, first published in handwritten form with Ramanujan's lost notebook [44, pp. 135–177]. Equivalent forms of (14) and (16) are offered by Ramanujan in his first paper on congruences for $p(n)$ [40], [42, pp. 210–213]. Hardy extracted some of the material from this unpublished manuscript for Ramanujan's posthumously published paper [41], [42, pp. 232–238]. A typed version of Ramanujan's unpublished manuscript, together with proofs and commentary, has been prepared by Berndt and K. Ono [12]. The latter paper and the new edition of Ramanujan's *Collected Papers* [42, pp. 372–375] contain several references to proofs of (14)–(17). H. H. Chan [19] has shown the equivalence of (14) and (15), as well as the equivalence of (16) and (17). To the best of our knowledge, (17) was first proved by N. J. Fine [22]. See also a paper by O. Kolberg [26] for another proof.

We provide here one example from [12] to illustrate how Ramanujan used Eisenstein series to establish the congruence $p(7n+5) \equiv 0 \pmod{7}$, $n \geq 0$. Further examples can be found in [12] and [41].

Ramanujan [39, Table I] evidently was the first to observe that

$$Q^2 = 1 + 480 \sum_{n=1}^{\infty} \frac{n^7 q^n}{1 - q^n}. \tag{18}$$

The identity (18) also follows readily from the theory of modular forms, since both E_4^2 and E_8 are modular forms of weight 8 on the full modular group, and the vector space of these forms has dimension 1. It is easy to see from (18) and (2)–(4) that

$$Q^2 = P + 7J \qquad \text{and} \qquad R = 1 + 7J, \tag{19}$$

where J (not the same at each occurrence) is a power series in q with integral coefficients. It follows from (19) that

$$\left(1728 q(q;q)_\infty^{24}\right)^2 = (Q^3 - R^2)^2 = P^3 - 2PQ + R + 7J. \tag{20}$$

Now [39, Tables II and III, resp.]

$$PQ - R = 720 \sum_{n=1}^{\infty} n\sigma_3(n) q^n \tag{21}$$

and

$$P^3 - 3PQ + 2R = -1728 \sum_{n=1}^{\infty} n^2 \sigma_1(n) q^n. \tag{22}$$

Furthermore, we easily see that

$$(q;q)_\infty^{48} = \frac{(q;q)_\infty^{49}}{(q;q)_\infty} = \frac{(q^{49};q^{49})_\infty}{(q;q)_\infty} + 7J. \tag{23}$$

It follows from (20)–(23) that

$$q^2 \frac{(q^{49};q^{49})_\infty}{(q;q)_\infty} = \sum_{n=1}^{\infty} \left\{ n^2\sigma_1(n) - n\sigma_3(n) \right\} q^n + 7J.$$

In other words,

$$(q^{49};q^{49})_\infty \sum_{n=0}^{\infty} p(n)q^{n+2} = \sum_{n=1}^{\infty} \left\{ n^2\sigma_1(n) - n\sigma_3(n) \right\} q^n + 7J.$$

It immediately follows that, for $n \geq 1$,

$$p(7n - 2) \equiv 0 \, (\text{mod} \, 7),$$

and, as a bonus,

$$p(n - 2) - p(n - 51) - p(n - 100) + p(n - 247)$$
$$+p(n - 345) - \cdots \equiv n^2\sigma_1(n) - n\sigma_3(n) \, (\text{mod} \, 7),$$

where $2, 51, 100, 247, \ldots$ are the numbers of the form $\frac{1}{2}(7\nu + 1)(21\nu + 4)$ and $\frac{1}{2}(7\nu - 1)(21\nu - 4)$.

4. Congruences for the Coefficients of Quotients of Eisenstein Series

In calculating the coefficients of the quotients of the Eisenstein series which appear in [6] and [14], and which we briefly discussed in Section 2, we noticed that for some quotients the coefficients in certain arithmetic progressions are divisible by prime powers, usually a power of 3. In view of Ramanujan's famous congruences for $p(n)$, one of which we proved in the previous section, it seemed natural for us to systematically investigate congruences of this type for Eisenstein series. In some cases, it was very easy to establish our observations, but in other cases, the task was considerably more difficult. We summarize what we have accomplished [15] in the following table. For each quotient, set $F(q) = \sum_{n=0}^{\infty} a_n q^n$.

$F(q)$	$n \equiv 2 \pmod 3$	$n \equiv 4 \pmod 8$
$1/P(q)$	$a_n \equiv 0 \pmod{3^4}$	
$1/Q(q)$	$a_n \equiv 0 \pmod{3^2}$	
$1/R(q)$	$a_n \equiv 0 \pmod{3^3}$	$a_n \equiv 0 \pmod{7^2}$
$P(q)/Q(q)$	$a_n \equiv 0 \pmod{3^3}$	
$P(q)/R(q)$	$\alpha_n \equiv 0 \pmod{3^2}$	$\alpha_n \equiv 0 \pmod 7$
$Q(q)/R(q)$	$\alpha_n \equiv 0 \pmod{3^3}$	
$P^2(q)/R(q)$	$a_n \equiv 0 \pmod{3^5}$	

5. Representations of Eisenstein Series as Quotients of Dedekind Eta-functions

In Section 3, we offered some identities for Eisenstein series that are useful for the study of $p(n)$. The eight identities we present in this section are found on pages 44, 50, 51, and 53 in Ramanujan's lost notebook [44] and were first proved by S. Raghavan and S. S. Rangachari [37]. Since they used the theory of modular forms, it seemed desirable to construct proofs in the spirit of Ramanujan to gain a better insight into how Ramanujan originally discovered them and to also seek a better understanding of the identities themselves. To that end, Berndt, Chan, J. Sohn, and S. H. Son [11] found proofs which depend *only* on material found in Ramanujan's notebooks [43]. All these identities for Eisenstein series are connected with modular equations of either degree 5 or degree 7. Z.–G. Liu [29] has recently constructed proofs of Ramanujan's septic identities and some new septic identities as well using the theory of elliptic functions, but employing methods from complex analysis. The methods in [11] do not use complex analysis.

We now record the eight identities in four separate theorems, after which we offer a few words about proofs.

Theorem 4 (p. 50) *For $Q(q)$ and $f(-q)$ defined by* (3) *and* (8), *respectively*,

$$Q(q) = \frac{f^{10}(-q)}{f^2(-q^5)} + 250qf^4(-q)f^4(-q^5) + 3125q^2\frac{f^{10}(-q^5)}{f^2(-q)}$$

and

$$Q(q^5) = \frac{f^{10}(-q)}{f^2(-q^5)} + 10qf^4(-q)f^4(-q^5) + 5q^2\frac{f^{10}(-q^5)}{f^2(-q)}.$$

Theorem 5 (p. 51) *For $f(-q)$ and $R(q)$ defined by* (8) *and* (4), *respectively*,

$$R(q) = \left(\frac{f^{15}(-q)}{f^3(-q^5)} - 500qf^9(-q)f^3(-q^5) - 15625q^2f^3(-q)f^9(-q^5)\right)$$
$$\times\sqrt{1 + 22q\frac{f^6(-q^5)}{f^6(-q)} + 125q^2\frac{f^{12}(-q^5)}{f^{12}(-q)}}$$

and

$$R(q^5) = \left(\frac{f^{15}(-q)}{f^3(-q^5)} + 4qf^9(-q)f^3(-q^5) - q^2f^3(-q)f^9(-q^5)\right)$$
$$\times\sqrt{1 + 22q\frac{f^6(-q^5)}{f^6(-q)} + 125q^2\frac{f^{12}(-q^5)}{f^{12}(-q)}}.$$

Theorem 6 *For* $|q| < 1$,

$$
\begin{aligned}
Q(q) &= \left(\frac{f^7(-q)}{f(-q^7)} + 5 \cdot 7^2 q f^3(-q) f^3(-q^7) + 7^4 q^2 \frac{f^7(-q^7)}{f(-q)} \right) \\
&\times \left(\frac{f^7(-q)}{f(-q^7)} + 13q f^3(-q) f^3(-q^7) + 49 q^2 \frac{f^7(-q^7)}{f(-q)} \right)^{1/3}
\end{aligned}
$$

and

$$
\begin{aligned}
Q(q^7) &= \left(\frac{f^7(-q)}{f(-q^7)} + 5q f^3(-q) f^3(-q^7) + q^2 \frac{f^7(-q^7)}{f(-q)} \right) \\
&\times \left(\frac{f^7(-q)}{f(-q^7)} + 13q f^3(-q) f^3(-q^7) + 49 q^2 \frac{f^7(-q^7)}{f(-q)} \right)^{1/3}.
\end{aligned}
$$

Theorem 7 *For* $|q| < 1$,

$$
\begin{aligned}
&R(q) \\
&= \left(\frac{f^7(-q)}{f(-q^7)} - 7^2 \left(5 + 2\sqrt{7}\right) q f^3(-q) f^3(-q^7) - 7^3 (21 + 8\sqrt{7}) q^2 \frac{f^7(-q^7)}{f(-q)} \right) \\
&\times \left(\frac{f^7(-q)}{f(-q^7)} - 7^2 (5 - 2\sqrt{7}) q f^3(-q) f^3(-q^7) - 7^3 (21 - 8\sqrt{7}) q^2 \frac{f^7(-q^7)}{f(-q)} \right)
\end{aligned}
$$

and

$$
\begin{aligned}
&R(q^7) \\
&= \left(\frac{f^7(-q)}{f(-q^7)} + (7 + 2\sqrt{7}) q f^3(-q) f^3(-q^7) + (21 + 8\sqrt{7}) q^2 \frac{f^7(-q^7)}{f(-q)} \right) \\
&\times \left(\frac{f^7(-q)}{f(-q^7)} + (7 - 2\sqrt{7}) q f^3(-q) f^3(-q^7) + (21 - 8\sqrt{7}) q^2 \frac{f^7(-q^7)}{f(-q)} \right).
\end{aligned}
$$

Two different methods were used in [11] to prove Theorems 4–7, but both employ modular equations discovered by Ramanujan and found in his notebooks [43]. The second method seems more amenable to generalization.

Ramanujan recorded several further, related identities in his lost notebook for P, Q, and R. Some of these involve differential equations satisfied by P. See the paper by Berndt, Chan, Sohn, and Son [11] for statements and proofs.

6.   Eisenstein Series and a Series Related to the Pentagonal Number Theorem

Page 188 of Ramanujan's lost notebook, in the pagination of [44], is devoted to the series

$$T_{2k}(q) := 1 + \sum_{n=1}^{\infty} (-1)^n \left\{ (6n - 1)^{2k} q^{n(3n-1)/2} + (6n + 1)^{2k} q^{n(3n+1)/2} \right\},$$

where $|q| < 1$. Note that the exponents $n(3n \pm 1)/2$ are the generalized pentagonal numbers. Ramanujan recorded formulas for $T_{2k}(q), k = 1, 2, \ldots, 6$, in terms of the Eisenstein series, P, Q, and R. The first three are given by

(i)
$$\frac{T_2(q)}{(q; q)_\infty} = P,$$

(ii)
$$\frac{T_4(q)}{(q; q)_\infty} = 3P^2 - 2Q,$$

(iii)
$$\frac{T_6(q)}{(q; q)_\infty} = 15P^3 - 30PQ + 16R.$$

Ramanujan's work on this page can be considered as a continuation of his study of representing certain kinds of series as polynomials in Eisenstein series in [39], [42, pp. 136–162].

The first formula, (i), has an interesting arithmetical interpretation. For $n \geq 1$, let $\sigma(n) = \sum_{d|n} d$, and define $\sigma(0) = -\frac{1}{24}$. Let n denote a nonnegative integer. Then

$$-24 \sum_{j+k(3k\pm1)/2=n} (-1)^k \sigma(j) = \begin{cases} (-1)^r (6r - 1)^2, & \text{if } n = r(3r - 1)/2, \\ (-1)^r (6r + 1)^2, & \text{if } n = r(3r + 1)/2, \\ 0, & \text{otherwise,} \end{cases}$$

$$(24)$$

where the sum is over all nonnegative pairs of integers (j, k) such that $j + k(3k \pm 1)/2 = n$.

There are many identities involving the divisor sums $\sigma_k(n) := \sum_{d|n} d^k$ in the literature, but we have not previously seen (24). Besides (24), other identities of Ramanujan can be reformulated in terms of divisor sums. In particular, see [2, pp. 326–329] and the references cited there. The most thorough study of identities of this sort has been undertaken by J. G. Huard, Z. M. Ou, B. K. Spearman, and K. S. Williams [25], where many references to the literature can also be found.

In general, $T_{2k}(q)$ can be represented as a polynomial in terms of the form $P^{2a} Q^{4b} R^{6c}$, where $2a + 4b + 6c = 2k$. It seems to be extremely difficult to find a general formula for this polynomial, but on page 188, Ramanujan does give the "first" five terms. Define the polynomials $f_{2k}(P, Q, R), k \geq 1$, by

$$f_{2k}(P, Q, R) := \frac{T_{2k}(q)}{(q; q)_\infty}.$$

Then, for $k \geq 1$,

$$f_{2k}(P, Q, R) = 1 \cdot 3 \cdots (2k - 1) \left\{ P^k - \frac{k(k - 1)}{3} P^{k-2} Q \right.$$
$$+ \frac{8k(k - 1)(k - 2)}{45} P^{k-3} R$$
$$- \frac{11k(k - 1)(k - 2)(k - 3)}{210} P^{k-4} Q^2$$
$$\left. + \frac{152k(k - 1)(k - 2)(k - 3)(k - 4)}{14175} P^{k-5} QR + \cdots \right\}.$$

Proofs of all the claims in this section can be found in a paper by Berndt and Yee [14].

7.   Eisenstein Series and Approximations to π

On page 211 in his lost notebook, in the pagination of [44], Ramanujan listed eight integers, 11, 19, 27, 43, 67, 163, 35, and 51 at the left margin. To the right of each integer, Ramanujan recorded a linear equation in Q^3 and R^2. The arguments q in these identities were not revealed by Ramanujan, but $q = -\exp(-\pi\sqrt{n})$, where n is the integer at the left margin. To the right of each equation in Q^3 and R^2, Ramanujan entered an equality involving π and square roots. (For the integer 51, the linear equation and the equality involving π, in fact, are not recorded by Ramanujan.) The equalities in the third column lead to approximations of π that are remindful of approximations given by Ramanujan in his famous paper on modular equations and approximations to π [38], [42, p. 33].

We offer a general theorem from which all the equalities in the third column follow. Considerable notation is first needed, however.

Let

$$P_n := P(-e^{-\pi\sqrt{n}}), \quad Q_n := Q(-e^{-\pi\sqrt{n}}), \quad \text{and} \quad R_n := R(-e^{-\pi\sqrt{n}}). \quad (25)$$

Recall that the modular j–invariant $j(\tau)$ is defined by

$$j(\tau) = 1728 \frac{Q^3(q)}{Q^3(q) - R^2(q)}, \quad q = e^{2\pi i \tau}, \quad \text{Im } \tau > 0.$$

In particular, if n is a positive integer,

$$j_n := j\left(\frac{3 + \sqrt{-n}}{2}\right) = 1728\frac{Q_n^3}{Q_n^3 - R_n^2}, \tag{26}$$

where Q_n and R_n are defined by (25). Furthermore, set

$$J_n = -\frac{1}{32}\sqrt[3]{j_n}. \tag{27}$$

Next, define

$$b_n = \{n(1728 - j_n)\}^{1/2} \tag{28}$$

and

$$a_n = \frac{1}{6}b_n\left\{1 - \frac{Q_n}{R_n}\left(P_n - \frac{6}{\pi\sqrt{n}}\right)\right\}. \tag{29}$$

The numbers a_n and b_n arise in series representations for $1/\pi$ proved by D. V. and G. V. Chudnovsky [21] and J. M. and P. B. Borwein [17].
We now have sufficient notation to state our first theorem.

Theorem 8 *If P_n, b_n, a_n, and J_n are defined by (25), (28), (29), and (27), respectively, then*

$$\frac{1}{\sqrt{Q_n}}\left(\sqrt{n}P_n - \frac{6}{\pi}\right) = \sqrt{n}\left(1 - 6\frac{a_n}{b_n}\right)\left(\frac{\left(\frac{8}{3}J_n\right)^3 + 1}{\left(\frac{8}{3}J_n\right)^3}\right)^{1/2}. \tag{30}$$

To illustrate Theorem 8, we offer Ramanujan's eight examples.

Corollary 9 *We have*

$$\frac{1}{\sqrt{Q_{11}}}\left(\sqrt{11}P_{11} - \frac{6}{\pi}\right) = \sqrt{2},$$

$$\frac{1}{\sqrt{Q_{19}}}\left(\sqrt{19}P_{19} - \frac{6}{\pi}\right) = \sqrt{6},$$

$$\frac{1}{\sqrt{Q_{27}}}\left(\sqrt{27}P_{27} - \frac{6}{\pi}\right) = 3\sqrt{\frac{6}{5}},$$

$$\frac{1}{\sqrt{Q_{43}}}\left(\sqrt{43}P_{43} - \frac{6}{\pi}\right) = 6\sqrt{\frac{3}{5}},$$

$$\frac{1}{\sqrt{Q_{67}}}\left(\sqrt{67}P_{67} - \frac{6}{\pi}\right) = 19\sqrt{\frac{6}{55}},$$

$$\frac{1}{\sqrt{Q_{163}}}\left(\sqrt{163}P_{163} - \frac{6}{\pi}\right) = 362\sqrt{\frac{3}{3335}},$$

$$\frac{1}{\sqrt{Q_{35}}}\left(\sqrt{35}P_{35} - \frac{6}{\pi}\right) = (2 + \sqrt{5})\sqrt{\frac{2}{\sqrt{5}}},$$

$$\frac{1}{\sqrt{Q_{51}}}\left(\sqrt{51}P_{51} - \frac{6}{\pi}\right) = .$$

Theorem 8 leads to the following theorem giving approximations of π.

Theorem 10 *We have*

$$\pi \approx \frac{6}{\sqrt{n} - r_n} =: A_n,$$

with the error approximately equal to

$$144\frac{\sqrt{n} + 5r_n}{(\sqrt{n} - r_n)^2}e^{-\pi\sqrt{n}},$$

where r_n is the algebraic expression on the right side in (30), i.e.,

$$r_n = \sqrt{n}\left(\frac{\left(\frac{8}{3}J_n\right)^3 + 1}{\left(\frac{8}{3}J_n\right)^3}\right)^{1/2}.$$

Theorem 10 with the examples above lead to approximations of π which we offer in the next table. Let N_n denote the number of digits of π which agree with the decimal expansion of A_n.

n	A_n	N_n
11	3.1538...	1
19	3.1423...	2
27	3.1416621...	3
43	3.141593...	5
67	3.14159266...	7
163	3.14159265358980...	12
35	3.141601...	3
51	3.14159289...	6

Although not mentioned by Ramanujan on page 211 in [44], the ideas needed to prove the results on this page lead to a very general infinite series expansion for $1/\pi$ in the spirit of those given by Ramanujan [38], and later by the Chudnovskys [21] and Borweins [17]. To state this expansion, we need further definitions.

Let

$$\mathbf{J}(q) := \frac{1728}{j\left(\dfrac{3+\tau}{2}\right)}$$

and

$$\mathbf{J}_n = \mathbf{J}(e^{-\pi\sqrt{n}}), \qquad n > 0. \tag{31}$$

Furthermore, let

$$t_n = \frac{Q_n}{R_n}\left(P_n - \frac{6}{\sqrt{n}\pi}\right), \tag{32}$$

where P_n, Q_n, and R_n are defined by (25).

We now state a general series representation for $1/\pi$.

Theorem 11 *If $\mathbf{J}_n$ and t_n are defined by (31) and (32), respectively, then*

$$\frac{6}{\sqrt{n}\sqrt{1-\mathbf{J}_n}}\frac{1}{\pi} = \sum_{k=0}^{\infty}\frac{\left(\frac{1}{6}\right)_k\left(\frac{5}{6}\right)_k\left(\frac{1}{2}\right)_k}{(k!)^3}\mathbf{J}_n^k(6k+1-t_n), \tag{33}$$

where $(a)_0 := 1$ and $(a)_n := a(a+1)(a+2)\cdots(a+n-1)$, for each positive integer n.

If j_n is defined by (26), then

$$j_{3n} = -27\frac{(\lambda_n^2-1)(9\lambda_n^2-1)^3}{\lambda_n^2}, \tag{34}$$

where

$$\lambda_n = \frac{e^{\pi\sqrt{n}/2}}{3\sqrt{3}}\frac{f^6(e^{-\pi\sqrt{n/3}})}{f^6(e^{-\pi\sqrt{3n}})}$$

and f is defined by (8). The numbers λ_n were first defined by Ramanujan on page 212 of his lost notebook, in the pagination of [44]. He recorded an extensive table of values which were verified by Berndt, Chan, S.-Y. Kang, and L.-C. Zhang [10] by a variety of methods.

In [8], Berndt and Chan used Theorem 11 to achieve a new "World Record" for the number of digits of $1/\pi$ per term. They used the value,

$$\lambda_{1105} = \left(\frac{\sqrt{5}+1}{2}\right)^{12} (4+\sqrt{17})^3 \left(\frac{15+\sqrt{221}}{2}\right)^3 (8+\sqrt{65})^3,$$

in the formula (34) to calculate J_{3315} (defined by (31)). Substituting this value for J_{3315} and the value of t_{3315} (defined by (32)) in (33), Berndt and Chan obtained a series for $1/\pi$ which yields about 73 or 74 digits of π per term. All the results in this section were established in a paper by Berndt and Chan [8].

8. Integrals of Eisenstein Series

In the Introduction, we gave one beautiful example of an integral of an Eisenstein series found in the lost notebook. Here we present a different kind of example. At first glance, it does not appear that there is any connection with Eisenstein series. Recall that the functions $f(-q)$ and $\eta(\tau)$ are defined in (8).

Theorem 12 *Suppose that $0 < q < 1$. Then*

$$q^{1/9} \prod_{n=1}^{\infty} (1-q^n)^{\chi(n)n} = \exp\left(-C_3 - \frac{1}{9}\int_q^1 \frac{f^9(-t)}{f^3(-t^3)} \frac{dt}{t}\right), \qquad (35)$$

where

$$C_3 = \frac{3\sqrt{3}}{4\pi} L(2,\chi) = L'(-1,\chi). \qquad (36)$$

Here, $L(s,\chi)$ denotes the Dirichlet L-function defined for Re $s > 1$ by $L(s,\chi) = \sum_{n=1}^{\infty} \chi(n)/n^s$. Furthermore, in (35) and (36), $\chi(n) = \left(\frac{n}{3}\right)$, the Legendre symbol.

This result, found on page 207 of Ramanujan's lost notebook, was first proved by Son [46], except that he did not establish Ramanujan's formula (36) for C_3. Berndt and Zaharescu [16] found another proof of Theorem 12, which included a proof of (36).

We now explain the connection of Theorem 12 with Eisenstein series. A key to the proof of (35) is the identity

$$\frac{f^9(-q)}{f^3(-q^3)} = 1 - 9\sum_{n=1}^{\infty}\sum_{d|n} \left(\frac{d}{3}\right) d^2 q^n, \qquad (37)$$

which was first proved by L. Carlitz [18] and then by Son [46]. The series on the right side of (37) is an example of an Eisenstein series $E_{k,\chi}(\tau)$ of weight k, defined on the congruence subgroup $\Gamma_0(N)$ by

$$E_{k,\chi}(\tau) := 1 - \frac{2k}{B_{k,\chi}} \sum_{n=1}^{\infty} \sum_{d|n} \chi(d) d^{k-1} q^n, \qquad q = e^{2\pi i \tau}, \qquad (38)$$

where χ is a Dirichlet character of modulus N, and $B_{k,\chi}$ denotes the kth generalized Bernoulli number associated with χ.

Ramanujan's identity (17), when written in the form,

$$E_{3,\chi}(z) = 1 - \frac{7}{8} \sum_{n=1}^{\infty} \sum_{d|n} \chi(d) d^2 q^n = \frac{\eta^7(\tau)}{\eta(7\tau)} + \frac{49}{8} \eta^3(\tau) \eta^3(7\tau), \qquad (39)$$

provides another example, where now $\chi(n)$ denotes the Legendre symbol modulo 7. Then (39) leads to the identity

$$q^{8/7} \prod_{n=1}^{\infty} (1 - q^n)^{\chi(n)n}$$

$$= \exp\left(-C_7 - \frac{8}{7} \int_q^1 \left\{ \frac{f^7(-t)}{f(-t^7)} + \frac{49}{8} t f^3(-t) f^3(-t^7) \right\} \frac{dt}{t} \right), \qquad (40)$$

where

$$C_7 = L'(-1, \chi). \qquad (41)$$

Clearly (40) and (41) are analogues of (35) and (36), respectively.

S. Ahlgren, Berndt, Yee, and Zaharescu [1] have proved a general theorem from which (35) and (40) follow as special cases. The integral appearing in these authors' general theorem contains the Eisenstein series (38). It is only in certain instances that we have identities of the type (37) and (39), which enable us to reformulate the identities in terms of integrals of eta-functions, as we have in (35) and (40).

References

[1] S. Ahlgren, B. C. Berndt, A. J. Yee, and A. Zaharescu, *Integrals of Eisenstein series, derivatives of L-functions, and the Dedekind eta-function*, International Mathematics Research Notices **2002**, No. 32, 1723–1738.

[2] B. C. Berndt, *Ramanujan's Notebooks, Part II*, Springer–Verlag, New York, 1989.

[3] B. C. Berndt, *Ramanujan's Notebooks, Part III*, Springer–Verlag, New York, 1991.

[4] B. C. Berndt, *Ramanujan's Notebooks, Part IV*, Springer–Verlag, New York, 1994.

[5] B. C. Berndt, *Ramanujan's Notebooks, Part V*, Springer–Verlag, New York, 1998.

[6] B. C. Berndt and P. R. Bialek, *On the power series coefficients of certain quotients of Eisenstein series*, submitted for publication.

[7] B. C. Berndt, P. Bialek, and A. J. Yee, *Formulas of Ramanujan for the power series coefficients of certain quotients of Eisenstein series*, International Mathematics Research Notices **2002**, No. 21, 1077–1109.

[8] B. C. Berndt and H. H. Chan, *Eisenstein series and approximations to π*, Illinois J. Math. **45** (2001), 75–90.

[9] B. C. Berndt, H. H. Chan, and S.-S. Huang, *Incomplete elliptic integrals in Ramanujan's lost notebook*, in *q–series from a Contemporary Perspective*, M. E. H. Ismail and D. Stanton, eds., American Mathematical Society, Providence, RI, 2000, pp. 79–126.

[10] B. C. Berndt, H. H. Chan, S.-Y. Kang, and L.-C. Zhang, *A certain quotient of eta-functions found in Ramanujan's lost notebook*, Pacific J. Math. **202** (2002), 267–304.

[11] B. C. Berndt, H. H. Chan, J. Sohn, and S. H. Son, *Eisenstein series in Ramanujan's lost notebook*, Ramanujan J. **4** (2000), 81–114.

[12] B. C. Berndt and K. Ono, *Ramanujan's unpublished manuscript on the partition and tau functions with proofs and commentary*, Sém. Lotharingien de Combinatoire **42** (1999), 63 pp; The Andrews Festschrift, D. Foata and G.-N. Han, eds., Springer–Verlag, Berlin, 2001, pp. 39–110.

[13] B. C. Berndt and R. A. Rankin, *Ramanujan: Letters and Commentary*, American Mathematical Society, Providence, RI, 1995; London Mathematical Society, London, 1995.

[14] B. C. Berndt and A. J. Yee, *A page on Eisenstein series in Ramanujan's lost notebook*, Glasgow Math. J., to appear.

[15] B. C. Berndt and A. J. Yee, *Congruences for the coefficients of quotients of Eisenstein series*, Acta Arith. **104** (2002), 297–308.

[16] B. C. Berndt and A. Zaharescu, *An integral of Dedekind eta-functions in Ramanujan's lost notebook*, J. Reine Angew. Math., to appear.

[17] J. M. Borwein and P. M. Borwein, *More Ramanujan-type series for $1/\pi$*, in *Ramanujan Revisited*, G. E. Andrews, R. A. Askey, B. C. Berndt, K. G. Ramanathan, and R. A. Rankin, eds., Academic Press, Boston, 1988, pp. 359–374

[18] L. Carlitz, *Note on some partition formulae*, Quart. J. Math. (Oxford) (2) **4** (1953), 168–172.

[19] H. H. Chan, *On the equivalence of Ramanujan's partition identities and a connection with the Rogers–Ramanujan continued fraction*, J. Math. Anal. Applics. **198** (1996), 111–120.

[20] H. H. Chan and K. S. Chua, *Representations of integers as sums of 32 squares*, Ramanujan J., to appear.

[21] D. V. Chudnovsky and G. V. Chudnovsky, *Approximation and complex multiplication according to Ramanujan*, in *Ramanujan Revisited*, G. E. Andrews, R. A. Askey, B. C. Berndt, K. G. Ramanathan, and R. A. Rankin, eds., Academic Press, Boston, 1988, pp. 375–472.

52

[22] N. J. Fine, *On a system of modular functions connected with the Ramanujan identities*, Tohoku Math. J. (2) **8** (1956), 149–164.

[23] G. H. Hardy and S. Ramanujan, *Asymptotic formulae in combinatory analysis*, Proc. London Math. Soc. (2) **17** (1918), 75–118.

[24] G. H. Hardy and S. Ramanujan, *On the coefficients in the expansions of certain modular functions*, Proc. Royal Soc. A **95** (1918), 144–155.

[25] J. G. Huard, Z. M. Ou, B. K. Spearman, and K. S. Williams, *Elementary evaluation of certain convolution sums involving divisor functions*, in *Number Theory for the Millennium*, Vol. 2, M. A. Bennett, B. C. Berndt, N. Boston, H. G. Diamond, A. J. Hildebrand, and W. Philipp, eds., A K Peters, Natick, MA, 2002, pp. 229–273.

[26] O. Kolberg, *Note on the Eisenstein series of* $\Gamma_0(p)$, Arb. Univ. Bergen, Mat.-Naturv. Ser., No. 6 (1968), 20 pp.

[27] J. Lehner, *The Fourier coefficients of automorphic forms on horocyclic groups, II*, Mich. Math. J. **6** (1959), 173–193.

[28] J. Lehner, *The Fourier coefficients of automorphic forms on horocyclic groups, III*, Mich. Math. J. **7** (1960), 65–74.

[29] Z.-G. Liu, *Some Eisenstein series identities related to modular equations of the seventh order*, Pacific J. Math., to appear.

[30] Z.-G. Liu, *On the representation of integers as sums of squares*, in *q-Series with Applications to Combinatorics, Number Theory, and Physics*, B. C. Berndt and K. Ono, eds., Contemp. Math. No. 291, American Mathematical Society, Providence, RI, 2001, pp. 163–176.

[31] S. Milne, *Infinite families of exact sums of squares formulas, Jacobi elliptic functions, continued fractions, and Schur functions*, Ramanujan J. **6** (2002), pp. 7–149.

[32] K. Ono, *Representations of integers as sums of squares*, J. Number Thy., to appear.

[33] H. Petersson, *Konstruktion der Modulformen und der zu gewissen Grenzkreisgruppen gehörigen automorphen Formen von positiver reeller Dimension und die vollständige Bestimmung ihrer Fourierkoeffzienten*, S.-B. Heidelberger Akad. Wiss. Math. Nat. Kl. (1950), 415–474.

[34] H. Petersson, *Über automorphe Orthogonalfunktionen und die Konstruktion der automorphen Formen von positiver reeller Dimension*, Math. Ann. **127** (1954), 33–81.

[35] H. Petersson, *Über automorphe Formen mit Singularitäten im Diskontinuitätsgebiet*, Math. Ann. **129** (1955), 370–390.

[36] H. Poincaré, *Oeuvres*, Vol. 2, Gauthiers–Villars, Paris, 1916.

[37] S. Raghavan and S. S. Rangachari, *On Ramanujan's elliptic integrals and modular identities*, in *Number Theory and Related Topics*, Oxford University Press, Bombay, 1989, pp. 119–149.

[38] S. Ramanujan, *Modular equations and approximations to* π, Quart. J. Math. **45** (1914), 350–372.

[39] S. Ramanujan, *On certain arithmetical functions*, Trans. Cambridge Philos. Soc. **22** (1916), 159–184.

[40] S. Ramanujan, *Some properties of p(n), the number of partitions of n* Proc. Cambridge Philos. Soc. **19** (1919), 207–210.

[41] S. Ramanujan, *Congruence properties of partitions*, Math. Z. **9** (1921), 147–153.

[42] S. Ramanujan, *Collected Papers*, Cambridge University Press, Cambridge, 1927; reprinted by Chelsea, New York, 1962; reprinted by the American Mathematical Society, Providence, RI, 2000.

[43] S. Ramanujan, *Notebooks* (2 volumes), Tata Institute of Fundamental Research, Bombay, 1957.

[44] S. Ramanujan, *The Lost Notebook and Other Unpublished Papers*, Narosa, New Delhi, 1988.

[45] R. A. Rankin, *Modular Forms and Functions*, Cambridge University Press, Cambridge, 1977.

[46] S. H. Son, *Some integrals of theta functions in Ramanujan's lost notebook*, in *Fifth Conference of the Canadian Number Theory Association*, R. Gupta and K. S. Williams, eds., CRM Proc. and Lecture Notes, Vol. 19, American Mathematical Society, Providence, RI, 1999, pp. 323–332.

NEW APPLICATIONS OF A RESULT OF GALOCHKIN ON LINEAR INDEPENDENCE

Peter Bundschuh

Mathematisches Institut der Universität, Weyertal 86-90, 50931 Köln, Germany

pb@math.uni-koeln.de

Abstract   We propose a slightly improved version of a criterion of Galochkin for the linear independence of the values of functions which are analytic in some neighborhood of the origin. Whereas Galochkin himself applied his criterion only to functions satisfying very special Mahler functional equations, we present new applications to solutions of certain Poincaré type functional equations. In particular, we propose a generalization of a recent irrationality result due to Duverney, and we give a new proof of the Tschakaloff-Skolem theorem on the linear independence of values, at appropriate points, of the Tschakaloff function T_q which is intimately connected with the Jacobian theta function ϑ_3, in the usual notation.

Keywords: linear independence, Poincaré functional equations, q-functions

1.   Introduction

A few years ago, Galochkin [5] published an elegant and very general-looking result reducing the linear independence of *numbers*, under suitable technical conditions, to the linear independence of *functions*. Usually, this second kind of linear independence can be checked much easier. The title of Galochkin's article "Linear independence of values of functions satisfying Mahler functional equations" clearly indicates the type of applications he had in mind. And indeed, the *only* application of his general theorem presented in his paper concerns certain values of Mahler functions from a special class. On the other hand, in the domain of these functions, the use of Mahler's method for transcendence and algebraic independence leads to much stronger arithmetical assertions (compare [9]).

It is the main aim of the present paper to show that Galochkin's theorem leads in another domain essentially to the best qualitative results available nowadays. Thereby we mean irrationality and linear independence questions on the values of so-called q-functions, or more precisely, of solutions f of certain linear Poincaré type functional equations of first order, i.e. of the shape

$$f(qz) = P(z)f(z) + Q(z) \tag{1}$$

where P, Q are rational functions and q is a fixed complex parameter with $|q| > 1$.

2. Results

First we state Galochkin's linear independence criterion in a slightly sharpened version as follows. Let K be either $\mathbb{Q}$ or an imaginary quadratic number field, and O_K the ring of integers of K. Suppose that the power series

$$f_\mu(z) := \sum_{\nu=0}^{\infty} c_{\mu\nu} z^\nu \in K[[z]] \quad (\mu = 1, \ldots, M) \tag{2}$$

converge in some neighborhood U of $z = 0$ and put $\underline{f} := {}^T(f_1, \ldots, f_M)$.

Theorem 1. *Suppose 1 and the $f_1(z), \ldots, f_M(z) \in K[[z]]$ from (2) to be linearly independent over $K(z)$. Let $(\zeta_k)_{k=0,1,\ldots}$ be a sequence in $U \cap K^\times$ such that the equation*

$$\underline{f}(\zeta_0) = A_k \underline{f}(\zeta_k) + \underline{b}_k \tag{3}$$

holds for any $k \in \mathbb{N}_0 := \{0, 1, 2, \ldots\}$ with some $A_k \in GL_M(K)$ and $\underline{b}_k \in K^M$. Further, assume that there is a sequence $(\gamma_k)_{k=0,1,\ldots}$ in $K^\times$ such that $\gamma_k A_k$ and $\gamma_k \underline{b}_k$ have all entries in O_K, and that all entries of $\gamma_k A_k$ are bounded in absolute value by Γ_k. Suppose finally that the condition

$$\lim_{k \to \infty} \Gamma_k (|\zeta_k|^{1+1/M} d(\zeta_k))^N = 0 \tag{4}$$

holds for some $N \in \mathbb{N} := \{1, 2, \ldots\}$ which is independent of k. Then the numbers $1, f_1(\zeta_0), \ldots, f_M(\zeta_0)$ are linearly independent over K.

Here, for $\zeta \in K$, we denote $d(\zeta) := \min\{|\tau| \mid \tau \in O_K \setminus \{0\}, \zeta\tau \in O_K\}$. It should be observed that we have, in general, the inequality $d(\zeta) \leq \mathrm{den}(\zeta)$ between our $d(\zeta)$ and the usual denominator of ζ as introduced for algebraic numbers. Plainly, for $\zeta \in \mathbb{Q}$, both notions coincide.

Evidently, condition (4) implies

$$\lim_{k \to \infty} \zeta_k = 0. \tag{5}$$

It should be remarked that in Galochkin's original version of Theorem 1 condition (4) is slightly stronger than ours in so far as there the exponent of $|\zeta_k|$ is $1+1/M-\varepsilon$ with arbitrarily small $\varepsilon \in \mathbb{R}_+$. Admittedly, this small difference is not important for our applications but just an esthetical question. On the other hand, we have to point out the fact that Γ_k has only to bound the absolute values of the entries of the matrix $\gamma_k A_k$ but *not* those of the vector $\gamma_k \underline{b}_k$. This observation will be crucial for all of our applications, the first of which is the following slight generalization of a recent irrationality result of Duverney (see [4], Théorème 2).

Theorem 2. *Let $f \in K[[z]]$ be the solution of the functional equation*

$$z^g f(z) = P(z)f(qz) + Q(z) \tag{6}$$

with fixed $g \in \mathbb{N}$, $q \in O_K$, $|q| > 1$ and polynomials $P, Q \in K[z]$ satisfying $P(0) \neq 0$ and $\deg P \leq g$. Then f defines a function, meromorphic in the complex plane, with the following property: If $f \notin K(z)$, and if $\alpha \in K^\times$ is not a pole of f, then $f(\alpha) \notin K$.

A simple consequence of Theorem 2 concerning the so-called Tschaka-loff-function

$$T_q(z) := \sum_{n=0}^{\infty} q^{-n(n-1)/2} z^n \tag{7}$$

satisfying the functional equation

$$z T_q(z) = \frac{1}{q} T_q(qz) - \frac{1}{q} \tag{8}$$

reads as follows (compare [13 I] and [3], Korollar 5).

Corollary 1. *If $q \in O_K$, $|q| > 1$ and $\alpha \in K^\times$, then $T_q(\alpha) \notin K$.*

To describe a more general application of Theorem 2, let us suppose $\beta, q \in \mathbb{C}$, $|q| > 1$. Then the series

$$\sum_{n=0}^{\infty} (1-\beta)\cdots(1-\beta q^{n-1}) q^{-n(n-1)/2} z^n \tag{9}$$

is a polynomial for $\beta \in \{1, q^{-1}, q^{-2}, \ldots\}$ and coincides with T_q for $\beta = 0$. If $\beta \in \mathbb{C}^\times \setminus \{1, q^{-1}, q^{-2}, \ldots\}$ the series (9) has radius of convergence $1/|\beta|$ and defines in $|z| < 1/|\beta|$ a holomorphic function $F_q(z; \beta)$. This function satisfies the functional equation

$$z f(z) = (\beta z + \frac{1}{q}) f(qz) - \frac{1}{q}, \tag{10}$$

and therefore can be analytically continued over $\mathbb{C} \setminus \{-1/\beta, -q/\beta, -q^2/\beta, \ldots\}$, and, under the last hypotheses on β, all these exceptional points are poles of the function. Denoting this continuation again by $F_q(z; \beta)$, we have the following generalization of Corollary 1 which is essentially due to Matala-aho [8].

Corollary 2. *Let* $q \in O_K$, $|q| > 1$, *and* $\beta \in K \setminus \{1, q^{-1}, q^{-2} \ldots\}$. *Then* $F_q(\alpha; \beta) \notin K$ *if* $\alpha \in K^\times \setminus \{-1/\beta, -q/\beta, -q^2/\beta, \ldots\}$.

Another important q-function is the q-exponential function E_q defined by

$$E_q(z) := \prod_{j=1}^{\infty} (1 + zq^{-j}) = \sum_{n=0}^{\infty} z^n \Big/ \prod_{\nu=1}^{n} (q^\nu - 1) \qquad (11)$$

satisfying the functional equation

$$E_q(qz) = (1 + z)E_q(z) \qquad (12)$$

which is obviously *not* of the form (6). Therefore we cannot deduce *directly* from Theorem 2 the following well-known result on E_q (compare, e.g., [6], [3], Korollar 1).

Corollary 3. *If* $q \in O_K$, $|q| > 1$ *and* $\beta \in K^\times \setminus \{-q, -q^2, \ldots\}$, *then* $E_q(\beta) \notin K$.

In section 4 we will explain a beautiful trick due to Amou, Katsurada and Väänänen [1] to derive Corollary 3 *still* from Theorem 2.

Next we present truly linear independence results on functions f satisfying slightly different functional equations of type (1) than they did in Theorem 2, compare (6).

Theorem 3. *Let* $f \in K[[z]]$ *be the entire transcendental solution of the functional equation*

$$f(qz) = az^g f(z) + Q(z) \qquad (13)$$

with $g \in \mathbb{N}$, $a \in K^\times$, $q \in O_k$, $|q| > 1$, *and* $Q \in K[z]$. *Suppose that* $\alpha_1, \ldots, \alpha_m \in K^\times$ *satisfy the conditions* $(\alpha_\mu/\alpha_{\mu'})^g \notin q^{\mathbb{Z}}$ *for* $\mu \neq \mu'$. *Then, for any* $\ell \in \mathbb{N}_0$, *the* $(\ell + 1)m + 1$ *numbers*

$$1, f(\alpha_1), \ldots, f(\alpha_m), \ldots, f^{(\ell)}(\alpha_1), \ldots, f^{(\ell)}(\alpha_m) \qquad (14)$$

are linearly independent over K.

It is easily seen that the unique solution $f \in K[[z]]$ of (13) converging in some neighborhood of $z = 0$ is automatically an entire function. This is not necessarily so in the case of the solutions of (6) as we have seen before Corollary 2.

Since the Tschakaloff function T_q from (7) satisfies (13) with $g = 1$, $a = q$, and $Q(z) = 1$, we deduce from Theorem 3 immediately the following

Corollary 4. *Suppose $q \in O_K$, $|q| > 1$, and $\alpha_1, \ldots, \alpha_m \in K^\times$ such that $\alpha_\mu / \alpha_{\mu'} \notin q^{\mathbb{Z}}$ holds if $\mu \neq \mu'$. Then, for any $\ell \in \mathbb{N}_0$, the following numbers are linearly independent over K*

$$1, T_q(\alpha_1), \ldots, T_q(\alpha_m), \ldots, T_q^{(\ell)}(\alpha_1), \ldots, T_q^{(\ell)}(\alpha_m).$$

This result was first proved in the case $K = \mathbb{Q}$ by Skolem [11], Theorem 10, using the Hilbert-Perron method based essentially on divisibility considerations. The particular case $\ell = 0$ of Corollary 4, i.e. without derivatives, dates back to Tschakaloff [13 II] who used Padé approximations of the first kind, plus some non-vanishing arguments for certain determinants. But it is true that in Tschakaloff's proof the number q has not necessarily to be an *integer* in K. Both these methods allow quantitative refinements of Corollary 4 which can be found in the literature.

Whereas the before-mentioned two methods were based on more or less explicit constructions of appropriate diophantine approximations, Bézivin [2] developed from 1988 on a new and more function-theoretic method to get linear independence results on the numbers (14) for quite general entire functions f. We remark that his key argument is to apply an appropriate criterion à la Kronecker or Borel-Dwork for the rationality of a suitable auxiliary function. But in spite of the relative generality of his results, his method has one fundamental lack: It does not allow, at least until now, to deduce quantitative refinements. The same remark is true of our present method via Galochkin's criterion, too.

We conclude this section with the observation that Corollary 4 does not hold, even in the case $\ell = 0$, without the condition $\alpha_\mu / \alpha_{\mu'} \notin q^{\mathbb{Z}}$ for $\mu \neq \mu'$ (if $m \geq 2$). Clearly, this follows easily from the functional equation (8) for T_q.

3.   A few comments on the proof of Theorem 1

This proof depends decisively on the following Lemma which, for its parts (i) and (ii), uses the non-trivial solubility of a system of less linear homogeneous equations than unknowns as in Mahler's method for transcendence and algebraic independence.

Lemma. *Suppose that the functions $1, f_1(z), \ldots, f_M(z) \in K[[z]]$ from Theorem 1 are linearly independent over $K(z)$. Then, for each $N \in$*

N_0, there exist $n_0, \ldots, n_M \in \mathbb{N}$ with $N \le n_0 < n_1 < \cdots < n_M$, and polynomials $P_{\lambda\mu} \in O_K[z]$ $(\lambda, \mu = 0, \ldots, M)$ such that
(i) $\deg P_{\lambda\mu} \le n_\lambda$,
(ii) $\mathrm{ord}_0(P_{\lambda 0} f_\mu - P_{\lambda\mu}) > (1 + \frac{1}{M}) n_\lambda$,
(iii) $\det(P_{\lambda\mu})_{\lambda, \mu = 0, \ldots, M} \ne 0$.

Proof. For each $n \in \mathbb{N}_0$ there exists $P_0 \in O_K[z] \setminus \{0\}$ with $\deg P_0 \le n$ such that the coefficients of the power series

$$P_0(z) f_\mu(z) = \sum_{\nu=0}^{\infty} \Lambda_{\mu\nu} z^\nu \quad (\mu = 1, \ldots, M)$$

satisfy the conditions

$$\Lambda_{\mu\nu} = 0 \quad \text{for} \quad 1 \le \mu \le M, \ n < \nu \le (1 + \frac{1}{M})n. \tag{15}$$

For this it is enough to write down P_0 with unknown coefficients, and to consider the equations (15) as a system of $p := M[n/M]$ $(\le n)$ linear homogeneous equations in the $q := n + 1$ coefficients of P_0. Since $p < q$, this system has a non-trivial solution in O_K^{n+1}. Defining

$$P_\mu(z) := \sum_{\nu=0}^{n} \Lambda_{\mu\nu} z^\nu \quad (\mu = 1, \ldots, M)$$

ensures the simultaneous validity of (i) and (ii) for each $n \in \mathbb{N}_0$.

The corresponding lemma of Galochkin [5] has
(ii') $\mathrm{ord}_0(P_{\lambda 0} f_\mu - P_{\lambda\mu}) \ge (1 + \frac{1}{M} - \varepsilon) n_\lambda$
instead of our (ii). But it is easily checked that, after this minor improvement, Galochkin's arguments go through to yield our above Lemma.

With this Lemma one is in a position to prove Theorem 1 along the lines of [5]. Plainly, our exponent $1 + 1/M$ in condition (4) instead of Galochkin's $1 + 1/M - \varepsilon$ comes just from using (ii) instead of (ii'). $\square$

4. Proof of Theorem 2 and its corollaries

To begin with, we observe first that, under the conditions on P, Q, q, g of Theorem 2, (6) has a unique solution $f \in K[[z]]$ with convergence radius $R_f \in]0, +\infty]$. Next we assert that the functional equation

$$f(z) \prod_{\kappa=1}^{k} q^{g\kappa} P\left(\frac{z}{q^\kappa}\right)$$

$$= z^{gk} f\left(\frac{z}{q^k}\right) - \sum_{j=0}^{k-1} z^{gj} q^{g(j+1)} Q\left(\frac{z}{q^{j+1}}\right) \prod_{\kappa=j+2}^{k} q^{g\kappa} P\left(\frac{z}{q^\kappa}\right) \tag{16}$$

holds for each $k \in \mathbb{N}_0$, of course, under the usual conventions that empty products or sums have to be interpreted as 1 or 0 respectively. Clearly, the proof of (16) is by induction on k using the functional equation (6) (which is equivalent with (16) for $k = 1$). If the above R_f is finite, then (16) can be used to extend f from $|z| < R_f$ meromorphically into the whole complex plane. If this extension is a rational function, i.e. $\in \mathbb{C}(z)$, then we may even suppose $f \in K(z)$; this is a standard argument to be found, e.g., in [10], p. 149 for the case $K = \mathbb{Q}$, or in [7], pp. 33-35 more generally.

Suppose now $f \notin K(z)$ or equivalently that $1, f(z)$ are linearly independent over $K(z)$. Let $\alpha \in K^\times$ first satisfy $|\alpha| < \vartheta \le R_f$, where $\vartheta \in \mathbb{R}_+$ is such that $P(z) \ne 0$ in $|z| < \vartheta$, compare the hypothesis $P(0) \ne 0$. Defining $f_1(w) := f(\alpha w)$ we see from (16)

$$
f_1(1) = \frac{(\pi\sigma^g)^k}{\prod\limits_{\kappa=1}^{k}\left\{\pi\tau^g q^{g\kappa} P\left(\frac{\sigma}{\tau}q^{-\kappa}\right)\right\}} f_1(q^{-k})
$$
$$
- \sum_{j=0}^{k-1} \frac{\pi^{j+1}\tau^g\sigma^{gj}q^{g(j+1)}\omega Q(\frac{\sigma}{\tau}q^{-j-1})}{\omega \prod\limits_{\kappa=1}^{j+1}\left\{\pi\tau^g q^{g\kappa} P(\frac{\sigma}{\tau}q^{-\kappa})\right\}} \tag{17}
$$

where $\sigma, \tau, \pi, \omega \in O_K \setminus \{0\}$ are such that $\alpha = \sigma/\tau$ and $\pi P(z), \omega Q(z) \in O_K[z]$. Since $1, f_1(w)$ are linearly independent over $K(w)$, we may apply Theorem 1 with $\zeta_k := q^{-k}$ for all $k \in \mathbb{N}_0$, U containing the closed unit disc of the w-plane, $A_k \in K^\times$ the factor of $f_1(q^{-k})$ in (17), and $-b_k$ the sum on the right-hand side of (17). Clearly, putting $\rho := \max\{0, \deg Q - g\}$,

$$
\gamma_k := \omega(\tau q^k)^\rho \prod_{\kappa=1}^{k}\left\{\pi\tau^g q^{g\kappa} P(\frac{\sigma}{\tau}q^{-\kappa})\right\} \in O_K \setminus \{0\} \tag{18}
$$

is enough to ensure $\gamma_k A_k, \gamma_k b_k \in O_K$ since $\deg P \le g$. Therefore we may take

$$
\Gamma_k := |\omega\tau^\rho| \, |\pi\sigma^g q^\rho|^k \tag{19}
$$

for any $k \in \mathbb{N}_0$, and since $d(q^{-k}) = |q|^k$, condition (4) is equivalent with $(|\pi\sigma^g q^\rho| \, |q|^{-N})^k \to 0$ as $k \to \infty$, and this holds if we fix $N \in \mathbb{N}$ in such a way that $|q|^N > |\pi\sigma^g q^\rho|$. Therefore we get $f(\alpha) = f_1(1) \notin K$ from Theorem 1, if $\alpha \in K^\times$ satisfies the additional above condition $|\alpha| < \vartheta$.

Let now $\alpha \in K^\times$ be not a pole of f. Then, by (16), no αq^{-k}, $k = 1, 2, \ldots$, is a pole of f, and the assumption $f(\alpha) \in K$ would imply $f(\alpha q^{-k}) \in K$ for any $k \in \mathbb{N}$ contradicting the fact we proved just before.

Remark. As one can easily see from (17) and (18), the absolute value of $\gamma_k b_k$ can, in general, have size $|q|^{ck^2 + O(k)}$ for some $c \in \mathbb{R}_+$ which cannot be bounded by (19) for large k.

Proof of Corollary 2. In virtue of the obvious Corollary 1, we may suppose $\beta \in K^\times \setminus \{1, q^{-1}, q^{-2}, \ldots\}$ in Corollary 2. By the observations before Corollary 2 its statement follows immediately from Theorem 2 since $F_q(z; \beta) \notin K(z)$ for β as above. □

Proof of Corollary 3. We have $E_q(\beta) = \sum\limits_{n=0}^{\infty} \beta^n q^{-n(n+1)/2} \Big/ \prod\limits_{\nu=1}^{n} (1 - q^{-\nu})$ for fixed $\beta \in \mathbb{C}$, by (11), and we consider the auxiliary function

$$E(z; \beta) := \sum_{n=0}^{\infty} \frac{\beta^n q^{-n(n+1)/2} z^n}{\prod\limits_{\nu=1}^{n} (1 - zq^{-\nu})}.$$

This function is meromorphic in $\mathbb{C}$, with poles at most in the points $q, q^2, \ldots$, and satisfies the functional equation

$$\beta z E(z; \beta) = (1 - z) E(qz; \beta) + z - 1 \tag{20}$$

of type (6) with $g = 1$, $P(z) = (1 - z)/\beta$, $Q(z) = (z - 1)/\beta$ if $\beta \neq 0$. Furthermore we know

$$E(1; \beta) = E_q(\beta), \tag{21}$$

by construction. If we can show $E(z; \beta) \notin K(z)$ for $\beta \in K^\times \setminus \{-q, -q^2, \ldots\}$, then Theorem 2, applied with $\alpha = 1$, yields the assertion of Corollary 3, compare (21). □

If, for a certain $\beta \neq 0$, the function $E(z; \beta)$ has a pole (contained in $q^{\mathbb{N}}$, as we saw above), then, by (20), it has infinitely many poles. Therefore the assumption $E(z; \beta) \in K(z)$ leads immediately to $E(z; \beta) \in K[z]$, i.e. to $E(z; \beta) = c_0 + \ldots + c_s z^s$ with $c_s \neq 0$. Using (20) we find $\beta = -q^s$ $(s \in \mathbb{N})$ in the same way as before. Thus we get $E(z; \beta) \notin K(z)$ for $\beta \in \mathbb{C}^\times \setminus \{-q, -q^2, \ldots\}$.

5. Proof of Theorem 3, arithmetical part

Using induction on k we get from (13)

$$f(z) = a^k z^{gk} q^{-gk(k+1)/2} f(zq^{-k}) + Q_k(z)$$

with

$$Q_k(z) := \sum_{\kappa=0}^{k-1} a^\kappa z^{\kappa g} q^{-g\kappa(\kappa+1)/2} Q(zq^{-\kappa-1}) \tag{22}$$

for $k = 0, 1, \ldots$. Differentiating λ times and replacing then z by $\alpha_\mu z$ leads to

$$f^{(\lambda)}(\alpha_\mu z) \;=\; \frac{a^k}{q^{gk(k+1)/2}} \sum_{\sigma=0}^{\lambda} \frac{\lambda!}{\sigma!} \binom{gk}{\lambda-\sigma} (\alpha_\mu z)^{gk-\lambda+\sigma} q^{-\sigma k} f^{(\sigma)}\left(\frac{\alpha_\mu z}{q^k}\right)$$

$$+ \, Q_k^{(\lambda)}(\alpha_\mu z) \tag{23}$$

for $\mu = 1, \ldots, m;\ \lambda = 0, \ldots, \ell$.

We try to apply Theorem 1, with $M := (\ell + 1)m$ to the functions

$$f_{\lambda m + \mu}(z) \;:=\; f^{(\lambda)}(\alpha_\mu z) \quad (\mu = 1, \ldots, m;\ \lambda = 0, \ldots, \ell) \tag{24}$$

with $\zeta_k := q^{-k}$ $(k = 0, 1, \ldots)$. From (23) and (24) we find

$$f_{\lambda m + \mu}(1) \;=\; \frac{a^k}{q^{gk(k+1)/2}} \sum_{\sigma=0}^{\lambda} \frac{\lambda!}{\sigma!} \binom{gk}{\lambda-\sigma} \alpha_\mu^{gk-\lambda+\sigma} q^{-\sigma k} f_{\sigma m + \mu}(q^{-k})$$

$$+ \, Q_k^{(\lambda)}(\alpha_\mu). \tag{25}$$

This means that (3) is indeed satisfied with

$$\underline{b}_k \;=\; {}^T(Q_k(\alpha_1), \ldots, Q_k(\alpha_m), \ldots, Q_k^{(\ell)}(\alpha_1), \ldots, Q_k^{(\ell)}(\alpha_m))$$

and with an $M \times M$ matrix A_k of the form

$$a^k q^{-gk(k+1)/2} \begin{pmatrix} A_k^{(0,0)} & \cdots & A_k^{(0,\ell)} \\ \vdots & & \vdots \\ A_k^{(\ell,0)} & \cdots & A_k^{(\ell,\ell)} \end{pmatrix} \tag{26}$$

where the $A_k^{(\lambda,\sigma)}$ themselves are $m \times m$ diagonal matrices with $A_k^{(\lambda,\sigma)} = O$ for $\sigma > \lambda$, and with

$$A_k^{(\lambda,\sigma)} \;=\; \frac{\lambda!}{\sigma!} \binom{gk}{\lambda-\sigma} q^{-\sigma k} \operatorname{diag}\left(\alpha_1^{gk-\lambda+\sigma}, \ldots, \alpha_m^{gk-\lambda+\sigma}\right) \tag{27}$$

for $\sigma \leq \lambda$, by (25). Clearly, $\det A_k \neq 0$ for any $k \in \mathbb{N}$.

Let $\tau, \eta \in O_K \setminus \{0\}$ be such that $\tau\alpha_1, \ldots, \tau\alpha_m,\ \eta a \in O_K$, and define

$$\gamma_k \;:=\; \eta^k q^{gk(k+1)/2 + \ell k} \tau^{gk} \left(\omega \tau^t q^{tk}\right) \tag{28}$$

with $t := \deg Q$, and $\omega \in O_K \setminus \{0\}$ such that $\omega Q(z) \in O_K[z]$. From (26) and (27) we see that $\gamma_k A_k$ has all entries in O_K, even if we disregard the factor $(\ldots)$ in (28). From (22) we find

$$Q_k^{(\lambda)}(\alpha_\mu) = \sum_{\kappa=0}^{k-1} \frac{a^\kappa}{q^{g\kappa(\kappa+1)/2}} \sum_{\sigma=0}^{\lambda} \frac{\lambda!}{\sigma!} \binom{\kappa g}{\lambda-\sigma} \alpha_\mu^{\kappa g-\lambda+\sigma} q^{-\sigma(\kappa+1)} Q^{(\sigma)}\left(\frac{\alpha_\mu}{q^{\kappa+1}}\right),$$

and therefore we get $\gamma_k Q_k^{(\lambda)}(\alpha_\mu) \in O_K$ which means that $\gamma_k \underline{b}_k$ has all entries in O_K. Because of (26), (27), (28) it is obvious that all entries of $\gamma_k A_k$ can be bounded by

$$\Gamma_k := BC^k$$

with some constants $B, C > 1$ independent of k. Thus (4) is satisfied if we choose N with $|q|^N > C^{(\ell+1)m}$. Clearly, Theorem 3 follows from Theorem 1, if we can show that 1 and the functions in (24) are linearly independent over $K(z)$. For this we shall need the conditions $(\alpha_\mu/\alpha_{\mu'})^g \notin q^{\mathbb{Z}}$ for $\mu \neq \mu'$ in Theorem 3.

6.   Proof of Theorem 3, analytic part

This proof will be carried out inductively beginning with the case $\ell = 0$. Clearly, 1 and $f_1(z) = f(\alpha_1 z)$ are linearly independent over $K(z)$ since $\alpha_1 \neq 0$ and f is transcendental by hypothesis. Suppose now $m > 1$ and select the smallest $\mu \in \{2, \ldots, m\}$ such that $1, f_1, \ldots, f_{\mu-1}$ are linearly independent, but $1, f_1, \ldots, f_\mu$ are linearly dependent over $K(z)$. This means that with certain appropriate $A_0, \ldots, A_{\mu-1} \in K(z)$ we have

$$f(\alpha_\mu z) = A_0(z) + \sum_{\rho=1}^{\mu-1} A_\rho(z) f(\alpha_\rho z)$$

where not all $A_1, \ldots, A_{\mu-1}$ vanish. Replacing here z by qz, and then using the functional equation (13) leads to

$$f(\alpha_\mu z) = A_0^*(z) + \sum_{\rho=1}^{\mu-1} (\alpha_\rho/\alpha_\mu)^g A_\rho(qz) f(\alpha_\rho z)$$

with $A_0^* \in K(z)$. By our assumption on μ we conclude from the last two equations $A_0^* = A_0$ and for $\rho = 1, \ldots, \mu - 1$

$$\alpha_\mu^g A_\rho(z) = \alpha_\rho^g A_\rho(qz). \tag{29}$$

Selecting ρ with $A_\rho \neq 0$ let δz^ω with $\delta \neq 0$, $\omega \in \mathbb{Z}$ be the lowest term in the Laurent expansion of A_ρ about the origin. Then (29) yields $(\alpha_\mu/\alpha_\rho)^g = q^\omega$, contradicting one of our hypotheses in Theorem 3. Therefore $1, f_1, \ldots, f_m$ are linearly independent over $K(z)$.

Before continuing we note that (13) (or (22), (23) with $k = 1$, $\alpha_\mu = 1, z \mapsto qz$) leads to

$$q^\lambda f^{(\lambda)}(qz) = a \sum_{\kappa=0}^{\lambda} \binom{\lambda}{\kappa} \binom{g}{\kappa} \kappa! z^{g-\kappa} f^{(\lambda-\kappa)}(z) + Q^{(\lambda)}(z). \tag{30}$$

Let us now suppose that for some $(\ell, n) \in \mathbb{N} \times \{1, \ldots, m\}$ it is yet proved that the functions

$$1, f(\alpha_1 z), \ldots, f(\alpha_m z), \ldots, f^{(\ell-1)}(\alpha_1 z), \ldots, f^{(\ell-1)}(\alpha_m z),$$
$$f^{(\ell)}(\alpha_1 z), \ldots, f^{(\ell)}(\alpha_{n-1} z) \tag{31}$$

are linearly independent over $K(z)$ (if $n = 1$ this means just that the ℓth derivatives are missing) whereas these functions together with $f^{(\ell)}(\alpha_n z)$ are linearly dependent over $K(z)$. Therefore we have a relation

$$f^{(\ell)}(\alpha_n z) = A_0(z) + \sum_{\lambda=0}^{\ell} \sum_{\mu}{}' A_{\lambda,\mu}(z) f^{(\lambda)}(\alpha_\mu z) \tag{32}$$

where not all $A_{\lambda,\mu} \in K(z)$ vanish. Here $\sum_{\mu}'$ indicates summation over $\mu = 1, \ldots, m$ if $\lambda < \ell$, and over $\mu = 1, \ldots, n-1$ if $\lambda = \ell$. Replacing z by qz in (32) and eliminating from the arising equation all $f^{(\lambda)}(\alpha_\mu q z)$ via (30) in favour of the $f^{(\lambda-\kappa)}(\alpha_\mu z)$ we find

$$(\alpha_n z)^g \sum_{\lambda=0}^{\ell} \sum_{\mu}{}' A_{\lambda,\mu}(z) f^{(\lambda)}(\alpha_\mu z)$$

$$+ \sum_{\kappa=1}^{\ell} \binom{\ell}{\kappa} \binom{g}{\kappa} \kappa! (\alpha_n z)^{g-\kappa} f^{(\ell-\kappa)}(\alpha_n z)$$

$$= \sum_{\lambda=0}^{\ell} q^{\ell-\lambda} \sum_{\mu}{}' A_{\lambda,\mu}(qz) \sum_{\kappa=0}^{\lambda} \binom{\lambda}{\kappa} \binom{g}{\kappa} \kappa! (\alpha_\mu z)^{g-\kappa} f^{(\lambda-\kappa)}(\alpha_\mu z). \tag{33}$$

Now we are in a position to compare coefficients since the functions in (31) are linearly independent over $K(z)$. Indeed, comparing the factors of $f^{(\ell-1)}(\alpha_n z)$ on both sides of (33) we are led to

$$A_{\ell-1,n}(z) + \frac{\ell g}{\alpha_n z} = q A_{\ell-1,n}(qz)$$

and thus $A_{\ell-1,n} \neq 0$ since $\ell \in \mathbb{N}$. If δz^ω is again the lowest Laurent series term of this rational function, then $(\ell g)/(\alpha_n z) = \delta(q^{\omega+1} - 1)z^\omega +$ terms of higher order, and this is a contradiction in both cases $\omega = -1$ and $\omega \neq -1$. Thereby the proof of Theorem 3 is complete.

Remark. Of course, our proof in this section gives that 1 and the $(\ell+1)m$ functions in (24) are linearly independent over $\mathbb{C}(z)$ for non-zero complex

$\alpha_1, \ldots, \alpha_m$ satisfying $(\alpha_\mu / \alpha_{\mu'})^g \notin q^{\mathbb{Z}}$ for $\mu \neq \mu'$. In particular, taking $m = 1$, and $\alpha_1 = 1$ we find again the well-known fact that the entire transcendental solution f of the functional equation (13) cannot satisfy any non-trivial linear differential equation with polynomial coefficients.

References

[1] M. Amou, M. Katsurada and K. Väänänen, "Arithmetical properties of the values of functions satisfying certain functional equations of Poincaré", *Acta Arith.* **99** (2001) 389-407.

[2] J.-P. Bézivin, "Indépendance linéaire des valeurs des solutions transcendantes de certaines équations fonctionnelles", *Manuscripta Math.* **61** (1988) 103-129.

[3] P. Bundschuh, "Ein Satz über ganze Funktionen und Irrationalitätsaussagen", *Invent. Math.* **9** (1970) 175-184.

[4] D. Duverney, "Propriétés arithmétiques des solutions de certaines équations fonctionnelles de Poincaré", *J. Théor. Nombres Bordeaux* **8** (1996) 443-447.

[5] A.I. Galochkin, "Linear independence of values of functions satisfying Mahler functional equations", *Vestnik Moskov. Univ. Ser.I Mat. Mekh.* **1997**, 14-17, 72, Engl. transl.: *Moscow Univ. Math. Bull.* **52** (1997) 14-17.

[6] A.V. Lototsky, "Sur l'irrationalité d'un produit infini", *Mat. Sb.* **12** (54) (1943) 262-272.

[7] K. Mahler, *Lectures on Transcendental Numbers*, Springer, Berlin et al., LNM 546, 1976.

[8] T. Matala-aho, "Remarks on the arithmetic properties of certain hypergeometric series of Gauss and Heine", *Acta Univ. Oulu. Ser. A Sci. Rerum Natur.* **219** (1991) 1-112.

[9] Ku. Nishioka, *Mahler Functions and Transcendence*, Springer, Berlin et al., LNM 1631, 1996.

[10] G. Pólya und G. Szegö, *Aufgaben und Lehrsätze aus der Analysis II* (4. Aufl.), Springer, Berlin et al., 1971.

[11] T. Skolem, "Some theorems on irrationality and linear independence", in: Den 11te Skand. Math. Kongr. Trondheim (1949) 77-98.

[12] T. Stihl und R. Wallisser, "Zur Irrationalität und linearen Unabhängigkeit der Werte der Lösungen einer Funktionalgleichung von Poincaré", *J. Reine Angew. Math.* **341** (1983) 98-110.

[13] L. Tschakaloff, "Arithmetische Eigenschaften der unendlichen Reihe $\sum_{\nu=0}^{\infty} a^{-\nu(\nu-1)/2} x^\nu$", I: *Math. Ann.* **80** (1921) 62-74, II: *Math. Ann.* **84** (1921) 100-114.

[14] R. Wallisser, "Über die arithmetische Natur der Werte der Lösungen einer Funktionalgleichung von H. Poincaré", *Acta Arith.* **25** (1973) 81-92.

PARTITIONS MODULO PRIME POWERS AND BINOMIAL COEFFICIENTS

Tianxin Cai

Department of Mathematics, Zhejiang University, Hangzhou, 310028, P.R.China

tcai@matematicas.udea.edu.co

Abstract   In this paper, we show some identities which connect binomial coefficients with the number of solutions of congruences of some type.

Keywords: partitions, modulo prime powers, binomial coefficients.

1.   Introduction

Suppose p is a prime, $n \geq 1$, $1 \leq k \leq p$, define $f_s(k)$ to be the number of solutions of the congruence

$$k \equiv i_1 + i_2 + \cdots + i_s \ (mod \ p) \,|\, 1 \leq i_1 < i_2 < \cdots < i_s \leq np - 1.$$

We show in [1] that

$$f_s(0) - (-1)^{s-[\frac{s}{p}]} \binom{n-1}{[\frac{s}{p}]} = f_s(1) = \cdots = f_s(p-1) \qquad (1)$$

for all $1 \leq s \leq np - 1$, this result plays key role in generalizing the Wolstenholme Theorem on q-binomial coefficients. In this note, we assume that p^t $(t \geq 1)$ is a prime power, $t \geq 1$, define $f_s(k, t)$ to be the number of solutions of the congruence

$$k \equiv i_1 + i_2 + \cdots + i_s \ (mod \ p^t) \,|\, 1 \leq i_1 < i_2 < \cdots < i_s \leq p^t - 1. \qquad (2)$$

We try to show some properties of $f_s(k, t)$, especially the relationship with binomial coefficients. For convenience, we define $f_0(0, t) = f_0(p^t, t) = 1$; $f_0(k, t) = 0$, if $1 \leq k \leq p^t - 1$. The main result we obtain is the following,

Theorem 1. *Let p be any prime, then*

$$\binom{p^t - 1}{s} = f_s(1, t) \, p^t + \sum_{1 \leq j \leq t} \left(f_s(p^j, t) - f_s(p^{j-1}, t) \right) p^{t-j} \qquad (3)$$

for all $0 \leq s \leq p^t - 1, t \geq 1$.

Similarly, if we define $F_s(k,t)$ to be the number of solutions of the congruence

$$k \equiv i_1 + i_2 + \cdots + i_s \ (mod \ p^t) \,|\, 1 \leq i_1 < i_2 < \cdots < i_s \leq p^t$$

and $F_0(k,t) := f_0(k,t)$, we have the following,

Theorem 2. *Let* p *be any prime, then*

$$\binom{p^t}{s} = F_s(1,t)\,p^t + \sum_{1 \leq j \leq t} \left(F_s(p^j,t) - F_s(p^{j-1},t) \right) p^{t-j} \qquad (4)$$

for all $0 \leq s \leq p^t, t \geq 1$.

2. Proofs and propositions

Proof of Theorem 1. If $(k_1, p^t) = (k_2, p^t)$, it's easy to verify that $f_s(k_1, t) = f_s(k_2, t)$, since one could establish a bijection between Y_{k_1} and Y_{k_2} by

$$\{i_1, i_2, \ldots, i_s\} \to \{k_2'\overline{k_1'}i_1, k_2'\overline{k_1'}i_2, \ldots, k_2'\overline{k_1'}i_s\},$$

here $k_1 = p^a k_1'$, $k_2 = p^a k_2'$, $a \geq 0$, $(k_1'k_2', p) = 1$, $\overline{k_1'}$ is an associate of k_1' modulo p^t, i.e., $k_1'\overline{k_1'} \equiv 1 \ (mod \ p^t)$, and $Y_k = \{y_k\}$ is the set of all solutions of (2). Hence

$$\binom{p^t - 1}{s} = \sum_{1 \leq k \leq p^t} f_s(k,t)$$

$$= f_s(1,t)\,\phi(p^t) + f_s(p,t)\,\phi(p^{t-1}) + \cdots + f_s(p^t,t)\,\phi(1)$$

$$= f_s(1,t)\,p^t + \sum_{1 \leq j \leq t} \left\{ f_s(p^j,t) - f_s(p^{j-1},t) \right\} p^{t-j}$$

here $\phi(n)$ is the Euler's function. This completes the proof of Theorem 1. $\qquad \square$

Proof of Theorem 2. Theorem 2 could be proved in the same way as Theorem 1. However, we want to make use of Theorem 1, noting that

$F_s(k,t) = f_s(k,t) + f_{s-1}(k,t)$, we have from Theorem 1,

$$\binom{p^t}{s} = \binom{p^t - 1}{s} + \binom{p^t - 1}{s - 1}$$

$$= f_s(1,t)p^t + \sum_{1 \le j \le t} \left\{ f_s(p^j, t) - f_s(p^{j-1}, t) \right\} p^{t-j}$$

$$+ f_{s-1}(1,t)p^t + \sum_{1 \le j \le t} \left\{ f_{s-1}(p^j, t) - f_{s-1}(p^{j-1}, t) \right\} p^{t-j}$$

$$= F_s(1,t)p^t + \sum_{1 \le j \le t} \left\{ F_s(p^j, t) - F_s(p^{j-1}, t) \right\} p^{t-j},$$

here we use $\binom{n}{r} = \binom{n-1}{r} + \binom{n-1}{r-1}$ twice. This completes the proof of Theorem 2. $\qquad\square$

Proposition 1. Let p be any prime, then

$$\binom{p^t - 1}{s} = \sum_{j=0}^{t} (-1)^{s - [s/p^j]} f_{[s/p^j]}(1, t - j) p^{t-j} \tag{5}$$

for all $0 \le s \le p^t - 1$, here $f_s(1,0) := (-1)^s$.

Proof. Noting that for $0 \le j < p, 1 \le k \le p^t, 1 \le s \le p^t - 1$,

$$\sum_{\substack{1 \le k \le p^t \\ k \equiv j \,(\mathrm{mod}\, p)}} f_s(k,t) = f'_s(j,t), \tag{6}$$

here $f'_s(j,t)$ is the number of solutions of (2) modulo p instead of p^t. Combining (1) with (6), from the proof of Theorem 1 one has

$$\binom{p^t - 1}{s} = (-1)^{s - [\frac{s}{p}]} \binom{p^{t-1} - 1}{[\frac{s}{p}]} + f_s(1,t) \, p^t \tag{7}$$

for all $0 \le s \le p^t - 1$. From the well-known result $[\frac{n}{ab}] = [\frac{[n/a]}{b}]$, (5) follows by using (7) repeatedly. $\qquad\square$

Proposition 2. Let p be any prime, then

$$\sum_{0 \le s \le p^t - 1} (-1)^s f_s(1,t) = \begin{cases} -1, & \text{if } t = 1 \\ 0, & \text{if } t > 1 \end{cases},$$

$$\sum_{0 \le s \le p^t - 1} f_s(1,t) = \begin{cases} 2^{2^t - t - 1}, & \text{if } p = 2 \\ 2^{p^{t-1} - 1}(2^{p^{t-1}(p-1)} - 1)/p^t, & \text{if } p \ge 3 \end{cases}.$$

Proposition 2 is an immediate consequence of (7).

Proposition 3. Let p be any prime, then

$$\binom{p^t}{s} = \sum_{j=0}^{l}(-1)^{s-s/p^j}F_{s/p^j}(1,t-j)\,p^{t-j} \tag{8}$$

for all $0 \le s \le p^t$, here $0 \le l \le t$ is the largest integer that p^l divides s, $F_s(1,0) := (-1)^s$.

Proof. Let $F_s(k)$ be the number of solutions of the congruence,

$$k \equiv i_1 + i_2 + \cdots + i_s\,(mod\,p)\,|\,1 \le i_1 < i_2 < \cdots < i_s \le p^t,$$

the author shows in [1] that

$$F_s(0) - F_s(1) = cdots = F_s(0) - F_s(p-1)$$

$$= \begin{cases} 0 & \text{if } p \nmid s \\ (-1)^{s-s/p}\binom{p^{t-1}}{s/p} & \text{if } p \mid s. \end{cases} \tag{9}$$

Similarly as $f_s(k,t)$, we could prove that $F_s(k_1,t) = F_s(k_2,t)$, if $(k_1,p^t) = (k_2,p^t)$. Hence one has from (9),

$$\binom{p^t}{s} = F_s(1,t)\,p^t + \begin{cases} 0 & \text{if } p \nmid s \\ (-1)^{s-s/p}\binom{p^{t-1}}{s/p} & \text{if } p \mid s. \end{cases} \tag{10}$$

(8) follows directly by using (10) repeatedly. □

Proposition 4. Let p be any prime, $t \ge 1$, then

$$\sum_{0 \le s \le p^t}(-1)^s F_s(1,t) = 0,$$

$$\sum_{0 \le s \le p^t} F_s(1,t) = \begin{cases} 2^{2^t-t}, & \text{if } p = 2 \\ 2^{p^{t-1}}(2^{p^{t-1}(p-1)} - 1)/p^t, & \text{if } p \ge 3. \end{cases}$$

Proposition 4 is an immediate consequence of (10).

3.   One more result

Generally, for any prime p, $t \ge 1$, $0 \le j < p^t$, $0 \le s \le p^t - j$, $1 \le k \le p^t$, define $f_s^{(j)}(k,t)$ to be the number of solutions of the congruence

$$k \equiv i_1 + i_2 + \cdots + i_s\,(mod\,p^t)\,|\,1 \le i_1 < i_2 < \cdots < i_s \le p^t - j.$$

In particular $f_s^{(0)}(k,t) = F_s(k,t)$, $f_s^{(1)}(k,t) = f_s(k,t)$, for $j \geq 2$, define $f_0^{(j)}(k,t) = f_0(k,t)$, most of the above formulae do not hold any more. However, we have the following,

Proposition 5. Let p be any prime, $t \geq 1$, $1 \leq j < p^t$, $1 \leq k \leq p^t$, then

$$f_s(k,t) = \sum_{i=0}^{s} (-1)^i F_{s-i}(k,t), \tag{11}$$

and

$$f_s^{(j)}(k,t) = \sum_{0 \leq i_1 + i_2 + \cdots + i_{j-1} \leq s} (-1)^{i_1 + i_2 + \cdots + i_{j-1}}$$
$$\times f_{s-i_1-i_2-\cdots-i_{j-1}}(k + i_1 + 2i_2 + \cdots + (j-1)i_{j-1}) \tag{12}$$

for all $0 \leq s \leq p^t - j$.

Proof. For $l \geq 0$, let's consider $f_s^{(l)}(k,t) - f_s^{(l+1)}(k,t)$, taking $i_s = p^t - l$. From the definition of $f_s^{(l)}(k,t)$, one has $f_s^{(l)}(k,t) - f_s^{(l+1)}(k,t) = f_{s-1}^{(l+1)}(k+j,t)$, or

$$f_s^{(l+1)}(k,t) = f_s^{(l)}(k,t) - f_{s-1}^{(l+1)}(k+j,t). \tag{13}$$

By using (13) repeatedly,

$$f_s^{(l+1)}(k,t) = \sum_{i=0}^{s} (-1)^i f_{s-i}^{(l)}(k + il, t). \tag{14}$$

If $l = 0$, (14) is equivalent to (11). Now, we want to prove (12) by induction on j. If $j = 1$, (12) is trivial, if $j = 2$, (12) follows from (14)($l = 1$). Suppose that (12) is true for all $j \geq l$ ($l \geq 2$), again from (14),

$$f_s^{(l+1)}(k,t) = \sum_{i=0}^{s} (-1)^i f_{s-i}^{(l)}(k + il, t)$$
$$= \sum_{i=0}^{s} (-1)^i \sum_{0 \leq i_1 + i_2 + \cdots + i_{l-1} \leq s-i} (-1)^{i_1 + i_2 + \cdots + i_{l-1}}$$
$$\times f_{s-i_1-i_2-\cdots-i_{l-1}}(k + i_1 + 2i_2 + \cdots + (l-1)i_{l-1}).$$

By the assumption of induction, (12) follows. $\qquad\square$

Remark. Either (3) or (4) reminds us of the well-known formula,

$$\binom{x}{n} = a_{n,1}x + a_{n,2}x^2 + \cdots + a_{n,n}x^n,$$

where $a_{n,i} = n!S(n,i)(1 \leq i \leq n)$, S(n,i) is the Stirling number of the first kind. Comparing (3) with (5), there might be hopefully a formula like

$$f_s(p^j, t) - f_s(p^{j-1}, t) = (-1)^{s-[s/p^j]} f_{[s/p^j]}(1, t-j) \tag{15}$$

for all $1 \leq j \leq t$, but I cannot prove it except for $t \leq 2$. It's easy to verify from (15) that

$$f_s(p^j, t) = \sum_{1 \leq i \leq j} (-1)^{s-[s/p^i]} f_{[s/p^i]}(1, t-i). \tag{16}$$

Combining (16) with (7), one derive a reverse formula for (3), i.e.,

$$f_s(p^i, t) = \left\{ \binom{p^t - 1}{s} + \sum_{1 \leq j \leq i} (-1)^{s-[s/p^j]} \binom{p^{t-j} - 1}{[s/p^j]} \right.$$

$$\left. \times p^{j-1}(p-1) - (-1)^{s-[s/p^{i+1}]} \binom{p^{t-i-1} - 1}{[s/p^{i+1}]} p^i \right\} \bigg/ p^t$$

for all $0 \leq s \leq p^t - 1, i \geq 0$, here the sum vanishes when $i = 0$.

References

[1] Tianxin Cai, A generalization of 'Wolstenholme-type' on q-binomial coefficients, *preprint.*

INFINITE SUMS, DIOPHANTINE EQUATIONS AND FERMAT'S LAST THEOREM *

Henri DARMON
CICMA, Mathematics Dept., McGILL University, Montréal, Canada H3A 2K6
darmon@math.mcgill.ca

Claude LEVESQUE
CICMA, Dép. de Mathématiques et de Statistique, Université LAVAL, Québec, Canada G1K 7P4
cl@mat.ulaval.ca

Abstract   Thanks to the results of Andrew Wiles, we know that Fermat's last theorem is true. As a matter of fact, this result is a corollary of a major result of Wiles: *every semi-stable elliptic curve over* **Q** *is modular*. The modularity of elliptic curves over **Q** is the content of the Shimura-Taniyama conjecture, and in this lecture, we will restrain ourselves to explaining in elementary terms the meaning of this deep conjecture.

Keywords: diophantine equation, elliptic curve, Fermat's last theorem, Fermat–Pell equation, modular form, Pythagoras' equation.

1.   Introduction

A few years ago, the New York Times highlighted the proof of Fermat's last theorem by Andrew Wiles, completed in collaboration with his former Ph.D. student Richard Taylor. This was the last chapter in an epic initiated around 1630, when Pierre de Fermat wrote in the margin of his Latin version of Diophantus' ARITHMETICA the following enigmatic lines, unaware of the passions they were about to unleash:

Cubum autem in duos cubos, aut quadrato-quadratum in duos quadrato-quadratos, et generaliter nullam in infinitum ultra quadratum, olloque

---

*Written English version of a lecture given in French by Henri Darmon on October 14, 1995, at CEGEP de Lévis-Lauzon on the occasion of the *Colloque des Sciences Mathématiques du Québec* and which appeared in French in the *Comptes Rendus du 38ᵉ Congrès de l'Association Mathématique du Québec*.

*potestatem in duos ejusdem nominis fas est dividere. Cujus rei ermat's
demonstrationem mirabilem sane detexi. Hanc marginis exiguitas non
caperet.*

In plain English, for those unfamiliar with Latin:

*One cannot write a cube as a sum of two cubes, a fourth power as a sum
of two fourth powers, and more generally a perfect power as a sum of
two like powers. I have found a quite remarkable proof of this fact, but
the margin is too narrow to contain it.*

The sequel is well-known: Fermat never revealed his alleged proof.
Thousands of mathematicians (from amateurs to most famous scholars)
working desperately hard at refinding this proof were baffled for more
than three centuries.

Fermat's Last Theorem. *The equation*

$$\boxed{x^n + y^n = z^n} \quad (n \geq 3) \tag{1.1}$$

has no integral solution with $xyz \neq 0$.

Using his so-called *method of infinite descent*, Fermat himself proved
the theorem when $n = 4$. Euler is credited for the proof of the case
$n = 3$ (though his proof was incomplete). The list of mathematicians
who worked on this problem of Fermat reads like a Pantheon of number
theory: Dirichlet, Legendre, Cauchy, Lamé, Sophie Germain, Lebesgue,
Kummer, Wieferich, to name but the most famous. Their results secured
the proof of Fermat's last theorem for all exponents $n \leq 100$.

Though the importance of the theorem looks like being mostly symbolic, this problem of Fermat was extraordinarily fruitful for modern
mathematics. Kummer's efforts generated huge bulks of mathematical
theories: algebraic number theory, cyclotomic fields. In 1985, the theory
of elliptic curves and modular forms threw an unexpected light on the
problem. This point of view was initiated by Gerhard Frey and led ten
years later to the proof of Wiles.

Here is (at last!) this famous proof of Fermat's last theorem which
was so keenly sought for. Roughly! (With references quoted from the
appendix.)

> **Proof of Fermat's Last Theorem.**
> *By K. Ribet [R], the Shimura–Taniyama conjecture (for
> semi-stable elliptic curves) implies the truth of Fermat's
> last theorem.*
> *Thanks to the works of Wiles [W] and Taylor–Wiles [T–
> W], we know that the Shimura–Taniyama conjecture is
> true for semi-stable elliptic curves.* Q.E.D.

This is a very short proof and it could possibly fit in that famous margin
of the book of Diophantus. Hence Fermat's proof, if it existed, was
different...

Readers will point out that this last proof lacks some details! The papers of Wiles and Taylor-Wiles cover more than 130 pages of the prestigious journal *"Annals of Mathematics"*, and rely on numerous previous papers which could hardly be summarized in less than one thousand pages addressed to initiated readers.

So Wiles did not succeed in making his proof contained in some narrow margin of any manuscript. In August 1995, the organizers of a conference held in Boston on F last theorem got off with printing the proof on a tee-shirt, put on by the first author during his lecture at the *C des Sciences mathématiques du Québec*, and whose content is reproduced in the appendix.

In this lecture, we will refrain from dealing with the existing link between Fermat's last theorem and the Shimura–Taniyama conjecture; we refer interested readers to papers listed in the bibliography. We shall restrain ourselves to explaining in elementary terms the meaning of the Shimura–Taniyama conjecture. As a matter of fact, we would like to make readers aware of the importance of this conjecture, which goes much beyond Fermat's last theorem, and is tied to some of the deepest and most fundamental questions of number theory.

2. Pythagoras' equation

Let us start with Pythagoras' equation

$$\boxed{x^2 + y^2 = 1} \qquad (2.1)$$

whose non-zero *rational solutions* $(x, y) = (\frac{a}{c}, \frac{b}{c})$ give birth to Pythagoras' triples (a, b, c) verifying the equation $a^2 + b^2 = c^2$. This equation was highlighted in Diophantus' treatise and led Fermat to consider the case where the exponents are greater than 2. (So our starting point is the same as Fermat's one, even if we will not deal with his last theorem...)

The rational solutions of Pythagoras' equation are given in a parametric way by

$$(x, y) = \left(\frac{1 - t^2}{1 + t^2}, \frac{2t}{1 + t^2} \right), \quad t \in \mathbf{Q} \cup \{\infty\}, \qquad (2.2)$$

which provides the classification of Pythagoras' triples and leads to the complete solution of Fermat's equation for $n = 2$. Integral solutions (with $x, y \in \mathbf{Z}$) are still simpler to describe. There are 4 of them, namely $(1, 0)$, $(-1, 0)$, $(0, 1)$, $(0, -1)$; hence we write

$$N_\mathbf{Z} = 4. \qquad (2.3)$$

We can also study the equation $x^2 + y^2 = 1$ on fields other than the rational numbers; for instance, the field $\mathbf{R}$ of real numbers, or the fields $\mathbf{F}_p = \{0, 1, 2, \ldots, p-1\}$ of congruence classes modulo p, where p is a prime number.

Solutions in real numbers of the equation $x^2 + y^2 = 1$ correspond to points on a circle of radius 1. Let us give the set of real solutions a quantitative measure by writing

$$N_{\mathbf{R}} = 2\pi, \tag{2.4}$$

the circonference of the circle.

The solutions of $x^2 + y^2 = 1$ on $\mathbf{F}_p$ form a finite set, and we set

$$N_p = \#\{(x, y) \in \mathbf{F}_p^2 : x^2 + y^2 = 1\}. \tag{2.5}$$

To calculate N_p, we let x run between 0 and $p-1$ and look for solutions whose first coordinate is x. There will be 0, 1, or 2 solutions according to whether $1 - x^2$ is not a square modulo p, is equal to 0, or is a non-zero square modulo p, respectively. Since half of the non-zero integers modulo p are squares, it is expected that N_p is roughly equal to p; this prompts us to define a_p as the "error term" of this rough estimate:

$$a_p = p - N_p. \tag{2.6}$$

In so doing, we arrive at the main problem which, as will be seen later, leads directly to the Shimura–Taniyama conjecture.

Problem 1. *Does there exist a simple formula for the numbers N_p as a function of p (or, which in the same, for the numbers a_p)?*

Experimental methods play an important role in the theory of numbers, probably to a greater extent than in other fields of pure mathematics. Gauss was a prodigious calculator, and found his quadratic reciprocity law in some empiric way, before giving it many rigorous proofs. Following in the footsteps of the master, let us give a list of the values of N_p for some values of p.

p	N_p	a_p
2	2	0
3	4	-1
5	4	1
7	8	-1
11	12	-1
13	12	1
17	16	1
19	20	-1
23	24	-1
29	28	1
31	32	-1
37	36	1
41	40	1
$\vdots$	$\vdots$	$\vdots$
10007	10008	-1
$\vdots$	$\vdots$	$\vdots$

Table 1: $x^2 + y^2 = 1$

A look at the table leads at once to the following conjecture.

Conjecture 2. *The value of N_p is 2 if $p = 2$ and we have*

$$
N_p = \begin{cases} p - 1 & \text{if } p \equiv +1 \pmod 4, \\[2mm] p + 1 & \text{if } p \equiv -1 \pmod 4. \end{cases} \tag{2.7}
$$

(In particular, we see that $p \neq N_p$, which might be of interest to our computer science colleagues: **P $\neq$ NP!**)

How can we prove Conjecture 2? Let us come back to the parametrization

$$
(x, y) = \left(\frac{1 - t^2}{1 + t^2}, \frac{2t}{1 + t^2} \right). \tag{2.8}
$$

The values $t = 0, 1, \dots, p - 1, \infty$ give birth to a complete list of $p + 1$ distinct solutions, excepted when -1 is a square j^2 modulo p. In the latter case, the denominator vanishes for the two values $t = j, -j$, so these values are not admissible. Therefore, when p is odd,

$$
N_p = \begin{cases} p - 1 & \text{if } -1 \text{ is a square modulo } p, \\[2mm] p + 1 & \text{if } -1 \text{ is not a square modulo } p. \end{cases} \tag{2.9}
$$

The condition that -1 be a square modulo p may *a priori* look subtle, but we are fortunate to be able to count on the following theorem proved by Fermat.

Theorem 3 (Fermat). *The integer* -1 *is a square modulo p if and only if* $p = 2$ *or* $p \equiv 1$ (mod 4).

Here is a proof, slightly different from that of Fermat. The multiplicative group $\mathbf{F}_p^\times$ is cyclic of order $p - 1$, and the element -1 of order 2 has a square root if and only if $\mathbf{F}_p^\times$ possesses some elements of order 4.

Theorem 3 (that we just proved) together with formula (2.9) provides a proof of Conjecture 2 about the value of N_p. What is the purpose of such an explicit formula for N_p? Let us consider, for instance, the following infinite product (taken over all the primes p):

$$\prod_p \frac{p}{N_p} \quad = \quad \prod_p \left(1 - \frac{a_p}{p}\right)^{-1} \tag{2.10}$$

$$``=" \quad \left\{\prod_{p \equiv 1(4)} \left(1 - \frac{1}{p}\right)^{-1}\right\} \cdot \left\{\prod_{p \equiv -1(4)} \left(1 + \frac{1}{p}\right)^{-1}\right\}$$

$$``=" \quad \left\{\prod_{p \equiv 1(4)} \left(1 + \frac{1}{p} + \frac{1}{p^2} + \frac{1}{p^3} + \cdots\right)\right\}$$

$$\cdot \left\{\prod_{p \equiv -1(4)} \left(1 - \frac{1}{p} + \frac{1}{p^2} - \frac{1}{p^3} + \cdots\right)\right\}$$

$$``=" \quad 1 - \frac{1}{3} + \frac{1}{5} - \frac{1}{7} + \frac{1}{9} - \frac{1}{11} + \frac{1}{13} - \cdots \tag{2.11}$$

$$= \quad \frac{\pi}{4} \qquad \text{(by Leibniz's formula),} \tag{2.12}$$

where the equality (2.11) is (formally) a consequence of the unique factorization of integers as products of powers of primes. We then deduce

$$\prod_p \frac{N_p}{p} = \frac{4}{\pi}. \tag{2.13}$$

To tell the truth, our proof of the equality (2.13) is a fallacy, because of the off-hand way the convergence questions were dealt with (this contempt would give analysts the shivers). This is why some equalities were used within inverted commas. Eighteenth century mathematicians like Euler were quite at ease with such formal series manipulations, guided by their instinct to reach the right conclusion by avoiding traps. As a matter of fact, it is true that

$$\prod_{p} \frac{N_p}{p} \quad \text{converges to} \quad \frac{4}{\pi},$$

though the convergence is very slow.

Recalling that $N_{\mathbf{R}} = 2\pi$ and that $N_{\mathbf{Z}} = 4$, we conclude that

$$\left(\prod_{p} \frac{N_p}{p} \right) \cdot N_{\mathbf{R}} = 2N_{\mathbf{Z}}. \tag{2.14}$$

This magical formula unveils a mysterious relation between the solutions of the equation $x^2 + y^2 = 1$ on finite fields $\mathbf{F}_p$, on the real numbers $\mathbf{R}$, and on the ring $\mathbf{Z}$ of integers. In particular, the numbers N_p which depend only on the solutions of the equation $x^2 + y^2 = 1$ on $\mathbf{F}_p$, "know" the behaviour of the equation over the real numbers: thanks to these numbers N_p, we recover the number π, related to the circumference of the circle. Fundamentally, this is only a simple reinterpretation of Leibniz's formula, but in fact this is quite a fruitful one. At the beginning of the twenty-first century, number theory had not yet digested the deep meaning of this formula and of its generalizations, as will be seen later.

3.   The Fermat–Pell equation

In his abundant correspondence with his colleagues from Europe, Fermat liked to send them mathematical challenges. By doing so, he invited the English mathematicians Wallis and Brouncker to find the integer solutions of the equation

$$\boxed{x^2 - 61y^2 = 1}. \tag{3.1}$$

This is a particular case of the so-called Fermat–Pell equation $x^2 - Dy^2 = 1$. Fermat had a crush for this equation and had developed a general method to solve it, based on continued fractions. When $D = 61$, the smallest non-trivial solution is

$$(x, y) = (1766319049, 226153980). \tag{3.2}$$

It is the odd size of this smallest solution that led Fermat to take $D = 61$, although he pretended (with a bit of maliciousness) that this value of D was taken at random. This Fermat–Pell equation, of degree 2, is a conic in the plane, as is Pythagoras' equation. Let us denote by N_p the number of solutions modulo p, and let us give once more the list of the numbers N_p for some values of p.

p	N_p	a_p
2	2	0
3	2	1
5	4	1
7	8	-1
11	12	-1
13	12	1
17	18	-1
19	18	1
23	24	-1
29	30	-1
31	32	-1
37	38	-1
41	40	1
43	44	-1
47	46	1
53	54	-1
59	60	-1
61	122	-61
67	68	-1
71	72	-1
73	72	1
$\vdots$	$\vdots$	$\vdots$
10007	10006	1
10009	10008	1
$\vdots$	$\vdots$	$\vdots$

Table 2: $x^2 - 61y^2 = 1$

Using the parametrization

$$(x, y) = \left(\frac{1 + 61t^2}{1 - 61t^2}, \frac{2t}{1 - 61t^2} \right), \quad t \in \mathbf{Q} \cup \{\infty\}, \tag{3.3}$$

of the conic (3.1), we find as before that $N_2 = 2$, that $N_p = 2p$ if $p = 61$, and that otherwise

$$N_p = \begin{cases} p - 1 & \text{if 61 is a square modulo } p, \\ \\ p + 1 & \text{if 61 is not a square modulo } p. \end{cases} \tag{3.4}$$

Let us now use Gauss reciprocity law which for our purposes asserts that for p odd, 61 is a square modulo p if and only if p is a square modulo 61. So for $p \neq 2, 61$, we find

$$N_p = \begin{cases} p - 1 & \text{if } p \text{ is a square modulo } 61, \\ \\ p + 1 & \text{if } p \text{ is not a square modulo } 61. \end{cases} \tag{3.5}$$

This simple formula (which is periodic since it depends only on p modulo 61) for the numbers N_p allows to deduce, with formal calculations closely copied on those of equations (2.10) to (2.12), the identity

$$\prod_p \frac{p}{N_p} \quad \text{``}=\text{''} \quad \frac{1}{2}\sum_n \frac{a_n}{n}, \tag{3.6}$$

where

$$a_n = \begin{cases} 0 & \text{if } 61|n, \text{ or if } n \text{ is even,} \\ +1 & \text{if } n \text{ odd is a non-zero square modulo } 61, \\ -1 & \text{if } n \text{ odd is not a square modulo } 61. \end{cases} \tag{3.7}$$

One verifies (with the help of Abel's summation formula, for instance) that the infinite sum in (3.6) converges (conditionally). Some kind of heroic calculations (which we invite the readers to do) lead to an identity analogous to the formula (2.12) of Leibniz,

$$\sum_n \frac{a_n}{n} = \frac{\log(1766319049 + 226153980\sqrt{61})}{2\sqrt{61}}. \tag{3.8}$$

One recognizes in this expression the coefficients which appeared in the solution (3.2) of (3.1). In conclusion, the knowledge of the numbers N_p allowed us to "recover" a (fundamental) solution of a Fermat–Pell equation.

As a matter of fact, the identity (3.6) can be formally rewritten as

$$\left(\prod_p \frac{N_p}{p}\right) \cdot N_{\mathbf{R}} \quad \text{``}=\text{''} \quad 4\sqrt{61}N_{\mathbf{Z}}. \tag{3.9}$$

The quantities $N_{\mathbf{R}}$ and $N_{\mathbf{Z}}$ are both infinite, since the hyperbola defined by the equation $x^2 - 61y^2 = 1$ has no finite length and the Fermat–Pell equation possesses an infinity of integral solutions. It is all the same natural to define the quotient $\frac{N_{\mathbf{R}}}{N_{\mathbf{Z}}}$ as

$$\frac{N_{\mathbf{R}}}{N_{\mathbf{Z}}} := \log(1766319049 + 226153980\sqrt{61}), \tag{3.10}$$

namely, as the quantity appearing in the numerator of the right hand side of (3.8). As a matter of fact, the set of integral solutions of (3.1) is an abelian group isomorphic to $\mathbf{Z} \times \mathbf{Z}/2\mathbf{Z}$ and the application

$$(x, y) \mapsto \log(|x + y\sqrt{61}|) \tag{3.11}$$

sends this group into a discrete subgroup G of $\mathbf{R}$ which is isomorphic to $\mathbf{R}$. It is therefore natural to define $N_{\mathbf{R}}/N_{\mathbf{Z}}$ as the volume of $\mathbf{R}$, *i.e.*, as in (3.10).

After a few months, Wallis and Brouncker gave an answer to Fermat's question, sending him the solution (3.2) of (3.1), together with a general method (essentially similar to the method of Fermat based on continued fractions) to solve the Fermat–Pell equation $x^2 - Dy^2 = 1$. We do not know what was the reaction of the Toulouse mathematician, but one can imagine he felt some secret resentment... This shows that Wiles and Taylor are not the first two English mathematicians to brilliantly take up Fermat's challenges.

4.   The equation $x^3 + y^3 = 1$

Let us keep the same momentum, and after having dealt with conics let us switch to equations of degree 3. As a tribute to Fermat, let us study for instance

$$\boxed{x^3 + y^3 = 1}. \tag{4.1}$$

Does there exist as before a simple formula for the number N_p of solutions of this equation modulo p? Once more, let us give a table.

p	N_p	a_p
2	2	0
3	3	0
5	5	0
7	6	1
11	11	0
13	6	7
17	17	0
19	24	−5
23	23	0
29	29	0
31	33	−2
37	24	13
41	41	0
43	10	33
47	47	0
53	53	0
⋮	⋮	⋮
10007	10007	0
10009	9825	184

Table 3: $x^3 + y^3 = 1$

Contrary to the case of the degree 2 equations, the integers a_p are not all 0 or ± 1, and seem to behave rather randomly. However, one may guess by inspection a few properties of these integers a_p. For example, it looks like a_p always vanishes when 3 divides $p + 1$. But what is going on when $p \equiv 1 \pmod 3$? Once more, it is Gauss himself who provided the answer by proving the following theorem.

Theorem 4 (Gauss). (1) *If* $p \equiv -1 \pmod 3$, *then* $a_p = 0$.

(2) *If* $p \equiv 1 \pmod 3$, *then the number $4p$ can be written as* $4p = A^2 + 27B^2$ *with* $A \equiv -1 \pmod 3$, *which makes A unique, so we have* $a_p = A + 2$.

The following table allows us to verify this theorem for a few values of p:

p	N_p	a_p	$4p = A^2 + 27B^2$
2	2	0	$- - -$
3	3	0	$- - -$
5	5	0	$- - -$
7	6	1	$28 = (-1)^2 + 27 \cdot 1^2$
11	11	0	$- - -$
13	6	7	$52 = 5^2 + 27 \cdot 1^2$
17	17	0	$- - -$
19	24	-5	$76 = (-7)^2 + 27 \cdot 1^2$
23	23	0	$- - -$
29	29	0	$- - -$
31	33	-2	$124 = (-4)^2 + 27 \cdot 2^2$
37	24	13	$148 = 11^2 + 27 \cdot 1^2$
41	41	0	$- - -$
43	10	33	$172 = 8^2 + 27 \cdot 2^2$
47	47	0	$- - -$
53	53	0	$- - -$
$\vdots$	$\vdots$	$\vdots$	$\vdots$
10007	10007	0	$- - -$
10009	9825	184	$40036 = 182^2 + 27 \cdot 16^2$
$\vdots$	$\vdots$	$\vdots$	$\vdots$

Table 4: $x^3 + y^3 = 1$ (sequel)

5. Elliptic curves

An elliptic curve is a diophantine equation of degree 3 having at least one rational solution. For example, the equation $x^3 + y^3 = 1$. One can prove that any elliptic curve over the rational numbers $\mathbf{Q}$ may be written, after a proper change of variables, in the form

$$y^2 = x^3 + ax + b, \tag{5.1}$$

where a, b are rational numbers.

As before, denote by N_p the number of solutions of the equation (5.1) over the finite field $\mathbf{F}_p$ of p elements.

Question 5. *Is there an explicit formula for the numbers N_p associated to an elliptic curve like the equation $x^3 + y^3 = 1$?*

Said otherwise, we would like to generalize the result of Gauss for the equation $x^3 + y^3 = 1$ to the case of any given elliptic curve. This is exactly the scope of the Shimura–Taniyama conjecture proved by Wiles for a very large class of elliptic curves.

Before giving explicit statements, let us see how the land lies by considering the elliptic curve

$$y^2 + y = x^3 - x^2 \qquad (5.2)$$

studied by Eichler. Here are some values of N_p as calculated by a computer:

p	N_p	a_p
2	4	−2
3	4	−1
5	4	1
7	9	−2
11	10	1
13	9	4
17	19	−2
19	19	0
23	24	−1
29	29	0
31	24	7
⋮	⋮	⋮
10007	9989	18
⋮	⋮	⋮

Table 5: $y^2 + y = x^3 - x^2$

This time, it is more difficult to guess a structure for the values of the integers a_p which again seem to behave rather randomly. Hasse proved the deep inequality

$$|a_p| \leq 2\sqrt{p} \qquad (5.3)$$

(valid for all elliptic curves), but this is far from providing an *exact formula* for the numbers N_p.

Eichler, building on deep results of Hecke, was however successful in obtaining an exact formula. The starting point is to extend the definition of the coefficient a_p (valid for the prime index p) to any index n by setting

$$\begin{cases} a_1 & = & 1, \\[2mm] a_p & = & p - N_p, \\[2mm] a_{p^r} & = & a_p a_{p^{r-1}} - p a_{p^{r-2}}, \\[2mm] a_n & = & \prod_{i=1}^{r} a_{p_i^{e_i}}, \quad \text{where} \quad n = \prod_{i=1}^{r} p_i^{e_i}. \end{cases} \qquad (5.4)$$

We notice that this extension is a rather natural one: if we denote by N_{p^r} the number of solutions of the elliptic curve over the finite field $\mathbf{F}_{p^r}$ of p^r elements, then we have

$$a_{p^r} = p^r - N_{p^r}. \tag{5.5}$$

Theorem 6 (Eichler). *The formal series* $\sum\limits_{n=1}^{\infty} a_n q^n$ *is given by the formula:*

$$q \prod_{n=1}^{\infty} (1 - q^n)^2 \cdot (1 - q^{11n})^2 = q - 2\mathbf{q^2} - q^3 + 2q^4 + \mathbf{q^5} + 2q^6 - 2\mathbf{q^7}$$

$$- 2q^9 - 2q^{10} + \mathbf{q^{11}} - 2q^{12} + 4\mathbf{q^{13}} + 4q^{14}$$

$$- q^{15} - 4q^{16} - 2\mathbf{q^{17}} + 4q^{18} + 2q^{20} + 2q^{21}$$

$$- 2q^{22} - \mathbf{q^{23}} - 4q^{25} - 8q^{26} + 5q^{27} - 4q^{28}$$

$$+ 2q^{30} + 7\mathbf{q^{31}} + \cdots + 18\mathbf{q^{10007}} + \cdots$$

The reader can at leisure verify the truth of Eichler's theorem for a few values of p, by comparing the coefficients of q^p written in boldface, with the values from Table 5.

The Shimura–Taniyama conjecture, proved by Wiles, is a direct generalization of Eichler's theorem, in the sense that Wiles gave a *very precise description* of the generating function $\sum\limits_n a_n q^n$, where the integers a_n are the coefficients associated to any given elliptic curve.

More precisely, let

$$f(z) = \sum_{n=1}^{\infty} a_n e^{2\pi i n z} \tag{5.6}$$

be a Fourier series with coefficients $a_n \in \mathbf{R}$, and let N be a positive integer. We say that $f(z)$ is a *modular form of level* N if the following conditions are satisfied:

(1) The series defining f converges for $Im(z) > 0$, i.e., when $|e^{2\pi i z}| < 1$. The series f then represents a holomorphic function on the Poincaré upper half plane of complex numbers having a strictly positive imaginary part.

(2) For all $\begin{pmatrix} a & b \\ Nc & d \end{pmatrix} \in SL_2(\mathbf{Z})$, we have

$$f\left(\frac{az + b}{Ncz + d}\right) = (Ncz + d)^2 f(z), \tag{5.7}$$

where $SL_2(\mathbf{Z})$ is the group of 2×2 matrices of determinant 1 with coefficients in $\mathbf{Z}$.

Here is at last the famous Shimura–Taniyama conjecture.

Conjecture 7 (Shimura–Taniyama). *Let $y^2 = x^3 + ax + b$ be an elliptic curve over the rational numbers $\mathbf{Q}$, and let a_n $(n = 1, 2, \dots)$ be the integers defined for this curve by the equations of (5.4). Then the generating function*

$$f(z) = \sum_{n=1}^{\infty} a_n e^{2\pi i n z} \tag{5.8}$$

is a modular form.

In fact, the conjecture is more precise:

(1) It predicts the value of the level N of the modular form associated to the elliptic curve. This level would be equal to the *arithmetic conductor* of the curve, which depends only on the primes having "bad reduction". The exact definition of N will not be used in our treatment.

(2) The space of modular forms of a given level N is a vector space over $\mathbf{R}$ whose dimension, a finite number, can easily be calculated out of the value of N. This space is equipped with certain natural linear operators defined by Hecke. The conjecture also states that the modular form f is an eigenform (*i.e.*, a characteristic vector) for all Hecke operators.

One shows that there is but a finite number of modular forms of level N which are eigenforms for all Hecke operators, and whose first Fourier coefficient a_1 is equal to 1. So once the conductor N of an elliptic curve has been calculated, we are led to a finite list of possibilities for the sequence $\{a_n\}_{n \in \mathbf{N}}$ associated to this curve. From this point of view, the Shimura–Taniyama conjecture gives an explicit formula for the numbers N_p of rational points on the elliptic curve modulo p.

Thanks to the works of Wiles and Taylor–Wiles, we now know that the Shimura–Taniyama conjecture is true for a very large class of elliptic curves. As a matter of fact, Diamond proved, improving upon the results of Wiles and Taylor-Wiles, that it suffices that the elliptic curve has good reduction, or in the worst case has only one double point modulo 3 or 5.

The formula of Wiles for the integers N_p associated to an elliptic curve looks at first less explicit than that of Fermat (Conjecture 2) for the equation $x^2 + y^2 = 1$, or than that of Theorem 4 of Gauss for the equation $x^3 + y^3 = 1$. Nevertheless it allows one to give a meaning to the expression $\prod_p \frac{p}{N_p}$, or to be more precise[1], to the quantities

$$\prod_p \frac{p}{N_p + 1}.$$

This is achieved by introducing the L-series associated to the elliptic curve E:

$$L(E, s) = \prod_p \left(1 - \frac{a_p}{p^s} + \frac{1}{p^{2s-1}}\right)^{-1} = \sum_n \frac{a_n}{n^s}. \qquad (5.9)$$

One notes that formally,

$$L(E, 1) \quad " = " \quad \prod_p \frac{p}{N_p + 1}, \qquad (5.10)$$

though the series defining $L(E, s)$ converges only for $\mathrm{Re}(s) > \frac{3}{2}$. In order to make $L(E, 1)$ meaningful, one needs to know that the series defining $L(E, s)$ admits an analytic continuation at least up to the value $s = 1$.

The following fundamental result of Hecke will then prove useful.

Theorem 8 (Hecke). *If the sequence $\{a_n\}_{n \in \mathbb{N}}$ comes from a modular form, then the function $L(E, s)$ admits an analytic continuation to the whole complex plane, and in particular, the value of $L(E, 1)$ is well defined.*

If one knows that the elliptic curve E is modular, then the result of Hecke allows one to define

$$\prod_p \frac{p}{N_p + 1} := L(E, 1). \qquad (5.11)$$

As in the previous example, one may expect some useful pieces of arithmetic information about the curve E from the value of $L(E, 1)$ (or more generally, from the behaviour of $L(E, s)$ at the neighborhood of $s = 1$).

This is exactly the content of the Birch–Swinnerton-Dyer conjecture, of which a particular case is the following.

---

[1]In our naïve definition of N_p, we systematically omitted to count the solution which corresponds to the "point at infinity" and which naturally comes into play when one considers an equation of the elliptic curve in the Desargues projective plane. It is therefore natural to replace N_p by $N_p + 1$.

Weak Birch–Sinnerton-Dyer conjecture. *The elliptic curve E possesses a finite number of rational points if and only if $L(E, 1) \neq 0$.*

This conjecture is far from being proved, and is still one of the most important open questions in the theory of elliptic curves. One can count although on some partial results, for instance, the following one, which is a consequence of the works of Gross–Zagier, Kolyvagin, together with an analytic result due to Bump–Friedberg–Hoffstein and Murty–Murty.

Theorem 9 (Gross–Zagier, Kolyvagin). *Let E be a modular elliptic curve. If the function $L(E, s)$ possesses a zero of order 0 or 1 at $s = 1$, then the weak Birch–Swinnerton-Dyer conjecture is true for E.*

The case where the function $L(E, s)$ has a zero of order > 1 still remains very mysterious. One expects in this case that the equation of the curve E has always rational solutions, but we still ignore how to find (or build) them in a systematic way, or even whether or not there is an algorithm to determine in all cases the set of all rational solutions. Despite spectacular progresses over the past few years, several number theorists, in love with elliptic curves, will be kept very busy.

Appendix: The t-shirt of the Boston University Conference

On the front of the above-mentioned t-shirt, one can read the following.

FERMAT'S LAST THEOREM: *Let $n, a, b, c \in \mathbf{Z}$ with $n > 2$. If $a^n + b^n = c^n$ then $abc = 0$.*

Proof. The proof follows a program formulated around 1985 by Frey and Serre [F,S]. By classical results of Fermat, Euler, Dirichlet, Legendre and Lamé, we may assume that $n = p$, an odd prime ≥ 11. Suppose $a, b, c \in \mathbf{Z}, abc \neq 0$, and $a^p + b^p = c^p$. Without loss of generality we may assume $2|a$ and $b \equiv 1 \pmod{4}$. Frey [F] observed that the elliptic curve $E : y^2 = x(x - a^p)(x + b^p)$ has the following "remarkable" properties:
(1) E is semistable with conductor $N_E = \prod_{\ell | abc} \ell$; and
(2) $\rho_{f,p} = \rho_{E,p}$ is unramified outside $2p$ and is flat at p.
By the modularity theorem of Wiles and Taylor–Wiles [W,T–W], there is an eigenform $f \in S_2(\Gamma_0(N_E))$ such that $\rho_{f,p} = \bar{\rho}_{E,p}$. A theorem of Mazur implies that $\bar{\rho}_{E,p}$ is irreducible, so Ribet's theorem [R] produces a Hecke eigenform $g \in S_2(\Gamma_0(2))$. But $X_0(2)$ has genus zero, so $S_2(\Gamma_0(2)) = 0$. This is a contradiction and Fermat's Last Theorem follows. Q.E.D.

On the back of the t-shirt, one finds the following bibliography.

References

[**F**] Frey, G.: Links between stable elliptic curves and certain Diophantine equations. *Ann. Univ. Sarav.* **1** (1986), 1–40.

[**R**] Ribet, K.: On modular representations of Gal($\bar{\mathbf{Q}}/\mathbf{Q}$)) arising from modular forms. *Invent. Math.* **100** (1990), 431–476.

[**S**] Serre, J.-P.: Sur les représentations modulaires de degré 2 de *Gal*($\bar{\mathbf{Q}}/\mathbf{Q}$), *Duke Math. J.* **54** (1987), 179–230.

[**T–W**] Taylor, R.L., Wiles, A.: Ring-theoretic properties of certain Hecke algebras. *Annals of Math.* **141** (1995), 553–572.

[**W**] Wiles, A.: Modular elliptic curves and Fermat's Last Theorem. *Annals of Math.* **141** (1995), 443–551.

Annoted bibliography

The references appear under seven headings, each one dealing with a given theme. Readers interested only by easily understood survey papers will appreciate references 1 to 4, 8 to 11, 14 to 18 of Section B.

(A) Fermat's last theorem

The following references provide historic informations about Fermat's last theorem or about methods not dealing with elliptic curves

1. E.T. Bell, *The Last Problem*, 2e édition, MAA Spectrum, Mathematical Association of America, Washington, DC, 1990, 326 pages.

2. H.M. Edwards, *Fermat's Last Theorem: A Genetic Introduction to Algebraic Number Theory*, Graduate Texts in Math. **50**, Springer–Verlag, New York, Berlin, Heidelberg, 1977, 410 pages.

3. C. Houzel, *De Diophante à Fermat*, in *Pour la Science* **220**, January 1996, 88–96.

4. P. Ribenboim, *13 Lectures on Fermat's Last Theorem*, Springer–Verlag, New York, Berlin, Heidelberg, 1979, 302 pages.

5. L.C. Washington, *Introduction to Cyclotomic Fields*, Graduate Texts in Math. **83**, Springer–Verlag, New York Berlin 1982, 389 pages.

(B) Elliptic curves and Fermat's last theorem

To learn more on the links between Fermat's last theorem and elliptic curves, we suggest the following references.

1. N. Boston, *A Taylor-made Plug for Wiles' Proof*, College Math. J. **26**, No. 2, 1995, 100–105.

2. B. Cipra, *"A Truly Remarkable Proof"*, in *What's happening in the Mathematical Sciences*, AMS Volume **2**, 1994, 3–7.

3. J. Coates, *Wiles Receives NAS Award in Mathematics*, Notices of the AMS **43**, 7, 1994, 760–763.

4. D.A. Cox, *Introduction to Fermat's Last Theorem*, Amer. Math. Monthly **101**, No. 1, 1994, 3–14.

5. B. Edixoven, *Le rôle de la conjecture de Serre dans la preuve du théorème de Fermat*, Gazette des mathématiciens **66**, Oct. 1995, 25–41. Addendum: idem **67**, Jan. 1996, 19.

6. G. Faltings, *The Proof of Fermat's Last Theorem by R. Taylor and A. Wiles*, Notices AMS **42**, No. 7, 743–746.

7. G. Frey, *Links Between Stable Elliptic Curves and Certain Diophantine Equations*, Ann. Univ. Sarav. **1**, 1986, 1–40.

8. G. Frey, *Links Between Elliptic Curves and Solutions of $A-B=C$*, Indian Math. Soc. **51**, 1987, 117–145.

9. G. Frey, *Links Between Solutions of $A-B=C$ and Elliptic Curves*, in *Number Theory, Ulm, 1987, Proceedings*, Lecture Notes in Math. **1380**, Springer–Verlag, New York, 1989, 31–62.

10. D. Goldfeld, *Beyond the last theorem*, in *The Sciences* **1996**, March/April, 34–40.

11. C. Goldstein, *Le théorème de Fermat*, La Recherche **263**, Mars 1994, 268–275.

12. C. Goldstein, *Un théorème de Fermat et ses lecteurs*, Presses Universitaires de Vincennes, 1995.

13. F.Q. Gouvêa, *A Marvelous Proof*, Amer. Math. Monthly **101**, No. 3, 1994, 203–222.

14. B. Hayes and K. Ribet, *Fermat's Last Theorem and Modern Arithmetic*, Amer. Scientist **82**, 1994, 144–156.

15. Y. Hellegouarch, *Points d'ordre $2p^h$ sur les courbes elliptiques*, Acta Arith. **26**, 1974/75, 253–263.

16. Y. Hellegouarch, *Fermat enfin démontré*, in *Pour la Science* **220**, February 1996, 92–97.

17. S. Lang, *Old and New Conjectured Diophantine Inequalities*, Bull. AMS (New Series) **23**, No. 1, 1990, 37–75.

18. B. Mazur, *Number Theory as Gadfly*, Amer. Math. Monthly **98**, No. 7, 1991, 593–610.

19. B. Mazur, *Questions about Number*, in *New Directions in Mathematics*, Cambridge Univ. Press, Cambridge, à paraître.

20. M.R. Murty, *Fermat's Last Theorem: an Outline*, Gazette Sc. Math. Québec, Vol. **XVI**, No. 1, 1993, 4–13.

21. M.R. Murty, *Reflections on Fermat's Last Theorem*, Elem. Math. **50** (1995) no. 1, 3–11.

22. J. Oesterlé, *Nouvelles approches du "théorème" de Fermat"*, Séminaire Bourbaki No. **694** (1987-88), Astérisque **161–162**, 1988, 165–186.

23. K. Ribet, *On Modular Representations of Gal($\bar{\mathbf{Q}}$/$\mathbf{Q}$) Arising from Modular Forms*, Invent. Math. **100**, 1990, 431–476.

24. K. Ribet, *From the Taniyama-Shimura Conjecture to Fermat's Last Theorem*, Ann. Fac. Sci. Toulouse (5) **11** (1990) no. 1, 116–139.

25. K. Ribet, *Wiles Proves Taniyama's Conjecture; Fermat's Last Theorem Follows*, Notices Amer. Math. Soc. **40**, 1993, 575–576.

26. K. Ribet, *Galois Representations and Modular Forms*, Bull. AMS (New Series) **32**, No. 4, 1995, 375–402.

27. M. Rosen, *New Results on the Arithmetic of Elliptic Curves*, Gazette Sc. Math. Québec, Vol. **XIV**, No. 1, 1993, 30–43.

28. K. Rubin and A. Silverberg, *A Report on Wiles' Cambridge Lectures*, Bull Amer. Math. Soc. (New Series) **31**, 1994, 15–38.

29. R. Schoof, *Proof of Taniyama-Weil Conjecture for Semi-stable Elliptic Curves over $\mathbf{Q}$*, Duke Math. J. **54**, 1987, 179–230.

30. J-P. Serre, *Sur les représentations modulaires de degré 2 de Gal($\bar{\mathbf{Q}}$/$\mathbf{Q}$)*, Duke Math. J. **54**, 1987, 179–230.

31. J-P. Serre, *Lettre à J.-F. Mestre*, in *Current Trends in Arithmetical Algebraic Geometry*, ed. by K. Ribet, Contemporary Mathematics **67**, AMS, 1987.

92

32. A. van der Poorten, *Notes on Fermat's Last Theorem*, Canadian Math. Society Series of Monographs and Advanced Texts, Wiley Interscience, Jan. 1996.

33. A. Wiles, *Modular Forms, Elliptic Curves, and Fermat's Last Theorem*, Proc. International Congress of Math., 1994, Birkhauser Verlag, Basel, 1995, 243–245.

(C) About the works of Wiles and Taylor

The following references concentrate on the work of Wiles and his *per se* proof of the Shimura–Taniyama conjecture.

1. J. Coates and S.T. Yau, *Elliptic Curves and Modular Forms*, in Proceedings of a conference in Hong Kong in 1993, International Press, Cambridge (MA) and Hong Kong, 1995.

2. H. Darmon, F. Diamond and R. Taylor, *Fermat's Last Theorem*, Current Developments in Math. **1**, International Press, 1995, 1–154.

3. H. Darmon, *The Shimura–Taniyama Conjecture, (d'après Wiles)*, (en Russe) Uspekhi Mat. Nauk **50** (1995), no. 3(303), pages 33–82. (English version in Russian Math Surveys).

4. V.K. Murty, ed., *Elliptic Curves, Galois Representations and Modular Forms*, CMS Conference Proc., AMS, Providence RI, 1996.

5. J. Oesterlé, *Travaux de Wiles (et Taylor...), Partie II*, Séminaire Bourbaki 1994–95, exposé No. **804**, 20 pages.

6. K. Ribet, *Galois Representations and Modular Forms*, Bull. AMS (New Series) **32**, 1995, No. 4, 375–402.

7. J-P. Serre, *Travaux de Wiles (et Taylor...), Partie I*, Séminaire Bourbaki 1994–95, exposé No. **803**, 13 pages.

8. R.L. Taylor and A. Wiles, *Ring Theoretic Properties of Certain Hecke Algebras*, Annals of Math. **141**, 1995, 553–572.

9. A. Wiles, *Modular Elliptic Curves and Fermat's Last Theorem*, Annals of Math. **141**, 1995, 443–551.

(D) Videos

Some readers may enjoy the numerous videos dealing with Fermat's last theorem and its proof.

1. Fermat Fest, *Fermat's Last Theorem. The Theorem and Its Proof: an Exploration of Issues and Ideas.* Shown on the occasion of a "Fermat Fest" in San Francisco, CA, on July 28, 1993, Video, *Selected Lectures in Mathematics*, AMS, Providence, RI, 1994, (98 min.)

2. B. Mazur, *Modular Elliptic Curves and Fermat's Last Theorem*, CMS meeting in Vancouver, August 1993, Video, *Selected Lectures in Mathematics*, AMS, Providence, RI, 1995, (50 min.)

3. K. Ribet, *Modular Elliptic Curves and Fermat's Last Theorem*, Lecture given at George Washington U. , Washington DC, 1993, Video, *Selected Lectures in Mathematics*, AMS, Providence, RI, 1993, (100 min.)

(E) Fermat and Gauss

To learn more on the works of Fermat and Gauss, in particular on the proof of Theorem 3 of Fermat, on the Fermat–Pell equation, and on the equation $x^3 + y^3 = 1$:

1. L.E. Dickson, *History of the Theory of Numbers*, Vol. II, Chelsea Publ. Co., New York, 1971.

2. K. Ireland and M. Rosen, *A Classical Introduction to Modern Number Theory*, 2nd edition, Graduate Texts in Math. **84**, Springer-Verlag, New York, 1990, 389 pages.

3. W. Scharlau and H. Opolka, *From Fermat to Minkowski. Lectures on the Theory of Numbers and Its Historical Development*, Translated from the German by Walter K. Bühler and G. Cornell, Undergraduate Texts in Math., Springer–Verlag, New York-Berlin, 1985, 184 pages.

4. A. Weil, *Fermat et l'équation de Pell*, in *Collected Papers*, Vol. III, Springer–Verlag, New York, 1979, 413–420.

5. A. Weil, *Number Theory. An Approach Through History. From Hammurapi to Legendre*, Birkhauser Boston Inc., Boston, MA, 1984, 375 pages.

(F) Elliptic curves

There is plenty of choice for the readers keen to learn more on elliptic curves.

1. J.W.S. Cassels, *Lectures on Elliptic Curves*, London Math. Society Student Texts **24**, Cambridge University Press, 1991, 137 pages.

2. H. Darmon, *Wiles' Theorem and the Arithmetic of Elliptic Curves*, in *Modular Forms and Fermat's Last Theorem*, Springer–Verlag, New York, 1997, 549-569.

3. D. Husemöller, *Elliptic Curves*, Graduate Texts in Math. **111**, Springer–Verlag, New York, 1987, 350 pages.

4. H. Kisilevsky and M.R. Murty, *Elliptic Curves and Related Topics*, CRM Proceedings and Lecture Notes, AMS, 1994, 195 pages.

5. A.W. Knapp, *Elliptic Curves*, Mathematical Notes **40**, Princeton U. Press, Princeton, NJ, 1992, 427 pages.

6. S. Lang, *Elliptic Curves: Diophantine Analysis*, Springer–Verlag, New York, 1978, 261 pages.

7. M.R. Murty and V.K. Murty, *Lectures on Elliptic Curves*, Lectures given at Andhra U., India, 1989, 92 pages.

8. M.R. Murty, *Topics in Number Theory*, Lectures given at the Mehta Research Institute, India, 1993, 117 pages.

9. J.H. Silverman and J. Tate, *Rational Points on Elliptic Curves*, Undergraduate Texts in Math., Springer–Verlag, New York, 1992, 281 pages.

10. J.H. Silverman, *The Arithmetic of Elliptic Curves*, Graduate Texts in Math. **106**, Springer–Verlag, New York, 1992, 400 pages.

11. J.H. Silverman, *Advanced Topics in the Arithmetic of Elliptic Curves*, Graduate Texts in Math., vol. **151**, Springer–Verlag, New York, 1994, 525 pages.

12. J. Tate, *Rational Points on Elliptic Curves*, Philips Lectures, Haverford College, 1961, unpublished notes.

(G) Modular forms and functions and the Shimura–Taniyama conjecture

1. T. Apostol, *Modular Functions and Dirichlet Series in Number Theory*, Graduate Texts in Math. **41**, Springer–Verlag, New York, 1976, 248 pages.

2. J. Cremona, *Algorithms for Modular Elliptic Curves*, Cambridge Univ. Press, Cambridge, 1992, 343 pages.

3. N. Koblitz, *Introduction to Elliptic Curves and Modular Forms*, 2nd edition, Graduate Texts in Math. **97**, Springer–Verlag, New York, 1993, 248 pages.

4. S. Lang, *Introduction to Modular Forms*, Springer–Verlag, New York, 1976, 261 pages.

5. T. Miyake, *Modular Forms*, Springer–Verlag, New York, 1989.

6. M.R. Murty, *Elliptic Curves and Modular Forms*, Can. Math. Bull. **34** (3), 1991, 375–384.

7. A. Ogg, *Modular Forms and Dirichlet Series*, Benjamin, New York, 1969.

8. J-P. Serre, *A Course in Arithmetic*, 2nd edition, Graduate Texts in Math. **7**, Springer–Verlag, New York, Berlin, Heidelberg, 1973, 115 pages.

Added in proof. *In the December* 1999 *issue of the Notices of the AMS, H. Darmon reported on the recent proof by C. Breuil, B. Conrad, F. Diamond and R. Taylor of the full Shimura–Taniyama–Weil conjecture for all elliptic curves over* **Q**.

ON THE NATURE OF THE "EXPLICIT FORMULAS" IN ANALYTIC NUMBER THEORY — A SIMPLE EXAMPLE

Christopher Deninger

Mathematisches Institut, WWU Münster, Einsteinstr. 62, 48149 Münster, Germany

deninge@math.uni-muenster.de

Abstract   We interpret the "explicit formulas" in the sense of analytic number theory for the zeta function of an elliptic curve over a finite field as a transversal index theorem on a 3-dimensional laminated space.

Keywords: Explicit formula, elliptic curve, transversal index, transversally elliptic operator, foliation, arithmetic topology

1.   Introduction

The "explicit formulas" in analytic number theory relate the prime numbers with the zeroes of the Riemann zeta function. They were first used by Riemann in his famous note which started analytic number theory. Various authors, notably A. Weil, subsequently developed generalizations of these formulas. A simple version of the "explicit formulas for the Riemann zeta function is the following. Given a test function $\alpha \in C_0^\infty(\mathbb{R})$ we have

$$\Phi(0) - \sum_\rho \Phi(\rho) + \Phi(1) = \sum_p \log p \sum_{k \geq 1} \alpha(k \log p)$$
$$+ \sum_p \log p \sum_{k \leq -1} p^k \alpha(k \log p) + W_\infty(\alpha) . \quad (1)$$

Here ρ runs over the non-trivial zeroes of $\zeta(s)$ and Φ is the entire function defined by the integral

$$\Phi(s) = \int_{\mathbb{R}} e^{ts} \alpha(t) \, dt \ .$$

Moreover we have set

$$W_\infty(\alpha) = \alpha(0) \log \pi + \int_0^\infty (\alpha(t) + e^{-t}\alpha(-t))(1 - e^{-2t})^{-1}$$
$$ - \alpha(0)t^{-1}e^{-2t} \, dt \ .$$

Formula (1) and generalizations can be found in [3] for example. We will view (1) as the "explicit formula" for the function

$$\hat{\zeta}(s) = \zeta(s)\zeta_\infty(s) \ , \quad \text{where} \quad \zeta_\infty(s) = \pi^{-s/2}\Gamma\left(\frac{s}{2}\right) \ .$$

In a sense $\hat{\zeta}(s)$ is an Euler product over *all* valuations of $\mathbb{Q}$. Note that $\hat{\zeta}(s)$ has simple poles at $s = 0, 1$ and that its zeroes are exactly the non-trivial zeroes of $\zeta(s)$.

There are good reasons to think that formula (1) has conceptual origins rooted in analysis on certain foliated spaces. These spaces however remain to be found e.g. [4], [5].

In this note we will be concerned with "explicit formulas" for certain function fields.
A function field F (of transcendence degree one) over the finite field $\mathbb{F}_q$ with $q = p^r$ elements is a finite extension of the field $\mathbb{F}_q(T)$ of rational functions over $\mathbb{F}_q$ such that $\mathbb{F}_q$ is algebraically closed in F. Since the time of E. Artin's thesis it is well known that there are many analogies between the arithmetic of number fields and such function fields. In particular a zeta function can be defined for F,

$$\hat{\zeta}_F(s) = \prod_w (1 - Nw^{-s})^{-1} \ .$$

Here w runs over the valuations of F and Nw is the number of elements in the (finite) residue field of w. This Euler product converges for $\operatorname{Re} s > 1$ and $\hat{\zeta}_F(s)$ can be meromorphically continued to the entire plane. In fact, it is a rational function of q^{-s}. The function $\hat{\zeta}_F(s)$ has simple poles at the following points on $\operatorname{Re} s = 0$ resp. $\operatorname{Re} s = 1$:

$$\frac{2\pi i\nu}{\log q} \quad \text{resp.} \quad 1 + \frac{2\pi i\nu}{\log q} \quad \text{for } \nu \in \mathbb{Z} \ .$$

Using geometric methods A. Weil was able to prove that the Riemann hypotheses holds for $\hat{\zeta}_F(s)$ i.e. that all zeroes ρ of $\hat{\zeta}_F(s)$ lie on the line $\mathrm{Re}\, s = 1/2$.

The usual methods give the following analogue of (1) if F has genus g:

$$\sum_{\nu \in \mathbb{Z}} \Phi\left(\frac{2\pi i \nu}{\log q}\right) - \sum_{\rho} \Phi(\rho) + \sum_{\nu \in \mathbb{Z}} \Phi\left(1 + \frac{2\pi i \nu}{\log q}\right)$$

$$= \alpha(0)(2 - 2g)\log q + \sum_w \log Nw \sum_{k \geq 1} \alpha(k \log Nw)$$

$$+ \sum_w \log Nw \sum_{k \leq -1} Nw^k \alpha(k \log Nw) \,. \tag{2}$$

Note that there is no term like $W_\infty(\alpha)$ which corresponds to the Archimedian valuation of $\mathbb{Q}$. In the function field case all valuations are non-Archimedian.

Now let E_0 be an elliptic curve over $\mathbb{F}_q$ with function field $F = \mathbb{F}_q(E_0)$. In this note we will interpret formula (2) for $F = \mathbb{F}_q(E_0)$ as a transversal index theorem for an $\mathbb{R}$-action on a suitable 3-dimensional space X which is not a manifold. This is the content of section 3.

Related spaces for elliptic curves have been considered by Ihara but not published. See [8], [9], [10], [11] however for his constructions in the cyclotomic, modular and Shimura curve cases.

In section 2 we provide some background on classical transversal index theory on manifolds. Finally in the last section we give another example of a transversal index calculation on our space X above.

In the field of arithmetic topology, see e.g. [18], certain analogies are studied between number fields and 3-manifolds. The constructions of the present note make some of these analogies concrete in the easy case where the number field is replaced by $F = \mathbb{F}_q(E_0)$. See for instance the explicit bijection in proposition 3.3 between valuations of F and certain knots in the 3-space X.

I would like to thank Y. Ihara, Y. Kordyukov, P. Schneider and M. Volkov for helpful comments and discussions. I am grateful to S. Kanemitsu, Professor of mathematics and head of the international Kanemitsu corporation [16] for the invitation to the joint Japanese–Chinese meeting in analytic number theory. I would also like to thank K. Matsumoto for kindly supporting my stay in Japan. Finally I am very grateful to the referee for valuable suggestions.

2. A short introduction to transversal Index theory

Transversal index theory is concerned with (pseudo-)differential operators or complexes of such operators, which are elliptic in the directions transversal to the orbits of an action by a Lie group G.

For compact groups the theory was initiated in Atiyah's lecture note [2]. For non-compact groups the definition of the transversal index as a distribution on G is due to Hörmander [19], [14]. Connes and Moscovici have greatly generalized the theory using non-commutative methods.

Let us recall Hörmander's definition of the transversal index. Let X be a compact manifold and consider a complex of differential operators acting on the smooth sections of vector-bundles $E_0, \ldots, E_N$ over X:

$$0 \longrightarrow C^\infty(E_0) \xrightarrow{D_0} C^\infty(E_1) \xrightarrow{D_1} \ldots \longrightarrow C^\infty(E_N) \longrightarrow 0 . \qquad (3)$$

Let us assume that a Lie group G acts smoothly on X and that we are given smooth (co-)actions $\tilde{g}_j : g^* E_j \to E_j$ on the bundles E_j. Furthermore the differential operators should commute with the induced G-actions on the $C^\infty(E_j)$. Let $\pi : T^*X \setminus 0 \to X$ denote the projection. Then associated to (3) there is the symbol sequence – a complex of vector bundles over $T^*X \setminus 0$:

$$0 \longrightarrow \pi^* E_0 \xrightarrow{\sigma(D_0)} \pi^* E_1 \longrightarrow \ldots \longrightarrow \pi^* E_N \longrightarrow 0 . \qquad (4)$$

Define the characteristic variety at the j-th place $C_j \subset T^*X \setminus 0$ to be the set of ξ in $T^*X \setminus 0$ over which (4) is not exact.

The complex (3) is called transversally elliptic with respect to the G-action, if all C_j are disjoint with $T^*_G X \setminus 0$. Here $T^*_G X$ is the closed G-invariant subspace of T^*X consisting of those $\xi \in T^*X$ that annihilate all vector fields along the G-orbits [2] p. 7.

Choose hermitian metrics on E_j and a volume form on X (or density if X is non-orientable). This gives a scalar product on each $C^\infty(E_j)$. Let $L^2(E_j)$ be the L^2-completion. We extend the operator

$$D_j : C^\infty(E_j) \longrightarrow C^\infty(E_{j+1})$$

to a closed densely defined operator

$$\tilde{D}_j : L^2(E_j) \longrightarrow L^2(E_{j+1})$$

by setting $\tilde{D}_j = (D_j^\dagger)^*$ where $D_j^\dagger$ is the formal adjoint of D_j. The domain of definition of $\tilde{D}_j$ consists of those $f \in L^2(E_j)$ for which $D_j f$, taken in the distributional sense is in $L^2(E_{j+1})$. The operator $\tilde{D}_j$ is known as the weak (or maximal) closure of D_j.

For simplicity assume that G acts conformally on the $L^2(E_j)$. Then G commutes with $D_j^\dagger, \tilde{D}_j$ and hence it acts on

$$\mathrm{Harm}_{L^2}^j(X) = \mathrm{Ker}\, \tilde{D}_j \cap \mathrm{Ker}\,(\tilde{D}_j)^* \ .$$

As a closed subspace of $L^2(E_j)$ this is a Hilbert space. For $\alpha \in C_0^\infty(G)$ define an endomorphism of $\mathrm{Harm}_{L^2}^j(X)$ by the formula:

$$S_j(\alpha) = \int_G \alpha(g)g^* \, dg \ .$$

Then Hörmander proved, c.f. [19] p. 29, [14]:

Theorem 2.1. $S_j(\alpha)$ *is of trace class and the map $\alpha \mapsto \mathrm{Tr}(S_j(\alpha))$ is a distribution T_j on G. The representation of G on $\mathrm{Harm}_{L^2}^j(X)$ decomposes into a discrete direct sum of irreducible representations.*

The transversal index of (3) can now be defined as the distribution:

$$\mathrm{Ind}_t(D) = \sum_{j=0}^N (-1)^j T_j \quad \text{in } \mathcal{D}'(\mathbb{R}) \ .$$

We go on to describe two computations of a transversal index for a $G = \mathbb{R}$-action due to Álvarez López, Kordyukov [1] and Lazarov [12].

Consider a compact $(d+1)$-dimensional manifold X with a smooth one-codimensional foliation $\mathcal{F}$ and a smooth $\mathbb{R}$-action whose orbits are everywhere transversal to the leaves of $\mathcal{F}$ and which maps leaves to leaves.

Then the de Rham complex along the leaves of $\mathcal{F}$:

$$0 \longrightarrow C^\infty(\Lambda^0 T^* \mathcal{F}) \xrightarrow{d_\mathcal{F}} C^\infty(\Lambda^1 T^* \mathcal{F}) \xrightarrow{d_\mathcal{F}} \cdots \longrightarrow C^\infty(\Lambda^d T^* \mathcal{F}) \longrightarrow 0$$

is transversally elliptic to the $\mathbb{R}$-action. Under a certain technical condition – $\mathbb{R}$ has to act isometrically on $T\mathcal{F}$ – the transversal index $\mathrm{Ind}_t(d_\mathcal{F})$ is defined as above.

We assume that all compact orbits γ of the $\mathbb{R}$-action are non-degenerate in the sense that for all $x \in \gamma$ and $k \in \mathbb{Z} \setminus 0$ the endomorphism $T_x \phi^{kl(\gamma)}$

has the eigenvalue 1 with algebraic multiplicity one. Here $l(\gamma)$ is the length of γ. Note that the tangent vector Y_x to the $\mathbb{R}$-orbit through x is kept fixed by $T_x \phi^{kl(\gamma)}$.

We set

$$\varepsilon_\gamma(k) = \operatorname{sgn} \det(1 - T_x \phi^{kl(\gamma)} \,|\, T_x X / \mathbb{R} \cdot Y_x) = \operatorname{sgn} \det(1 - T_x \phi^{kl(\gamma)} \,|\, T_x \mathcal{F}) \,.$$

The following theorem is proved in [1], [12] and (on $\mathbb{R}^*$) in [6]:

Theorem 2.2. $\operatorname{Ind}_t(d_{\mathcal{F}}) = \chi_{\mathrm{Co}}(\mathcal{F}, \mu)\delta_0 + \sum_\gamma l(\gamma) \sum_{k \in \mathbb{Z} \setminus 0} \varepsilon_\gamma(k)\delta_{kl(\gamma)}$ in $\mathcal{D}'(\mathbb{R})$.

Here $\chi_{\mathrm{Co}}(\mathcal{F}, \mu)$ is Connes' Euler characteristic of the foliation with respect to a certain transversal measure μ.

REMARKS 1) The version of this theorem in [1] uses a more general definition of the transversal index and does not require that $\mathbb{R}$ acts isometrically on $T\mathcal{F}$.
2) We will not discuss the term $\chi_{\mathrm{Co}}(\mathcal{F}, \mu)$ because in the situation of the next section its analogue vanishes.

Let us now assume that $d = 2g$ is even and that $\mathcal{F}$ is actually a foliation by complex manifolds i.e. that there exists a smooth almost complex structure J on $T\mathcal{F}$ whose restriction to every leaf is integrable. Moreover every $t \in \mathbb{R}$ should induce (bi-)holomorphic maps between leaves. Then we can consider the Dolbeault complex along the leaves of $\mathcal{F}$:

$$0 \longrightarrow C^\infty(\Lambda_{\mathbb{C}}^0 T_c^* \mathcal{F}) \xrightarrow{\bar{\partial}_{\mathcal{F}}} C^\infty(\Lambda_{\mathbb{C}}^1 T_c^* \mathcal{F}) \xrightarrow{\bar{\partial}_{\mathcal{F}}} \ldots \longrightarrow C^\infty(\Lambda_{\mathbb{C}}^g T_c^* \mathcal{F}) \longrightarrow 0 \,.$$

The following result is a special case of Lazarov's theorem 2.10 in [12]:

Theorem 2.3. *In $\mathcal{D}'(\mathbb{R})$ the following equality holds:*

$$\operatorname{Ind}_t(\bar{\partial}_{\mathcal{F}}) = \chi_{\mathrm{Co}}(\mathcal{F}, \mathcal{O}, \mu) \cdot \delta_0$$
$$+ \sum_\gamma l(\gamma) \sum_{k \in \mathbb{Z} \setminus 0} \det{}_{\mathbb{C}}(1 - T_{cx}\phi^{kl(\gamma)} \,|\, T_{cx}\mathcal{F})^{-1}\delta_{kl(\gamma)} \,.$$

Here x is any point on the relevant orbit γ and $\chi_{\mathrm{Co}}(\mathcal{F}, \mathcal{O}, \mu)$ is the holomorphic Connes' Euler characteristic of the foliation with respect to μ.

3.   A transversal index calculation on a generalized solenoid

Consider an elliptic curve E_0 over $\mathbb{F}_q$, where $q = p^r$, together with its q-th power Frobenius endomorphism $\varphi_0 : E_0 \to E_0$ over $\mathbb{F}_q$.

We consider a lift of the pair (E_0, φ_0) to characteristic zero as follows.

First assume that E_0 is ordinary and let $R = W(\mathbb{F}_q)$ denote the ring of Witt vectors of $\mathbb{F}_q$ with maximal ideal $\mathfrak{m} = pR$ and quotient field L. Note that $R/\mathfrak{m} = \mathbb{F}_q$. Up to isomorphism there is then a unique elliptic curve $\mathcal{E}$ over spec R such that $\mathcal{E} \otimes \mathbb{F}_q = E_0$ and such that

$$\operatorname{End}_R(\mathcal{E}) \xrightarrow{\sim} \operatorname{End}_{\mathbb{F}_q}(E_0)$$

is an isomorphism. This $\mathcal{E}$ is called the canonical lift of E_0.

In particular, there is a unique endomorphism $\varphi : \mathcal{E} \to \mathcal{E}$ lifting φ_0 i.e. such that

$$(\mathcal{E}, \varphi) \otimes \mathbb{F}_q = (E_0, \varphi_0) .$$

Now let E_0 be supersingular. There exist the following:

- a complete local integral domain R with field of fractions L a finite extension of $\mathbb{Q}_p$ such that $R/\mathfrak{m} = \mathbb{F}_q$ if $\mathfrak{m}$ is the maximal ideal of R.

- an elliptic curve $\mathcal{E}$ over spec R together with an endomorphism $\varphi : \mathcal{E} \to \mathcal{E}$ such that

$$(\mathcal{E}, \varphi) \otimes \mathbb{F}_q = (E_0, \varphi_0) .$$

The pair $(\mathcal{E}, \varphi)$ is not canonically determined. In the following, any choice will do if E_0 is supersingular. Proofs or references for these facts can be found in [15]. Up to a finite extension of $\mathbb{F}_q$ they follow from [7] p. 259–263.

Let $E_0/\mathbb{F}_q$ be arbitrary and consider $(\mathcal{E}, \varphi)$ as above. We denote by $E = \mathcal{E} \otimes_R L$ the generic fibre. Then

$$\operatorname{End}_L^0(E) = \operatorname{End}_L(E) \otimes \mathbb{Q}$$

is a field K which is either $\mathbb{Q}$ or an imaginary quadratic extension of $\mathbb{Q}$.

Fixing an embedding $L \subset \mathbb{C}$ we can consider the complex analytic elliptic curve $E(\mathbb{C})$. Let ω be a non-zero holomorphic one-form on $E(\mathbb{C})$ and let

$\Gamma \subset \mathbb{C}$ be its period lattice. Then the Abel–Jacobi map

$$E(\mathbb{C}) \xrightarrow{\sim} \mathbb{C}/\Gamma , \quad p \longmapsto \int_0^p \omega \bmod \Gamma$$

defines an isomorphism.

There is a homomorphism:

$$\Theta : K \cong \operatorname{End}_L^0(E) \longrightarrow \operatorname{End}^0(\mathbb{C}/\Gamma) .$$

Take the embedding $K \subset \mathbb{C}$ for which $\Theta(\alpha)$ induces multiplication by α on the Lie algebra $\mathbb{C}$ of $\mathbb{C}/\Gamma$.

Let

$$\xi \in \Theta^{-1}(\operatorname{End}_L(E)) \subset K \subset \mathbb{C}$$

be the unique element with $\Theta(\xi) = \varphi \otimes L$. According to a theorem of Hasse – the Riemann conjecture for elliptic curves over function fields – we have $|\xi|^2 = q$.

It is clear that $\xi\Gamma \subset \Gamma$. Hence we may consider the ξ-adic "Tate modules":

$$T_\xi\Gamma = \varprojlim \Gamma/\xi^n\Gamma \quad \text{and} \quad V_\xi\Gamma = T_\xi\Gamma \otimes \mathbb{Q} .$$

Note that $V_\xi\Gamma$ is a $\mathbb{Q}_p$-vector space of dimension one or two.

We define an additive subgroup $V \subset \mathbb{C}$ by

$$V = \Gamma[\xi^{-1}] = \bigcup_{\nu \geq 0} \xi^{-\nu}\Gamma .$$

Then V acts by translation on $V_\xi\Gamma$ since multiplication by ξ is invertible on $V_\xi\Gamma$. We now introduce actions of V and $\Lambda = (\log q)\mathbb{Z} \subset \mathbb{R}$ on the space

$$\tilde{X} = \mathbb{C} \times V_\xi\Gamma \times \mathbb{R} .$$

They are defined by the formulas:

$$(z, \hat{v}, t) \cdot v = (z + v, \hat{v} - v, t) \quad \text{for } v \in V$$

and

$$(z, \hat{v}, t) \cdot \lambda = (\xi^{-\nu}z, \xi^{-\nu}\hat{v}, t + \lambda) \quad \text{for } \lambda = \nu \log q \in \Lambda .$$

Together we get a free right action on $\tilde{X}$ of the semidirect product

$$H = V \rtimes \Lambda = V \cdot \Lambda$$

where $\lambda \cdot v = v^\lambda \cdot \lambda$ with $v^\lambda = \xi^\nu v$.

The 3-dimensional space on which we want to do transversal index theory is the compact quotient $X = \tilde{X}/H$.

The Lie group in question is $\mathbb{R}$ which acts on X by translation. For $s \in \mathbb{R}$ we set $\phi^s(z, \hat{v}, t) = (z, \hat{v}, t + s)$. In order to define the relevant differential operators we make the following observations. The space X is a smooth foliated space in the sense of [13] in two ways. First there is the foliation $\mathcal{L}$ of X by its 3-dimensional path components. They are non-compact and $\mathbb{R}$-equivariantly diffeomorphic to either $\mathbb{C} \times \mathbb{R}$ or a suitable circle bundle over $\mathbb{C}$, see the end of this section for details.

The transversals for $\mathcal{L}$ are totally disconnected. Thus X becomes a generalized solenoid in the sense of [20].

Secondly there is the foliation $\mathcal{FL}$ of X by the surfaces $\tau(\mathbb{C} \times \{\hat{v}\} \times \{t\})$ where $\tau : \tilde{X} \to X$ is the natural projection. Its leaves are all diffeomorphic to $\mathbb{C}$. Any leaf L of $\mathcal{L}$ is foliated as a 3-dimensional manifold by the leaves F of $\mathcal{FL}$ that are contained in L. The resulting foliation $\mathcal{F}_L$ of L is one-codimensional and everywhere transversal to the orbits of the $\mathbb{R}$-action restricted to L.

The union of the tangent spaces to the $\mathcal{FL}$-leaves is naturally a rank 2 vector bundle on X denoted by $T\mathcal{FL}$. As in [13] let $\mathcal{A}^i_{\mathcal{FL}}(X)$ denote the space of those sections of $\Lambda^i T^* \mathcal{FL}$ which are smooth on the $\mathcal{FL}$-leaves and continuous transversally. We then have the de Rham complex along the $\mathcal{FL}$-leaves [13]:

$$0 \longrightarrow \mathcal{A}^0_{\mathcal{FL}}(X) \xrightarrow{d_{\mathcal{FL}}} \mathcal{A}^1_{\mathcal{FL}}(X) \xrightarrow{d_{\mathcal{FL}}} \mathcal{A}^2_{\mathcal{FL}}(X) \longrightarrow 0 . \qquad (5)$$

Because of the equation

$$(d_{\mathcal{FL}}\omega)|_L = d_{\mathcal{F}_L}(\omega|_L)$$

this complex can be viewed as the family of de Rham complexes along the leaves L of $\mathcal{L}$

$$0 \longrightarrow \mathcal{A}^0_{\mathcal{F}_L}(L) \xrightarrow{d_{\mathcal{F}_L}} \mathcal{A}^1_{\mathcal{F}_L}(L) \xrightarrow{d_{\mathcal{F}_L}} \mathcal{A}^2_{\mathcal{F}_L}(L) \longrightarrow 0 . \qquad (6)$$

These complexes are transversally elliptic with respect to the $\mathbb{R}$-action on L.

We now define the transversal index for the complex (5) similarly as in the case of manifolds.

Let μ_ξ denote a Haar measure on the locally compact abelian group $V_\pi \Gamma$. Then the measure

$$\tilde{\mu} = dx\, dy \otimes \mu_\xi \otimes dt$$

on $\tilde{X}$ is H-invariant and hence induces a measure μ on X. In this regard, note that

$$\mu_\xi(\xi^\nu A) = |\xi|^{-2\nu}\mu_\xi(A) = q^{-\nu}\mu_\xi(A) .$$

It suffices to check this for $\nu \geq 0$ where it follows from the facts that $(\xi^{-\nu})_*\mu$ is a Haar measure and that

$$T_\xi\Gamma/\xi^\nu T_\xi\Gamma \cong \Gamma/\xi^\nu\Gamma$$

has $|\xi|^{2\nu}$ elements.

It is clear that the $\mathbb{R}$-action on X preserves the measure μ. Next we need a Riemannian metric on $T\mathcal{FL}$. Consider the Riemannian metric on the bundle

$$T\mathbb{C} \times V_\xi\Gamma \times \mathbb{R} = \mathbb{C} \times \tilde{X}$$

over $\tilde{X}$ given by:

$$\tilde{g}_{(z,\hat{v},t)}(\xi,\eta) = e^t\mathrm{Re}\,(\xi\overline{\eta}) .$$

It is H-invariant and induces a Riemannian metric g on $T\mathcal{FL}$ and hence on $\Lambda^\bullet T^*\mathcal{FL}$. Together with μ we get a scalar product on $\mathcal{A}^\bullet_{\mathcal{FL}}(X)$. With respect to this scalar product $d_{\mathcal{FL}}$ has an adjoint (the formal adjoint)

$$d^\dagger_{\mathcal{FL}} : \mathcal{A}^\bullet_{\mathcal{FL}}(X) \longrightarrow \mathcal{A}^{\bullet-1}_{\mathcal{FL}}(X) .$$

It is given by

$$d^\dagger_{\mathcal{FL}} = e^t d^\dagger_\mathbb{C} .$$

Here $d^\dagger_\mathbb{C}$ is the formal adjoint of $d_\mathbb{C} : \mathcal{A}^\bullet(\mathbb{C}) \to \mathcal{A}^{\bullet+1}(\mathbb{C})$ applied to forms in $\mathcal{A}^{\bullet+1}_{\mathcal{FL}}(X)$.

Explicitly we have the following formulas:

$$d^0_{\mathcal{FL}}f = \frac{\partial f}{\partial x}dx + \frac{\partial f}{\partial y}dy \quad \text{for } f \in \mathcal{A}^0_{\mathcal{FL}}(X) = C^\infty(X) \tag{7}$$

$$d^{0\dagger}_{\mathcal{FL}}\omega = -e^t\left(\frac{\partial\alpha}{\partial x} + \frac{\partial\beta}{\partial y}\right) \quad \text{for } \omega = \alpha\,dx + \beta\,dy \in \mathcal{A}^1_{\mathcal{FL}}(X) \tag{8}$$

$$d^1_{\mathcal{FL}}\omega = \left(\frac{\partial\beta}{\partial x} - \frac{\partial\alpha}{\partial y}\right)dx\,dy \quad \text{for } \omega = \alpha\,dx + \beta\,dy \in \mathcal{A}^1_{\mathcal{FL}}(X) \tag{9}$$

$$d^{1\dagger}_{\mathcal{FL}}\eta = e^t\left(\frac{\partial\gamma}{\partial y}\,dx - \frac{\partial\gamma}{\partial x}\,dy\right) \quad \text{for } \eta = \gamma\,dx\,dy \in \mathcal{A}^2_{\mathcal{FL}}(X) . \tag{10}$$

Note that dx and dy do not define global forms (along $\mathcal{FL}$) on X. The coefficients α, β, γ are smooth functions on $\tilde{X}$ which satisfy suitable invariance properties with respect to the H-action, so that ω, η descend to forms on X.

Let $\mathcal{A}^{\bullet}_{\mathcal{FL},L^2}(X)$ be the L^2-completion of $\mathcal{A}^{\bullet}_{\mathcal{FL}}(X)$ and view $d_{\mathcal{FL}}$ as an unbounded operator. We define the weak closure of $d_{\mathcal{FL}}$ to be the closed unbounded operator

$$\tilde{d}_{\mathcal{FL}} = d^{\dagger *}_{\mathcal{FL}} .$$

Set

$$\mathrm{Harm}^{\bullet}_{L^2}(X) = \mathrm{Ker}\, \tilde{d}_{\mathcal{FL}} \cap \mathrm{Ker}\, (\tilde{d}_{\mathcal{FL}})^* .$$

This is a closed subspace of $\mathcal{A}^{\bullet}_{\mathcal{FL},L^2}(X)$. For later calculations note that $(\tilde{d}_{\mathcal{FL}})^* \subset \widetilde{d^{\dagger}_{\mathcal{FL}}}$ and therefore:

$$\mathrm{Harm}^{\bullet}_{L^2}(X) \subset \widetilde{\mathrm{Harm}^{\bullet}_{L^2}}(X) = \mathrm{Ker}\, \tilde{d}_{\mathcal{FL}} \cap \mathrm{Ker}\, \widetilde{d^{\dagger}_{\mathcal{FL}}} .$$

The metric g on $T\mathcal{FL}$ has the property that

$$g(T_p\phi^t(\xi), T_p\phi^t(\eta)) = e^t g(\xi, \eta)$$

for all $p \in X$. Since μ is ϕ^t-invariant, it follows that for $\omega, \omega' \in \mathcal{A}^j_{\mathcal{FL}}(X)$ we have:

$$(\phi^{t*}\omega, \phi^{t*}\omega') = e^{jt}(\omega, \omega') .$$

Thus $\exp(-\frac{1}{2}jt)\phi^{t*}$ is an orthogonal operator on the real Hilbert space $\mathcal{A}^j_{\mathcal{FL}}(X)$. It follows that ϕ^{t*} has a unique extension to $\mathcal{A}^j_{\mathcal{FL},L^2}(X)$ such that $\exp(-\frac{1}{2}jt)\phi^{t*}$ is orthogonal. It follows that ϕ^{t*} commutes not only with $\tilde{d}_{\mathcal{FL}}$ but also with $\tilde{d}^*_{\mathcal{FL}}$. In particular ϕ^{t*} leaves $\mathrm{Harm}^{\bullet}_{L^2}(X)$ invariant and the representation U_j of $\mathbb{R}$ on the real Hilbert space $\mathrm{Harm}^j_{L^2}(X)$ given by

$$U_j(t) = \exp\left(-\frac{1}{2}jt\right)\phi^{t*}$$

is orthogonal.

Let $\alpha \in C^{\infty}_0(\mathbb{R})$ be a test function on $\mathbb{R}$. Consider the bounded operator

$$S_j(\alpha) = \int_{\mathbb{R}} \alpha(t)\phi^{t*}\, dt = \int_{\mathbb{R}} \exp\left(\frac{1}{2}jt\right)\alpha(t)U_j(t)\, dt$$

on $\mathrm{Harm}^j_{L^2}(X)$.

Theorem 3.1. *For every* $\alpha \in C_0^\infty(\mathbb{R})$ *the operator* $S_j(\alpha)$ *is of trace class on* $\mathrm{Harm}_{L^2}^j(X)$ *and its trace is given for* $j = 0, 1, 2$ *by:*

$$\mathrm{Tr}(S_0(\alpha) \,|\, \mathrm{Harm}_{L^2}^0(X)) = \sum_{\nu \in \mathbb{Z}} \Phi\left(\frac{2\pi i \nu}{\log q}\right)$$

$$\mathrm{Tr}(S_1(\alpha) \,|\, \mathrm{Harm}_{L^2}^1(X)) = \sum_{\rho} \Phi(\rho)$$

$$\mathrm{Tr}(S_2(\alpha) \,|\, \mathrm{Harm}_{L^2}^2(X)) = \sum_{\nu \in \mathbb{Z}} \Phi\left(1 + \frac{2\pi i \nu}{\log q}\right).$$

Here ρ *runs over the zeroes of* $\hat\zeta_E(s)$. *The eigenvalues of* ϕ^{t*} *on* $\mathrm{Harm}_{L^2}^j(X)$ *are the numbers*

$$\exp\left(\frac{2\pi i \nu t}{\log q}\right) \qquad \text{for } \nu \in \mathbb{Z} \text{ if } j = 0$$

$$\exp(t\rho) \qquad \text{for } \hat\zeta_E(\rho) = 0 \text{ if } j = 1$$

$$\exp\left(1 + \frac{2\pi i \nu t}{\log q}\right) \qquad \text{for } \nu \in \mathbb{Z} \text{ if } j = 2 \, .$$

Before giving the proof of the theorem we make the following observation. The maps

$$\alpha \longmapsto \mathrm{Tr}(S_j(\alpha) \,|\, \mathrm{Harm}_{L^2}^j(X))$$

define distributions T_j. As in the case of manifolds we define the transversal index $\mathrm{Ind}_t(d_{\mathcal{FL}})$ of the de Rham complex (5) along $\mathcal{FL}$ to be the alternating sum:

$$\mathrm{Ind}_t(d_{\mathcal{FL}}) = T_0 - T_1 + T_2 \, .$$

Corollary 3.2. *For the function field* F *of an elliptic curve over* $\mathbb{F}_q$ *the left hand side of the explicit formula (2) equals* $\mathrm{Ind}_t(d_{\mathcal{FL}})(\alpha)$.

Proof of Theorem 3.1. We discuss the trace on $\mathrm{Harm}_{L^2}^1(X)$. The other cases are similar but easier. A 1-form $\omega = \alpha\,dz + \beta\,d\bar z$ in $\mathcal{A}_{\mathcal{FL},L^2}^1(X)_{\mathbb{C}}$ is given by (classes of) V-invariant $\mathbb{C}$-valued L_{loc}^2-functions on $\tilde X$ such that for $\lambda = \nu \log q \in \Lambda$ we have

$$\lambda^*\alpha = \xi^\nu \alpha \quad \text{and} \quad \lambda^*\beta = \xi^\nu \beta \, .$$

It follows from equations (8) and (9) that ω is in $\widetilde{\mathrm{Harm}}_{L^2}^1(X)$ precisely when the distributional (partial) derivatives of suitable representatives α

and β with respect to both x and y vanish. The argument uses Fubini's theorem. Thus α and β can be chosen to be independent of the $\mathbb{C}$-coordinates.

Fourier theory shows that the action of V by translation on $V_\xi\Gamma$ is ergodic with respect to μ_ξ. It follows that α, β are independent of the $V_\xi\Gamma$-coordinate as well.

Let Log_q denote the branch of the complex logarithm for the basis q determined by $-\pi < \mathrm{Arg}\,\alpha \leq \pi$. According to what we have seen, we may view:

$$\alpha \cdot \exp(-t\mathrm{Log}_q\xi) \quad \text{and} \quad \beta \cdot \exp(-t\mathrm{Log}_q\bar{\xi}) \tag{11}$$

as elements of $L^2(\mathbb{R}/\Lambda; \mathbb{C})$.

The continuous functions being dense in $L^2(\mathbb{R}/\Lambda; \mathbb{C})$ it follows that ω is orthogonal to the image of $\tilde{d}_{\mathcal{FL}}$ i.e. that $\tilde{d}^*_{\mathcal{FL}}\omega = 0$. Hence we have:

$$\mathrm{Harm}^1_{L^2}(X) = \widetilde{\mathrm{Harm}}^1_{L^2}(X) \,.$$

It follows that $\mathrm{Harm}^1_{L^2}(X)_{\mathbb{C}}$ has an orthogonal basis consisting of the 1-forms

$$\exp t\left(\mathrm{Log}_q\xi + \frac{2\pi i\nu}{\log q}\right) dz \quad \text{and} \quad \exp t\left(\mathrm{Log}_q\bar{\xi} + \frac{2\pi i\nu}{\log q}\right) d\bar{z} \quad \text{for } \nu \in \mathbb{Z}\,.$$

These are eigenvectors with eigenvalue $\exp(t_0\rho)$ for the operator ϕ^{t_0*}. Here:

$$\rho = \mathrm{Log}_q\xi + \frac{2\pi i\nu}{\log q} \quad \text{resp.} \quad \rho = \mathrm{Log}_q\bar{\xi} + \frac{2\pi i\nu}{\log q} \quad \text{for } \nu \in \mathbb{Z}\,.$$

The zeta function of E is known to equal the following rational function in q^{-s} c.f. [17] Ch. V:

$$\hat{\zeta}_E(s) = \frac{(1 - \pi q^{-s})(1 - \bar{\pi}q^{-s})}{(1 - q^{-s})(1 - q^{1-s})}\,.$$

Hence the ρ's are exactly the zeroes of $\hat{\zeta}_E(s)$. The assertions on the operator $S_1(\alpha)$ follow from straightforward estimates. $\qquad \square$

According to corollary 3.2 the explicit formula (2) may be viewed as a formula for the transversal index $\mathrm{Ind}_t(d_{\mathcal{F}})$. We will now check that its right hand side can be described in terms of the closed orbits of the $\mathbb{R}$-action, similarly as in theorem 2.2.

Proposition 3.3. *There is a natural bijection between the set of valuations w of $F = \mathbb{F}_q(E_0)$ and the set of compact $\mathbb{R}$-orbits on X. It has the following property: If w corresponds to $\gamma = \gamma_w$, then we have*

$$l(\gamma_w) = \log N(w) .$$

Proof. Let V act on $\mathbb{C} \times V_\xi \Gamma$ by the formula

$$(z, \hat{v}) \cdot v = (z + v, \hat{v} - v)$$

and let

$$\overline{M} = \mathbb{C} \times_V V_\xi \Gamma$$

be the quotient. Then Λ acts on $\overline{M}$ via

$$(z, \hat{v}) \cdot \lambda = (\xi^{-\nu} z, \xi^{-\nu} \hat{v}) \quad \text{for } \lambda = \nu \log q$$

and we may write X as the suspension

$$X = \overline{M} \times_\Lambda \mathbb{R} .$$

Identify $\overline{M}$ with the image of $\overline{M} \times 0$ in X. Then the map

$$\gamma \longmapsto \gamma \cap \overline{M}$$

gives a bijection between the compact $\mathbb{R}$-orbits of X of length $n \log q$ and the $\lambda = (-\log q)$-orbits on $\overline{M}$ of order n. Note that $(-\log q) \in \Lambda$ operates by diagonal multiplication with ξ on $\overline{M}$.

The natural map

$$\mathbb{C} \times_\Gamma T_\xi \Gamma \stackrel{\sim}{\longrightarrow} \mathbb{C} \times_V V_\xi \Gamma$$

is an isomorphism and equivariant with respect to diagonal multiplication with ξ on both sides.

Under the projection

$$\mathbb{C} \times_\Gamma T_\xi \Gamma \longrightarrow \mathbb{C}/\Gamma$$

the orbits of order n of diagonal ξ-multiplication on $\mathbb{C} \times_\Gamma T_\xi \Gamma$ are mapped bijectively onto the ξ-orbits of order n on $\mathbb{C}/\Gamma$. The inverse map sends the orbit of $z + \Gamma$ to the orbit of $[z, \hat{\gamma}]$ where

$$\hat{\gamma} = (1 - \xi^n)^{-1} \gamma \quad \text{if} \quad \gamma = \xi^n z - z \in \Gamma .$$

Note that the endomorphism $1 - \xi^n$ of $T_\xi \Gamma$ is invertible with inverse

$$(1 - \xi^n)^{-1} = \sum_{\mu=0}^{\infty} \xi^{n\mu} .$$

It remains to construct a bijection between the $\Theta(\xi)$-orbits on $E(\mathbb{C})$ of order n with the φ_0-orbits on $E_0(\overline{\mathbb{F}}_q)$ of the same order. The latter are in bijection with the closed points of E_0 of degree n i.e. with the valuations w of $F = \mathbb{F}_q(E_0)$ of degree n. Note that such bijections have been constructed by Ihara in the cyclotomic, modular and Shimura curve cases [8], [9], [10], [11]. However for elliptic curves there does not seem to be a reference in the literature although this case is certainly known to Ihara as well.

Let $\overline{L}$ be the algebraic closure of L in $\mathbb{C}$ and let $\overline{R}$ be the integral closure of R in $\overline{L}$. Fix a maximal ideal $\overline{\mathfrak{m}}$ over $\mathfrak{m}$ in $\overline{R}$. Then

$$\overline{R}/\overline{\mathfrak{m}} = \overline{\mathbb{F}}_q.$$

Since $\mathcal{E}$ is proper over spec R the natural map

$$\mathcal{E}(\overline{R}) \xrightarrow{\sim} E(\overline{L})$$

is an isomorphism. The torsion points of $E(\mathbb{C})$ being algebraic over L, the inclusion

$$E(\overline{L}) \hookrightarrow E(\mathbb{C})$$

induces an isomorphism on the torsion subgroups.

Consider the reduction map obtained by composition:

$$\text{red} : E(\mathbb{C})_{\text{tors}} \cong \mathcal{E}(\overline{R})_{\text{tors}} \longrightarrow \mathcal{E}(\overline{R}/\overline{\mathfrak{m}})_{\text{tors}} = E_0(\overline{\mathbb{F}}_q) .$$

It is equivariant with respect to the actions by $\Theta(\xi)$ on the left and by the q-Frobenius φ_0 on the right. The group

$$(E(\mathbb{C})_{\text{tors}})^{\Theta(\xi^n)=1} = E(\mathbb{C})^{\Theta(\xi^n)=1} = (\mathbb{C}/\Gamma)^{\xi^n=1} = (\xi^n - 1)^{-1}\Gamma/\Gamma$$

has order $|\xi^n - 1|^2$. The group

$$E_0(\overline{\mathbb{F}}_q)^{\varphi_0^n=1} = E_0(\mathbb{F}_{q^n})$$

has the same order, since it is known that

$$|\xi^n - 1|^2 = q^n + 1 - (\xi^n + \overline{\xi}^n) = \text{card } E_0(\mathbb{F}_{q^n}) .$$

For any integer $N \geq 1$ prime to p the restriction

$$\text{red} : E(\mathbb{C})_N \xrightarrow{\sim} E_0(\overline{\mathbb{F}}_q)_N$$

is an isomorphism, [17] IV, prop. 2.3 and VII prop. 3.1.

Now assume that E_0 is supersingular. Then $E_0(\overline{\mathbb{F}}_q)$ has no p-torsion. Hence $N = |\xi^n - 1|^2 = \operatorname{card} E_0(\mathbb{F}_{q^n})$ is prime to p. It follows that reduction induces an isomorphism:

$$\operatorname{red} : E(\mathbb{C})^{\Theta(\xi^n)-1} = (E(\mathbb{C})_N)^{\Theta(\xi^n)-1} \xrightarrow{\sim} (E_0(\overline{\mathbb{F}}_q)_N)^{\varphi_0^n=1} = E_0(\mathbb{F}_{q^n}) .$$

Therefore red also induces a bijection between the $\Theta(\xi)$-orbits on $E(\mathbb{C})$ of order n and the φ_0-orbits on $E_0(\overline{\mathbb{F}}_q)$ of the same order.

We now assume that E_0 is ordinary. In this case we had taken $\mathcal{E}$ to be the canonical lifting of E_0 to $R = W(\mathbb{F}_q)$. Then the canonical sequence of p-divisible groups

$$0 \longrightarrow \mathbb{G}_m(p) \longrightarrow \mathcal{E}(p) \longrightarrow \mathbb{Q}_p/\mathbb{Z}_p \longrightarrow 0$$

over R is split. Hence we get a commutative diagram with compatible splittings:

$$
\begin{array}{ccccccccc}
0 & \longrightarrow & \mu_{p^n}(\overline{R}) & \longrightarrow & \mathcal{E}_{p^n}(\overline{R}) & = & \mathbb{Z}/p^n & \longrightarrow & 0 \\
& & \big\downarrow & & \big\downarrow{\scriptstyle\text{red}} & & \big\| & & \\
0 & \longrightarrow & \mu_{p^n}(\overline{\mathbb{F}}_q) & \longrightarrow & (E_{0p^n})(\overline{\mathbb{F}}_q) & = & \mathbb{Z}^{p^n} & \longrightarrow & 0.
\end{array}
$$

This in turn gives a split exact sequence:

$$0 \longrightarrow \mu_{p^n}(\overline{R}) \longrightarrow E(\overline{L})_{p^n} \xrightarrow{\text{red}} E_0(\overline{\mathbb{F}}_q)_{p^n} \longrightarrow 0$$

and hence, by the above, a split exact sequence for any $N \geq 1$

$$0 \longrightarrow \mu_{N_p}(\overline{R}) \longrightarrow E(\overline{L})_N \xrightarrow{\text{red}} E_0(\overline{\mathbb{F}}_q)_N \longrightarrow 0 .$$

Here N_p is the largest power of p dividing N.

Taking

$$N = |\xi^n - 1|^2 = \operatorname{card} E_0(\mathbb{F}_{q^n})$$

as above and passing to $\Theta(\xi)^n - 1$ resp. $\varphi_0^n - 1$ fixed-modules we get the exact (red was split!) sequence:

$$0 \longrightarrow \mu_{N_p}(\overline{R})^{\Theta(\xi)^n-1} \longrightarrow E(\overline{L})^{\Theta(\xi)^n-1} \xrightarrow{\text{red}} E_0(\mathbb{F}_{q^n}) \longrightarrow 0 .$$

Since the middle and right hand group have the same order it follows again that

$$\operatorname{red} : E(\mathbb{C})^{\Theta(\xi)^n-1} = E(\overline{L})^{\Theta(\xi)^n-1} \xrightarrow{\sim} E_0(\mathbb{F}_{q^n})$$

is an isomorphism. We conclude as before.

The fact that $\mu_{N_p}(\overline{R})^{\Theta(\xi)^n - 1}$ vanishes can be seen directly.

Namely $\Theta(\xi)$ acts on $\zeta \in \mu_{p^\infty}$ by raising ζ to the q-th power. But if $\zeta^{q^n} = \zeta$ for some $n \geq 1$, if follows that $\zeta = 1$.   □

Combining corollary 3.2 and proposition 3.3 we arrive at the following index theoretic way to write the explicit formula (2) for an ordinary elliptic curve:

Corollary 3.4. *The following equality holds in $\mathcal{D}'(\mathbb{R})$:*

$$\mathrm{Ind}_t(d_{\mathcal{F}\mathcal{L}}) = \sum_\gamma l(\gamma) \sum_{k \geq 1} \delta_{kl(\gamma)} + \sum_\gamma l(\gamma) \sum_{k \leq -1} e^{kl(\gamma)} \delta_{kl(\gamma)} \, .$$

This looks exactly like the transversal index formula 2.2, except for the factor $e^{kl(\gamma)}$ in front of the Dirac distribution $\delta_{kl(\gamma)}$ for $k \leq -1$. Geometrically this factor results from the conformal behaviour of our metric g under the flow – a behaviour which is linked to the solenoidal structure of X.

In our case $\chi_{\mathrm{Co}}(\mathcal{F}, \mu)$ can be shown to equal $\chi(E(\mathbb{C})) \log q$ which is zero.

It is possible to prove 3.4 directly using e.g. the Poisson summation formula as in the next section. Our example suggests that under suitable conditions transversal index theory generalizes to solenoids or even more general laminated spaces instead of manifolds. This is also confirmed by the example in the next section.

One cannot help wondering whether the explicit formula (1) for the Riemann zeta function might have a similar interpretation as an index theorem. The arguments in [4], [5] suggest that in this case the $\mathbb{R}$-action will have a fixed point and that a generalization of transversal index theory may be required where the operator is transversally elliptic except in isolated points.

We end this section by a closer inspection of how the periodic orbits are distributed within the leaves of the $\mathcal{L}$-foliation on X.

For $\hat{v} \in V_\xi \Gamma$ one checks that

$$\Lambda_{\hat{v}} = \{\nu \log q \,|\, (\xi^{-\nu} - 1)\hat{v} \in V\} \subset \Lambda$$

is a subgroup. It acts on $\mathbb{C} \times \mathbb{R}$ by the formula:

$$(z, t) \cdot \lambda = (\xi^{-\nu} z + (\xi^{-\nu} - 1)\hat{v}, t + \lambda) \quad \text{for } \lambda = \nu \log q \in \Lambda_{\hat{v}} \, .$$

Here we use the embeddings

$$V_\xi \Gamma \hookleftarrow V \hookrightarrow \mathbb{C}$$

constructed above. Set $\mathbb{C} \times_{\Lambda_{\hat{v}}} \mathbb{R} = (\mathbb{C} \times \mathbb{R})/\Lambda_{\hat{v}}$. The map

$$\mathbb{C} \times_{\Lambda_{\hat{v}}} \mathbb{R} \hookrightarrow X \ , \ [z,t] \longmapsto [z,\hat{v},t]$$

is an $\mathbb{R}$-equivariant diffeomorphism onto the $\mathcal{L}$-leaf obtained by project-ing $\mathbb{C} \times \{\hat{v}\} \times \mathbb{R}$ to X.

For $\Lambda_{\hat{v}} = 0$ we have $\mathbb{C} \times_{\Lambda_v} \mathbb{R} = \mathbb{C} \times \mathbb{R}$. For $\Lambda_{\hat{v}} \neq 0$ we may view $\mathbb{C} \times_{\Lambda_{\hat{v}}} \mathbb{R}$ as a complex line bundle over the circle $\mathbb{R}/\Lambda_{\hat{v}}$ with an $\mathbb{R}$-action.

Note that since the set

$$\bigcup_{\nu \in \mathbb{Z}\backslash 0} (\xi^{-\nu} - 1)^{-1}V$$

ist countable there are at most countably many $\hat{v}$ with $\Lambda_{\hat{v}} \neq 0$. Hence only countably many $\mathcal{L}$-leaves are diffeomorphic to $\mathbb{C} \times_{\Lambda_{\hat{v}}} \mathbb{R}$. The space of $\mathcal{L}$-leaves is $V_\xi \Gamma / H$.

It is clear that $\mathcal{L}$-leaves corresponding to $\hat{v}$ mod H with $\Lambda_{\hat{v}} = 0$ do not contain compact orbits of the $\mathbb{R}$-action.

Assume therefore that $\Lambda_{\hat{v}} \neq 0$ and let $\lambda_0 = \nu_0 \log q$ be a generator of $\Lambda_{\hat{v}}$. Thus $\nu_0 = [\Lambda : \Lambda_{\hat{v}}]$.

The two embeddings $V_\xi \Gamma \hookleftarrow V \hookrightarrow \mathbb{C}$ being K-equivariant, it follows that for all $k \in \mathbb{Z} \backslash 0$ we have:

$$(\xi^{-\nu_0 k} - 1)^{-1}(\xi^{-\nu_0 k} - 1)\hat{v} = (\xi^{-\nu_0} - 1)^{-1}(\xi^{-\nu_0} - 1)\hat{v} \ .$$

It follows that the only finite orbits of the $\Lambda_{\hat{v}}$-action on $\mathbb{C}$ given by

$$z \cdot \lambda = \xi^{-\nu}z + (\xi^{-\nu} - 1)\hat{v} \quad \text{for } \lambda = \nu \log q \in \Lambda_{\hat{v}}$$

consists of the single point $z_0 = -(\xi^{-\nu_0} - 1)\hat{v}$.

Hence the suspended flow, which is just the $\mathbb{R}$-action on $\mathbb{C} \times_{\Lambda_{\hat{v}}} \mathbb{R}$ has precisely one compact orbit $\gamma_{\hat{v}}$ and it has length

$$l(\gamma_{\hat{v}}) = \lambda_0 = [\Lambda : \Lambda_{\hat{v}}] \log q = \text{vol}\,\mathbb{R}/\Lambda_{\hat{v}} \ .$$

4. A holomorphic transversal index formula on X

We keep the notations from the last section. The foliation $\mathcal{FL}$ of X carries an obvious complex structure and we can consider the Dolbeault complex along $\mathcal{FL}$

$$0 \longrightarrow A^{0,0}_{\mathcal{FL}}(X) \xrightarrow{\bar{\partial}_{\mathcal{F}}} A^{0,1}_{\mathcal{FL}}(X) \longrightarrow 0 .$$

As the hermitian metric on $T_c \mathcal{FL}$ etc. we take the metric induced by

$$h_{(z,\hat{v},t)}(\xi,\eta) = e^t \xi \bar{\eta} .$$

Looking at Lazarov's result 2.3 and taking the factor $e^{kl(\gamma)}$ in 3.4 into account makes the statement of the next result plausible.

Note also that in our case $\chi_{\mathrm{Co}}(\mathcal{F}, \mathcal{O}, \mu)$ can be shown to equal $\chi(E(\mathbb{C}), \mathcal{O}) \log q$ which is zero.

The precise definition of $\mathrm{Ind}_t(\bar{\partial}_{\mathcal{FL}})$ is given in the proof of 4.1.

Theorem 4.1. *In $\mathcal{D}'(\mathbb{R})$ we have:*

$$
\begin{aligned}
\mathrm{Ind}_t(\bar{\partial}_{\mathcal{FL}}) \;=\; & \sum_{\gamma} l(\gamma) \sum_{k \geq 1} \det_{\mathbb{C}}(1 - T_{cx}\phi^{kl(\gamma)} \,|\, T_{cx}\mathcal{FL})^{-1} \delta_{kl(\gamma)} \\
& + \sum_{\gamma} l(\gamma) \sum_{k \leq -1} e^{kl(\gamma)} \det_{\mathbb{C}}(1 - T_{cx}\phi^{kl(\gamma)} \,|\, T_{cx}\mathcal{FL})^{-1} \delta_{kl(\gamma)} .
\end{aligned}
$$

Here γ runs over the compact $\mathbb{R}$-orbits on X and in the k-sums, x is any point on γ.

Proof. For $f \in A^{0,0}_{\mathcal{FL}}(X)_{\mathbb{C}} = C^\infty(X, \mathbb{C})$ we have

$$\bar{\partial}_{\mathcal{FL}} f = \frac{\partial f}{\partial \bar{z}} d\bar{z} \tag{12}$$

and for $\omega = \beta \, d\bar{z}$ in $A^{0,1}_{\mathcal{FL}}(X)$

$$\bar{\partial}^\dagger_{\mathcal{FL}} \omega = -2e^t \frac{\partial \beta}{\partial z} . \tag{13}$$

Define $\tilde{\bar{\partial}}_{\mathcal{FL}}$ and $(\tilde{\bar{\partial}}_{\mathcal{FL}})^*$ as closed unbounded operators on $A^{0,0}_{\mathcal{FL}, L^2}(X)_{\mathbb{C}}$ and $A^{0,1}_{\mathcal{FL}, L^2}(X)_{\mathbb{C}}$ as before.

Let f be in the kernel of $\tilde{\bar{\partial}}_{\mathcal{FL}}$. By elliptic regularity f can be represented by a $\mathbb{C}$-valued H-invariant L^2_{loc}-function on $\tilde{X}$ which we also denote by f. The function f_t on $\overline{M} = \mathbb{C} \times_V V_\xi \Gamma$ defined by

$$f_t[z, \hat{v}] = [z, \hat{v}, t]$$

is in $\mathcal{L}^2(\overline{M})$ for Lebesgue almost all t in $\mathbb{R}$. This follows from Fubini's theorem since f is in $\mathcal{L}^2$. We now observe that $\overline{M}$ is a compact abelian group with character group

$$\overline{M}^\vee = \{\chi \otimes \chi' \,|\, \chi \in \mathbb{C}^\vee, \chi' \in (V_\xi \Gamma)^\vee \text{ such that } \chi\,|_V = \chi'\,|_V\}\,.$$

For f_t in $\mathcal{L}^2(\overline{M})$ we may therefore write

$$f_t(z, \hat{v}) = \sum_{\chi|_V = \chi'|_V} a_{\chi,\chi'} \chi(z) \chi'(\hat{v}) \quad \text{in } L^2(\overline{M})\,. \tag{14}$$

Every character χ of $\mathbb{C}$ can be written in the form

$$\chi(z) = \chi_w(z) = \exp(wz - \overline{w}\overline{z})$$

for some unique $w \in \mathbb{C}$. By assumption $\overline{\partial}_z f_t = 0$ in the distributional sense. Since distributional derivatives commute with convergent series of distributions it follows that

$$\overline{\partial}_z f_t(z, \hat{v}) = \sum_{\chi|_V = \chi'|_V} a_{\chi,\chi'} \frac{\partial \chi(z)}{\partial \overline{z}} \chi'(\hat{v})\, d\overline{z} = 0\,.$$

Since

$$\frac{\partial \chi_w}{\partial \overline{z}} = -\overline{w}\chi_w\,,$$

it follows that in (17) we have $a_{\chi,\chi'} = 0$ for all $\chi \neq 1$. But a character χ' of $V_\xi \Gamma$ with $\chi' = 1$ on V is trivial since V is dense in $V_\xi \Gamma$. It follows that f_t is constant in $L^2(\overline{M})$.

Hence we have seen that we have an $\mathbb{R}$-equivariant isomorphism:

$$\mathrm{Ker}\, \widetilde{\overline{\partial}}_{\mathcal{FL}} = L^2(\mathbb{R}/\Lambda, \mathbb{C})\,.$$

It follows that for every $\alpha \in C_0^\infty(\mathbb{R})$ the operator

$$S_0(\alpha) = \int_\mathbb{R} \alpha(t)\phi^{t*}\, dt \quad \text{on } \mathrm{Ker}\, \widetilde{\overline{\partial}}_{\mathcal{FL}}$$

is of trace class. Its trace is given by the formula

$$T_0(\alpha) = \mathrm{Tr}(S_0(\alpha)\,|\,\mathrm{Ker}\, \widetilde{\overline{\partial}}_{\mathcal{FL}}) = \sum_{\nu \in \mathbb{Z}} \Phi\left(\frac{2\pi i \nu}{\log q}\right)\,.$$

Similar arguments as before show that $\mathrm{Ker}\, \widetilde{\overline{\partial}}^\dagger \supset \mathrm{Ker}\, \widetilde{\overline{\partial}}^*_{\mathcal{FL}}$ may be identified with

$$L^2(\mathbb{R}/\Lambda, \mathbb{C}) \cdot \exp(t \mathrm{Log}_q \overline{\xi})\, d\overline{z}$$

which in fact lies in the kernel of $\widetilde{\bar{\partial}}^*$. Hence we have

$$\operatorname{Ker}\widetilde{\bar{\partial}}^{\dagger} = \ker\widetilde{\bar{\partial}}^*$$

in our case. Moreover the operator

$$S_1(\alpha) = \int_{\mathbb{R}} \alpha(t)\phi^{t*}\,dt \quad \text{on } \operatorname{Ker}\widetilde{\bar{\partial}}^*_{\mathcal{FL}}$$

is of trace class and

$$T_1(\alpha) = \operatorname{Tr}(S_1(\alpha)\,|\,\operatorname{Ker}\widetilde{\bar{\partial}}^*_{\mathcal{FL}}) = \sum_{\nu\in\mathbb{Z}}\Phi\left(\operatorname{Log}_q\bar{\xi} + \frac{2\pi i\nu}{\log q}\right).$$

Both T_0 and T_1 are distributions on $\mathbb{R}$ and we set

$$\operatorname{Ind}_t(\bar{\partial}_{\mathcal{FL}}) = T_0 - T_1.$$

For $\psi \in C_0^\infty(\mathbb{R})$ and

$$\hat{\psi}(x) = \int_{\mathbb{R}} e^{-2\pi i x t}\psi(t)\,dt$$

the Poisson summation formula asserts that

$$\sum_{n\in\mathbb{Z}}\psi(n) = \sum_{n\in\mathbb{Z}}\hat{\psi}(n).$$

It implies that

$$\operatorname{Ind}_t(\bar{\partial}_{\mathcal{FL}}) = \log q\sum_{n\in\mathbb{Z}}(1 - \bar{\xi}^n)\delta_{n\log q}.$$

Using the formulas valid for $n \geq 1$:

$$\operatorname{card} E_0(\mathbb{F}_{q^n}) = (1 - \xi^n)(1 - \bar{\xi}^n)$$

and

$$\operatorname{card} E_0(\mathbb{F}_{q^n}) = \sum_{\deg w\,|\,n}\deg w$$

we find after a short calculation:

$$\begin{aligned}
\operatorname{Ind}_t(\bar{\partial}_{\mathcal{FL}}) &= \sum_w \log Nw \sum_{k\geq 1}(1 - \xi^{k\deg w})^{-1}\delta_{k\log N(w)} \\
&+ \sum_w \log Nw \sum_{k\leq -1} q^{k\deg w}(1 - \xi^{k\deg w})^{-1}\delta_{k\log N(w)}.
\end{aligned}$$

Now if x is a point on a closed orbit γ then the map induced by $\phi^{kl(\gamma)}$ on the complex tangent space $T_{cx}\mathcal{FL}$ is multiplication by $\xi^{k(\log q)^{-1}l(\gamma)}$.

Using proposition 3.3 we therefore get the assertion. $\qquad\square$

References

[1] J.A. Álvarez López, Y. Kordyukov, Distributional Betti numbers of transitive foliations of codimension one. Preprint 2000

[2] M.F. Atiyah, Elliptic operators and compact groups. Springer LNM **401**, 1974

[3] K. Barner, On A. Weil's explicit formula. J. Reine Angew. Math. **323** (1981), 139–152

[4] C. Deninger, Some analogies between number theory and dynamical systems on foliated spaces. Doc. Math. J. DMV. Extra Volume ICM I (1998), 23–46

[5] C. Deninger, Number theory and dynamical systems on foliated spaces. In: Jber. d. dt. Math.-Verein **103** (2001), 79–100

[6] C. Deninger, W. Singhof, A note on dynamical trace formulas. In: M.L. Lapidus, M. van Frankenhuysen (eds.), Dynamical Spectral and Arithmetic Zeta-Functions. In: AMS Contemp. Math. **290**, 41–55

[7] M. Deuring, Die Typen der Multiplikatorenringe elliptischer Funktionenkörper. Abh. Math. Sem. Hamburg **14** (1941), 197–272

[8] Y. Ihara, On $(\infty \times p)$-adic coverings of curves (the simplest example). Trudy Mat. Inst. Steklov **132** (1973), 133-148

[9] Y. Ihara, On congruence monodromy problems. Vols 1,2, Lecture Notes, nos. 1,2, Dept. of Mathematics, Univ. of Tokyo, Tokyo 1968, 1969. MR # 6706, # 6707

[10] Y. Ihara, Non-abelian classfields over function fields in special cases. Proc. Internat. Congress Math. (Nice 1970), vol 1, Gauthier-Villars, Paris 1971, 381–390

[11] Y. Ihara, Congruence relations and Shimura curves. Proceedings of Symp. Pure Math. **33** (1979) part 2, 291–311

[12] C. Lazarov, Transverse index and periodic orbits. GAFA **10** (2000), 124–159

[13] C.C. Moore, C. Schochet, Global analysis on foliated spaces. MSRI Publications **9**, Springer 1988

[14] A. Neske, F. Zickermann, The index of transversally elliptic complexes. Proceedings of the 13th winter school on abstract analysis (Srni, 1985). Rend. Circ. Mat. Palermo (2) Suppl. No. **9** (1986), 165–175

[15] F. Oort, Lifting an endomorphism of an elliptic curve to characteristic zero. Indag. Math. **35** (1973), 466–470

[16] Robocop 3, Orion pictures 1993

[17] J.H. Silverman, The arithmetic of elliptic curves. Springer GTM **106**, 1986

[18] A.S. Sikora, Analogies between group actions on 3-manifolds and number fields. Preprint arXiv:math. GT/0107210, 29. Juli 2001

[19] I.M. Singer, Index theory for elliptic operators, Proc. Symp. Pure Math. **28** (1973), 11–31

[20] D. Sullivan, Linking the universalities of Milnor–Thurston Feigenbaum and Ahlfors–Bers. In: Topological methods in modern mathematics. Publish or perish 1993, 543–564

PRODUCT REPRESENTATIONS BY RATIONALS

P.D.T.A. Elliott*
University of Colorado, Boulder, Colorado 80309-0385
pdtae@euclid.colorado.edu

Abstract   A survey is made of the representability of arbitrary rationals by products of given rationals and their reciprocals. Rationals of number theoretic interest are emphasised. There are connections with group theory, harmonic analysis and the theory of algorithms. Outstanding problems of the discipline are identified.

Keywords: Product representations; rationals; polynomials; abelian groups; harmonic analysis.

1.   The central problem

Given a sequence of positive rationals $r_1, r_2, \ldots$, what can we say about those rationals that may be represented in the form

$$r_1^{a_1} r_2^{a_2} \cdots r_k^{a_k}$$

with integers a_i, $i = 1, \ldots, k$, positive, negative or zero? This appears to be a difficult question.

We may appreciate this question by recasting it algebraically: If Q^* denotes the multiplicative group of positive rationals and Γ its subgroup generated by the r_i, then what is the quotient group Q^*/Γ?

I shall denote a typical quotient group of this type by G.

The fundamental theorem of arithmetic asserts that Q^* is freely generated by the positive rational primes. If we interpret the equations $r_i = 1$, $i = 1, 2, \ldots$, as a sequence of relations between them, then we may realise every denumerable abelian group as such a quotient group G.

As a first example of this procedure consider a free group with generators $g_2, g_3, g_5, \ldots$, enumerated by the primes. Corresponding to the

*Partially supported by NSF contract DMS 0070496.

sequence of rationals $(k+1)/k$, $k = 1, 2, \ldots$, we acquire the relations

$$e^{-1}g_2 = e, \quad g_2^{-1}g_3 = e, \quad g_3^{-1}g_2^2 = e, \quad g_2^{-2}g_5 = e, \ldots$$

and so on, where e denotes the identity of the group. Successive concatenations

$$(1^{-1}2)(2^{-1}3)(e^{-1}4) \cdots ((k-1)^{-1}k)$$

show that every generator is in fact the identity, and G is trivial.

As a second example we choose the sequence of rationals r_i to be the shifted primes $p + 1$, $p = 2, 3, 5, \ldots$. Our relations are now

$$g_3 = e, \quad g_2g_3 = e, \quad g_2^3 = e, \ldots$$

Trial and error discovers the representations

$$g_2 = (g_2g_3)g_e^{-1}, \quad g_5 = (g_2^2g_5)g_2^{-2}, \quad g_7 = (g_2g_7)(g_2g_3)^{-1}g_3$$

corresponding to

$$2 = (5+1)(2+1)^{-1}, \quad 5 = (19+1)(3+1)^{-1}, \quad 7 = (13+1)(5_1)^{-1}(2+1)$$

and so on, but there is no known algebraic identity that would enable us to assert the triviality of G, although that assertion is almost certainly true.

Unfortunately, this second example portrays more nearly the nature of things. The word problem for groups is well studied, cf. Stilwell, [32]. It follows from a result of Britten, cf. [3] Lemma 2.31, that there is a sequence of positive rationals $f_1, r_2, \ldots$, for which no recursive algorithm algorithm, recursive can be given to decide whether an arbitrary positive integer n has a representation of the form

$$n = r_1^{a_1} \cdots r_k^{1_k}$$

where a_i are integers, positive, negative, or zero. Without further information concerning the sequence r_j we cannot decide our initial question by computation. It is not at all clear what that information might be.

For comparison, consider two celebrated problems concerning the addition of integers. Waring's conjecture, from 1770 and about which is an enormous literature, cf. Vaughan, [33], asserted that for each positive integer k there is a further integer s so that every positive integer n has a representation as s integral k^{th}-powers:

$$n = x_1^k + \cdots + x_s^k.$$

Goldbach's conjectures, from 1742 and about which there is an almost equally extensive literature, cf Pan Chengdong and Pan Chengbiao [31], asserts that every even integer above 3 is the sum of two primes, every odd integer above 5 the sum of three primes.

Whilst we expect that theoretical considerations will be required to completely settle these questions, the representability of any particular integer in the desired manner may be ascertained in a finite number of steps. There can be no more than a finite number of different representations and a main difficulty is that there may be an insufficient supply of integers of a specified type.

For product representations the situation is quite different. There may be infinitely many representations of the type

$$t = r_1^{a_1} \cdots r_k^{a_k},$$

and they may have varying k. The factorisation of an individual r_k with k large may be unfeasible in practice, and the possible cancellation of a particular large prime between the various r_j difficult to ascertain. It is not clear that those rationals t which possess a product representation of the desired type possess one for which the value of $|a_1| + \cdots + |a_k|$ lies below a specified uniform bound. To proceed further I limit the sequences f_j to those that are arithmetically interesting. Indeed, I shall view various rational functions from the standpoints of a series of problems involving products.

Let $F(x)$ be a rational function of x with rational coefficients, positive for all sufficiently large integer values of x. Let Γ or $\Gamma(F(n))$ be the subgroup of Q^* generated by the positive values of $F(n)$, $n = 1, 2, \ldots$, and denote the quotient group Q^*/Γ by $G(F)$ or $G(F(n))$.

Problem 1 *Determine the group $G(F)$.*

I indicate four approaches to this question.

2.   Elementary argument by induction

In our first example $F(x) - x^{-1}(x+1)$ and the quotient group was trivial. Note that if we replace F by $x^{-1}(x+k)$ with a positive integer k, then for each positive integer n, $n^{-1}(n+1) = (kn)^{-1}(kn+k)$ so that Γ and the group $G(F)$ are unaffected.

The group $G(F)$ need not be finite: in the case $F(x) = x^k$ it is a direct sum of denumerably many cyclic groups of order k, one for each prime.

A more interesting example is $F(x) = x^2 + 1$. If a prime p is of the form $4m + 1$, then the congruence $y^2 \equiv -1 \pmod{p}$ has solution with

$1 \leq y \leq (p-1)/2$. Thus $(y^2 + 1 = pw$, where the integer w does not exceed $p^{-1}[((p-1)/2)^2 + 1]$ and so $p/4$. An argument by induction shows that $G(F)$ is free on the primes of the form $4m + 3$, and a celebrated theory of Dirichlet guarantees that there are infinitely many such primes.

Various quadratic polynomials similar to $x^2 + 1$ are considered in the exercises of [9], Chapter 23.

If $F(x)$ is a polynomial of degree three or higher, then argument by induction alone becomes more difficult. A second approach can help.

3. Parametric solutions

We may try to parameterize representations

$$t = F(n_1)^{\varepsilon_1} \cdots F(n_k)^{\varepsilon_k},$$

with $\varepsilon_j = 1$ or -1, by seeking polynomials $K_j(x)$, $j = 1, \ldots, s$, with integer coefficients, leading coefficient positive, for which the product

$$F(K_1(x))^{\varepsilon_1} \cdots F(K_j(x))^{\varepsilon_k}$$

in some sense simplifies.

More generally, let $K(x)$, P_i denote typical polynomials in $\mathbb{Z}[x]$ with positive leading coefficients. Let $Q(x)^*$ be the multiplicative group generated by the rational functions P_1/P_2. For a given rational function $F(x)$ in $Q(x)^*$ let $\Delta(F(x))$ be the subgroup of $Q(x)^*$ which is generated by the $F(K(x))$ for varying $K(x)$. Define the quotient group $H(F) = Q(x)^*/\Delta(F(x))$.

Problem 2 *Determine the group $H(F)$.*

With this question we seem to have made matters worse. To illustrate an advantage of the wider viewpoint consider Problem 1 when $F(x) = x^3 - 1$. The identity $x^3 + 1 = F(x^2)F(x)^{-1}$ shows that in constructing group products in Q^* using the integers $F(n)$, $n = 1, 2, \ldots$, we may employ the integers $n^3 + 1$, $n = 1, 2, dots$, as well.

As an example, using rings of shift operators it is shown in [12] that for any nonzero integer a, the group $H(x(x^2 + a))$ is cyclic or order 3, generated by the image of x. In particular, there are polynomials $K_j(x)$, $j = 1, \ldots, s$, for which an identity

$$x^3 = \prod_{j=1}^{s} (K_j(x)(K_j(z)^2 + a))^{\varepsilon_j},$$

with each $\varepsilon_j = +1$ or -1, is valid.

Viewed in $Q(x)^*$, the polynomials $x, x - 1, x - 2$ cannot be distinct mod $\Delta(x(x^2 + a)))$. If x and $x - 2$ belong to the same coset, then $(2x)^{-1}(2x - 2)$ and so $x^{-1}(x - 1)$ belong to $\Delta(x(x^2 + a))$. The remaining alternatives lead to the same conclusion. As a consequence we may deduce that the group $G(n(n^2 + 1))$ is trivial.

As an alternative to algebraic manipulation we may dualise and appeal to the methods of analysis. This gives a third approach to Problem 1.

4.   Harmonic analysis

The group dual to Q^* is a direct product of denumerably many copies of $\mathbb{R}/\mathbb{Z}$, the additive group of reals (mod 1). It is somewhat large. In the parlance of number theory, a typical character on Q^* is a completely multiplicative function g, with values in the complex unit circle, and domain of definition extended to the positive rationals m/n by $g(m/n) = g(m)\overline{g(n)}$.

If the multiplicative function g satisfies only the traditional requirement that $g(ab) = g(a)g(b)$ whenever a and b are mutually prime, then we may extend it to the rationals by confining ourselves to representations m/n with $(m, n) = 1$.

If Γ is a subgroup of Q^*, then the characters on Q^*/Γ are those characters on Q^* that assume the value 1 on every element of Γ. In fact, to determine the nature of the quotient group Q^*/Γ it is only necessary to study homomorphisms of Q^*/Γ into the multiplicative group of the roots of unity. Thus, the group $G(G)$ is trivial and only if every completely multiplicative function g with values in the roots of unity and which satisfies

$$g(F(n)) = 1, \qquad n = 1, 2, \ldots, \qquad F(n) > 0$$

assumes the value 1 on all the positive integers, cf. [9], Chapter 15, Lemma 15.6.

In accordance with standard practice in number theory, it is therefore natural to pose another question.

Problem 3 *For what multiplicative functions g with values in the complex unit disc does the asymptotic mean-value*

$$\lim_{y \to \infty} y^{-1} \sum_{\substack{n \leq y \\ F(n) > 0}} g(F(n))$$

exist? In particular, when is the limit 1?

With this question we begin to appreciate the difficulty of Problem 1. A satisfactory answer to Problem 3 is currently known only when F is a linear polynomial or its reciprocal.

For the purposes of studying Problem 1 a full solution to Problem 3 is not necessary. The torsion properties of a quotient group Q^*/Γ may be determined by characterising the group homomorphisms of Q^*/Γ into the additive group of real numbers.

For example, let $A > 0, B, a > 0, b$ be four integers for which $Ab = aB \neq 0$. Let Γ be the subgroup of Q^* generated by the rationals $(An+B)/(an+b)$ with n running through all positive integers exceeding $\max(-B/A, -b/a)$. Composed with the from Q^*, any homomorphism from Q^*/Γ to the additive reals gives rise to a real-valued completely additive function

$$f : Q^* \to Q^*/\Gamma \to (\mathbb{R}, +)$$

which satisfies

$$f(An + B) - f(an + b) = 0$$

for all appropriately large integers n.

As for multiplicative functions, an additive function that satisfies only the requirement $f(ab) = f(a) + f(b)$ whenever a and b are mutually prime, may be extended to the rationals m/n with $(m, n) = 1$ by $f(m/n) = f(m) - f(n)$. Completely additive and completely multiplicative functions are homomorphisms of the group Q^* that differ only according to the groups in which they take values. The notation, which is hallowed by tradition, is chosen to reflect the differing methodologies with which the homomorphisms are treated.

Additive functions of the above type may be studied under the rubric of Probabilistic Number Theory. Moreover, in 1969, Kátai [28] [29], asked for a characterisation of those real-valued additive functions f for which

$$\lim_{n \to \infty} (f(An + B) - f(an + b))$$

exists.

By 1980 I had settled Kátai's question: such functions must be of the form $D \log n$ on the integers prime to $aA(Ab - Ba)$, cd. [7], [8]. The torsion properties of $G((An + B)/(an + b))$ could then be determined, cf. [9], Chapter 16. These may be described as follows.

A prime p will be called *exceptional* if there are integers $h_1 > h_2 \geq 0$, $t_1 > t_2 \geq 0$, so that $p^{h_1} \parallel a, p^{h_2} \parallel b; p^{t_1} \parallel A, p^{t_2} \parallel B$. Here $p^s \parallel k$ denotes that p^s divides k but p^{s+1} does not.

Let ρ be the ratio $(a, b)/(A, B)$ of highest common factors, ρ_0 that part of ρ made up of powers of exceptional primes, ρ_1 the positive rational for which $\rho_0 = \rho_1^m$ with m as large as possible. For example, if $a = 2^5 3^2 5^2 7$,

$b = 2^4 3.5.7$, $A = 2.3^2 t^4 7^3$, $B = 3.5^3 7^2$, then the exceptional primes are 2, 3, and 5, so that $\rho = 2^4 5^{-2} 7$, $\rho_0 = 2^4 5^{-2}$ and $\rho_1 = 2^2 5^{-1}$.

Let k be a positive integer.

In [9], Lemma 16.1, I prove that a positive rational t has a representation

$$t^v = \prod_{i=1}^{m} \left(\frac{an_i + b}{An_i + B} \right)^{d_i}$$

with integers d_i, $v > 0$ and $n_i > k$ if and only if it is the product of a rational which is not divisible by an exceptional prime, and of a power of ρ_1. The Lemma is stated only for integral t but the argument works generally. Note that the multiplicative condition to be satisfied by t does not depend upon k.

In that same reference I employ an estimate for a Kloosterman trigonometric sum to show that $G((An + B)/(an + b))$ is finitely generated.

Let T denote the subgroup of $G((An + B)/(an + b))$ formed by the elements of finite order—the torsion subgroup. It follows from the general theory of abelian groups that there is a direct product decomposition

$$G((An + B)/(an + b)) = T \otimes K$$

where K is a free subgroup. From the previous result T is generated by the non-exceptional primes, then a little further argument shows that the K has free-group rank r is no exceptional prime divides ρ_1, and $r - 1$ otherwise.

As a particular example, corresponding to the rational function $(15x + 13)/(30x + 7)$ there are two exceptional primes, 3 and 5. Here $\rho_1 = 1$ and the free subgroup of $G((15n + 13)/(30n + 7))$ has rank 2. The torsion group T is not trivial. If

$$n = \prod_{n=1}^{w} \left(\frac{15n_i + 13}{30n_i + 7} \right)^{d_i}.$$

then there is an integer d so that $n \equiv (\bar{7}13)^d \pmod 5$, where $\bar{7}7 \equiv 1 \pmod 5$. Since the possibilities of $(\bar{7}13)^d$ are only $\pm 1 \pmod 5$, no integer of the form $15m + 2$ will belong to Γ, although some power of it must.

It is not difficult to show that the torsion subgroups of the various groups $G((An + B)/(an_b))$ can have arbitrarily high order. Because the dual group of $G((An + B)(an + b))$ is currently not well understood, the exact nature of the torsion group T is not known.

An exotic application of group homomorphisms of Q^* into the additive reals may be found in [9], Chapter 23, problems 8–14, where an argument supporting the following result is implicitly sketched. Let the real t be given. Then every prime $p \equiv 1 \pmod 3$ has a representation

$$p^v = \prod_i \left(\frac{[n_i^2 + 9n_i + 21]^4 [n_i^2 - n_i + 1]}{[n_i^2 + 5n_i + 7]^2} \right)^{\varepsilon_i}$$

with integers $v > 0$, $\varepsilon_i \pm 1$, $n_i > t$.

When the homomorphisms $Q^*/\Gamma \to (\mathbb{R}, +)$ are understood but not those from Q^*/Γ into $\mathbb{R}/\mathbb{Z}$, we may mediate by choosing a prime ℓ and considering the composition

$$Q^* \to Q^*/\Gamma \to G/G^\ell \to \mathbb{Z}/\ell\mathbb{Z}$$

instead, the first two homomorphisms canonical, the third arbitrary. In particular, the group G/G^ℓ, of G by its ℓth-powers, may be viewed as a vector space over the finite field $\mathbb{A}/\ell\mathbb{Z}$.

For example, let $J(x)$ have the form $(x + a_1)^{b_1} \cdots (x + a_h)^{b_h}$ with integer roots $-a_i \leq 0$, and non-zero integral exponents b_j, $1 \leq j \leq h$, possibly negative but whose highest common factor is 1. For a given integer $k \geq 3$ form the subgroup Γ of Q^* generated by the $J(n)$ with $n \geq k$. Then the group Q^*/Γ is trivial, cf. [9], Theorem 17.1.

Moreover, there is a constant c so that every positive integer n has a representation

$$n = \prod_j J(n_j)^{\varepsilon_j}$$

with $\varepsilon_j = \pm 1$ and $k \leq n_j \leq cn$ for each j. Apparently a representation of this type was sought by Burgess. If χ is a non-principal Dirichlet character $\pmod D$, and $\chi(J(n)) = 1$ for $k \leq n \leq y$, then $\chi(m) = 1$ for $a \leq m \leq c^{-1}y$. In pursuit of a small integer n for which $\chi(J(n)) \neq 1$ we may essentially restrict ourselves to the simpler case $J(x) = x$. It should be mentioned that this case remains very difficult.

In view of these results the following questions suggest themselves.

Problem 4 *If $F(x)$ is a rational function for which $G(F(n))$ is finitely generated, then can one give explicit bounds for a set of generators? More generally, can one give explicit bounds for a set of coset representatives for the torsion subgroup of $G(F(n))$?*

Let $a_1, a_2, \ldots$ be a sequence of positive integers. It is shown in [9], Chapter 15, that if a representation

$$m^v = a_1^{d_1} \cdots a_t^{d_t}$$

is known to exist with integers a_j constrained by $1 \leq a_j \leq M$, and some positive integer v, then an algorithm can be given to determine a representation of the same form with explicit values of the integer powers d_j and the minimal value of v. A theoretical localisation of the a_j appearing in a product representation has practical consequences.

5.   Groups generated by primes

Representing an odd integer as the sum of three odd integers, $N = x_1 + x_2 + x_3$, is an easy question of Waring type. Requiring the x_j to be primes, as Goldbach did, increases the difficulty very seriously; the local distribution of the primes is not known with precision. Likewise, the behavior of even simple rational functions on the primes is much less immediate than their behavior on the integers.

Problem 5 *With $F(s)$ a rational function of the type considered in Problem 1, determine the group $G(F(p))$.*

There is an analogue of Problem 3.

Problem 6 *For what multiplicative functions does the asymptotic mean-value*

$$\lim_{y \to \infty} \frac{\log y}{y} \sum_{p \leq y} g(F(p))$$

exist? In particular, when is the limit 1?

Consider the function $x^{-1}(x+1)$ from our very first example. Each odd prime p belongs to the same coset of Γ as $p+1$, and $p+1 = 2((p+1)/2)$. The group $G(p^{-1}(p+1))$ is generated by the prime 2. Since

$$2 = 3^{-1}(3+1)2^{-1}(2+1),$$

$G(p^{-1}(p+1))$ is also trivial.

The case $F(x) = x^{-1}(x+k)$ with k a positive integer exceeding 1 is not quite so simple. Any prime p belongs to the same coset of Γ as $p+k$. If $p+k$ is again a prime, so that the factorising argument in the case $k = 1$ cannot proceed, then $p+2k$ also belongs to $p(\text{mod}\,\Gamma)$. This inductive argument cannot continue indefinitely.

Let ℓ be a prime that does not divide pk. then one of the integers $p+jk$, $j = 1, \ldots, \ell - 1$, is divisible by ℓ. The group $G(P^{-1}(p+k))$ is generated by the primes not exceeding $(\ell - 1)k$.

The prime number theorem shows the product of the first m rational primes to be $\exp((1 + o(1))m)$ as $m \to \infty$. The smallest prime not dividing k does not exceed $c_1 \log k$ for some constant c_1, and a set of generators for $G(p^{-1}(p+k))$ can be found in the interval $(1, c_1 k \log k)$.

The third example illustrating Problem 1 has $F(s) = x^k$. Here $G(p^k)$ coincides with $G(n^k)$.

We continue with $F(x) = x^2 + 1$. If q_j, $1 \le a \le h$, are distinct primes of the form $4m + 3$, and $\bar{q}_j$ denotes a typical image under the canonical map $Q^* \to Q^*/\Gamma$, then any relation $\bar{q}_1^{b_1} \cdots \bar{q}_h^{b_h} = \bar{1}$ implies a factorisation

$$q_1^{b_1} \cdots q_h^{b_h} \prod_{i=1}^{u} (p_i^2 + 1)^{e_i} = \prod_{i=u+1}^{v} (p_i^2 + 1)^{e_i}$$

with p_i prime, the e_i positive integers. No prime factor of the integers $p_i^2 + 1$ is of the form $4m + 3$. Hence $q_1^{b_1} \cdots q_h^{b_h} = 1$ in Q^*, and every $b_j = 0$, $1 \le j \le h$. the images of the primes q of the form $4m + 3$ are free in the group $G(p^2 + 1)$. Nothing further seems to be know.

It seems likely that the groups $G(p^2 + 1)$ and $G(n^2 + 1)$ again coincide.

I return to consider the second example, $F(x) = x + 1$ in some detail. the nature of $G(n + 1)$ is evident; that of $G(p + 1)$ is not.

In 1904, Dickson [4], conjectured that unless prevented by residue class conditions, any finite collection of linear forms in $\mathbb{Z}[x]$ with positive leading coefficients simultaneous representation primes for infinitely many integer values of the variable. He regarded this as a natural generalisation of Dirichlet's theorem on primes in an arithmetic progression. An immediate consequence of Dickson's conjecture would be that every positive rational r had an infinite number of representations $r = (p + 1)/(q + 1)$ by shifted primes. The case $r = 2$ would assert that for primes q, $2q + 1$ is infinitely often prime, an analogue of the celebrated prime-pair conjecture.

We therefore expect $G(p + 1)$ to be trivial.

In 1968, Kátai [26], conjectured that any real-valued additive function that vanished on the shifted primes $p + 1$ vanished identically. Towards this conjecture he gave an argument that implicitly guaranteed the finite generation of the corresponding group $Q^*/\Gamma(p+1)$, Kátai [27]. Although his argument did not lend itself to the construction of an algorithm to provide a set of generators, it could be made to do so, cf. [9], Chapter 23, problems 78–84. Unfortunately, the bound so obtained on the largest generator is beyond the practical reach of current computer power.

I settled Kátai's conjecture in the affirmative [6]. From our earlier remarks concerning harmonic analysis on groups, $G(p + 1)$ must be a torsion group, and in view of the result of Kátai, finite. This was not realised at the time, cf. Dress and Volkmann [5], Wolke [35], Meyer [30], Elliott [9], Chapter 15.

The groups $G = Q^*/\Gamma$ appear explicitly in [9]. In Chapter 23 of that volume an argument delivering the bound $|G(p+1)| \leq 8$ is indicated.

There are currently two methods to estimate the order of $G(p+1)$ analytically. The first of these, [15], depends upon a bound for the spectral radius of the self-adjoint operator underlying the hermitian form

$$\sum_{p+1 \leq x} \left| \sum_{j=1}^{k} c_j g_j(p+1) \right|^2$$

where the g_j, $j = 1, \ldots, k$, are extensions to Q^* of group characters on $G(p+1)$:

$$g_j : Q^* \to Q^*/\Gamma(p+1) \to U,$$

with U the unit circle in the complex plane. Selberg's sieve method and the good mean-square distributions of multiplicative functions on residue classes play important rôles. Enhanced with estimates for bilinear forms, derived in turn from a spectral decomposition by the hyperbolic Laplacian, cf. Fouvry [25], the argument shows that the dual group of $G(p+1)$ and hence that group itself has order at most 3.

Duality transfers every operator norm estimate for the above hermitian form to an estimate for the form

$$\sum_{j=1}^{k} \left| \sum_{p+1 \leq x} d_p g_j(p+1) \right|^2 .$$

As a by-product of the harmonic treatment of $G(p+1)$ we obtain nontrivial estimates for the sum

$$\sum_{p+1 \leq x} g(p+1),$$

where g is a multiplicative function with values in the complex unit disc. Within this wide generality these estimates are currently the best known, but in strength far from what one would believe to be true, cf. [15], Theorems 5 and 6.

The second method for treating $G(p+1)$ provides a fourth general approach to Problem 1: Estimate the asymptotic frequency of integers belonging to a typical coset (mod Γ).

For each set of positive rationals E, define the Lower asymptotic density of E to be

$$d(E) = \liminf_{y \to \infty} y^{-1} \sum_{n \leq y, n \in E} 1,$$

the sum counting integers in E. Clearly $0 \leq d(E) \leq 1$, and for disjoint sets A and B, $d(A \cup B) \geq d(A) + d(B)$.

For a rational γ let γE denote the set of rationals in E each multiplied by γ.

Our second approach to $G(p+1)$ rests upon the uniformity of the following result, cf. [19].

For every positive integer k, and positive rational γ, the density of the integers of the form $(p+1)(\gamma(q+1))^{-1}$ with p, q primes exceeding k, is at least $11/40$.

A result of this nature may be obtained by a variant of the following argument.

Let $r(n)$ denote the number of representations of n as a quotient $(p+1)(\gamma(q+1))^{-1}$ with the primes p, q chosen to belong to a certain region that may depend upon parameters, say α, $0 < \alpha < 1$, and $x \geq 2$. An example is $p + 1 \leq \gamma(q+1)$ with $q \leq x^{\alpha}$. In particular, $n \leq x$. Some restriction is necessary in order that $r(n)$ be well defined.

We choose a bound for $r(n)$, say $z(n)$, so that

$$\sum_{n \leq x, r(n) \neq 0} 1 \geq \sum_{n \leq x} z(n)^{-1} r(n)$$

$$= \sum_{q} \sum_{p} z \left(\frac{p+1}{\gamma(q+1)} \right)^{-1},$$

the innersum taken over those primes not exceeding $\gamma(q+1)x - 1$ that satisfy $p \equiv -1 (\mathrm{mod}\, \gamma(q+1))$. As far as possible $z(n)$ is chosen to have an arithmetic nature convenient for calculation and to render the quotients $r(n)/z(n)$ equal. The primes q are chosen so that $\gamma(q + 1)$ is integral.

In this approach some knowledge of the distribution of primes over residue classes is required.

Let $E(\gamma)$ denote the set of integers of the form $(p+1)(\gamma(q+1))^{-1}$ with p, q prime, so that $k = 1$. If $\gamma_1, \ldots, \gamma_g$ are representatives of distinct cosets $(\mathrm{mod}\, \Gamma(p+1))$, then

$$\frac{11g}{40} \leq \sum_{j=1}^{g} d(E(\gamma_j)) \leq 1.$$

We gain a second proof that $|G(p+1)| \leq 3$, without the intervention of harmonic analysis on $G(p+1)$ or Q^*, but without the generation of estimates for the sums involving $g(p+1)$.

These two approaches to the study of the group $G(p+1)$ are flexible enough to address a problem characteristic of the study of representation by products.

6.   Multiplicity of representation

Let $r_1, r_2, \ldots$, be a sequence of positive rationals in which no value repeats infinitely often. We say that a rational t has infinitely many distinct representations

$$t = r_1^{d_1} \cdots r_w^{d_2},$$

with the d_j integers, if for each positive integer k there is a representation with $d_j = 0$ whenever $j < k$.

If Γ_k denotes the subgroup of Q^* generated by the r_i with $i \geq k$, then $\Gamma_1 \supseteq \Gamma_2 \supseteq \cdots$. The positive rationals which possess infinitely many group product representations by the r_j form a subgroup of Q^*, namely

$$\bigcap_{w=1}^{\infty} \Gamma_w.$$

We form the groups $G_k = Q^*/\Gamma_k$. It is natural to ask the following question.

Problem 7 *When are the groups G_k eventually isomorphic from some point on?*

I conjecture that when the r_j are given by the positive values of a rational function $F(x)$ on the integers or the primes, this is always the case. More generally:

Problem 8 *Determine the group of those rationals possessing infinitely many product representations.*

It is easy to see that, in an obvious notation, the groups $G_k(n^{-1}(n+1))$ are all trivial.

The corresponding proposition for the groups $G_k(n^2 + 1)$ is not immediately accessible to the elementary inductive argument that decided the structure of $G_1(n^2 + 1)$.

In this matter the derivation of parametric solutions via the groups $Q(x)^*/\Delta(F(x))$ can be very helpful. For example, when $F(x)$ is the cubic $x(x^2 + a)$ with $a \neq 0$, the product representation of x^3 given earlier enables representations

$$h^3 = x^{-3}(hx)^3 = \prod_{j=1}^{s} (F(x)^{-1}F(hx))^{\varepsilon_j},$$

affording at once infinitely many representations of h^3 by the $F(n)$.

We established earlier that $x^{-1}(x+1)$ belongs to $\Delta(F(x))$ in this case and so, therefore, does $x^{-1}(x + z)$ for any positive integer z. Setting $x = mk$, $z = k$ we see that, again in an obvious notation, $m^{-1}(m + 1)$

belongs to $\Gamma_k(F(n))$ for every $m \geq 1$. Choosing h^3 to be the first cube at least as large as m, we may represent

$$\prod_{y=m}^{h^3-1} y(y+1)^{-1}$$

by a member of $\Gamma_k(F(n))$ and deduce that m has infinitely many representations. All the groups $G_k(n(n^2 + a))$ are trivial.

A group theoretical argument reduces Problem 7 to a question with a more practical aspect: *When are the orders of the groups G_k uniformly bounded from some point on?*

Let $|L|$ denote the order of the group L. If $s \geq t$, then as a set Γ_s is contained in Γ_t; G_t may be viewed as a subgroup of G_s and $G_s/G_t \simeq \Gamma_t/\Gamma_s$. In particular,

$$|G_s| = |G_t| \prod_{i=1}^{s-t} |\Gamma_{s-i}/\Gamma_{s-i+1}|,$$

where each term in the product is an integer. If the orders of the groups G_s are ultimately bounded, then at most finitely many of the quotient groups Γ_k/Γ_{k+1} can be non-trivial. For all sufficiently large t the groups G_t are isomorphic.

This phenomenon is exemplified by the groups $G_k(p+1)$. According to the coset density result of the previous section, the order of the groups $G_k(p+1)$ is uniformly bounded by 3. Here the value of the bound is unimportant. Any bound would have also ensured that for all but finitely many positive k, the groups $G_k(p+1)$ are isomorphic.

This last result may also be obtained from the harmonic treatment of $G(p+1)$ by replacing the summation $p+1 \leq x$ in the associated hermitian forms with $k < p+1 \leq x$, this scarcely affecting the argument. Once an analytic treatment is obtained it is often stable under perturbations of boundary conditions.

We have proved that $\bigcap_{w=1}^{\infty} \Gamma_w(p+1) = \Gamma_{w_0}(p+1)$ for some w_0, but not that we may take $w_0 = 1$, although that is almost certainly the case.

In comparison with Waring's problem there remains a further question, and its answer for products appears complex.

7.   Number of representing terms

Problem 9 *In product representations*

$$r = \prod_{j=1}^{s} F(n_j)^{\varepsilon_j}$$

what can be said about the minimal value of s as a function of r? When can it be uniformly bounded, or even a fixed value?

There need not be a bounded value of s. In the example with $F(x) = (ax + b)/(Ax + B)$, let v be the order of the torsion group of $G(F(n))$, and m (≥ 2) an integer whose image $(\bmod \Gamma)$ has torsion. There will be a representation

$$m^v = \prod_{i=1}^{s} \left(\frac{an_i + b}{An_i + B} \right)^{\varepsilon_i}.$$

Since

$$\max_{n \neq -b/a, -B/A} \max_{\varepsilon = \pm 1} \left| \left(\frac{an + b}{An + B} \right)^{\varepsilon} \right| \leq e^{\alpha}$$

for some positive α, s is at least as large as $\alpha^{-1} v \log m$.

It is shown in [9], that for each $\beta > \log 3 (\log 4/3)^{-1}$, there is a constant $c(\beta)$ so that a representation of the above type for m^v exists with $s \leq c(\beta)(\log m)^{\beta}$. It seems likely that the lower bound on β may be reduced, but what the best possible would be is quite unclear.

Possession of a function-field analogue of the group $G(F(n))$ is again helpful. The determination of the group $H(F)$ when $F = x(x^2 + a)$ leads at once to an infinitude of representations for cubes of integers that employ a fixed number of terms in the product.

Combined with the iteration scheme $n \mapsto n/2$ if n is even, $n \mapsto (n+1)/2$ if n is odd, this shows that an integer m can be represented by a group product of integers $n(n^2 + a)$ using $O(\log m)$ terms. Whether a bounded or a fixed number of terms suffices for every positive integer is not clear.

A lower bound on the density of coset representatives $(\bmod \Gamma)$ helps here, too. Let k be a positive integer. For each j, let S_j denote the set of rationals in $\Gamma(p+1)$ that may be represented by a product of at most j shifted primes, each prime exceeding k. The density bound in §6 may be reinterpreted as $d(\gamma S_2) \geq 11/40$ uniformly in positive rationals γ.

Clearly $S_2 \subseteq S_3 \subseteq \cdots$. Suppose that $S_9 \neq S_{10}$. Then $S_j \neq S_{j+1}$ for $2 \leq j \leq 9$. Let α belong to S_5 but not S_4, and β belong to S_{10} but not S_9. The sets $S_2, \alpha S_2, \beta S_2, (\alpha\beta) S_2$ are mutually disjoint. Hence

$$11/10 \leq d(S_2) + d(\alpha S_2) + d(\beta S_2) + d(\alpha\beta S_2) \leq 1,$$

which is impossible. Therefore $S_9 = S_{10} = \cdots$.

We have shown that for each k, the members of $\Gamma_k(p+1)$ have a group product representation employing no more than 9 terms.

It is straightforward to prove that every member of $\Gamma_1(p+1)$ has a group representation with w terms provided $w \geq 9$.

Moreover, there is a pattern $(\varepsilon_1, \ldots, \varepsilon_t)$ for some $t \leq 9$, so that for some k_0 every member of Γ_k with $k \geq k_0$ has infinitely many product representations of the fixed form

$$(p_1 + 1)^{\varepsilon_1} \cdots (p_t + 1)^{\varepsilon_t},$$

with p_i prime. In particular, with g the (constant) order of the groups $G_k(p+1)$, so that $g \leq 3$, every power r^g of a rational r has such representations.

That this result is part of a general phenomenon was shown by Berrizbeitia and Elliott [1], [2]. In the notation of §1, let S_j denote the set of elements in Γ with product representations using at most j members of the sequence r_i. The following three propositions are equivalent.

There is a positive integer j and a positive c_1 so that $d(m^{-1}S_j) \geq c_1$ uniformly in positive integers m.

There is a positive integer k and a positive c_2 so that $d(\gamma S_k) \geq c_2$ uniformly in positive rationals γ.

The group Q^*/Γ is finite and as sets $\Gamma = S_\ell$ for some ℓ.

Moreover, granted a value for c_1 (or c_2), a value for ℓ may be computed.

I conclude this section with a more elaborate example illustrating the efficacy of counting coset representatives.

Let $f(x)$ be a polynomial with integer coefficients, leading coefficient positive, of degree at least one.

A prime ℓ is said to be singular with respect to a polynomial f if there is an integer m so that the congruence $mf(n) \equiv 1 \pmod{\ell}$ has $\ell - 1$ solutions in reduced classes $n \pmod{\ell}$. No singular prime exceeds $1 + \deg(\text{ree}) f$.

Let Δ be the product of the primes singular with respect to f. There are classes $u_j \pmod{\Delta}$, $j = 1, \ldots, \phi(\Delta)$, to which an integer m belongs if and only if no congruence $mf(n) \equiv 1 \pmod{\ell}$ has $\ell - 1$ reduced solutions. This collection always contains the class $0 \pmod{\Delta}$. For example, if $f(x) = x^3 + 5$, then only $\ell = 2$ or $\ell = 3$ might be singular, and they are not. If $f(x) = x^2 + 3$, then the prime 3 is singular and no other. We may set $\Delta = 3$, $u_1 = 0$, $u_2 = 2$.

We attach to the polynomial f a density defined on the sets of rationals E by

$$d_f(E) = \liminf_{y \to \infty} \left(\frac{\phi(\Delta)y}{\Delta} \right)^{-1} \sum_{j=1}^{\phi(\Delta)} \sum_{\substack{n \leq y, n \equiv u_j (\text{mod} \Delta) \\ n \in E}} 1,$$

the innersum counting only positive integers. For polynomials without singular primes this is the standard lower asymptotic density defined in §**5**. Note that if $\bar{\Delta}$ is the product of primes singular for at least one of a collection of polynomials f_j, $j = 1, \ldots, t$, then each $\bar{\Delta}f_j$ has no singular primes.

In terms of this density we have the following generalisation of the case $f(x) = x + 1$, cf. [21], Theorem 4.

Let $0 < 3\alpha \deg f < 1$. The integers representable in the form $(p + 1)f(q)^{-1}$ with p, q primes that satisfy $(2\sqrt{2})^{-1} \leq (p + 1)^\alpha q^{-(1+\alpha)} \leq \sqrt{2}$ have density (attached to f) at least $\frac{1}{4}(1 + \alpha)(1 + \alpha \deg f)^{-1}$.

In particular, those integers unrestrictedly representable in the form $(p + 1)f(q)^{-1}$ with p, q prime, have density at least $1/4$.

The polynomial $x + 1$ has no singular primes.

For $k \geq 1$, let Γ_k be the subgroup of Q^* generated by the integers representable in the form $(p + 1)f(q)^{-1}$ with p, q primes as large as k. Let G_k be the quotient group Q^*/Γ_k.

If $E(m)$ denotes the set of integers representable in the form $(p + 1)(mf(q))^{-1}$ with an integer m, then

$$\sum_{u=1}^{5} d_f(E(m^u)) \geq 5/4.$$

At least one pair $E(m^u) \cap E(m^v)$ with $1 \leq u < v \leq 5$ is non-empty. There is a corresponding representation

$$m^{v-u} = \frac{p_1 + 1}{f(q_1)} \cdot \frac{f(q_2)}{p_2 + 1)}$$

with primes $p_i, q_i \geq k$. Every integer $(\text{mod} \, \Gamma_k)$ has torsion. Each coset $(\text{mod} \, \Gamma_k)$ has an integer representative.

Let $m_1, \ldots, m_g$ be representatives of distinct cosets $(\text{mod} \, \Gamma_k)$. Then

$$1 \geq \sum_{j=1}^{g} d_f(E(m_j)) \geq g/4.$$

and $|G_k| \leq 4$. Since this bound is uniform in k, from some point on the groups G_k are isomorphic. The common group G has order at most 4.

As for the polynomials $x + 1$, we define Y_j to be the set of rationals in Γ_k that may be represented by a group product of at most j integral ratios $(p + 1)f(q)^{-1}$ with primes $p, q \geq k$.

Again $Y_1 \subseteq Y_2 \subseteq \cdots$. In this more general case, and in an obvious notation, the assumption $Y_{11} \neq Y_{12}$ provides members β_1 in $Y_3 - Y_2$, β_2 in $Y_6 - Y_5$, β_3 in $Y_9 - Y_8$ and β_4 in $Y_{12} - Y_{11}$. Let D be a common denominator for the β_j, $1 \leq j \leq 4$, and let Δ be the product of the primes singular for f. For each j we have

$$d_{\Delta f}((\beta_j D)^{-1} Y_1) \geq 1/4,$$

and the contradiction

$$5/4 \leq d_{\Delta f}(D^{-1} Y_1) + \sum_{j=1}^{4} d_{\Delta f}((\beta_j D)^{-1} Y_1) \leq 1$$

renders our assumption untenable. As sets $\Gamma_k = Y_{11}$ for each $k \geq 1$.

In particular, there is a fixed pattern $\varepsilon_1, \ldots, \varepsilon_t$, with each $\varepsilon_j = +1$ or -1, and $t \leq 11$, so that every positive rational r has a representation

$$r^{|G|} = \left(\frac{p_1 + 1}{f(q_1)}\right)^{\varepsilon_1} \cdots \left(\frac{p_t + 1}{f(q_1)}\right)^{\varepsilon_t}$$

with the p_i, q_i prime and the $(p_i + 1)f(q_i)^{-1}$ integral.

Doubtless $|G|$ should be 1, but the example $f(x) = x^2$ with r an integral square shows that without a restriction upon f, the value $t = 2$ would be the best possible.

I conjecture that every positive rational r has infinitely many representations

$$r = \frac{p_1 + 1}{f(q_1)} \cdot \frac{f(q_2)}{p_2 + 1}$$

with p_i, q_i prime, $(p_i + 1)f(q_i)^{-1}$ integral, $i = 1, 2$.

For individual polynomials f and rationals r, better may be achieved, cf. [21], Theorems 1 and 2. There are primes p_1, q_1 so that

$$2\left(\frac{p_1 + 1}{q_1^2 + 3}\right)^{-1} = \frac{p_2 + 1}{q_2^2 + 3} \cdot \frac{q_3^2 + 3}{p_3 + 1}$$

has infinitely many solutions with primes p_j, q_j, and integral quotients $(p_j + 1)(q_j^2 + 3)^{-1}$, $j = 2, 3$. Moreover, the pair (p_1, q_1) may be assumed to belong to the list $(223, 5)$, $(383, 3)$, $(3583, 5)$ and $(3583, 2)$.

Likewise, there are primes p_1, q_1 so that

$$2\left(\frac{p_1 + 1}{q_1^3 + 5}\right)^{-1} = \frac{p_2 + 1}{q_2^3 + 5} \cdot \frac{q_3^3 + 5}{p_3 + 1}$$

has infinitely many solutions with primes p_j, q_j and integral quotients $(p_j + 1)(q_j^3 + 5)^{-1}$, $j = 2, 3$. The pair (p_1, q_1) may then be assumed to belong to the list $(31, 3)$, $(173, 7)$, $(7, 3)$ and $(3, 3)$.

In the foregoing argument no use has been made of the parameter α. If $c > 0$ and we choose α suitably small, then we may require that in the representations by $(p + 1)f(q)^{-1}$ the localisation

$$(2\sqrt{2})^{-1} \leq (p + 1)^{\alpha} q^{-(1+\alpha)} \leq \sqrt{2}$$

be satisfied, and so show that for any positive rational r one of the equations

$$(x_1 + 1)f(x_2) - r^k(x_3 + 1)f(x_4) = 0, \qquad k = 1, 2, 3, 4,$$

has an infinity of solutions with $x_2 > x_4^c$ and all the x_j prime. Presumably each of the equations has such solutions.

8.   Dense cosets

Much of the argument in the previous section depended upon possession of cosets containing many integer representatives. Harmonic analysis allows us to measure how dense these integer representatives need to be, cf. [9], Chapter 22, [11].

Let $a_1 < a_2 < \cdots$ be a sequence of positive integers, of positive upper asymptotic density d;

$$d = \limsup_{y \to \infty} y^{-1} \sum_{a_j \leq y} 1.$$

Then *there is a set of primes q for which $\sum q^{-1}$ converges to a sum bounded in terms of d, so that the subgroup of Q^*/Γ generated by the integers not divisible by one of these q is finite, of order not exceeding $1/d$.*

Here, as in §1, Γ is the subgroup of Q^* generated by the a_j. Thus, every integer n coprime to each q has a group product representation

$$n^v = \prod_{j=1}^{m} a_j^{d_j}$$

with v and the d_j integers, $1 \leq v \leq 1/d$. The occurrence of a thin set over which one has no control is a characteristic of methods employing harmonic analysis. Here it is appropriate, since the sequence of integers not divisible by any of a sequence of chosen primes q for which $\sum q^{-1}$ converges has positive asymptotic density $\prod(1 - q^{-1})$. It is also characteristic of harmonic analysis that we can measure the size of the set of exceptional primes q but we cannot readily locate individual members.

Fortunately, arithmetic properties of the a_j combined with a local sieve often enable the difficulties caused by the existence of a thin set of exceptional primes to be finessed, cf. [9], Chapter 12, entitled *From L^2 to L^∞*. An example of this procedure will be considered in the next section.

9.   Localised representations

Define the height $h(r)$ of a rational $r = a/b$, in lowest terms, to be $\max(|a|, |b|)$. We shall say that two rationals are mutually prime rationals if their canonical representations in Q^* possess no common primes.

It seems likely that if the positive integer N is large enough, then any rational r mutually prime to N and of a height small compared to that of N will have a representation

$$r = \frac{N - p}{N - q}$$

with p, q primes less than N. In the case N odd, $r = 2$, this would be equivalent to a representation $N = 2q - p$, of Goldbach type.

Such representations being currently out of reach, we might try for a group product representation

$$r = \prod_{i=1}^{s} (N - p_i)^{\varepsilon_i}$$

with $\varepsilon_i = \pm 1$, p_i prime and s bounded independently of N and all r of height up to a fixed power of $\log N$. This also being out of reach, we might allow s to vary with r and/or N.

In algebraic terms, let N be a positive integer, and let Q_1^* be the multiplicative group of rationals generated by the primes less than N and mutually prime to it. Let P be a subset of the primes in Q_1^* and let $\Gamma(P)$ be the subgroup of Q_1^* generated by the $N - p$ with p in P.

Problem 10 *Determine, as far as possible, the group $Q_1^*/\Gamma(P)$.*

The group is at once finite, but its order may vary with P and N.

If the set P is constrained to lie in the interval $(1, N/2)$, then every term $N - p$ in a product representation will exceed $N/2$. It is not difficult to prove that most integers $N - p$ with $p < N/2$ possess prime factors decidedly large compared to any fixed power of $\log N$. In any group product representation of a rational with height small compared to N, say $r = 2$, all such large prime factors must cancel away.

For sets P whose dependence upon N is well defined, there is an associated question.

Problem 11 *In the notation of problem 10, for which multiplicative functions g with values in the complex unit disc does*

$$\lim_{N \to \infty} \frac{\log N}{N} \sum_{p \in P} g(N - p)$$

exist? In particular, when is it 1?

If the set P is comprised of all primes strictly between $N/3$ and $2N/3$, then each $N - p$ belongs to the same range. Every ratio $(N - p_1)(N - p_2)^{-1}$ with p_1, p_2 in P lies in the interval $(1/2, 2)$. In general there may not be any integers of the form $(N - p_1)(m(N - p_2))^{-1}$ with which to generate coset representatives $\bmod \, \Gamma(P)$ in Q_1^*. However, harmonic analysis remains viable.

Suppose that P contains at least $\delta \pi(N) > 0$ primes, with $0 < \delta < 1$. Then *for $N \geq N_0(\delta)$ there is a set of primes q with $\sum q^{-1} \leq c_1(\delta)$ such that G_1, the subgroup of $Q_1^*/\Gamma(P)$ generated by the rationals in Q_1^* with no q-factor, satisfies $|G_1| \leq c_2(\delta)$.*

Moreover, let Q_2^* be the subgroup of Q_1^* when the q factors are removed. There is then a subgroup L of G_1, $|L| \leq 4/\delta$, and an integer D in Q_1^*, so that the diagram

$$(\mathbb{Z}/D\mathbb{Z})^*$$

$$\nearrow \qquad \searrow$$

$$Q_2^* \quad \longrightarrow \quad G_1/L$$

commutes, cf [18].

Here $(\mathbb{Z}/D\mathbb{Z})^*$ is the multiplicative group of reduced residues $(\bmod \, D)$, the maps $Q_2^* \to (\mathbb{Z}/D\mathbb{Z})^*$, $Q_2 \to G_1 \to G_1/L$ are canonical. In the language of [9], Chapter 22, the group G_1/L is *arithmic group*.

Explicit values for D and the $c_j(\delta)$ may be given, but not for the individual exceptional primes q.

One may say that the representability of an integer m, not divisible by any of the exceptional primes q, largely depends upon the residue class $(\bmod \, D)$ to which m belongs. If $1 \leq m < N$, $m \equiv 1 (\bmod \, D)$, $(m, q) = 1$ for all q and $(m, N) = 1$, then there is a representation

$$m^{|L|} = \prod_{p \in P} (N - p)^{d_p}$$

with d_p integral.

It seems likely that the bound on the order of L should be $1/\delta$ rather than $4/\delta$. Then $\delta > 1/2$ would force L to be trivial, and the group G_1 itself would be arithmic.

Given a positive integer d, small compared to N and coprime to it, we cannot determine whether d is divisible by any exceptional prime q. Were the q to include all the primes in an interval $(N^\varepsilon, N]$, $0 < \varepsilon < 1$, then the natural approach of looking for a prime p so that $N - p = md$ with $(m, q) = 1$ for all q would amount to representing N as the sum $p + t$ with all the prime factors of t less than N^ε. For small values of ε this is a problem difficult of itself.

If the primes in P are better localised, then the exceptional primes q may be dealt with using sieves, as indicated in the previous section. Thus, cf. [17], *there is an integer k so that if $c > 0$, $N > N_0(c)$, then every integer in the range $1 \leq m \leq (\log N)^c$, $(m, N) = 1$ has a representation*

$$m^k = \prod_{N/4 < p \leq N/2} (N - p)^{d_p}$$

with p prime, d_p integral.

Allowing the k to vary with individual m we may assert that $1 \leq k \leq 4$, with a uniform value if $1 \leq n \leq \log \log N$, cf. [22]. We employ a local density to count integers with a representation of the form $w = (N - p)(mt^g)^{-1}$, where t has no prime factor among the exceptional q, satisfies $t \equiv 1 \pmod{D}$ and we choose $g = |L|$; or t is unrestricted $\pmod{D}$ and we choose $g = |G_1|$. If $0 < \alpha < 1$, then a typical local density

$$\Delta_m = \left(\sum_{\substack{2M/3 < s \leq M \\ (s, N) = 1}} 1 \right)^{-1} \sum_{2M/3 < w \leq M} 1,$$

with $M = N^{1-\alpha}$, satisfies $\Delta_m \geq (1 + o(1))/4$ as $N \to \infty$ and, if $c > 0$, uniformly for integers m in the interval $[1, (\log N)^c]$. As a consequence, there will be a representation

$$m^u = \frac{(N - p_1)}{t_1^g} \cdot \frac{t_2^g}{(N - p_2)}$$

with $1 \leq u \leq 4$. The t_j^g will inherit membership of $\Gamma(P)$ from the earlier result obtained by harmonic analysis.

To this end it is appropriate to study the distribution in residue classes of integers t, $1 \leq t \leq N$, that do not possess prime factors q taken from a set concerning which it is known only that $\sum q^{-1}$ is bounded uniformly in N. This is a problem interesting in its own right. Corresponding to a non-principal Dirichlet character $(\bmod D)$, the natural Dirichlet series associated with the integers t is

$$\sum_{t=1}^{\infty} \chi(t)t^{-s} = L(s,\chi) \prod_{q}(1 - \chi(q)q^{-s}), \quad \mathrm{Re}\,(s) > 1.$$

If χ is quadratic, the q run through the primes in $(N^{1/2}, N]$ that satisfy $\chi(q) = -1$, and s has a real-value σ in the interval $1/2 < \sigma < 1$, then subject to the availability of an estimate

$$\sum_{\substack{p \le y \\ p \equiv a (\bmod D)}} 1 = (1 + o(1))\frac{y}{\phi(D) \log y}, \quad y \to \infty,$$

uniformly for $N^{1/2} \le y \le N$, we have

$$\prod_{q}(1 - \chi(q)q^{-\sigma}) = \exp\left(\frac{(1 + o(1))N^{1-\sigma}}{(1 - \sigma)\phi(D) \log N}\right), \quad N \to \infty,$$

a factor that would overpower any standard estimate for the Dirichlet series $L(s, \chi)$ employed in a classical Riemannian study of primes using analytic continuation. This is certainly the case if $1 \le D \le \log N$.

A completely different approach is needed. The following result serves, cf. [13], [23].

Let D be an integer, $2 \le D \le x$, $\varepsilon > 0$. Let g be a multiplicative function with values in the complex unit disc.

There is a character $\chi_1 (\bmod D)$, real if g is real, so that when $0 < \gamma < 1$,

$$\sum_{\substack{n \le y \\ n \equiv a (\bmod D)}} g(n) - \frac{1}{\phi(D)} \sum_{\substack{n \le y \\ (n,D)=1}} g(n) \quad -\frac{\overline{\chi_1(a)}}{\phi(D)} \sum_{n \le y} g(n)\chi_1(n)$$

$$\ll \frac{y}{\phi(D)}\left(\frac{\log D}{\log y}\right)^{1/4-\varepsilon}$$

uniformly for $(a, D) = 1$, $D \le y$, $x^{\gamma} \le y \le x$, the implied constant depending at most upon ε, γ.

In this generality the extra term involving a character χ_1 is appropriate, since g itself may be a character $(\bmod D)$

When g is the characteristic function of the integers t not divisible by any of a set of primes q with $\sum q^{-1} \le K$, at the expense of a factor $\exp(K)$, the sum involving χ_1 may be deleted and the exponent $1/4 - \varepsilon$ in the error term improved to $1/2 - \varepsilon$.

An inductive argument of Erdös and Ruzsa [24], shows that there are at least $x \exp(-\exp(c_3 K))$, $c_3 > 0$, such integers t in the interval $[1, x]$, provided x is sufficiently large. Moreover, the double exponential in this lower bound is best possible. The desired uniformity with respect to the location of the exceptional primes q perhaps militates against the existence of an asymptotic estimate with a readily comprehensible main term. The above result guarantees the t to be well distributed in residue classes without appeal to such a asymptotic estimate.

Choosing g to be a suitable restriction of the Möbius μ-function, and employing sieve results of the 'fundamental lemma' type derived for purposes in probabilistic number theory, we arrive at the following result; cf. [13] Appendix, [23].

Let D be an integer, $2 \leq D \leq x^{1/2}$, $0 < \varepsilon < 1$. There is a real character $\chi_1(\bmod D)$ so that then $0 < \gamma < 1$,

$$
\sum_{\substack{p \leq y \\ p \equiv a(\bmod D)}} \log p - \frac{1}{\varphi(D)} \sum_{\substack{p \leq y \\ (p,D)=1}} \log p \;-\; \frac{\chi_1(a)}{\varphi(D)} \sum_{p \leq y} \chi_1(p) \log p
$$
$$
\ll \frac{y}{\phi(D)} \left(\frac{\log D}{\log y} \right)^{1/2 - \epsilon}
$$

uniformly for $(a, D) = 1$, $D \leq y^{1-\varepsilon}$, $x^\gamma \leq y \leq x$, the implied constant depending at most upon ε and γ.

It is shown in [20] that a weaker version of this result may be employed as the cornerstone of a new proof of the celebrated theorem of Linnik that for some c every reduced residue class $(\bmod D)$ contains a prime representative not exceeding D^c. The proof does not involve the use of the zeros of Dirichlet L-series (nor therefore of the Deuring–Heilbronn phenomenon),analytic continuation or appeal to the Pólya–Vinogradov inequality.

10.   An ideal method?

The arguments of the previous section decide the existence of group product representations

$$
m^v = \prod_{p \in P} (N - p)^{d_p},
$$

but give only limited control over the number of terms $\sum |d_p|$, $p \in P$. To gain the best from the harmonic approach it would be desirable to employ integration.

The group $\widehat{Q}_1^*$ dual to Q_1^* is $U^{\pi(N) - \omega(N)}$, the direct product of $\pi(N) - \omega(N)$ copies of the complex unit circle. Here $\omega(N)$ is the number of distinct prime factors of N. With the topology on U induced by the standard topology on $\mathbb{C}^2$, the product topology on $\widehat{Q}_1^*$ renders the group compact. Lebesgue measure on $(\mathbb{R},+)$ induces a measure on $\mathbb{R}/\mathbb{Z}$. The appropriate product measure formed from the measure on $\mathbb{R}/\mathbb{Z}$ gives the Haar measure ν, on $\widehat{Q}_1^*$, that satisfies $\nu\widehat{Q}_1^* = 1$.

In practice, a typical character $g : Q_1^* \to U$ is determined by its value on each prime p, say $g(p) = \exp(\pi i \alpha_p)$, $0 \le \alpha_p < 1$. The integration $d\nu$ is then $d\alpha_2 d\alpha_3 \cdots d\alpha_p \cdots$ with $d\alpha_p$ Lebesgue measure, and the suffices $2, 3, \ldots$ running through the primes up to N that do not divide N.

If

$$S(g) = \sum_{\substack{p \le N \\ p \in P}} (g(N-p) + g(\overline{N-p})),$$

then the number $\mathfrak{N}$ of representations of the integer n in the form

$$n = \prod_{i=1}^{k} (N-p)^{\varepsilon_i}$$

with $\varepsilon_i = 1$ or -1, has itself a representation

$$\mathfrak{N} = \int_{\widehat{Q}_1^*} \overline{g(n)}\, S(g)^k d\nu(g).$$

Here k may be large compared to height of n.

Define a (translation invariant) metric ϱ on $\widehat{Q}_1^*$ by

$$\varrho(g,h) = \left(\sum_{\substack{p \le N \\ (p,N)=1}} \frac{1}{p} |g(p) - h(p)|^2 \right)^{1/2}.$$

Bearing in mind experience in probabilistic number theory, an analogue of the circle method developed by Hardy and Littlewood to study additive problems of the Waring type might proceed as follows.

We decompose $\widehat{Q}_1^*$ into major and minor cells. For an appropriate positive δ, the *major cells* would be the tubular neighbourhood

$$\left(g;\, \inf_{|\tau| \le T} \varrho(g, h_\tau) \le \delta \right)$$

where h_τ is the completely multiplicative function given by $h_\tau(p) = p^{i\tau}\chi(p)$ for a real τ, and primitive Dirichlet character $\chi(\mathrm{mod}\, D)$. To ensure membership of $\widehat{Q}_1^*$, it is understood that contrary to classical practice, on the primes dividing the modulus D, χ is to be given the value 1. Moreover, the maximum of D and T, which may be regarded as in some sense the order of the character h_T, is to be reasonably 'small' compared to N.

What remains of $\widehat{Q}_1^*$ after the major cells have been removed, we call the *minor cells*.

The expectation is that if k is large enough, then the contribution to the integral from the major cells overwhelms that from the minor cells.

Similarly the group $\widehat{Q}^*$, dual to Q^*, is the direct product of denumerably many copies of the complex unit circle. Again it is compact, and the product measure ν derived from Lebesgue measure restricted to $\mathbb{R}/\mathbb{Z}$ gives Haar measure 1 to $\widehat{Q}$. If the series of non-negative weights $\sum w_p$ converges, and we define

$$L(g) = \sum_p w_p g(p+1),$$

then there is a representation

$$\sum_{n=(p+1)/(q+1)} w_p w_q = \int_{\widehat{Q}^*} g(n)|L(g)|^2 d\nu(g),$$

and so on. For study of $\widehat{Q}$ a family of metrics

$$\left(\sum p^{-\lambda}|g(p) - h(p)|^2\right)^{1/2}, \quad \lambda > 1,$$

seems appropriate.

The additive problems studied by Hardy, Littlewood and Ramanujan concerned the group $(\mathbb{Z}, +)$. The dual of $\mathbb{Z}$ is $\mathbb{R}/\mathbb{Z}$. In their case(s) a natural metric on $\mathbb{R}/\mathbb{Z}$ is

$$\|\alpha - \beta\| = \min_{m \in \mathbb{Z}} |\alpha - \beta - m|.$$

A typical character $\mathbb{Z} \to U$ is given by $n \mapsto \exp(2\pi i \alpha n)$, where α is an element of $\mathbb{R}/\mathbb{Z}$. Localised solutions to (say) Goldbach's problems then require non-trivial estimates for sums of the form

$$\sum_{p \leq N} \exp(2\pi i \alpha p).$$

To obtain such estimates is a serious matter, cf. Vinogradov [34], Vaughan [33].

In our present circumstances $(\mathbb{Z}, +)$ has been replaced by $(Q^*, \times)$ or one of its subgroups, and $\mathbb{R}/\mathbb{Z}$ by a direct product of denumerably many copies of $\mathbb{R}/\mathbb{Z}$. The exponential sums involving primes are typically replaced by

$$\sum_{p<N} g(N-p) \quad \text{or} \quad \sum_p w_p g(p+1),$$

for multiplicative functions g with values in the complex unit disc. A number of comments concerning the estimation of such sums are made in ref. [18]. All current methods to estimate these sums lead to results weaker than are believed to be true. However, in this approach new difficulties might be expected.

Consider the problem of representing m^v, $m \geq 2$, in the form

$$m^v = \prod_{i=1}^{s} \left(\frac{an_i + b}{An_i + B} \right)^{\varepsilon_i},$$

where $aB \neq Ab$, as earlier. It is proved in ref. [9] that with a suitably chosen value of v, bounded in terms of only a, b, A and B, such a representation exists with each $n_i \leq c_1 m^2$ and $s \leq c_2(\log m)^2$. The values of c_1 and c_2 are uniform in all m for which any such a representation of m^v exists.

As we proved earlier, $s \geq c_3 \log m$ for some $c_3 > 0$ must hold.

In our harmonic approach, Q^* may be replaced by Q_2^*, the subgroup of Q^* constructed by removing the rationals whose canonical factorisation contains one of the finitely many exceptional primes described in §4. If we define

$$W(g) = 2\text{Re} \left(\sum_{\max(-b/a,-B/A)<n\leq c_1 m^2} g\left(\frac{an+b}{An+B} \right) \right),$$

then representations of the above type for m^v are counted by the integral

$$\mathfrak{N} = \int_{\widehat{Q}_2^*} g(m)^v W(g)^s d\nu(g),$$

where ν is projected from the Haar measure on $\widehat{Q}_1^*$. From what we know already, a direct analogue of the Hardy–Littlewood procedure must fail for small values of s, not because of insufficiently strong estimates for the sum $W(g)$, but because $\mathfrak{N} = 0$ for $s < c_3 \log m$. Nevertheless, $\mathfrak{N} > 0$ if $s \geq c_2(\log m)^2$. The number of terms needed in the representing product increases seriously with the size of the integer to be represented. This is a situation that does not arise in additive problems of Waring or Goldbach type.

A better appreciation of the application of harmonic analysis to the study of product representations would apparently require the routine consideration of large numbers of variables, and therefore likely the adoption of an aesthetic, not to say a methodology, from the study of the physical world.

11. Simultaneous representations

Let $F_j(x)$, $j = 1, \ldots, k$, be rational functions in $Q(x)$ that assume positive values for all integers $n \geq n_0$. Let $Q^{\otimes k}$ be the direct product of k copies of the group Q^*, and let Γ be the subgroup of $Q^{\otimes k}$ generated by the elements $F_1(n) \otimes F_2(n) \otimes \cdots \otimes F_k(n)$, $n \geq n_0$.

The quotient group $Q^{\otimes k}/\Gamma$ is a natural object with which to account the simultaneous representation of rationals:

$$t_1 = \prod_{i=1}^{s} F_1(n_i)^{\varepsilon_i}$$

$$\cdots$$

$$t_k = \prod_{i=1}^{s} F_k(n_i)^{\varepsilon_i}.$$

There are k-dimensional analogues of all eleven of the problems raised in §§1-9, as well as a k-dimensional analogue of the possible ideal method offered in §10. Elementary methods are less effective and progress so far has largely depended upon harmonic analysis.

Corresponding to each group homomorphism

$$g : Q^{\otimes k} \to Q^{\otimes k}/\Gamma \to U$$

there is a k-tuple of homomorphisms $g_j : Q^* \to U$, $j = 1, \ldots, k$ that satisfy

$$g_1(F_1(n))g_2(F_2(n)) \cdots g_k(F_k(n)) = 1$$

for all $n \geq n_0$. These are defined by $g_j(n) = g((1 \otimes \cdots \otimes n \otimes \cdots \otimes 1))$, where n lies in the jth coordinate, and 1 lies in all remaining $k - 1$ coordinates.

There is then a simultaneous product representation of the above kind if and only if whenever the g_j assume values in the group of roots of unity, $g_j(t_j) = 1$ for every j, $1 \leq j \leq k$, holds.

A natural analogue of Problem 3 would be

Problem 12 *For what multiplicative functions g_j, $j = 1, \ldots, k$, with values in the complex unit disc does the asymptotic mean-value*

$$\lim_{y \to \infty} y^{-1} \sum_{n \leq y} g_1(F_1(n)) \cdots g_k(F_k(n))$$

exist? In particular, when is the limit 1?

A strong version of Problem 10 would replace the $N - p$ by $F_1(p) \otimes \cdots \otimes F_k(p)$ where the coefficients of the rational functions $F_j(x)$ were allowed to depend in a suitably restricted manner upon N.

Rather than enumerate analogues to the earlier questions I offer two examples.

If once again

$$J(x) = (x + a_1)^{b_1} \cdots (x + a_h)^{b_h}$$

is the rational function in $Q(x)$ featured in the problem of Burgess considered in §4, and t is a positive integer, then *every pair of positive integers m_1, m_2 has a simultaneous representation*

$$m_1 = \prod_{j=1}^{s} R(n_j)^{\varepsilon_j}$$

$$m_2 = \prod_{j=1}^{s} R(n_j + t)^{\varepsilon_j}$$

with positive integers n_j and each $\varepsilon_j = +1$ or -1; cf. [9], Chapter 18. The proof features linear recurrences in modules, and asymptotic lower bounds for elliptic power sums.

Let $m_1 > m_2$. Since

$$\frac{m_1}{m_2} = \prod_{j=1}^{s} \left(\prod_{\ell=1}^{h} \left(\frac{n_j + a_\ell}{n_j + a_\ell + t} \right)^{b_\ell} \right)^{\varepsilon_j},$$

the number of terms needed to simultaneously represent the m_j is at least $c \log(m_1/m_2)$, where $c^{-1} = \sum_{\ell=1}^{h} |b_\ell| \log(t + 1)$.

Note that the case $m_1 = m_2$ is not trivial.

For the second example, let m_1 and m_2 be positive integers not divisible by 3. Then *there is a simultaneous representation*

$$m_1 = \prod_{i=1}^{s} (3n_i - 17)^{\varepsilon_i}$$

$$m_2 = \prod_{i=1}^{s} (3n_i + 19)^{\varepsilon_i}.$$

with integers $n_i \geq 7$, $\varepsilon_i = +1$ or -1, if and only if $m_1 \equiv m_2 \equiv 1, 4$ or $7 (mod\, 9)$ is satisfied; cf. [9] Theorem (19.2) and [10]. In this case $10^s \geq \max(m_1/m_2, m_2/m_1)$.

148

A number of unsolved problems are offered in [9], Chapter 23. Problem 8 of that chapter implicitly conjectures that every pair of positive integers m_1, m_2 has a simultaneous representation

$$m_1 = \prod_{i=1}^{s}(p_i + 1)^{\varepsilon_i}$$

$$m_2 = \prod_{i=1}^{s}(p_i + 3)^{\varepsilon_i}$$

with the p_i prime. At the moment, a method to estimate the corresponding character sum

$$\sum_{p \leq y} g_1(p+1)g_2(p+3)$$

is nowhere in sight. With this comment I close my short survey of a challenging and, to me, most interesting area. Further remarks may be found in the concluding section of my American Mathematical Society memoir, [14], and in the last chapter of my book, [16].

References

[1] Berrizbeitia, P., Elliott, P.D.T.A. (1998). On products of shifted primes. *The Ramanujan Journal* (Erdös Memorial issue), 2:219–223.

[2] Berrizbeitia, P., Elliott, P.D.T.A. (1999). Product bases for the rationals. *Canadian Math. Bull.*, 42:441–444.

[3] Britten, J.L. (1956–58). Solution of the word problem for certain types of groups, I. *Proc. Glasgow Math. Soc.*, 3:45–56.

[4] Dickson, L.E. (1904). A new extension of Dirichelt's theorem on prime numbers. *Messenger of Mathematics*, 33:155–161.

[5] Dress, F., Volkmann B. (1974). Ensembles d'unicité pour les fonctions arithmétiques additives ou multiplicatives. *C.R. Acad. Sci. Paris, Sér. A*, 287:43–46.

[6] Elliott, P.D.T.A. (1974). A conjecture of Kátai. *Acta Arith*, 26:11-20.

[7] Elliott, P.D.T.A. (1983a). On the distribution of the roots of certain congruences and a problem for additive functions. *J. Number Theory*, 16:267–282.

[8] Elliott, P.D.T.A. (1983b). On additive functions $f(n)$ for which $f(an + b) - f(cn + d)$ is bounded. *J. Number Theory*, 16:285–310.

[9] Elliott, P.D.T.A. (1985). Arithmetic Functions and Integer Products. *Grund. der math. Wiss.*, 272, Springer-Verlag, New York, Berlin, Heidelberg, Tokyo.

[10] Elliott, P.D.T.A. (1986a). Linear recurrences in modules. *Bull. London Math. Soc.*, 18:1–4.

[11] Elliott, P.D.T.A. (1986b). On the multiplicative group generated by a dense sequence of integers. *Monatshefte für Math.*, 102(1):3–6.

[12] Elliott, P.D.T.A. (1987). Persistence of form and the value group of reducible cubics. *Trans. American Math. Soc.*, 299(1):133–143.

[13] Elliott, P.D.T.A. (1993). Multiplicative functions on arithmetic progressions, VI: More Middle Moduli. *J. Number Theory*, 44(2):178–208.

[14] Elliott, P.D.T.A. (1994). *On the Correlations of Multiplicative and the Sum of Additive Arithmetic Functions. American Math. Soc. Memoir*, 538(112):88 pp.

[15] Elliott, P.D.T.A. (1995). The multiplicative group of rationals generated by the shifted primes, I. *J. reine angew. Math*, 463:169–216.

[16] Elliott, P.D.T.A. (1997). *Duality in Analytic Number Theory*. Cambridge Tracts in Mathematics, 122:350 pp. Cambridge Univ. Press.

[17] Elliott, P.D.T.A. (1998). Products of shifted primes. Multiplicative analogues of Goldbach's problems, II. *The Ramanujan Journal* (Erdös Memorial issue), 2:201–217.

[18] Elliott, P.D.T.A. (1999). Products of shifted primes: Multiplicative analogues of Goldbach's problem. *Acta Arithmetica*, 88(1):31–50.

[19] Elliott, P.D.T.A. (2000a). The multiplicative group of rationals generated by the shifted primes, II. *J. reine angew. Math*, 519:59–71.

[20] Elliott, P.D.T.A. (200b). The least prime primitive root and Linnik's theorem. in Number Theory for the Millenium, *Proc. Millenial Conference in Number Theory* (Urbana, IL, 2000). M. A. Bennett et al., editors. A. K. Peters, Boston.

[21] Elliott, P.D.T.A. Primes, products and polynomials. To appear in *Inventiones Mathematicae*.

[22] Elliott, P.D.T.A. Products of shifted primes: Multiplicative analogs of Goldbach's Problem, IV. *Preprint*.

[23] Elliott, P.D.T.A. Multiplicative functions on arithmetic progressions, VII: Large Moduli. To be published by the *London Math. Soc.*

[24] Erdös, P., Ruzsa, I.Z. (1980). On the small sieve, I: shifting by primes. *J. Number Theory*, 12:385–394.

[25] Fouvry, E. (1984). Autour du théorème de Bombieri–Vinogradov. *Acta Math*, 152:219–244.

[26] Kátai, I. (1968a). On sets characterizing number-theoretical functions. *Acta Arith*, 13:315–320.

[27] Kátai, I. (1968b). On sets characterizing number-theoretical functions, II. (The set of "prime plus one" 's is a set of quasi-uniqueness.) *Acta Arith*, 16:1–4.

[28] Kátai, I. (1969). Some results and problems in the theory of additive functions. *Acta Sc. Math. Szeged*, 30:305–311.

[29] Kátai, I. (1970). On number theoretical functions. *Colloquia Mathematica Societas Janos Bolyai*, 2:133–136. North–Holland, Amsterdam.

[30] Meyer, J. (1980). Ensembles d'unicité pour les fonctions additives. Étude analogue dans le cas des fonctions multiplicatives. *Journée de Théorie Analytique et Élémentaire des Nombres, Orsay, 2 et 3 Juin, 1980*. Publications Mathématiques d'Orsay, 50–66.

[31] Pan Chengdong, Pan Chengbaio. (1992). *Goldbach Conjecture*. Science Press, Beijing, China.

[32] Stilwell, J. (1982). The word problem and the isomorphism problem for groups. *Bull. American Math. Soc*, 6(1):33–56.

[33] Vaughan, R.C. (1981). *The Hardy-Littlewood method.* Cambridge Tracts in Mathematics, 80. Cambridge Univ. Press.

[34] Vinogradov, I.M. (1954). *The method of trigonometrical sums in the theory of numbers.* Translated from the Russian, revised and annotated by Davenport, A. and Roth, K.F. Interscience, New York.

[35] Wolke, D. (1978). Bemerkungen über Eindeutigkeitsmengen additiver Funktionen. *Elem. der Math,* 33:14–16.

ON THE DISTRIBUTION OF αp MODULO ONE

Chaohua Jia

Institute of Mathematics, Academia Sinica, Beijing 100080, P. R. China

jiach@math08.math.ac.cn

Abstract   In this article, some topics on the approximation to the real number by rational numbers are introduced. In particular, the author introduces the progress on the situation in which the denominators of rational numbers are prime numbers and describes some methods used.

Keywords: prime number, sieve method, trigonometric sum.

Early in ancient China, people had used $\frac{22}{7}$ or $\frac{355}{113}$ to approximate to π. For any real number, we have an approximation by the rational number as follows.

Theorem 1 (Dirichlet). *Let α be any given real number. Then for any $P > 1$, there is a positive integer $q(\leq P)$ and an integer a with $(a, q) = 1$ such that*

$$\left| \alpha - \frac{a}{q} \right| < \frac{1}{qP}.$$

As a consequence, if α is an irrational number, there is an infinite sequence of rational numbers $\frac{a}{q}$ such that

$$\left| \alpha - \frac{a}{q} \right| < \frac{1}{q^2}.$$

There is a yet better result.

Theorem 2 (Hurwitz). *For any irrational number α, there are infinitely many rational numbers $\frac{a}{q}$ such that*

$$\left| \alpha - \frac{a}{q} \right| < \frac{1}{\sqrt{5}q^2}.$$

151

This result is best possible. Taking $\alpha = \frac{1}{2}(\sqrt{5} - 1)$, we can show that for any constant $c < \frac{1}{\sqrt{5}}$, the inequality

$$\left| \alpha - \frac{a}{q} \right| < \frac{c}{q^2}$$

has finite solutions in a and q.

Therefore for further research, we have to restrict either α or $\frac{a}{q}$ respectively. To begin with restricting α, there is a classical result.

Theorem 3 (Liouville). *Assume that α is a real algebraic number of degree n. Then for any $\varepsilon > 0$, the inequality*

$$\left| \alpha - \frac{a}{q} \right| < \frac{1}{q^{n+\varepsilon}}$$

has finite solutions in a and q.

There is much progress on this problem. Finally Roth improved the exponent to $2 + \varepsilon$ which is best possible.

Then we assume α is an arbitrary irrational number and turn to restrict $\frac{a}{q}$, when many interesting topics arise. I take only one example.

Theorem 4 (D. R. Heath-Brown). *For any irrational number α, there are infinitely many square-free integers m and n such that*

$$\left| \alpha - \frac{m}{n} \right| < \frac{1}{n^{\frac{5}{3} - \varepsilon}}.$$

By Theorem 1, for any irrational number α and every $\varepsilon > 0$, there are positive integer q and integer a such that

$$|\alpha q - a| < \varepsilon.$$

Kronecker generalized this result.

Theorem 5 (Kronecker). *For any given irrational number α, real number γ and any $\varepsilon > 0$, there are positive integer q and integer a such that*

$$|\alpha q - a - \gamma| < \varepsilon.$$

If we would like to get more information on the distribution of αq in the neighborhood of any given real number γ, we should introduce the following definition.

Definition. *Let P_i $(i = 1, 2, \ldots)$ be a sequence of points in the interval $[0, 1)$. For any given a, b $(0 \leq a < b \leq 1)$, $Z[n; a, b]$ denotes the number of i $(1 \leq i \leq n)$ such that P_i falls into the interval $[a, b]$. If*

$$\lim_{n \to \infty} \frac{1}{n} \cdot Z[n; a, b] = b - a$$

always holds, then we say the sequence P_i is uniformly distributed.

In the case of a sequence of real numbers P_i ($i = 1, 2, \ldots$), we say P_i is uniformly distributed modulo one if their fractional parts $\{P_i\}$ is uniformly distributed.

H. Weyl gave a criterion on the uniform distribution. He connected the problem with trigonometric sums by the Fourier analysis so that the problem can be dealt with by the application of the estimate for trigonometric sums.

Theorem 6 (H. Weyl). *The sequence*

$$x_1, x_2, \ldots, x_n, \ldots, \qquad 0 \le x_i \le 1$$

is uniformly distributed, if and only if, for any integer $m \neq 0$, we have

$$\lim_{n \to \infty} \frac{1}{n} \sum_{i=1}^{n} e(mx_i) = 0,$$

where $e(y) = e^{2\pi i y}$.

In the following I shall sketch out the progress on the approximation

$$\left| \alpha - \frac{a}{p} \right| < \frac{1}{p^{1+\tau-\varepsilon}}$$

or the distribution of αp modulo one

$$\|\alpha p\| < \frac{1}{p^{\tau-\varepsilon}}, \tag{1}$$

where $\|x\|$ denotes the smallest distance from x to integers, p denotes the prime number and $\tau > 0$.

I. M. Vinogradov [7] first proved the inequality (1) with $\tau = \frac{1}{5}$ has infinitely many solutions of prime numbers p. He transformed the problem into the estimate for the trigonometric sum

$$\sum_{h=1}^{H} \left| \sum_{p \sim x} e(\alpha h p) \right| \tag{2}$$

by the Fourier analysis, where $p \sim x$ means there are positive constants c_1 and c_2 such that $c_1 x < p \le c_2 x$. Vinogradov applied his original ideas to give a non-trivial bound for the trigonometric sum

$$\sum_{p \sim x} e(\alpha h p).$$

He showed that the sum on the prime variables

$$\sum_{p\sim x} e(\beta p) \tag{3}$$

can be written as a collection of two kinds of sums.
   One is

$$\sum_{m\sim M} \sum_{n\sim N} a(m)b(n)e(\beta mn) \tag{4}$$

which is called the sum of type II, where $MN = x$, $a(m)$, $b(n) = O(1)$.
   The other is

$$\sum_{m\sim M} \sum_{n\sim N} a(m)e(\beta mn) \tag{5}$$

which is called the sum of type I, where $MN = x$, $a(m) = O(1)$. For
this sum, we have

$$\sum_{m\sim M} \sum_{n\sim N} a(m)e(\beta mn) \ll \sum_{m\sim M} \min\left(N, \frac{1}{\|\beta m\|}\right).$$

On average, the term

$$\min\left(N, \frac{1}{\|\beta m\|}\right)$$

is small so that we can get a non-trivial bound for the sum of type I.
   For the sum of type II, we apply Cauchy's inequality to transform it
into the sum of type I so that we can get a non-trivial bound. In this
way, Vinogradov deduced his conclusion.
   In 1977, Vaughan [6] got $\tau = \frac{1}{4}$. He still used Vinogradov's ideas but
treated the summation on h in a non-trivial way.
   In 1983, Harman [1] proved $\tau = \frac{3}{10}$. He introduced the sieve method
into this problem for the first time but had no improvement on the
estimate for trigonometric sums.
   Let

$$\mathcal{A} = \{n : \ x < n \le 2x, \ \|\alpha n\| < x^{-\tau+\varepsilon}\}, \qquad P(z) = \prod_{p<z} p.$$

We define the sieve function

$$S(\mathcal{A}, z) = \sum_{\substack{n\in\mathcal{A} \\ (n, P(z))=1}} 1.$$

Then

$$\#\{p : \ x < p \le 2x, \ \|\alpha p\| < x^{-\tau+\varepsilon}\} = S(\mathcal{A}, (2x)^{\frac{1}{2}}).$$

If we can prove

$$S(\mathcal{A}, (2x)^{\frac{1}{2}}) > 0, \tag{6}$$

then we deduce the conclusion.

By Buchstab's identity, we have

$$
\begin{aligned}
S(\mathcal{A}, (2x)^{\frac{1}{2}}) &= S(\mathcal{A}, x^{\varphi}) - \sum_{x^{\varphi} < p \le (2x)^{\frac{1}{2}}} S(\mathcal{A}_p, p) \\
&= S(\mathcal{A}, x^{\varphi}) - \sum_{x^{\varphi} < p \le (2x)^{\frac{1}{2}}} S(\mathcal{A}_p, x^{\varphi}) \\
&\quad + \sum_{\substack{x^{\varphi} < p \le (2x)^{\frac{1}{2}}}} \sum_{\substack{x^{\varphi} < q < p \\ q < (\frac{2x}{p})^{\frac{1}{2}}}} S(\mathcal{A}_{pq}, q), \tag{7}
\end{aligned}
$$

where φ is a certain positive constant.

We can transform sieve functions into trigonometric sums. By the estimate for the sums of type I and II, Harman got the asymptotic formula for the sum

$$\sum_{p \le P} S(\mathcal{A}_p, x^{\varphi}). \tag{8}$$

We hope to make φ as big as possible in order to get more asymptotic formulas. In the application of sieve method in some other problems, one has to take $\varphi = \varepsilon$ if there is only the estimate for the sum of type I but not for type II.

Now we can get asymptotic formulas for the first two sums on the right side of (7). For the third sum in (7), we divide it into three parts. In the first part, we use the estimate for the sum of type II to get asymptotic formulas, which is the best situation. In the second part, under some conditions, we can decompose the sums further. In the third part, we use the positivity of sieve functions to replace them by 0. Repeating this procedure, we can get a positive lower bound of $S(\mathcal{A}, (2x)^{\frac{1}{2}})$ so that the conclusion follows.

In 1993, Jia [4] improved some techniques in the above discussion to get $\tau = \frac{4}{13}$.

In 1996, Harman [2] made further innovation with the sieve method to get $\tau = \frac{7}{22}$. His major idea is that when $S(\mathcal{A}_p, p)$ can not be dealt with, one can write

$$\sum_{p \sim P} S(\mathcal{A}_p, p) = \sum_{p \sim P} \left(\sum_a 1 \right) = \sum_a \left(\sum_{p \sim P} 1 \right)$$

and then applies the sieve method to the sum $\sum_{p \sim P} 1$.

Harman mentioned in his paper that the exponent $\tau = \frac{1}{3}$ is the limit of all currently known methods unless one assumes very strong results on primes in arithmetic progressions.

In 2000, Jia [5] proved $\tau = \frac{9}{28}$.

In 2002, Heath-Brown and Jia [3] made a new progress toward $\tau = \frac{1}{3}$. We got $\tau = \frac{16}{49} = 0.3265\ldots$.

In that paper, a new breakthrough on the estimate for trigonometric sums was made so that we can deal with the tri-linear sum

$$\sum_{m \sim M} a(m) \sum_{r \sim R} \alpha(r) \sum_{s \sim S} \beta(s) f(m, rs), \tag{9}$$

where $f(m, n)$ is a suitable expression.

If $\beta(s) \equiv 1$, we apply the estimate for the Kloosterman sum in (9). For the estimate for general tri-linear sum, we transform it into the estimate for $N_k(A, B)$ which is the number of solutions of the Diophantine equation

$$\frac{a_1}{b_1} + \cdots + \frac{a_k}{b_k} = \frac{a_{k+1}}{b_{k+1}} + \cdots + \frac{a_{2k}}{b_{2k}}, \tag{10}$$

where $a_i \sim A$, $b_i \sim B$. By some theorems on the geometry of numbers, we get an upper bound

$$N_k(A, B) \ll B^\varepsilon (A^k B^k + A^{2k-1} B).$$

By Vaughan's identity, we transform the sum of type II into the tri-linear sum. Then the above estimate for the tri-linear sum works and a good bound for the sum of type II follows.

Combining this new estimate for trigonometric sums with some developing techniques on the sieve method, we get the exponent $\tau = \frac{16}{49}$ at last.

Recently, Mikawa announced that he had proved $\tau = \frac{1}{3}$. His research will open new prospects for the study on this topic.

Let us review the progress on this problem. Write

$$\tau = \frac{n}{3n + 1}. \tag{11}$$

Then

$n = \frac{1}{2}$, Vinogradov [7].

$n = 1$, Vaughan [6].

$n = 3$, Harman [1].

$n = 4$, Jia [4].

$n = 7$, Harman [2].

$n = 9$, Jia [5].

$n = 16$, Heath-Brown and Jia [3].

$n = \infty$, an old destination and a new starting point.

References

[1] G. Harman, On the distribution of αp modulo one, *J. London Math. Soc.* (2), **27** (1983), 9–18.

[2] G. Harman, On the distribution of αp modulo one II, *Proc. London Math. Soc.* (3), **72** (1996), 241–260.

[3] D. R. Heath-Brown and C. Jia, The distribution of αp modulo one, *Proc. London Math. Soc.* (3), **84** (2002), 79–104.

[4] C. Jia, On the distribution of αp modulo one, *J. Number Theory*, **45** (1993), 241–253.

[5] C. Jia, On the distribution of αp modulo one (II), *Sci. China, Ser. A*, **43** (2000), 703–721.

[6] R. C. Vaughan, On the distribution of αp modulo 1, *Mathematika*, **24** (1977), 135–141.

[7] I. M. Vinogradov, *The method of trigonometric sums in the theory of numbers*, translated from the Russian by K. F. Roth and A. Davenport (Wiley-Interscience, London, 1954).

RAMANUJAN'S FORMULA AND MODULAR FORMS

Shigeru Kanemitsu*
Graduate School of Advanced Technology, University of Kinki, Iizuka, Fukuoka, 820-8555, Japan
kanemitu@fuk.kindai.ac.jp

Yoshio Tanigawa*
Graduate School of Mathematics, Nagoya University, Nagoya, 464-8602, Japan
tanigawa@math.nagoya-u.ac.jp

Masami Yoshimoto*
Graduate School of Mathematics, Nagoya University, Nagoya, 464-8602, Japan
x02001n@math.nagoya-u.ac.jp

Dedicated to Professor Jonas Kubilius on his eightieth birthday

Abstract   In the theory of zeta-functions, which are defined wherever there are defined norms or substitutes thereof, the ingredients – modular relations, functional equations, incomplete gamma series, and the like – are placed like nodes on the woofs. Some of them are woven by warps as Hecke theory or Lavrik's theory. The former connects the modular relation to the functional equation, thus making it possible to go to and from between the more orderly world of automorphic forms and the less orderly one of zeta-functions while the latter relates functional equations and incomplete gamma series in the same vein, the idea originating from Riemann. We have found a warp stitching all of these nodes-ingredients, enabling us to warp from one node to another as well as providing us with a guiding principle to locate the exact position and direction of research, a guiding thread to give a clear picture of the whole scene through opaque mist of complexity. We shall illustrate the principle by examples of various zeta-functions satisfying Hecke's functional equation, *i.e.* the one with a single gamma factor, in which category many of

*The authors are supported by Grant-in-Aid for Scientific Research No. 14540051, 14540021 and 14005245 respectively.

the important zeta-functions are contained, notably, the Riemann zeta-, Dirichlet L-, Epstein zeta-, the automorphic zeta-functions, etc. In particular, we shall be concerned with the automorphic zeta-functions, the zeta functions arising from automorphic forms, evaluating their special values and obtaining incomplete gamma series.

Keywords: functional equation, Hecke theory, incomplete gamma series, modular relation, Ramanujan's formula, Riemann-Siegel integral formula

1. Introduction and the equivalent statements to functional equations

One of the most famous formulas of S. Ramanujan is decidedly the one connecting the special values of the Riemann zeta-function at odd, positive integer arguments to Lambert series, which are rapidly convergent and most amenable to numerical calculations (cf. §2). In the statement there appears the condition of the form $\alpha\beta = \pi^2$ ($\alpha > 0$, $\beta > 0$). We started these lines of research that we are stating partially in the present article, from the naive question as to why such a condition is needed, then went on to find that it is the very modular relation in disguised form (avatar), proceeding further on to discover that there is a guiding thread (of Ariadone or of the Fates as the case may be) which penetrates the labyrinth of the theory of zeta-functions. It is the line starting from the more orderly world of automorphic forms to less orderly world of zeta-functions by integration, known as Hecke theory, and then go down to the series expressions with incomplete gamma function coefficients which we refer to as the *incomplete gamma (function) series (expression)* and which involves already the Riemann-Siegel (integral) formula in some cases as well as approximate functional equations.

This last expression that incorporates both functional equations and modular relations has been considered as discovered by A. F. Lavrik [91], but there were important contributions to this theory by Russian mathematicians including R. O. Kuz'min [85, 86], and Ju. V. Linnik [92, 93] prior to Lavrik. As explained in §5.1, the results of Kuz'min and Linnik complementary to each other have been unified by Lavrik by the introduction of the extraneous function whose appropriate choice gives as extremal cases the results of Kuz'min and Linnik.

Our new principle is to the effect that we may go up and down on this route freely by differentiation and integration, so that we might call it "susmna principle", or "Wu-xin principle", or "the four elements principle" originating from the ancient Hindu or Chinese or Western principle:

automorphic forms

$\uparrow\downarrow$

$\varphi(s)$ – modular relation – $\psi(\delta - s)$

$\uparrow\downarrow$

incomplete gamma series

$\uparrow\downarrow$

Riemann-Siegel

$\uparrow\downarrow$

?

This scheme, though speculative, gives a very clear picture as to what one is working with, enables one to fit fragmental pieces scattered among literature over 200 years to their proper places, and renders one somehow perceptible of the meaning of many otherwise unpredictable aspects in zeta-function theory.

Moreover, it sets aside the functional equations as a side-effect, placing the modular principle in the central line, suggesting of the "niryana"–the true body of the entity.

Our objective in this paper is to illustrate this principle in the setting of cusp forms starting from a new interpretation of Ramanujan's formula, explaining the meaning of the title.

More specifically, we shall prove, within Hecke theory, that from the celebrated Ramanujan formula (2.1) for $n > 0$ follows the case $n < 0$ of the same by "differentiation", that the negative case of (2.1) is nothing but the automorphic property of the corresponding Eisenstein series $G_{2n+2}(z)$, and that the case $n = 0$ gives rise to an elucidation of the situation surrounding the Dedekind eta-function.

The above-mentioned transition from the positive case to the negative case is made by the Mellin transform with shifted argument (cf. Remark 2.3) in which the fact is incorporated that the differentiation of the Lambert series (2.4) corresponds essentially to the replacement of $\varphi(s)$ by $\varphi(s - 1)$ (cf. (2.16)). It turns our that this last line of argument was already pointed out by M. Razar [102] (p.6, ll.2–3) who proved the special case of Theorem 3.1 below in the case where $a = \kappa - 1$, as the evaluation of the Eichler integral, the $(\kappa - 1)$-fold integral of the cusp form f, of a somewhat more general nature than ours, in terms of the values of the corresponding Dirichlet series $L(s, f)$. Razar [102], however, did not make use of the functional equation and the procedure may be different from ours, which is indeed a reverse one to Razar's, performing differentiation and making use of the functional equation. A. Weil [126] has made similar work on Eichler integrals (cf. Remark 2.5). L. J. Goldstein [50] has introduced the notion of generalized Eichler integrals and obtained expressions for special values of ray class zeta-functions of real quadratic fields, which was used by Goldstein and Razar [52] to study the values $\zeta(2k-1)$, at the end of whose paper they remarked that their Corollary 3.4 reduces to the Ramanujan formula.

We might thus say that our standpoint is to try to extract as much elixir as possible from the Ramanujan formula and show that it can be an ideal model for building up similar theories in other fields. Thus in §3 we shall prove a rapidly convergent series expression for the zeta-function $L(s, f)$ associated to a cusp form f.

In §4 we shall go out of Hecke theory and prove the Riemann-Siegel integral formula for $L(s, f)$ above in two ways, one due to R. O. Kuz'min, the other to N. S. Koshlyakov in the spirit of A. F. Lavrik (§4 and §5.1).

It is possible to state a more general framework, but we shall confine ourselves to the case of Hecke's functional equation, *i.e.* the functional equation with a single gamma factor.

Notation. In this paper we shall use the following standard notation. Let $\Gamma(s, z)$ and $K_s(z)$ denote the incomplete gamma function of the second kind and the modified Bessel function given by

$$\Gamma(s, z) = \int_z^\infty e^{-u} u^{s-1} du, \quad \Re z > 0, \Re s > 0 \tag{1.1}$$

$$K_\nu(z) = \frac{1}{2}\left(\frac{z}{2}\right)^\nu \int_0^\infty e^{-t-\frac{z^2}{4t}} t^{-\nu-1} dt, \quad \Re\nu > -\frac{1}{2}, |\arg z| < \frac{\pi}{4}, \tag{1.2}$$

respectively. We also let $W_{\nu,\mu}(x)$ be the Whittaker function defined by, for example,

$$W_{\mu,\nu}(z) = \frac{e^{-\frac{z}{2}} z^\mu}{\Gamma\left(\nu - \mu + \frac{1}{2}\right)} \int_0^\infty e^{-t} t^{\nu-\mu-\frac{1}{2}} \left(1 + \frac{t}{z}\right)^{\nu+\mu-\frac{1}{2}} dt, \tag{1.3}$$

for $\Re\left(\nu - \mu + \frac{1}{2}\right) > 0$ and $|\arg z| < 3\pi/2$.

Definition. Let

$$0 < \lambda_1 < \lambda_2 < \cdots \to \infty, 0 < \mu_1 < \mu_2 < \cdots \to \infty$$

be increasing sequences of real numbers. For complex sequences $\{a_n\}$, $\{b_n\}$ form the Dirichlet series

$$\varphi(s) = \sum_{n=1}^\infty \frac{a_n}{\lambda_n^s} \quad \text{and} \quad \psi(s) = \sum_{n=1}^\infty \frac{b_n}{\mu_n^s}$$

which we assume are absolutely convergent for $\sigma > \sigma_a^*$ and $\sigma > \sigma_b^*$, respectively. Then $\varphi(s)$ and $\psi(s)$ are said to satisfy Hecke's functional equation

$$A^{-s}\Gamma(s)\varphi(s) = A^{-(\delta-s)}\Gamma(\delta-s)\psi(\delta-s) \tag{1.4}$$

if there exists a regular function $\chi(s)$ outside of a compact set S such that

$$\chi(s) = A^{-s}\Gamma(s)\varphi(s), \quad \sigma > \alpha \; (\geq \sigma_a^*)$$

and

$$\chi(s) = A^{-(\delta-s)}\Gamma(\delta - s)\psi(\delta - s), \quad \sigma < \beta \; (\leq \delta - \sigma_b^*)$$

and such that $\chi(s)$ is convex in the sense that

$$e^{-\varepsilon|t|}\chi(\sigma + it) = O(1), \quad 0 < \varepsilon < \frac{\pi}{2},$$

uniformly in σ, $\sigma_1 \leq \sigma \leq \sigma_2$, $|t| \to \infty$.

Following Bochner [26], define the residual function

$$P(x) = \frac{1}{2\pi i}\int_C \chi(s)x^{-s}ds,$$

where C encircles all the singularities of $\chi(s)$ in S.

The following theorem constitutes the basis of all that follows.

Theorem 1.1. *The functional equation (1.4), Bochner's modular relation (1.5) below, Lavrik's incomplete gamma series (1.6) below and Berndt's Bessel series (1.7) below are all equivalent:*

$$\sum_{n=1}^{\infty} a_n e^{-\lambda_n x} = \left(\frac{A}{x}\right)^{\delta}\sum_{n=1}^{\infty} b_n e^{-\mu_n \frac{A^2}{x}} + P\left(\frac{x}{A}\right) \tag{1.5}$$

with $\Re x > 0$,

$$A^{-s}\Gamma(s)\varphi(s) = A^{-s}\sum_{n=1}^{\infty}\frac{a_n}{\lambda_n^s}\Gamma(s, Aw\lambda_n)$$

$$+ A^{-(\delta-s)}\sum_{n=1}^{\infty}\frac{b_n}{\mu_n^{\delta-s}}\Gamma(\delta - s, Aw^{-1}\mu_n) + \int_0^w P(x)x^{s-1}dx \tag{1.6}$$

with $\Re w > 0$, *and*

$$A^{-s}\Gamma(s)\varphi(s,\alpha) = 2\alpha^{\frac{\delta-s}{2}}\sum_{n=1}^{\infty}b_n\mu_n^{-\frac{\delta-s}{2}}K_{s-\delta}(2A\sqrt{\alpha\mu_n})$$

$$+ A^{-s}\int_0^{\infty}e^{-\alpha u}u^{s-1}P\left(\frac{u}{A}\right)du \tag{1.7}$$

with $\alpha > 0, \sigma > \max\{\delta - \frac{1}{2}, -1\}, s \neq 0$, *where*

$$\varphi(s,\alpha) = \sum_{n=1}^{\infty}\frac{a_n}{(\lambda_n + \alpha)^s} \tag{1.8}$$

denotes the Hurwitz type (generalized) Dirichlet series associated to φ.

Proof. The equivalence of the functional equation and the modular relation is proved by Bochner [26]. The equivalence of the modular relation and the incomplete gamma series has been given in [77] complementing Lavrik's observation that differentiation of (1.7) implies (1.6), *i.e.* we have proved that (1.7) follows from (1.6) by integration with respect to x over $(0, Aw)$ $(w > 0)$ after multiplying both sides by x^{s-1}, plus analytic continuation.

The equivalence of (1.4) and (1.7) is first given by Berndt [11] and then a different proof is given in [13] and a generalization is given [14] (cf. §5.3).

We may deduce (1.7) quite easily either from (1.5) by integrating (1.5) over $(0, \infty)$ after multiplying $e^{-\alpha x} x^{s-1}$, plus analytic continuation, or from (1.6) by the Laplace transform.

Finally we note that (1.7) implies (1.4) by letting $\alpha \to 0+$ in the case $\Re s < \Re \alpha - \kappa$ whence the assertion of the theorem follows. $\qquad\square$

Remark 1.1. In its general form, with multiple gamma factors, the equivalence of three assertions (1.5), (1.6) and (1.7) is given in Lavrik [91], but as the title suggests, the most important parameter τ $(\Re \tau > 0)$ is an "extraneous" one which does not appear in (1.4) nor (1.5), which suggests that Hecke correspondence was not perceived as in our context, $i\tau$ being to be regarded as the variable of automorphic functions in the upper half-plane.

As is remarked in [77], Lavrik refers to the modular relation only in this paper [91] (see the lines 10–11, p.521, which reads "the general form of the zeta-function equation (3) is formulated for the first time here, ..."), he may have missed the works of Bochner et al. Thus, Lavrik's works may be considered as on the lines of (Riemann), Kuz'min, Linnik, missing the work of Koshlyakov (cf. §5.2).

We remark that Terras [118, 121, 122] develops the equivalence of the functional equation (1.4) and the incomplete gamma expression (in her terms) (1.6) with $w = 1$.

Although we did not state a more general case of (1.4) in which a factor, say c, appears on the right-side, we may incorporate it in the equivalent assertions (1.5), (1.6) and (1.7) by multiplying the right-side members by c (cf. [76], [74] and Example 1.2 below).

We shall now give some examples illustrating Theorem 1.1.

Example 1.1. The Riemann zeta-function

$$\zeta(s) = \prod_{p:\text{prime}} (1 - p^{-s})^{-1} = \sum_{n=1}^{\infty} n^{-s}, \quad \sigma > 1 \qquad (1.9)$$

viewed as one satisfying (1.4) (or rather (1.16) below) with

$$a_n = b_n = 1, \ \lambda_n = \mu_n = n^2, \ A = \pi, \ \delta = \frac{1}{2}$$

and

$$P(x) = \frac{1}{2}\left(x^{-\frac{1}{2}} - 1\right).$$

Equivalent assertions read

$$\pi^{-s}\Gamma(s)\zeta(2s) = \pi^{-\left(\frac{1}{2}-s\right)}\Gamma\left(\frac{1}{2} - s\right)\zeta(1 - 2s), \tag{1.10}$$

$$\sum_{n=1}^{\infty} e^{-\pi n^2 x} = x^{-\frac{1}{2}}\sum_{n=1}^{\infty} e^{-\frac{\pi n^2}{x}} + \frac{1}{2}\left(x^{-\frac{1}{2}} - 1\right), \quad \Re x > 0, \tag{1.11}$$

$$\pi^{-s}\Gamma(s)\zeta(2s) = \pi^{-s}\sum_{n=1}^{\infty}\frac{1}{n^{2s}}\Gamma(s, \pi n^2 w)$$

$$+ \pi^{s-\frac{1}{2}}\sum_{n=1}^{\infty}\frac{1}{n^{1-2s}}\Gamma\left(\frac{1}{2} - s, \frac{\pi n^2}{w}\right) + \frac{w^{s-\frac{1}{2}}}{2s-1} - \frac{w^s}{2s}, \quad \Re w > 0, \tag{1.12}$$

$$\pi^{-s}\Gamma(s)\sum_{n=1}^{\infty}\frac{1}{(n^2 + \alpha^2)^s} = 2\alpha^{\frac{1}{2}-s}\sum_{n=-\infty}^{\infty}|n|^{s-\frac{1}{2}}K_{s-\frac{1}{2}}(2\pi\alpha|n|), \tag{1.13}$$

where the term corresponding to $n = 0$ on the right-hand side is to be understood to mean

$$\lim_{u\to 0} u^{s-\frac{1}{2}}K_{s-\frac{1}{2}}(2\pi\alpha u) = \frac{\sqrt{\pi}}{2}\alpha^{\frac{1}{2}-s}\Gamma\left(s - \frac{1}{2}\right). \tag{1.14}$$

Only (1.13) needs some explanation. In accordance with $\lambda_n = n^2$, we write α^2 for α. Then $\varphi(s, \alpha^2)$ in (1.8) becomes $\sum_{n=1}^{\infty}\frac{1}{(n^2+\alpha^2)^s}$. Since

$$A^{-s}\int_0^\infty e^{-\alpha^2 u}u^{s-1}P\left(\frac{u}{\pi}\right)du = \frac{1}{2}\sqrt{\pi}\alpha^{1-2s}\Gamma\left(s - \frac{1}{2}\right) - \frac{1}{2}\alpha^{-2s}\Gamma(s),$$

we obtain for (1.7)

$$\pi^{-s}\Gamma(s)\sum_{n=1}^{\infty}\frac{1}{(n^2 + \alpha^2)^s} = 2\alpha^{\frac{1}{2}-s}\sum_{n=1}^{\infty}n^{s-\frac{1}{2}}K_{s-\frac{1}{2}}(2\pi\alpha n)$$

$$+ \frac{1}{2}\pi^{-s+\frac{1}{2}}\alpha^{1-2s}\Gamma\left(s - \frac{1}{2}\right) - \frac{1}{2}\pi^{-s}\alpha^{-2s}\Gamma(s). \tag{1.15}$$

Moving the last term on the right of (1.15) to the left and adopting the interpretation (1.14) completes the proof of (1.13).

Remark 1.2. (i) Functional equation (1.10) is satisfied by $\zeta(2s)$ not by $\zeta(s)$ which satisfies Riemann's functional equation

$$\pi^{-\frac{s}{2}}\Gamma\left(\frac{s}{2}\right)\zeta(s) = \pi^{-\frac{1-s}{2}}\Gamma\left(\frac{1-s}{2}\right)\zeta(1-s). \tag{1.16}$$

After Hecke, it has been universally adopted to think of the Riemann zeta-function as a cusp form zeta-function $\zeta(2s)$ satisfying (1.10), but as (1.13) reveals, there is rather a big difference between these two types of functional equation and zeta-functions satisfying them (for this, cf. 4.1 below).

(ii) The incomplete gamma series expression (1.12) for $\zeta(s)$ reads

$$\pi^{-\frac{s}{2}}\Gamma\left(\frac{s}{2}\right)\zeta(s) = \pi^{-\frac{s}{2}}\sum_{n=1}^{\infty}\frac{1}{n^s}\Gamma\left(\frac{s}{2}, \pi n^2 w\right)$$

$$+ \pi^{-\frac{1-s}{2}}\sum_{n=1}^{\infty}\frac{1}{n^{1-s}}\Gamma\left(\frac{1-s}{2}, \frac{\pi n^2}{w}\right) + \frac{w^{\frac{s-1}{2}}}{s-1} - \frac{w^{\frac{s}{2}}}{s}, \quad \Re w > 0. \tag{1.17}$$

For Kuz'min's important observation, which is markedly far-reaching, cf. 4.1, the basis of which is (1.17).

It amounts to saying that (1.17) is valid in the limiting case of $w = i$ ($\Re w = 0$), the infinite series being absolutely convergent, and that the Riemann-Siegel integral formula follows from it. Formula (1.17) with $w = i$ reads

$$\pi^{-\frac{s}{2}}\Gamma\left(\frac{s}{2}\right)\zeta(s) = \frac{i^{\frac{s-1}{2}}}{s-1} - \frac{i^{\frac{s}{2}}}{s} + \varepsilon^s \sum_{n=1}^{\infty}\int_1^{\infty} e^{-\pi n^2 x i}x^{\frac{s}{2}-1}dx$$

$$+ \varepsilon^{s-1}\sum_{n=1}^{\infty}\int_1^{\infty} e^{\pi n^2 x i}x^{\frac{1-s}{2}-1}dx, \tag{1.18}$$

where $\varepsilon = e^{\frac{\pi}{4}i}$.

Note that if we choose in (1.17)

$$w = \Delta^2 \exp\left\{i\left(\frac{\pi}{2} - \frac{1}{|t|}\right)\frac{t}{|t|}\right\}, \quad \Delta > 0.$$

following Lavrik [91], we obtain the approximate functional equation due to Hardy and Littlewood

$$\zeta(s) = \sum_{n \le x}\frac{1}{n^s} + \pi^{s-\frac{1}{2}}\frac{\Gamma\left(1 - \frac{s}{2}\right)}{\Gamma\left(\frac{s}{2}\right)}\sum_{n \le y}\frac{1}{n^{1-s}} + O\left(x^{-\sigma}\right) + O\left(|t|^{\frac{1}{2}-\sigma}y^{\sigma-1}\right) \tag{1.19}$$

for $0 < \sigma < 1$, $x > h$, $y > h$ with $2\pi xy = |t|$, for a fixed constant $h > 0$. For approximate functional equations, cf. §5.2.

(iii) For (1.13) confer [77] where Watson-Guinand's formula is deduced.

Example 1.2. Let $Q[\underline{x}]$ $(\underline{x} = (x_1, \ldots, x_m) \in \mathbb{R}^m)$ denote a positive definite quadratic form with coefficient matrix Q in m variables. Denoting the inner product of two vectors $\underline{x}$, $\underline{\delta}$ by $(\underline{x}, \underline{\delta})$, we introduce the Epstein zeta-function (of the Hurwitz-Lerch type) $Z \left| \begin{matrix} \gamma \\ \underline{\delta} \end{matrix} \right| (s)_Q$ by

$$Z \left| \begin{matrix} \gamma \\ \underline{\delta} \end{matrix} \right| (s)_Q = Z \left| \begin{matrix} \gamma_1 \ldots \gamma_m \\ \delta_1 \ldots \delta_m \end{matrix} \right| (s)_Q$$

$$= {\sum_{\underline{n} \in \mathbb{Z}^m}}' \frac{e^{2\pi i (\underline{n}, \underline{\delta})}}{Q[\underline{n} + \underline{\gamma}]^{\frac{s}{2}}}, \quad \sigma > m, \tag{1.20}$$

where $\underline{\gamma} = (\gamma_1, \ldots, \gamma_m)$ and $\underline{\delta} = (\delta_1, \ldots, \delta_m)$ are given vectors and the prime on the summation sign means that the case $\underline{n} + \underline{\gamma} = \underline{0}$ is to be omitted (which can occur only when $\underline{\gamma} \in \mathbb{Z}^m$).

The associated theta series is

$$\vartheta \left| \begin{matrix} \gamma \\ \underline{\delta} \end{matrix} \right| (0, z)_Q = \vartheta \left| \begin{matrix} \gamma_1 \ldots \gamma_m \\ \delta_1 \ldots \delta_m \end{matrix} \right| (0, z)_Q$$

$$= \sum_{\underline{n} \in \mathbb{Z}^m} e^{-\pi z Q[\underline{n} + \underline{\gamma}] + 2\pi i (\underline{n}, \underline{\delta})}. \tag{1.21}$$

With slight modification of the statements of Theorem 1, we may state equivalent assertions;

$$\pi^{-s} \Gamma(s) Z \left| \begin{matrix} \gamma \\ \underline{\delta} \end{matrix} \right| (2s)_Q$$

$$= e^{-2\pi i (\underline{\gamma}, \underline{\delta})} |Q|^{-\frac{1}{2}} \pi^{-\frac{m}{2} + s} \Gamma \left(\frac{m}{2} - s \right) Z \left| \begin{matrix} \underline{\delta} \\ -\underline{\gamma} \end{matrix} \right| (m - 2s)_{Q^{-1}}, \tag{1.22}$$

$$\vartheta \left| \begin{matrix} \gamma \\ \underline{\delta} \end{matrix} \right| (0, z)_Q = e^{-2\pi i (\underline{\gamma}, \underline{\delta})} |Q|^{-\frac{1}{2}} z^{-\frac{m}{2}} \vartheta \left| \begin{matrix} \underline{\delta} \\ -\underline{\gamma} \end{matrix} \right| (0, z^{-1})_{Q^{-1}}, \quad \Re z > 0, \tag{1.23}$$

$$\pi^{-s}\Gamma(s)Z\left|\begin{matrix}\underline{\gamma}\\\underline{\delta}\end{matrix}\right|(2s)_Q$$

$$= \pi^{-s}\sideset{}{'}\sum_{\underline{n}\in\mathbb{Z}^m}\frac{e^{2\pi i(\underline{n},\underline{\delta})}}{(Q[\underline{n}+\underline{\gamma}])^s}\Gamma(s,\pi wQ[\underline{n}+\underline{\gamma}])$$

$$+ |Q|^{-\frac{1}{2}}\pi^{-\frac{m}{2}+s}\sideset{}{'}\sum_{\underline{n}\in\mathbb{Z}^m}\frac{e^{-2\pi i(\underline{n}+\underline{\delta},\underline{\gamma})}}{(Q^{-1}[\underline{n}+\underline{\delta}])^{\frac{m}{2}-s}}\Gamma\left(\frac{m}{2}-s,\pi\frac{Q^{-1}[\underline{n}+\underline{\delta}]}{w}\right)$$

$$- \varepsilon(\underline{\gamma})\frac{w^s}{s} + \varepsilon(\underline{\delta})e^{-2\pi i(\underline{\gamma},\underline{\delta})}|Q|^{-\frac{1}{2}}\frac{w^{s-\frac{m}{2}}}{s-\frac{m}{2}}, \quad \Re w > 0, \tag{1.24}$$

and

$$\pi^{-s}\Gamma(s)\sum_{\substack{\underline{n}\in\mathbb{Z}^n\\Q[\underline{n}+\underline{\gamma}]\neq\underline{0}}}\frac{e^{2\pi i(\underline{n},\underline{\delta}}}{(Q[\underline{n}+\underline{\gamma}]+\alpha)^s}$$

$$= 2\alpha^{\frac{1}{2}(\frac{m}{2}-s)}\sum_{\substack{\underline{n}\in\mathbb{Z}^n\\Q^{-1}[\underline{n}+\underline{\delta}]\neq\underline{0}}}\frac{e^{-2\pi i(\underline{n},\underline{\gamma})}}{Q^{-1}[\underline{n}+\underline{\delta}]^{\frac{1}{2}(\frac{m}{2}-s)}}K_{\frac{m}{2}-s}(2\pi\sqrt{\alpha Q^{-1}[\underline{n}+\underline{\delta}]})$$

$$- \varepsilon(\underline{\gamma})(\pi\alpha)^{-s}\Gamma(s) + \varepsilon(\underline{\delta})(\pi\alpha)^{-s}e^{-2\pi i(\underline{\gamma},\underline{\delta})}|Q|^{-\frac{1}{2}}\Gamma\left(s-\frac{m}{2}\right), \tag{1.25}$$

where

$$\varepsilon(\underline{\gamma}) = \begin{cases}1 & \underline{\gamma}\in\mathbb{Z}^m\\0 & \underline{\gamma}\notin\mathbb{Z}^m.\end{cases}$$

The Bessel series expansion for $Z\left|\begin{matrix}\underline{\gamma}\\\underline{\delta}\end{matrix}\right|(s)_Q$ corresponding to (1.7) is rather complicated, and we do not state it here, referring the reader to Terras [115] and Berndt [17].

Remark 1.3. (i) The special case $\underline{\gamma}\in\mathbb{Z}^m$, $\underline{\delta}=\underline{0}$ and $w=1$ of (1.24) (multiplied by 2) coincide with Terras' Theorem 2.

(ii) Fundamental references on the (m-ary) Epstein zeta-function are Epstein [41], Berndt [17] and Terras [121, 122]. Epstein's original papers contain already a lot of information. As will be referred to in §5.3, Berndt's paper [17] is important but not duly mentioned in the existing vast literature. Terras' books [122] contain great many applications including Madelung constants (cf. [28], [31], [48]), but the general lines are developments of her papers [115, 121].

(iii) When Q is a positive definite matrix with integer entries, the Epstein zeta-function with Dirichlet character χ

$$Z(s,\chi,Q) = \sideset{}{'}\sum_{\underline{n}\in\mathbb{Z}^m}\frac{\chi(Q[\underline{n}])}{Q[\underline{n}]^{\frac{s}{2}}}$$

is another important object of study, and there have been results due to H. M. Stark and A. Terras [121] et al.

(iv) The special case of binary quadratic forms are most fundamental and have been studied rather extensively. We refer to the above papers of Berndt [17], Terras [121] and the papers of Chowla and Selberg (cf. Kurokawa [84]), Bateman-Grosswald, Rankin, Stark therein referred to.

Example 1.3. The Hurwitz-Lerch zeta-function is defined by

$$\phi(s; \alpha, \xi) = \sum_{n=0}^{\infty} \frac{e^{2\pi i n \xi}}{(n + \alpha)^s}.$$

The theory of this function is developed in [42] (cf. [71]) and the functional equation is well known and of ramified type, so that it is more convenient to work with the intermediate lattice zeta-function

$$\varphi_a(s) = \sum_{n \in \mathbb{Z}} \frac{e^{2\pi i n \xi}}{|n + \alpha|^{2s}} (\mathrm{sgn}(n + \alpha))^a, \tag{1.26}$$

where we impose the condition

$$0 < \alpha \leq \frac{1}{2} \tag{1.27}$$

and use the notation $a = 0$ or 1 and

$$\mathrm{sgn}\, x = \begin{cases} \frac{x}{|x|}, & x \neq 0, \\ 0, & x = 0. \end{cases}$$

We note that

$$\varphi_0(s) = Z \begin{vmatrix} \alpha \\ \xi \end{vmatrix} (2s)_Q, \quad Q = (n + \alpha)^2,$$

where $Z \begin{vmatrix} \alpha \\ \xi \end{vmatrix} (s)_Q$ is the Epstein zeta-function in Example 1.2.

The modular relation reads

$$\sum_{n \in \mathbb{Z}} (n + \alpha)^a e^{-(n+\alpha)^2 u + 2\pi i n \xi}$$

$$= i^a \left(\frac{\pi}{u}\right)^{\frac{1}{2}+a} \sum_{n \in \mathbb{Z}} (n + \xi)^a e^{-(n+\xi)^2 \frac{\pi^2}{u} - 2\pi i (n+\xi)}, \quad \Re u > 0. \tag{1.28}$$

The incomplete gamma series reads for $0 < \alpha \le \frac{1}{2}$, $0 < \xi < 1$, and $\Re w > 0$,

$$2\phi(s; \alpha, \xi) = 2 \sum_{n=0}^{\infty} \frac{e^{2\pi i n \xi}}{(n+\alpha)^s}$$

$$= \sum_{n=-\infty}^{\infty} \frac{e^{2\pi i n \xi}}{|n+\alpha|^s} \left\{ \frac{\Gamma\left(\frac{s}{2}, \pi w(n+\alpha)^2\right)}{\Gamma\left(\frac{s}{2}\right)} + \text{sgn}(n+\alpha) \frac{\Gamma\left(\frac{s+1}{2}, \pi w(n+\alpha)^2\right)}{\Gamma\left(\frac{s+1}{2}\right)} \right\}$$

$$+ \sqrt{2}\pi^{s-\frac{1}{2}} \sum_{n=-\infty}^{\infty} \frac{e^{-2\pi i \alpha(n+\xi)}}{|n+\xi|^{1-s}} \left\{ \frac{\Gamma\left(\frac{1-s}{2}, \frac{\pi(n+\xi)^2}{w}\right)}{\Gamma\left(\frac{s}{2}\right)} \right.$$

$$\left. + 2i\,\text{sgn}(n+\xi) \frac{\Gamma\left(1 - \frac{s}{2}, \frac{\pi(n+\xi)^2}{w}\right)}{\Gamma\left(\frac{s+1}{2}\right)} \right\}. \tag{1.29}$$

The transformation formula (1.28) follows from that of the theta series (cf. [43], p.370):

$$\vartheta_3(v, \tau) = e^{\frac{\pi i}{4}} \tau^{-\frac{1}{2}} e^{-\frac{\pi i v^2}{\tau}} \vartheta_3\left(\frac{v}{\tau}, -\frac{1}{\tau}\right),$$

where

$$\vartheta_3(v, \tau) = \sum_{n=-\infty}^{\infty} e^{\pi \tau i n^2 + 2\pi i v n}.$$

From (1.28) we first deduce the incomplete gamma series for $\varphi_a(s)$:

$$\varphi_a\left(\frac{s}{2}\right)$$

$$= \sum_{n=-\infty}^{\infty} \frac{e^{2\pi i n \xi}}{|n+\alpha|^s} (\text{sgn}(n+\alpha))^a \frac{\Gamma\left(\frac{s+a}{2}, \pi w(n+\alpha)^2\right)}{\Gamma\left(\frac{s+a}{2}\right)}$$

$$+ 2^{\frac{1}{2}+a} \pi^{s-\frac{1}{2}} i^a \sum_{n=-\infty}^{\infty} \frac{e^{-2\pi i \alpha(n+\xi)}}{|n+\xi|^{1-s}} (\text{sgn}(n+\xi))^a \frac{\Gamma\left(\frac{a+1-s}{2}, \frac{\pi(n+\xi)^2}{w}\right)}{\Gamma\left(\frac{s+a}{2}\right)}. \tag{1.30}$$

Adding (1.30) with $a = 0$ and 1 yields (1.29).

2. Ramanujan's formula and Guinand's formula

2.1. The following remarkable formula of Ramanujan is probably one of the most famous formulas named after him which is stated as Entry 21 (i), Chapter 14 of Ramanujan's Notebook II [23] (which is also stated as I, Entry 15 of Chapter 16 [22], cf. also IV, Entry 20 of [24]):

Let $\alpha > 0$, $\beta > 0$ satisfy the relation $\alpha\beta = \pi^2$ and let n be any positive integer. Then

$$\alpha^{-n}\left\{\frac{1}{2}\zeta(2n+1) + \sum_{k=1}^{\infty}\frac{k^{-2n-1}}{e^{2\alpha k}-1}\right\}$$

$$= (-\beta)^{-n}\left\{\frac{1}{2}\zeta(2n+1) + \sum_{k=1}^{\infty}\frac{k^{-2n-1}}{e^{2\beta k}-1}\right\}$$

$$- 2^{2n}\sum_{j=0}^{n+1}(-1)^j\frac{B_{2j}}{(2j)!}\frac{B_{2n+2-2j}}{(2n+2-2j)!}\alpha^{n+1-j}\beta^j, \qquad (2.1)$$

where B_m denotes the m-th Bernoulli number defined by

$$\frac{x}{e^x-1} = \sum_{m=0}^{\infty}\frac{B_m}{m!}x^m, \quad |x| < 2\pi.$$

Formula (2.1) is true for all integers $n \neq 0$ by interpreting a vacuous sum as 0. It looks like at a first glance that the case $n \geq 0$ counts more, giving an analytic expression for zeta values at odd positive integers, but the negative case is equally important in that it expresses the automorphic property of the corresponding Eisenstein series; the case $n = 0$ is no less important as we shall see below.

In order to include the case $n = 0$, we make use of Liouville's formula (cf. Koshlyakov [83])

$$\sum_{k=1}^{\infty}\frac{k^a}{e^{2\pi kx}-1} = \sum_{k=1}^{\infty}\sigma_a(k)e^{-2\pi kx}, \qquad (2.2)$$

where

$$\sigma_a(k) = \sum_{d|k}d^a \qquad (2.3)$$

denotes the sum of a-th powers of all divisors of k.

Incorporating (2.2) with $x = \alpha/\pi$, in (2.1), we may express Ramanujan's formula as

$$\sum_{k=1}^{\infty}\sigma_{-2n-1}(k)e^{-2\pi kx} + (-1)^{n+1}x^{2n}\sum_{k=1}^{\infty}\sigma_{-2n-1}(k)e^{-\frac{2\pi k}{x}} = P(x), \quad (2.4)$$

where

$$P(x) = \frac{(2\pi)^{2n+1}}{2x} \sum_{j=0}^{n+1} (-1)^j \frac{B_{2j}}{(2j)!} \frac{B_{2n+2-2j}}{(2n+2-2j)!} x^{2n+2-2j}$$

$$+ \begin{cases} -\dfrac{1}{2}\zeta(2n+1)\left\{1 + (-1)^{n+1}x^{2n}\right\} & \text{if } n \geq 1, \\ \dfrac{1}{2}\log x & \text{if } n = 0, \end{cases} \tag{2.5}$$

valid for $\Re x > 0$.

Indeed, $P(x)$ is a residual function given as the sum of the residues:

$$P(x) = \sum_{\xi \in R} \operatorname*{Res}_{s=\xi} (2\pi)^{-s}\Gamma(s)\zeta(s)\zeta(s + 2n + 1)x^{-s}, \tag{2.6}$$

where $R = \{-2n-1, -2n, -2n+1, -2n+3, \ldots, -3, -1, 0, 1\}$, and $s = 0$ is a double pole only when $n = 0$ (others are simple poles).

Stated in the form of (2.4), Ramanujan's formula (2.1) may be interpreted as a modular relation (cf. [77], Example 1 where the reason why we should have $\alpha\beta = \pi^2$ is elucidated) and therefore may include the case $n = 0$, in which case Liouville's formula (2.2) is nothing but the sum of the Taylor expansion of the logarithm function:

$$\sum_{k=1}^{\infty} \sigma_{-1}(k)q^k = \sum_{l,m=1}^{\infty} \frac{1}{l}q^{lm} = -\sum_{k=1}^{\infty} \log(1 - q^k)$$

$$= F(\tau), \tag{2.7}$$

say, where $q = e^{2\pi i \tau}$ denotes the local uniformization parameter and τ is in the upper-half plane $\mathcal{H}$, i.e. $\Im\tau > 0$.

We now put $x = \tau/i$ ($\Re x > 0$) in (2.4) and (2.5) with $n = 0$ to obtain

$$F(\tau) = F\left(-\frac{1}{\tau}\right) + \frac{\pi}{12}\left(\frac{i}{\tau} - \frac{\tau}{i}\right) + \frac{1}{2}\log\frac{\tau}{i}, \tag{2.8}$$

which is the formula deduced by A. Weil [125], p.401.

Recalling the definition of the Dedekind eta function

$$\eta(\tau) = q^{\frac{1}{24}} \prod_{k=1}^{\infty} (1 - q^k), \tag{2.9}$$

we see that

$$F(\tau) = \frac{\pi i}{12}\tau - \log\eta(\tau). \tag{2.10}$$

Rewriting (2.8) in terms of η on using (2.10), gives the most famous transformation formula for $\eta(\tau)$:

$$\log \eta \left(-\frac{1}{\tau} \right) = \log \eta(\tau) + \frac{1}{2} \log \frac{\tau}{i}, \qquad (2.11)$$

or

$$\eta \left(-\frac{1}{\tau} \right) = \sqrt{\frac{\tau}{i}} \eta(\tau). \qquad (2.12)$$

In (2.7) we summed the infinite sum $\sum_{l,m=1}^{\infty} \frac{1}{l} q^{lm}$ by grouping those l, m which give the same value n, say, of lm. If instead, we sum over m first, then we will get a geometric series, so that

$$F(\tau) = \sum_{l=1}^{\infty} \frac{1}{l(q^{-l} - 1)} = \frac{i}{2} \sum_{l=1}^{\infty} \frac{1}{l} (\cot \pi l \tau + i). \qquad (2.13)$$

Considering $F(\tau) - F(-\frac{1}{\tau})$, we are led to

$$F(\tau) - F\left(-\frac{1}{\tau}\right) = \frac{i}{2} \sum_{l=1}^{\infty} \frac{1}{l} \left(\cot \pi l \tau + \cot \frac{\pi l}{\tau} \right). \qquad (2.14)$$

Formula (2.13) suggests a definition of the Dedekind sum based on cotangent values, and (2.14) led C. L. Siegel [111] to apply the residue theorem to the function $z^{-1} f(\nu z)$, where $f(z) = \cot z \cot \frac{z}{\tau}$, $\nu = (n + \frac{1}{2})\pi$ ($n = 0, 1, \ldots$), giving a simple proof of (2.12). As Siegel's proof has its genesis in the modular relation, the calculation done in [111] is to find the residual function (and the right-hand side of (2.14)).

Thus we could say that the transformation formula (2.11) was within easy reach of Ramanujan.

In [77], after viewing Ramanujan's formula (2.4) as a modular relation in Example 1, we went on further to interpret A. P. Guinand's formula (Formula (9), [77]) in the same way as in Example 2. We note that the limiting case as $s \to -1$ of Guinand's formula in Example 1 is pointed out to be equivalent to (2.11) by Guinand. Then in Theorem 3 we deduced the special case of Guinand's formula pertaining to a rapidly convergent series for the Riemann zeta-values at odd, positive integers, as $2n$-times differentiated form of Ramanujan's formula (2.4). In the proof we differentiated a certain integral in Formula (21), [77], which represents the second sum on the right of (2.4).

We now restate the proof of that theorem in a much deeper perspective and illustrate a remarkable principle that differentiation of Ramanujan's formula, or the modular relation, gives rise to automorphy of the corresponding automorphic form.

To this end we introduce the integral $I_a(x)$ for a real and $x > 0$

$$I_a(x) = \frac{1}{2\pi i} \int_{(\kappa)} (2\pi)^{-s} \Gamma(s) \varphi(s - a) x^{-s} ds \qquad (2.15)$$

where

$$\varphi(s) = \zeta(s)\zeta(s + 2n + 1), \qquad (2.16)$$

with n a non-negative integer and $\kappa > \max\{1, 1 + a\}$. It is easily seen that

$$I_a(x) = \sum_{k=1}^{\infty} \sigma_{-2n-1}(k) k^a e^{-2\pi k x}. \qquad (2.17)$$

For simplicity we put

$$I(x) = I_0(x) = \frac{1}{2\pi i} \int_{(\kappa)} (2\pi)^{-s} \Gamma(s) \varphi(s) x^{-s} ds. \qquad (2.18)$$

Let a be a non-negative integer and differentiate $I(x)$ a-times with respect to x, whereby we perform differentiation under integral sign. Then we have the additional factor

$$\prod_{j=0}^{a-1} (-s - j),$$

which is $(-1)^a \frac{\Gamma(s+a)}{\Gamma(s)}$, whence we deduce the remarkable formula

$$\frac{d^a}{dx^a} I(x) = (-2\pi)^a I_a(x), \qquad (2.19)$$

i.e. a-times differentiation of the Lambert series (2.17) is effected by shifting the argument of $\varphi(s)$ by a in (2.18) and multiplying by $(-2\pi)^a$.

We shall now obtain a modular relation for $I_a(x)$, which is the a-times differentiated form of Ramanujan's formula (2.4), by incorporating the functional equation satisfied by $\varphi(s)$, an equivalent form of the modular relation (cf. §1)

$$(2\pi)^{-s} \Gamma(s) \varphi(s) = (-1)^n (2\pi)^{2n+s} \Gamma(-2n - s) \varphi(-2n - s). \qquad (2.20)$$

Moving the line of integration to $\sigma = -\kappa_1$ ($\kappa_1 > 2n + 1 - a$), whereby noting that the horizontal integrals vanish in the limit as $|t| \to \infty$, we have

$$I_a(x) = \frac{1}{2\pi i} \int_{(-\kappa_1)} \Gamma(s) \varphi(s - a) (2\pi x)^{-s} ds + P_a(x), \qquad (2.21)$$

where $P(x) = P_a(x)$ denotes the sum of residues of the integrand at its poles at $s = a - 2n - 1, a - 2n, a - 2n + 1, a - 2n + 3, \ldots, 0, a + 1$.

Substitute the formula

$$\varphi(s - a) = (-1)^n (2\pi)^{2n+2s-2a} \frac{\Gamma(a - 2n - s)}{\Gamma(s - a)} \varphi(a - 2n - s)$$

which follows from (2.20) in the integral, say, $J_a(x)$ on the right-hand side of (2.21). Then

$$
\begin{aligned}
J_a(x) &= \frac{(-1)^n}{2\pi i} \int_{(-\kappa_1)} \frac{\Gamma(s)\Gamma(a - 2n - s)}{\Gamma(s - a)} \varphi(a - 2n - s)(2\pi)^{2n+s-2a} x^{-s} ds \\
&= \frac{(-1)^n}{2\pi i} \int_{(a-2n+\kappa_1)} \frac{\Gamma(a - 2n - s)\Gamma(s)}{\Gamma(-2n - s)} \varphi(s)(2\pi)^{-s-a} x^{s-a+2n} ds.
\end{aligned}
$$

$$(2.22)$$

We note that by the reciprocal relation for the gamma function, we have

$$
\begin{aligned}
\frac{\Gamma(a - 2n - s)}{\Gamma(-2n - s)} &= \frac{\Gamma(1 + s + 2n)}{\Gamma(1 + s + 2n - a)} \frac{\sin \pi s}{\sin \pi(s - a)} \\
&= (-1)^a \frac{\Gamma(a + 2n + 1)}{\Gamma(s + 2n + 1 - a)}.
\end{aligned}
$$

$$(2.23)$$

Hence

$$J_a(x) = \frac{(-1)^{n+a}}{2\pi i} \int_{(a-2n+\kappa_1)} \frac{\Gamma(s + 2n + 1)\Gamma(s)}{\Gamma(s + 2n + 1 - a)} \varphi(s)(2\pi)^{-s-a} x^{s-a+2n} ds.$$

$$(2.24)$$

Since the gamma factor can be computed for $0 \leq a \leq 2n$ as,

$$\frac{\Gamma(s)\Gamma(s + 2n + 1)}{\Gamma(s + 2n + 1 - a)} = \sum_{k=0}^{a} \binom{a}{k} \frac{(2n - k)!}{(2n - a)!} \Gamma(s + k),$$

$$(2.25)$$

we conclude from (2.21) and (2.24) that

$$
\begin{aligned}
I_a(x) = {}&(-1)^{n+a}(2\pi)^{-a} x^{-a+2n} \sum_{k=0}^{a} \binom{a}{k} \frac{(2n - k)!}{(2n - a)!} \\
&\times \frac{1}{2\pi i} \int_{(a-2n+\kappa_1)} \Gamma(s + k)\varphi(s) \left(\frac{2\pi}{x}\right)^{-s} ds + P_a(x).
\end{aligned}
$$

$$(2.26)$$

Finally, we note that the integral on the right-hand side of (2.26) becomes

$$\frac{1}{2\pi i} \int_{(a-2n+\kappa_1+k)} \Gamma(s)\varphi(s - k) \left(\frac{2\pi}{x}\right)^{-s+k} ds,$$

which is identical with

$$\left(\frac{2\pi}{x}\right)^k I_k\left(\frac{1}{x}\right).$$

Thus we have proved

Theorem 2.1. *For the Mellin transform with shifted argument* $I_a(x)$ *as defined by (2.15), we have the modular relation for* $n \geq 0$ *and* $0 \leq a \leq 2n$,

$$I_a(x) = (-1)^{n+a}(2\pi)^{-a}x^{-a+2n}\sum_{k=0}^{a}\binom{a}{k}\frac{(2n-k)!}{(2n-a)!}\left(\frac{2\pi}{x}\right)^k I_k\left(\frac{1}{x}\right)$$

$$+ P_a(x), \tag{2.27}$$

where $P_a(x)$ *is the residual function (the sum of the residues).*

Particular cases are

$$P_0(x) = P(x),$$

given by (2.5), and

$$P_{2n}(x) = (2n)!\,\zeta(2n+2)(2\pi x)^{-2n-1} + \frac{1}{2}\zeta(-2n-1)(2\pi x)$$

$$+ \frac{(-1)^n}{2}(2\pi)^{-2n}(2n)!\zeta(2n+1), \quad n \geq 1. \tag{2.28}$$

On the other hand, for $a = 2n + 1$, we have

$$I_{2n+1}(x) = (-1)^{n+1}(2\pi)^{-2n-1}x^{-2n-2}I_{2n+1}\left(\frac{1}{x}\right) + P_{2n+1}(x), \tag{2.29}$$

where

$$P_{2n+1}(x) = (2n+1)!\zeta(2n+2)(2\pi x)^{-2n-2} - \frac{1}{2}\zeta(-2n-1) \ (n \geq 1). \tag{2.30}$$

The meaning of this equality is discussed below.

Remark 2.1. The case $a = 0$ is Ramanujan's formula (2.4), the case $a = 2n$ is Guinand's formula (cf. [77], Theorem 3), and the case $a = 2n + 1$ gives, on writing $m = -n - 1$, the negative case $(m < 0)$ of Ramanujan's formula (2.4).

The special case of (2.28) with $n = 1$, i.e. twice differentiated form of Ramanujan's formula, yields Terras' formula [116]

$$\zeta(3) = \frac{2}{45}\pi^3 - 4\sum_{k=1}^{\infty}e^{-2\pi k}\sigma_{-3}(k)\left(2\pi^2 k^2 + \pi k + \frac{1}{2}\right). \tag{2.31}$$

We now state a corollary to Theorem 2.1, which makes it possible to perform numerical calculation of zeta values in a remarkably efficient way.

Corollary 2.1 ([77], Corollary 1). *For $n \geq 1$ we have*

$$\frac{1}{2}(2n)!\,\zeta(2n+1) + \frac{(2\pi)^{2n+1}B_{2n+2}}{2(2n+2)}\left(\frac{1}{2n+1} - (-1)^n\right)$$

$$= (-1)^n (2\pi)^{2n} \sum_{r=1}^{2n+1} (r-1)!\,S(2n+1,r)A(1,r)$$

$$- \sum_{m=0}^{2n} \frac{(2n)!}{m!}(2\pi)^m \sum_{r=1}^{m+1} (r-1)!\,S(m+1,r)A(2n+1-m,r), \quad (2.32)$$

where

$$A(p,q) = \sum_{m=1}^{\infty} \frac{1}{m^p(e^{2\pi m} - 1)^q}$$

and $S(k,r)$ denote the Stirling number of the second kind defined by

$$x^k = \sum_{r=1}^{k} S(k,r)x(x-1)\cdots(x-r+1).$$

Example 2.1. We have

$$\text{(i)}\quad \zeta(3) = \frac{2}{45}\pi^3 - 8\pi^2 \left\{ \sum_{k=1}^{\infty} \frac{1}{k}\left(1 + \frac{1}{2\pi k} + \frac{1}{(2\pi k)^2}\right)\frac{1}{e^{2\pi k}-1} \right.$$

$$\left. + \sum_{k=1}^{\infty} \frac{1}{k}\left(3 + \frac{1}{2\pi k}\right)\frac{1}{(e^{2\pi k}-1)^2} + \sum_{k=1}^{\infty} \frac{2}{k}\frac{1}{(e^{2\pi k}-1)^3} \right\}$$

$$= 1.20205690315959428539973816151144999076498629234049888179227155534183820578631309018\ldots$$

correct up to 84 decimal places (by computing first 30 terms).

(ii) $\zeta(5) = \dfrac{2^2 \pi^5}{3^3 \cdot 5 \cdot 7} - \dfrac{8\pi^3}{3} \sum_{k=1}^{\infty} \dfrac{1}{k^2} \left(1 + \dfrac{3}{2\pi k} + \dfrac{6}{(2\pi k)^2} + \dfrac{6}{(2\pi k)^3}\right) \dfrac{1}{e^{2\pi k} - 1}$

$\qquad - \dfrac{8\pi^3}{3} \sum_{k=1}^{\infty} \dfrac{1}{k^2} \left(7 + \dfrac{9}{2\pi k} + \dfrac{6}{(2\pi k)^2}\right) \dfrac{1}{(e^{2\pi k} - 1)^2}$

$\qquad - 16\pi^3 \sum_{k=1}^{\infty} \dfrac{1}{k^2} \left(2 + \dfrac{1}{2\pi k}\right) \dfrac{1}{(e^{2\pi k} - 1)^3}$

$\qquad - 16\pi^3 \sum_{k=1}^{\infty} \dfrac{1}{k^2} \dfrac{1}{(e^{2\pi k} - 1)^4}$

$\qquad = 1.03692775514336992633136548645703416805708091950191281197419267790380358978628148456\,00\ldots$

Remark 2.2. In a recently published book [113] Srivastava and Choi record that by their formula, they were able to compute the value of $\zeta(3)$ accurate up to 7 decimal places by taking only first 50 terms.

It should be noted that their formula is easily derivable from [71] or [72] by differentiating (not integrating as it looks at a first glance) and forming linear combinations of resulting formulas and therefore that by their method they could never attain the acceleration of order $e^{2\pi}$,

$$e^{2\pi} = 535.54916555247647365030493\,26 \ldots,$$

and all endeavor of accelerating the series in this way will be pointless (cf. §5.5 at the end of this paper).

Remark 2.3. The method of proof of Theorem 2.1, which was not perceived in [77], the Mellin transform with shifted argument, is an additive version of the "pseudo modular relation principle" whose multiplicative analogues have been successfully applied in [74, 75, 76, 78]. We mean by a pseudo modular relation an integration process in which the relevant gamma factor is not the same as one appearing in the functional equation (cf. (2.20) and (2.21)). By this means we may extract the most important information and open a vast uncultivated field all over ahead. In this special case, however, as is remarked in the Introduction, the main formula (2.19) is the manifestation of the statement of Razar that the differentiation of Lambert series essentially corresponds to the shift of the argument of the associated Dirichlet series (cf. Remark 3.1).

Remark 2.4. In the beginning we have seen that the special case $n = 0$, $a = 0$ of Theorem 2.1, the Ramanujan-Guinand formula, is nothing other than the automorphic property of the Dedekind eta-function, which is an automorphic form of weight "1/2". This can be regarded as once differentiated form of Guinand's formula. We note that in the case of $n = 0$, the twice differentiated Guinand's formula, with a suitable modification, coincides with the automorphic property of the Eisenstein series E_2 (cf. Rankin [101], p.196). The E_2 is studied in [25] as Ramanujan's Eisenstein series P. These functions are related by

$$P(\tau) = 1 - 24 \sum_{k=1}^{\infty} \frac{kq^k}{1-q^k} = E_2(\tau) + \frac{3}{\pi y}$$

$$= 1 - \frac{12}{\pi i} F'(\tau)$$

$$= \frac{12}{\pi i} (\log \eta(\tau))',$$

where F is the function defined by (2.7).

2.2. In order to grasp the meaning of Formula (2.29) in the light of the transformation formula for Eisenstein series, we briefly review the theory of modular forms (there are many good books, but the most accessible ones for beginners would be Serre [106] or Doi-Miyake [37] or Shimura [108]).

Let κ be an even integer greater than or equal to 4. Let M_κ denote the space of modular forms $f(z)$ of weight κ with respect to the full modular group $SL_2(\mathbb{Z})$, where $f(z)$ is called a modular form with the above-mentioned description if it satisfies the following conditions:

M 1. $f(z)$ is a holomorphic function on the upper half-plane $\mathcal{H}$ and is holomorphic at infinity.

M 2. $f(z) = (cz + d)^{-\kappa} f\left(\dfrac{az+b}{cz+d}\right)$ for all $\begin{pmatrix} a & b \\ c & d \end{pmatrix} \in SL_2(\mathbb{Z})$.

A modular form $f(z)$ is called a cusp form if it vanishes at infinity.

The most important modular forms are Eisenstein series $G_\kappa(z)$ defined by

$$G_\kappa(z) = \sum_{m,n}{}' \frac{1}{(mz+n)^\kappa}, (\kappa \geq 4) \tag{2.33}$$

where the summation runs over all pairs (m, n) of integers other than $(0, 0)$. It is well known that G_κ is a modular form of weight κ with Fourier expansion

$$G_\kappa(z) = 2\zeta(\kappa) + 2\frac{(2\pi i)^\kappa}{(\kappa - 1)!} \sum_{m=1}^{\infty} \sigma_{\kappa-1}(m) e^{2\pi imz}, \tag{2.34}$$

and that M_κ has for its basis $\{G_4^a G_6^b\}$ with a, b non-negative integers satisfying $4a + 6b = \kappa$.

We are now in a position to interpret (2.29), once differentiated form of Guinand's formula($=2n+1$ times differentiated form of Ramanujan's formula=the negative case thereof), in terms of G_{2n+2}.

Stating (2.29) in explicit form

$$-2\sum_{k=1}^{\infty}\sigma_{2n+1}(k)e^{-2\pi kx} + \frac{B_{2n+2}}{2n+2}$$

$$= (-1)^{n+1}x^{-2n-2}\left\{-2\sum_{k=1}^{\infty}\sigma_{2n+1}(k)e^{-\frac{2\pi k}{x}} + \frac{B_{2n+2}}{2n+2}\right\}, \qquad (2.29)'$$

we see that it is nothing but

$$G_{2n+2}\left(-\frac{1}{ix}\right) = (ix)^{2n+2}G_{2n+2}(ix), \qquad (2.35)$$

i.e. the automorphy of $G_{2n+2}(z)$.

Since Guinand's formula (2.28) is interpreted as a modular relation in Example 2 [77], we may express what we have described above in the following scheme.

Principle

Automorphic Forms

diff. ↑   ↓ int.

$Z(s)$ — Modular Relation — $Z(\delta - s)$

⇕

Functional Equation

Remark 2.5. As a correspondence between modular forms and Dirichlet series $Z(s)$ satisfying functional equations with the gamma factor $\Gamma(s)$, the above principle has been known as Hecke correspondence (Hecke theory) [64]-[66], or the Riemann-Hecke-Bochner correspondence (cf. M. I. Knopp [82] for this terminology), antedated probably by Riemann [103], Hamburger [62], Siegel [111] and generalized by Weil [126], Bochner [26], Chandrasekharan and Narasimhan [33, 34], Terras [122] and Knopp [82]. In particular, Terras [122, I], Examples p.234 states the correspondence under Hecke theory between the Maass wave forms and the Dirichlet series as well as the Eisenstein series G_κ and the Dirichlet series $\zeta(s)\zeta(s+1-\kappa)$.

This most renowned Hecke correspondence (also [111]) can be described in short as "the (inverse) Mellin transform of the (completed) zeta-function $A^{-s}\Gamma(s)Z(s)$ is a constant times a certain modular form and vice versa."

Prominent as it is, it lacks e.g. a direct explanation why Knopp's "abundance principle" holds. Our point of view, though it works in a speculative way, gives a feeling that in stepping down from the more orderly world of modular forms to less orderly world of zeta-functions, one loses the order inherent in modular forms to allow more possible choices of zeta-functions, and that this can be thought of indefiniteness of integration constant (as sometimes perceived as (generalized) Eichler integrals), so to say. In the other direction, stepping up process is much more direct in our principle, *i.e.* just differentiation gives rise to the automorphic property of the corresponding modular form, thus making it possible to guess the form of the corresponding modular form.

Weil [126] calls the Hecke correspondence Hecke's lemma and states it in terms of the Mellin transforms of continuous functions on $\mathbb{R}_+^\times$ originating from modular forms (by rotating it by right-angle; similar process adopted by Grosswald [55]), thus associating Dirichlet series satisfying Hecke's type functional equation (with a single gamma factor) to every pair of cusps and every automorphic form for a Fuchsian group. He also deduced the same expression for Eichler integrals in terms of the Dirichlet series as Razar [102].

As remarked by Weil himself in the footnote of page 268, Bochner [26] developed rather broad generalization of the Hecke theory by extracting the group action $\tau \to -\tau^{-1}$ on translating it as $x \to x^{-1}$ on $\mathbb{R}_+^\times$, which was the view we have adopted in [77].

For a thorough list of zeta-functions whose functional equations follow from theta-relations, we refer the reader to F. Sato [104].

Our speculation suggests that the relation among modular forms (theta series) are inherited as relations among zeta-functions, and their very special case are the functional equations. It also allows us to speculate that the relations among multiple zeta-values are heritages of those among poly-logarithms (Okuda-Ueno), and likewise that relations among hypergeometric functions under a group action on variables may be inherent in the corresponding zeta-function-like objects.

3. Zeta-functions associated to automorphic forms

The purpose of this section is to prove an analogue of Theorem 2.1 for the L-functions associated to cusp forms and evaluate their special values, the periods, numerically.

Let $f(z) = \sum_{n=1}^{\infty} a_n e^{2\pi i n z}$ $(z \in \mathcal{H})$ be a cusp form of weight κ with respect to $SL_2(\mathbb{Z})$ (cf. Mod 1, 2 in §2) and let

$$L(s, f) = \sum_{n=1}^{\infty} \frac{a_n}{n^s}, \quad \Re s > \frac{\kappa + 1}{2} \tag{3.1}$$

be the corresponding Dirichlet series.

Theorem 3.1. *We have the modular relation for each integer $a \geq 1$ and $x > 0$,*

$$\sum_{n=1}^{\infty} \frac{a_n}{n^a} e^{-2\pi n x} = \sum_{j=0}^{a-1} \frac{(-2\pi x)^j}{j!} L(a - j, f)$$

$$+ (-2\pi x)^{-\frac{\kappa}{2}+a} \sum_{n=1}^{\infty} \frac{a_n}{n^{\frac{\kappa}{2}}} e^{-\frac{\pi n}{x}} W_{\frac{\kappa}{2}-a, \frac{\kappa-1}{2}} \left(\frac{2\pi n}{x}\right), \tag{3.2}$$

where $W_{\mu,\nu}$ denotes the Whittaker function introduced in §1.

Corollary 3.1. *For $a \in \mathbb{N}$, we have*

$$L(a, f) = \sum_{j=0}^{a-1} \frac{(2\pi)^j}{j!} \sum_{n=1}^{\infty} \frac{a_n}{n^{a-j}} e^{-2\pi n} + \frac{(-1)^{\frac{\kappa}{2}}}{(a-1)!} (2\pi)^{-\frac{\kappa}{2}+a}$$

$$\times \sum_{n=1}^{\infty} \frac{a_n}{n^{\frac{\kappa}{2}}} (2\pi n)^{\frac{a-1}{2}} e^{-\pi n} W_{\frac{\kappa-1-a}{2}, \frac{\kappa-a}{2}} (2\pi n). \tag{3.3}$$

In particular, for $a \leq \kappa - 1$, we have

$$L(a, f) = \sum_{j=0}^{a-1} \frac{(2\pi)^j}{j!} \sum_{n=1}^{\infty} \frac{a_n}{n^{a-j}} e^{-2\pi n} + \frac{(-1)^{\frac{\kappa}{2}}}{(a-1)!} (2\pi)^{a-1}$$

$$\times \sum_{n=1}^{\infty} \frac{a_n}{n} e^{-2\pi n} \left\{ 1 + \sum_{k=1}^{\kappa-a-1} \frac{(\kappa - a - 1)(\kappa - a - 2) \cdots (\kappa - a - k)}{(2\pi n)^k} \right\}. \tag{3.4}$$

Proof of Theorem 3.1. The proof goes almost word-for-word as that of Theorem 2.1, using the Mellin transform with shifted argument

$$T_a(x) = \frac{1}{2\pi i} \int_{(c)} \Gamma(s) L(s + a, f)(2\pi x)^{-s} ds \tag{3.5}$$

with $c > \frac{\kappa+1}{2}$, corresponding to (2.15). It is easily seen that

$$T_a(x) = \sum_{n=1}^{\infty} \frac{a_n}{n^a} e^{-2\pi n x}. \tag{3.6}$$

The functional equation remains the same form:

$$(2\pi)^{-s}\Gamma(s)L(s,f) = (-1)^{\kappa/2}(2\pi)^{-(\kappa-s)}\Gamma(\kappa-s)L(\kappa-s,f). \tag{3.7}$$

In exactly the same manner, we move the line of integration in (3.5) to the left: $\sigma = -c_1$ ($c_1 > a - 1$). Noting that $L(-n, f) = 0$, ($n \geq 0$) since $(2\pi)^{-s}\Gamma(s)L(s,f)$ is entire, we see that the sum $P_a(x)$ of residues is

$$P_a(x) = \sum_{j=0}^{a-1} \frac{(-1)^j}{j!}(2\pi x)^j L(a - j, f). \tag{3.8}$$

Hence, corresponding to (2.21), we have

$$T_a(x) = S_a(x) + P_a(x), \tag{3.9}$$

where

$$S_a(x) = \frac{1}{2\pi i} \int_{(-c_1)} \Gamma(s)L(s + a, f)(2\pi x)^{-s}ds. \tag{3.10}$$

Substituting the functional equation in the form

$$L(s + a, f) = (-1)^{\frac{\kappa}{2}}(2\pi)^{2(s+a)-\kappa} \frac{\Gamma(\kappa - s - a)}{\Gamma(s + a)} L(\kappa - s - a, f)$$

and making the change of variables $\kappa - s - a \to s$, we deduce

$$S_a(x) = (-1)^{\frac{\kappa}{2}}(2\pi)^a x^{-(\kappa-a)}$$
$$\times \frac{1}{2\pi i} \int_{(\kappa-a+c_1)} \frac{\Gamma(\kappa - a - s)\Gamma(s)}{\Gamma(\kappa - s)} L(s, f) \left(\frac{2\pi}{x}\right)^{-s} ds, \tag{3.11}$$

corresponding to (2.22).

Since we have, instead of (2.23),

$$\frac{\Gamma(\kappa - a - s)}{\Gamma(\kappa - s)} = (-1)^a \frac{\Gamma(s - \kappa + 1)}{\Gamma(s - \kappa + a + 1)}, \tag{3.12}$$

it follows that

$$S_a(x) = (-1)^{\frac{\kappa}{2}}(-2\pi)^a x^{a-\kappa}$$
$$\times \frac{1}{2\pi i} \int_{(\kappa-a+c_1)} \frac{\Gamma(s - \kappa + 1)\Gamma(s)}{\Gamma(s - \kappa + 1 + a)} L(s, f) \left(\frac{2\pi}{x}\right)^{-s} ds. \tag{3.13}$$

Now substituting the Dirichlet series expansion (3.1), changing the order of summation and integration, and writing $s - \frac{\kappa}{2}$ for s, we obtain

$$S_a(x) = (-1)^{\frac{\kappa}{2}+a}(2\pi)^a x^{a-\kappa}\left(\frac{2\pi}{x}\right)^{-\frac{\kappa}{2}}$$

$$\times \sum_{n=1}^{\infty} \frac{a_n}{n^{\frac{\kappa}{2}}} \frac{1}{2\pi i} \int\limits_{(\frac{\kappa}{2}-a+c_1)} \frac{\Gamma(s - \frac{\kappa}{2} + 1)\Gamma(s + \frac{\kappa}{2})}{\Gamma(s - \frac{\kappa}{2} + a + 1)}\left(\frac{2\pi n}{x}\right)^{-s} ds. \quad (3.14)$$

Finally we note that the integral appearing on the right-hand side is

$$e^{-\frac{\pi n}{x}} W_{\frac{\kappa}{2}-a,\frac{\kappa}{2}-\frac{1}{2}}\left(\frac{2\pi n}{x}\right)$$

by Formula 5.46 in [44, p.197], whence we conclude that (3.8), (3.9) and (3.14) lead to (3.2). The proof is complete. $\qquad\square$

Before turning to the proof of Corollary 3.1, we shall examine the case $a = 1$. By (3.2), we have

$$L(1, f) = \sum_{n=1}^{\infty} \frac{a_n}{n} e^{-2\pi n} + (-1)^{\frac{\kappa}{2}}(2\pi)^{-\frac{\kappa}{2}+1}\sum_{n=1}^{\infty} \frac{a_n}{n^{\frac{\kappa}{2}}} e^{-\pi n} W_{\frac{\kappa}{2}-1,\frac{\kappa-1}{2}}(2\pi n).$$

$$(3.15)$$

It is well known that

$$W_{\frac{\nu-1}{2},\frac{\nu}{2}}(z) = z^{-\frac{\nu-1}{2}} e^{\frac{z}{2}} \Gamma(\nu, z)$$

$$= z^{\frac{\nu-1}{2}} e^{-\frac{z}{2}}\left\{1 + \sum_{n=1}^{\nu-1} \frac{(\nu - 1)\cdots(\nu - n)}{z^n}\right\} \quad (3.16)$$

for $\nu \geq 1$. Hence (3.15) reduces to

$$L(1, f) = \sum_{n=1}^{\infty} \frac{a_n}{n} e^{-2\pi n}$$

$$+ (-1)^{\frac{\kappa}{2}}\sum_{n=1}^{\infty} \frac{a_n}{n} e^{-2\pi n}\left\{1 + \sum_{j=1}^{\kappa-2} \frac{(\kappa - 2)\cdots(\kappa - 1 - j)}{(2\pi n)^j}\right\},$$

$$(3.17)$$

i.e. Formula (3.4) with $a = 1$.

To prove Corollary 3.1, we need the following lemma.

Lemma 3.1. *For any $\kappa \in \mathbb{R}$, $2 \le a \in \mathbb{Z}$, we have*

$$(-1)^{a+1}(a-1)!\, W_{\frac{\kappa}{2}-a,\frac{\kappa-1}{2}}(z)$$

$$= z^{\frac{a-1}{2}} W_{\frac{\kappa-a-1}{2},\frac{\kappa-a}{2}}(z) + \sum_{j=1}^{a-1} (-1)^j \binom{a-1}{j} z^{\frac{a-j-1}{2}} W_{\frac{\kappa-a-1+j}{2},\frac{\kappa-a+j}{2}}(z).$$

$$(3.18)$$

Proof. We apply induction on a, starting from the case $a = 2$, in which case (3.18) results from the well known formula

$$W_{\mu,\nu}(z) = \sqrt{z}\,W_{\mu-\frac{1}{2},\nu-\frac{1}{2}}(z) + \left(\frac{1}{2} - \mu + \nu\right) W_{\mu-1,\nu}(z) \qquad (3.19)$$

[52, p.1090, 9.234.1] by putting $\mu = \frac{\kappa}{2} - 1$, $\nu = \frac{\kappa-1}{2}$.

Now we suppose (3.18) is true for a and any κ. Then in addition to (3.18), we have (3.18) with κ replaced by $\kappa - 1$:

$$S_{a-1,\kappa-1} := \sum_{j=1}^{a-1} (-1)^j \binom{a-1}{j} z^{\frac{a-j-1}{2}} W_{\frac{\kappa-a-2+j}{2},\frac{\kappa-a-1+j}{2}}(z)$$

$$= (-1)^{a+1}(a-1)!\, W_{\frac{\kappa-1}{2}-a,\frac{\kappa-2}{2}}(z) - z^{\frac{a-1}{2}} W_{\frac{\kappa-2-a}{2},\frac{\kappa-1-a}{2}}(z).$$

$$(3.20)$$

What we are to prove is

$$S_{a,\kappa} := \sum_{j=1}^{a} (-1)^j \binom{a}{j} z^{\frac{a-j}{2}} W_{\frac{\kappa-a-2+j}{2},\frac{\kappa-a-1+j}{2}}(z)$$

$$= (-1)^a a!\, W_{\frac{\kappa}{2}-a-1,\frac{\kappa-1}{2}}(z) - z^{\frac{a}{2}} W_{\frac{\kappa-a-2}{2},\frac{\kappa-a-1}{2}}(z). \qquad (3.21)$$

Separating the term with $j = 1$ in $S_{a,\kappa}$, writing $j + 1$ for j in the remaining sum and using the binomial identity $\binom{a}{k+1} = \binom{a-1}{k} + \binom{a-1}{k+1}$, we get

$$S_{a,\kappa} = -\, az^{\frac{a-1}{2}} W_{\frac{\kappa-a-1}{2},\frac{\kappa-a}{2}}(z) - S_{a-1,\kappa}$$

$$+ \sum_{j=1}^{a-2} (-1)^{j+1} \binom{a-1}{j+1} z^{\frac{a-1-j}{2}} W_{\frac{\kappa-a-1+j}{2},\frac{\kappa-a+j}{2}}(z). \qquad (3.22)$$

Substituting (3.18), we see that

$$S_{a,\kappa} = (-1)^a (a-1)! W_{\frac{\kappa}{2}-a, \frac{\kappa-1}{2}}(z) - (a-1) z^{\frac{a-1}{2}} W_{\frac{\kappa-a-1}{2}, \frac{\kappa-a}{2}}(z)$$

$$- \sum_{j=1}^{a-2} (-1)^j \binom{a-1}{j+1} z^{\frac{a-1-j}{2}} W_{\frac{\kappa-a-1+j}{2}, \frac{\kappa-a+j}{2}}(z)$$

$$= (-1)^a (a-1)! W_{\frac{\kappa}{2}-a, \frac{\kappa-1}{2}}(z)$$

$$- \sum_{j=0}^{a-2} (-1)^j \binom{a-1}{j+1} z^{\frac{a-1-j}{2}} W_{\frac{\kappa-a-1+j}{2}, \frac{\kappa-a+j}{2}}(z). \tag{3.23}$$

Since the sum on the right-hand side of the last equality in (3.23) is $-z^{\frac{1}{2}} S_{a-1,\kappa-1}$, it follows from (3.20) that

$$S_{a,\kappa} = (-1)^a (a-1)! \left(W_{\frac{\kappa}{2}-a, \frac{\kappa-1}{2}}(z) - z^{\frac{1}{2}} W_{\frac{\kappa-1-a}{2}, \frac{\kappa-2}{2}}(z) \right)$$

$$- z^{\frac{a}{2}} W_{\frac{\kappa-a-2}{2}, \frac{\kappa-a-1}{2}}(z). \tag{3.24}$$

The first term on the right-hand side of (3.24) is $a W_{\frac{\kappa}{2}-a-1, \frac{\kappa-1}{2}}(z)$, and so this proves (3.21), and the proof is complete. □

Proof of Corollary 3.1. Proof is based on induction on a. Suppose (3.3) is true for $1, 2, \ldots, a-1$, and substituting them in (3.2). Then we have

$$L(a,f) = \sum_{n=1}^{\infty} \frac{a_n}{n^a} e^{-2\pi n} - \sum_{j=1}^{a-1} \sum_{h=0}^{a-j-1} \frac{(-2\pi)^j}{j!} \frac{(2\pi)^h}{h!} \sum_{n=1}^{\infty} \frac{a_n}{n^{a-j-h}} e^{-2\pi n}$$

$$- (-1)^{\frac{\kappa}{2}} \sum_{j=1}^{a-1} \frac{(-2\pi)^j}{j!} \frac{(2\pi)^{-\frac{\kappa}{2}+a-j}}{(a-j-1)!}$$

$$\times \sum_{n=1}^{\infty} \frac{a_n}{n^{\frac{\kappa}{2}}} (2\pi n)^{\frac{a-j-1}{2}} e^{-\pi n} W_{\frac{\kappa-1-a+j}{2}, \frac{\kappa-a+j}{2}}(2\pi n)$$

$$- (-1)^{\frac{\kappa}{2}+a} (2\pi)^{-\frac{\kappa}{2}+a} \sum_{n=1}^{\infty} \frac{a_n}{n^{\frac{\kappa}{2}}} e^{-\pi n} W_{\frac{\kappa}{2}-a, \frac{\kappa-1}{2}}(2\pi n). \tag{3.25}$$

The first and the second terms on the right-hand side of (3.25) is equal to

$$\sum_{n=1}^{\infty} \frac{a_n}{n^a} e^{-2\pi n} - \sum_{k=1}^{a-1} \left(\sum_{j=1}^{k} \frac{(-1)^j}{j!(k-j)!} \right) (2\pi)^k \sum_{n=1}^{\infty} \frac{a_n}{n^{a-k}} e^{-2\pi n}$$

$$= \sum_{n=1}^{\infty} \frac{a_n}{n^a} e^{-2\pi n} + \sum_{k=1}^{a-1} \frac{(2\pi)^k}{k!} \sum_{n=1}^{\infty} \frac{a_n}{n^{a-k}} e^{-2\pi n},$$

coinciding with the first term on the right-hand side of (3.3).

On the other hand, the remaining terms on the right-hand side of (3.25) reduce to

$$\frac{(-1)^{\frac{\kappa}{2}}(2\pi)^{-\frac{\kappa}{2}+a}}{(a-1)!}\sum_{n=1}^{\infty}\frac{a_n}{n^{\frac{\kappa}{2}}}e^{-\pi n}\Big\{(-1)^{a+1}(a-1)!\,W_{\frac{\kappa}{2}-a,\frac{\kappa-1}{2}}(2\pi n)$$

$$-\sum_{j=1}^{a-1}(-1)^j\binom{a-1}{j}(2\pi n)^{\frac{a-j-1}{2}}W_{\frac{\kappa-1-a+j}{2},\frac{\kappa-a+j}{2}}(2\pi n)\Big\}$$

$$=\frac{(-1)^{\frac{\kappa}{2}}(2\pi)^{-\frac{\kappa}{2}+a}}{(a-1)!}\sum_{n=1}^{\infty}\frac{a_n}{n^{\frac{\kappa}{2}}}e^{-\pi n}(2\pi n)^{\frac{a-1}{2}}W_{\frac{\kappa-1-a}{2},\frac{\kappa-a}{2}}(2\pi n),$$

by Lemma 3.1, thereby completing the proof. □

We shall now state numerical examples supplied by Corollary 3.1 (see also Zagier [129] and Lang [87]).

Example   Let $\Delta(z) = \eta(z)^{24} = \sum_{n=1}^{\infty}\tau(n)e^{2\pi nz}$ be a cusp form of weight 12 with respect to $SL_2(\mathbb{Z})$, and let $L(s,\Delta)$ be the corresponding L-function. By (3.3), we get

$L(1,\Delta) = 0.0374412812685155417387703158411808705230498585 8289$
$\qquad 4411222067055966716462042166225657680708689684 8418$
$\qquad 5066718624644089408217748512$

and

$L(2,\Delta) = 0.1463745420912659894130009132749962159070673841 9062$
$\qquad 1201045597854577123050776453797047474867726507 2680$
$\qquad 11900851088117196024232255132$

which are correct up to 128 decimal places by computing first 50 terms.

By the theorems on periods of cusp forms (Eichler [40], Shimura [107] and Manin [95]), there exit two periods ω_+ and ω_- such that $L(a,f)$ $(1\le a\le \kappa-1)$ is a rational multiple of ω_+ if a is odd, and of ω_- if a is odd respectively. In the case $f(z) = \Delta(z)$, we have

$\omega_+ = 0.0214460667072841789427102865173197152673623795 5824$
$\qquad 8318782368217117721888181345657177138416514496 5125$
$\qquad 05706836607882239226051783 96$
$\omega_- = 0.0000482774800123446001888439306373712807603101 9956$
$\qquad 8016637476049430104693381532001893778229508716 1519$
$\qquad 3280456740060920451739038605$

and

$$(\Delta, \Delta) = \omega_+ \omega_-$$
$$=0.0000010353620568043209223478168122251645932 2490796$$
$$0950425299666957831367631065178091888477090812 5002$$
$$0176418696057835591055801221,$$

where (f, f) denotes the Petersson product of $f(z)$ (cf. [129]).

Remark 3.1. The functional equation (3.7) is equivalent to the modular relation

$$T_0(x) = (-1)^{\frac{\kappa}{2}} x^{-\kappa} T_0 \left(\frac{1}{x} \right).$$

Now a-times differentiation can be performed in the same way as in §2 to yield the formula

$$\frac{d^a}{dt^a} T_a(x) = (-2\pi)^a T_0(x).$$

As remarked in Remark 2.3, this pertains to the correspondence between the differentiation of Lambert series and the shift of the argument in the associated Dirichlet series. The difference between our Theorem 3.1 and Razar's results (Theorems 2 and 3 in [102]) is that his investigation is restricted to the case $a = \kappa - 1$ while ours is concerned with any $a \geq 1$, which enables us to express $L(a, f)$ in closed form as in Corollary 3.1.

4. The Riemann-Siegel integral formula for L-functions associated with cusp forms

In this section we shall establish the Riemann-Siegel integral formula for $L(s, f)$, for a cusp form f, as a generalization of the celebrated Riemann-Siegel integral formula for the Riemann zeta-function [109], expound in [38], [69], [123]. A. Guthman has recently worked on this subject energetically and obtained very interesting results [58, 59, 60, 61] including the Riemann-Siegel integral formula for $L(s, f)$ above and for $\zeta^2(s)$. But the modular relation was not perceived, and as a result proofs are rather complicated. In addition, antedating Guthman [58, 59, 60, 61], there are many papers by Russian mathematicians who got very important and hitherto best results, notably Kuz'min, Koshlyakov, Lavrik; save for the last, other two seem to be almost forgotten, the importance of their results notwithstanding.

The work still in progress, we wish to make this section rather of survey type, giving two proofs of the Riemann-Siegel integral formula, the first based on Kuz'min's method, the second on Koshlyakov's method, to restore the prominence of their achievements.

4.1 Kuz'min's method

This method is very closely related to the incomplete gamma function coefficients series for the zeta-function in question established in §1.

Recalling the functional equation (3.7) for $L(s, f)$, we see that Formula (1.6) now takes the form

$$
(2\pi)^{-s}\Gamma(s)L(s, f)
$$

$$
= (2\pi)^{-s} \sum_{n=1}^{\infty} \frac{a_n}{n^s} \Gamma(s, 2\pi n w)
$$

$$
+ (-1)^{\frac{\kappa}{2}} (2\pi)^{-(\kappa-s)} \sum_{n=1}^{\infty} \frac{a_n}{n^{\kappa-s}} \Gamma(\kappa - s, 2\pi n/w) \qquad (4.1)
$$

valid for $\Re w > 0$.

Kuz'min's argument lies in the observation that the series on the right of (1.12) corresponding to (4.1) are absolutely convergent not only for $\Re w > 0$ but also for $\arg w = \frac{\pi}{2}$, so that we may take w to be a purely imaginary number. It should be noted that although Kuz'min [86] treated the Dirichlet L-function, he restricted to the case where the gamma factor is $\Gamma(s/2)$ *i.e.* the case of Riemann's functional equation (cf. Lavrik [91]).

Our subsequent discussion is rather relevant to Akiyama and Tanigawa [1] (cf. Ferminger [46] also) in the sense that both follow by the choice $\arg w = \frac{\pi}{4}$ in the absolutely convergent region $\Re w > 0$, who gave the expression as an incomplete gamma function coefficients series for the L-function associated to a cusp form of weight k with respect to the congruence subgroup $\Gamma_0(N)$ of $SL_2(\mathbb{Z})$ (Lemma 1). They truncated the series at large M and made an estimate of the error term, proceeding to conduct numerical calculation of various zeta- and L-functions $L(s, E)$ for some range of $\Im s$. To make their calculation practical, Akiyama and Tanigawa made a delicate choice of the parameter w (r in their notation, they take $r = e^{i(\frac{\pi}{2} - \delta(t))}$, $0 < \delta(t) \le \frac{\pi}{2}$ and choose $\delta(t)$ in a delicate manner). The first paragraph of p.204 [1] is quite instructive when compared with the remark made by Lavrik [89].

Since we may prove the Riemann-Siegel integral formula for $L(s, E)$ also in the same vein as we prove it for $L(s, f)$, there is a hope that we may get an asymptotic formula, which then may enable us to avoid the use of incomplete gamma function values which they remark are difficult to calculate and are time-consuming (the first line of §4, p.1208 [1]). Compare this with Terras [121] and Terras [122].

Since Akiyama-Tanigawa's L-function $L(s, f)$ satisfies the same type of Hecke's functional equation

$$\left(\frac{2\pi}{\sqrt{N}}\right)^{-s} \Gamma(s)L(s,f) = \mu(-1)^{\frac{\kappa}{2}} \left(\frac{2\pi}{\sqrt{N}}\right)^{-(\kappa-s)} \Gamma(\kappa-s)L(\kappa-s,f) \quad (4.2)$$

$\mu = \pm 1$ as (3.7), we may state the result for this $L(s, f)$ (our case can be read off as the special case $N = 1$, $\mu = 1$).

Our aim is to prove the following Theorem 4.1 in two ways. To state it we introduce some notation.

For a cusp form $f(z) = \sum_{n=1}^{\infty} a_n e^{2\pi i n z}$ of weight κ with respect to $\Gamma_0(N)$, we put

$$\varphi(z, \xi) = \sum_{n=1}^{\infty} \frac{a_n}{z-n} e^{-\frac{2\pi n \xi}{\sqrt{N}}} \quad (4.3)$$

for $\Re\xi > 0$.

Theorem 4.1. *Let* $\varepsilon = e^{\frac{\pi i}{4}}$ *and* $0 < \Re s < \kappa$. *Then we have the Riemann-Siegel integral formula*

$$\left(\frac{2\pi}{\sqrt{N}}\right)^{-s} \Gamma(s)L(s,f) = \left(\frac{2\pi}{\sqrt{N}}\right)^{-s} \frac{\Gamma(s)}{2\pi i} \int_{0\swarrow 1} \varphi(z,\varepsilon) e^{\frac{2\pi i}{\sqrt{N}}\bar{\varepsilon}z} z^{-s} dz$$

$$- \mu(-1)^{\frac{\kappa}{2}} \left(\frac{2\pi}{\sqrt{N}}\right)^{-(\kappa-s)} \frac{\Gamma(\kappa-s)}{2\pi i} \int_{0\searrow 1} \varphi(z,\bar{\varepsilon}) e^{-\frac{2\pi i}{\sqrt{N}}\varepsilon z} z^{-(\kappa-s)} dz,$$

$$(4.4)$$

where the path $0 \swarrow 1$ *and* $0 \searrow 1$ *are well known ones due to Siegel, i.e. the directed lines making the angle* $\frac{5}{4}\pi$ *and* $\frac{3}{4}\pi$ *respectively, with the positive direction of the real axis and intersecting with it between the interval* $[0, 1]$.

Proof. In view of the form (4.2) of the functional equation, the expression (4.1) as series with incomplete gamma function coefficients takes the form

$$\left(\frac{2\pi}{\sqrt{N}}\right)^{-s} \Gamma(s)L(s,f) = \sum_{n=1}^{\infty} \frac{a_n}{\left(\frac{2\pi}{\sqrt{N}}n\right)^s} \Gamma\left(s, \frac{2\pi n w}{\sqrt{N}}\right)$$

$$+ \mu(-1)^{\frac{\kappa}{2}} \sum_{n=1}^{\infty} \frac{a_n}{\left(\frac{2\pi}{\sqrt{N}}n\right)^{\kappa-s}} \Gamma\left(\kappa - s, \frac{2\pi n w^{-1}}{\sqrt{N}}\right). \quad (4.5)$$

Since w is at our disposal as long as it remains in the sector $|\arg w| < \frac{\pi}{2}$, we may take $w = \varepsilon$ to get

$$\left(\frac{2\pi}{\sqrt{N}}\right)^{-s}\Gamma(s)L(s,f) = \left(\frac{2\pi}{\sqrt{N}}\right)^{-s}\sum_{n=1}^{\infty}\frac{a_n}{n^s}\Gamma\left(s,\frac{2\pi n\varepsilon}{\sqrt{N}}\right)$$

$$+ \mu(-1)^{\frac{\kappa}{2}}\left(\frac{2\pi}{\sqrt{N}}\right)^{-(\kappa-s)}\sum_{n=1}^{\infty}\frac{a_n}{n^{\kappa-s}}\Gamma\left(\kappa-s,\frac{2\pi n\bar{\varepsilon}}{\sqrt{N}}\right). \qquad (4.6)$$

As in [1], Formula (5), we have

$$\Gamma(s,w) = w^s\int_1^\infty e^{-wt}t^{s-1}dt$$

for $\Re w > 0$. Hence, as in Kuz'min [86], writing $t+1$ for t in the integral above, we have

$$\Gamma(s,w) = w^se^{-w}\int_0^\infty e^{-wt}(1+t)^{s-1}dt.$$

Suppose that $0 < \Re s < 1$ for the moment, and replace the factor $(1+t)^{s-1}$ by

$$\frac{1}{\Gamma(1-s)}\int_0^\infty e^{-(1+t)u}u^{-s}ds$$

to obtain

$$\Gamma(s,w) = w^ae^{-w}\frac{1}{\Gamma(1-s)}\int_0^\infty e^{-wt}dt\int_0^\infty e^{-(1+t)u}u^{-s}du. \qquad (4.7)$$

By absolute convergence, we may change the order of integration in (4.7), and therefore after integrating,

$$\Gamma(s,w) = w^se^{-w}\frac{\Gamma(s)\sin\pi s}{\pi}\int_0^\infty \frac{e^{-u}u^{-s}}{u+w}du, \qquad (4.8)$$

which holds for $\Re w > 0$ and $0 < \Re s < 1$.

Now putting $w = \frac{2\pi n\varepsilon}{\sqrt{N}}$ and $\frac{2\pi n\bar{\varepsilon}}{\sqrt{N}}$, and changing the variables $u \leftrightarrow \frac{2\pi x}{\sqrt{N}}$ in (4.8), we have

$$\Gamma\left(s,\frac{2\pi n\varepsilon}{\sqrt{N}}\right) = (\varepsilon n)^se^{-\frac{2\pi n\varepsilon}{\sqrt{N}}}\frac{\Gamma(s)\sin\pi s}{\pi}\int_0^\infty \frac{e^{-\frac{2\pi x}{\sqrt{N}}}}{x+\varepsilon n}x^{-s}dx \qquad (4.9)$$

and

$$\Gamma\left(s,\frac{2\pi n\bar{\varepsilon}}{\sqrt{N}}\right) = (\bar{\varepsilon} n)^se^{-\frac{2\pi n\bar{\varepsilon}}{\sqrt{N}}}\frac{\Gamma(s)\sin\pi s}{\pi}\int_0^\infty \frac{e^{-\frac{2\pi x}{\sqrt{N}}}}{x+\bar{\varepsilon} n}x^{-s}dx. \qquad (4.10)$$

Now we are to replace the last integral by a contour integral along the line from right to left intersecting with the imaginary axis between 0 and $-i$. In [85] Kuz'min did not give details, and we shall give some details.

Let $0 < \Re s < 1$ and $b, c > 0$. By Cauchy's theorem, we have

$$\int_{-\infty-ic}^{\infty-ic} \frac{e^{2\pi ibz}}{z-v} z^{-s} dz = -2ie^{\frac{\pi is}{2}} \sin \pi s \int_0^\infty \frac{e^{-2\pi bx}}{x+iv} x^{-s} dx \qquad (4.11)$$

for $\Im v < -c$, and

$$\int_{-\infty+ic}^{\infty+ic} \frac{e^{-2\pi ibz}}{z-v} z^{-s} dz = 2ie^{\frac{-\pi is}{2}} \sin \pi s \int_0^\infty \frac{e^{-2\pi bx}}{x-iv} x^{-s} dx \qquad (4.12)$$

for $\Im v > c$.

Choosing $b = 1/\sqrt{N}$, $0 < c < \frac{1}{\sqrt{2}}$ and putting $v = \bar{\varepsilon} n$ in (4.11) and $v = \varepsilon n$ in (4.12), we get

$$\sin \pi s \int_0^\infty \frac{e^{-\frac{2\pi}{\sqrt{N}}x}}{x+\varepsilon n} x^{-s} dx = \frac{\varepsilon^{-s}}{2i} \int_{0\diagup 1} \frac{e^{\frac{2\pi i}{\sqrt{N}}\bar{\varepsilon} z}}{z-n} z^{-s} dz \qquad (4.13)$$

and

$$\sin \pi s \int_0^\infty \frac{e^{-\frac{2\pi}{\sqrt{N}}x}}{x+\bar{\varepsilon} n} x^{-s} dx = -\frac{\varepsilon^s}{2i} \int_{0\diagdown 1} \frac{e^{-\frac{2\pi i}{\sqrt{N}}\varepsilon z}}{z-n} z^{-s} dz. \qquad (4.14)$$

Note that the integrals on the right-hand sides of the above two equations are absolutely convergent for $\Re s > 0$.

Substituting these results into (4.6), we get the assertion of Theorem 4.1. $\qquad \square$

For completeness' sake we give a proof of (4.11). (4.12) is proved in a similar way.

Consider the integral

$$I = \int_{-\infty-ic}^{\infty-ic} \frac{e^{2\pi ibz}}{z-v} z^{-s} dz,$$

which is absolutely convergent for $\sigma > 0$. Let $\mathcal{R}_{T,\varepsilon}$ be a rectangle which consist of vertices at $(T, 0), (-T, 0), (-T, -ic), (T, -ic)$ and where the origin is excluded by the lower semi-circle of radius ε. Integrating the function $\frac{e^{2\pi ibz}}{z-v} z^{-s}$ along the sides of $\mathcal{R}_{T,\varepsilon}$ and letting $T \to \infty$ and $\varepsilon \to 0$, we get

$$I + \int_\infty^0 \frac{e^{2\pi ibx}}{x-v} x^{-s} dx + \int_0^{-\infty} \frac{e^{2\pi ibx}}{x-v} x^{-s} dx = 0.$$

By rewriting the variable $x \to e^{-\pi i}x$ in the last integral, we deduce that

$$I = I_1 - e^{\pi i s}I_2,$$

where

$$I_1 = \int_0^\infty \frac{e^{2\pi i b x}}{x - v}x^{-s}dx, \quad I_2 = \int_0^\infty \frac{e^{-2\pi i b x}}{x + v}x^{-s}dx.$$

In I_1 we rotate the integration path by $\frac{\pi}{2}$ $(z = ix)$ to get

$$I_1 = e^{-\frac{\pi i}{2}s}\int_0^\infty \frac{e^{-2\pi b x}}{x + iv}x^{-s}dx,$$

which is absolutely convergent for $\sigma < 1$.

Similarly in I_2, rotating the path by $-\frac{\pi}{2}$ $(z = -ix)$, we get

$$I_2 = e^{-\frac{\pi i}{2}s}\int_0^\infty \frac{e^{-2\pi b x}}{x + iv}x^{-s}dx.$$

Thus, for $0 < \sigma < 1$ and $\Im v < -c$, we have

$$I = \left(e^{-\frac{\pi}{2}s} - e^{\frac{3\pi i}{2}s}\right)\int_0^\infty \frac{e^{-2\pi b x}}{x + iv}x^{=s}dx,$$

i.e. (4.11).

4.2   Koshlyakov's method

Koshlyakov [83] developed his theory of Dedekind zeta-function $\zeta_\Omega(s)$ of a rational or quadratic field Ω, real or imaginary. In the sense that he was successful in deriving the Riemann-Siegel integral formula for $\zeta_\Omega(s)$ for a real quadratic field Ω, Koshlyakov's achievement remains the record till now (cf. Guthman [61]).

The case of $\zeta_\Omega(s)$ with Ω an imaginary quadratic field is almost the same as the case of a cusp form zeta-function $L(s, f)$, and we shall prove here our Theorem 3.1 applying his reasoning in our set-up. We shall present it in such a way that by suitably modifying the proof, it can be interpreted as that of Koshlyakov's original argument.

Following him, we put

$$A = \frac{\sqrt{N}}{2\pi} \quad \left(\text{in his case} \quad A = \frac{\sqrt{|\Delta|}}{2^{r_2}\pi^{(r_1+r_2-1)/2}}\right).$$

Then the Dedekind zeta-function $\zeta_\Omega(s)$ of the field Ω or the cusp form zeta-function $L(s, f)$ satisfies the functional equation

$$\rho(s) = c\rho(r - s), \tag{4.15}$$

$c = 1$, $r = 1$ in Koshlyakov, $c = \mu(-1)^{\frac{\kappa}{2}}$, $r = \kappa$ in our case, where we put

$$\rho(s) = \begin{cases} A^s \Gamma(s/2)^{r_1} \Gamma(s)^{r_2} \zeta_\Omega(s) \\ A^s \Gamma(s) L(s, f) \end{cases} \tag{4.16}$$

as the case may be, and where r_1, r_2 are non-negative integers subject to $r_1 + r_2 \le 2$ (in the case in question $r_1 = 0$, $r_2 = 1$, see also (4.2)).

A remarkable feature of Koshlyakov's method is his ingenious use of the (solutions of) differential equations, foreseeing the 21st century interaction of number theory and differential equations. Only through his method, it is clearly understood why we should have cylinder functions.

Koshlyakov uses five functions K, L, X, Y and Z which are closely related to solutions of the Bessel differential equation, but we need X, Y and Z in what follows:

$$X(x) = X_{r_1, r_2}(x) = \frac{1}{2\pi i} \int_{(\alpha)} \Gamma\left(\frac{z}{2}\right)^{r_1} \Gamma(z)^{r_2} x^{-z} dz, \tag{4.17}$$

with $\Re x > 0$, $0 < \alpha < 2$ and

$$Y(x) = Y_{r_1, r_2}(x) = \frac{1}{2\pi i} \int_{(\alpha)} \frac{\pi}{2 \sin \frac{\pi z}{2} \Gamma\left(\frac{2-z}{2}\right)^{r_1} \Gamma(2-z)^{r_2}} x^{-z} dz \tag{4.18}$$

subject to the same restriction as for X.

Since in our case, $(r_1, r_2) = (0, 1)$ we have (cf. Table on p.122 of [83])

$$X(x) = e^{-x} \tag{4.19}$$

and

$$Y(x) = \frac{\sin x}{x}. \tag{4.20}$$

It is known ([83, (9.10),(9.11)]) that

$$\int_1^\infty X(ax) Y(bx) x \, dx = \frac{Z(a, b)}{a^2 + b^2}, \quad a > 0, b > 0, \tag{4.21}$$

where

$$Z(a, b) = \frac{X(a)}{2^{2-2r_2}} \left\{ \rho^{r_2} Y(b\rho) \right\}_{\rho=1}^{(r_1+2r_2-1)}$$

$$+ \frac{Y(b)}{2^{2-2r_2}} \left\{ (-1)^{(r_1+2r_2-1)} X(a\rho) \right\}_{\rho=1}^{(r_1+2r_2-1)}. \tag{4.22}$$

In our case, (4.21) and (4.22) together read

$$\int_1^\infty X(ax)Y(bx)x\,dx$$

$$= \frac{1}{a^2+b^2}\left\{ X(a)\frac{d}{dx}(xY(bx))_{x=1} - Y(b)\frac{d}{dx}(X(ax))_{x=1} \right\}$$

$$= \frac{e^{-a}}{a^2+b^2}\left(\cos b + a\frac{\sin b}{b} \right) = \frac{e^{-a}}{a^2+b^2}G(a,b), \qquad (4.23)$$

say.

We are now ready to another proof of Theorem 4.1 by Koshlyakov's method, featuring hereafter the case

$$\rho(s) = \rho_f(s) = A^s\Gamma(s)L(s,f).$$

Let $\alpha > \kappa$ in (4.18). With the above-mentioned notation (recall $\varepsilon = e^{\frac{\pi i}{4}}$, $A = \frac{\sqrt{N}}{2\pi}$ etc.), we calculate

$$\Phi_f(\varepsilon,s) = \varepsilon^s \sum_{n=1}^\infty a_n \int_1^\infty X\left(\frac{n\varepsilon x}{A}\right)x^{s-1}dx \qquad (4.24)$$

for $0 < \sigma < \kappa$.

Substituting (4.17) in (4.24) and changing the order of integration, which we may, we see that the inner integral is

$$\int_1^\infty x^{s-z-1}dz,$$

which is $1/(z-s)$.

Hence

$$\Phi_f(\varepsilon,s) = \frac{1}{2\pi i}\int_{(\alpha)} \rho_f(z)\frac{\varepsilon^{s-z}}{z-s}dz, \qquad (4.25)$$

for which we apply the same procedure, as in the proof of Theorems 2.1 and 3.1, of shifting the line of integration to the left to obtain

$$\Phi_f(\varepsilon,s) = \rho_f(s) + \frac{1}{2\pi i}\int_{(\kappa-\alpha)} \rho_f(z)\frac{\varepsilon^{s-z}}{z-s}dz, \qquad (4.26)$$

Changing the variables $z = \kappa - w$ and using the functional equation (4.15), we derive that

$$\frac{1}{2\pi i}\int_{(\kappa-\alpha)} \rho_f(z)\frac{\varepsilon^{s-z}}{z-s}dz = -c\bar{\varepsilon}^{\kappa-s}\sum_{n=1}^\infty a_n \int_1^\infty X\left(\frac{n\bar{\varepsilon}x}{A}\right)x^{\kappa-1-s}dx$$

$$= -c\Phi(\bar{\varepsilon},\kappa-s), \qquad (4.27)$$

by reversing the argument that led us from (4.24) to (4.25), whence that

$$\rho_f(s) = \Phi_f(\varepsilon, s) + c\Phi_f(\bar{\varepsilon}, \kappa - s) \tag{4.28}$$

for $0 < \sigma < \kappa$.

Now we shall prove an expression for the integral appearing on the right of (4.24), in terms of the function $G(a, b)$ introduced in (4.23).

First we note that from (4.18) we have for $0 < \sigma < 2$

$$x^{s-1} = \frac{2}{\pi} \sin \frac{\pi s}{2} \Gamma(s) \int_0^\infty Y(xt) x t^{1-s} dt. \tag{4.29}$$

We substitute (4.29) for x^{s-1} in the definition of the integral I_n in question

$$I_n = \int_1^\infty X\left(\frac{n\varepsilon x}{A}\right) x^{s-1} dx \tag{4.30}$$

and change the order of integration to infer that

$$
\begin{aligned}
I_n &= \frac{2}{\pi} \sin \frac{\pi s}{2} \Gamma(s) \int_0^\infty t^{1-s} \int_1^\infty X\left(\frac{n\varepsilon x}{A}\right) Y(xt) x \, dx \, dt \\
&= \frac{2}{\pi} \sin \frac{\pi s}{2} \Gamma(s) \int_0^\infty \left(\frac{u}{A}\right)^{1-s} \int_1^\infty X\left(\frac{n\varepsilon x}{A}\right) Y\left(\frac{ux}{A}\right) x \, dx \frac{1}{A} du
\end{aligned} \tag{4.31}
$$

by the change of variables $t = \frac{u}{A}$.

The inner integral on the right-hand side of the last equality in (4.31) is

$$\frac{e^{-\frac{n\varepsilon}{A}}}{\left(\frac{n\varepsilon}{A}\right)^2 + \left(\frac{u}{A}\right)^2} G_n(u), \tag{4.32}$$

in view of (4.23), where we put for simplicity

$$G_n(u) = G\left(\frac{n\varepsilon}{A}, \frac{u}{A}\right) = \frac{n\varepsilon}{u} \sin \frac{u}{A} + \cos \frac{u}{A}. \tag{4.33}$$

Hence

$$I_n = \frac{2}{\pi} A^s \sin \frac{\pi s}{2} \Gamma(s) e^{-\frac{n\varepsilon}{A}} J_n, \tag{4.34}$$

where

$$J_n = \int_0^\infty \frac{G_n(u)}{u^2 + in^2} u^{1-s} du. \tag{4.35}$$

It remains to express J_n as a contour integral, which can be done by the same procedure as we adopted in 4.1. Then, we have, for $0 < \xi < 1$,

$$\int_{\infty - i\xi}^{-\infty - i\xi} \frac{G_n(u)}{u^2 + in^2} u^{1-s} du$$

$$= -\int_0^\infty \frac{G_n(u)}{u^2 + in^2} u^{1-s} du + e^{\pi i s} \int_0^\infty \frac{G_n(u)}{u^2 + in^2} u^{1-s} du$$

$$= e^{\frac{\pi i s}{2}} 2i \sin \frac{\pi s}{2} \int_0^\infty \frac{G_n(u)}{u^2 + in^2} u^{1-s} du, \tag{4.36}$$

which leads, by the change of variables $u = \bar{\varepsilon} x$ on the left, to

$$\int_{0 \swarrow 1} \frac{G_n(\bar{\varepsilon} x)}{-i(x^2 - n^2)} (\bar{\varepsilon} x)^{1-s} \bar{\varepsilon} dx = \varepsilon^{2s} 2i \sin \frac{\pi s}{2} \int_0^\infty \frac{G_n(u)}{u^2 + in^2} u^{1-s} du. \tag{4.37}$$

Using (4.37) in (4.34), we conclude that

$$\varepsilon^s I_n = \frac{1}{\pi i} A^s \Gamma(s) e^{-\frac{n\varepsilon}{A}} \int_{0 \swarrow 1} \frac{G_n(\bar{\varepsilon} x)}{x^2 - n^2} x^{1-s} dx$$

$$= \frac{1}{2\pi i} A^s \Gamma(s) e^{-\frac{n\varepsilon}{A}} \int_{0 \swarrow 1} x^{-s} \left(\frac{1}{x - n} e^{\frac{i\bar{\varepsilon} x}{A}} + \frac{1}{x + n} e^{-\frac{i\bar{\varepsilon} x}{A}} \right) dx$$

$$= \frac{1}{2\pi i} A^s \Gamma(s) e^{-\frac{n\varepsilon}{A}} \int_{0 \swarrow 1} x^{-s} \frac{1}{x - n} e^{\frac{i\bar{\varepsilon} x}{A}} dx \tag{4.38}$$

whose left-hand side member is a holomorphic function of s in $\sigma > 0$ and whose right-hand side integral is convergent for $\sigma > 0$, and so (4.38) is valid in $\sigma > 0$.

Substituting (4.38) in (4.24), we finally obtain

$$\Phi_f(\varepsilon, s) = \frac{1}{2\pi i} A^s \Gamma(s) \int_{0 \swarrow 1} x^{-s} \left(\sum_{n=1}^\infty \frac{a_n e^{-\frac{n\varepsilon}{A}}}{x - n} \right) e^{\frac{i\varepsilon x}{A}} dx, \tag{4.39}$$

which coincides with the first integral on the right of (4.1).

We may adopt a similar reasoning to the counterpart

$$\Phi_f(\bar{\varepsilon}, \kappa - s) = \bar{\varepsilon}^{\kappa - s} \sum_{n=1}^\infty a_n I_n', \tag{4.40}$$

with

$$I_n' = \int_1^\infty X \left(\frac{n\bar{\varepsilon} x}{A} \right) x^{\kappa - 1 - s} ds. \tag{4.41}$$

Indeed I_n' can be expressed as

$$\frac{2}{\pi}A^{\kappa-s}\sin\frac{\pi(\kappa-s)}{2}\Gamma(\kappa-s)e^{-\frac{n\bar\varepsilon}{A}}J_n',$$

where

$$J_n' = \int_0^\infty \frac{\bar G_n(u)}{u^2-in^2}u^{1-\kappa+s}du$$

and where

$$\bar G_n(u) = \overline{G_n(u)} = \frac{n\bar\varepsilon}{u}\sin\frac{u}{A}+\cos\frac{u}{A}.$$

J_n' may be transformed into a contour integral by a similar way as above. In this way we get a counterpart of (4.38) valid in $0<\sigma<\kappa$:

$$\bar\varepsilon^{\kappa-s}\int_1^\infty X\left(\frac{n\bar\varepsilon x}{A}\right)x^{\kappa-1-s}dx$$

$$= -\frac{1}{2\pi i}A^{\kappa-s}\Gamma(\kappa-s)X\left(\frac{n\bar\varepsilon}{A}\right)\int_{0\searrow 1}x^{-(\kappa-s)}\frac{1}{x-n}e^{-\frac{i\varepsilon x}{A}}dx \qquad (4.42)$$

whence we deduce finally, corresponding to (4.39),

$$\Phi_f(\bar\varepsilon,\kappa-s) = -\frac{1}{2\pi i}A^{\kappa-s}\Gamma(\kappa-s)\int_{0\searrow 1}x^{-(\kappa-s)}\left(\sum_{n=1}^\infty \frac{a_n e^{-\frac{n\bar\varepsilon}{A}}}{x-n}e^{-\frac{i\varepsilon x}{A}}\right)dx,$$

$$(4.43)$$

which is seen to coincide with the second integral on the right of (4.1), and the proof is complete.

5. Research Notices

In the following we shall give a rambling survey of some of our recent research problems and thereto relevant works of other people.

5.1 Approximate functional equation

Since its introduction in 1921 [63] and subsequently by Hardy and Littlewood, the approximate functional equation has been one of the major sources of research on the behaviour of the zeta-function in the critical strip.

We refer to Lavrik [88] for the most complete list of references, one of which by Apostol and Sklar [4] deserves special attention as it contains an identity for the zeta-function of Hecke type in terms of the Riesz sum (Theorem 1). We have excluded the Riesz sum part in this paper, but this is significant not only from the point of view of application to analytic number theory but also of Eichler's integrals (see Weil [126]). This direction of research will be pursued in forthcoming papers.

With regard to approximate functional equations, we should add Jutila [70] which was reformed by Guthman [60] in the form of the Riemann-Siegel (asymptotic) formula.

Although only a passing remark is made in Lavrik, the papers of Linnik are to be worthy of special attention, where he proves an incomplete gamma function coefficient series expressions for Dirichlet L-functions (e.g. Formula (26.7) in [92] or (40.1) in Lemma 7 in [93]) which are suitable for deriving good estimate for the mean 6th power of Dirichlet L-function on the critical line for $|t| \ll 1$. In conjunction with Kuz'min's results which are suitable for treating the case $|t| \geq 1$, Linnik succeeded in proving the mean sixth power of Dirichlet L-function.

Lavrik elucidated the difference between these two cases of Kuz'min and Linnik as the choice of the argument of w in (1.12) and obtained a very general form of the functional relations involving an extraneous function $\Phi(z)$ (Lavrik [89]), which he further developed into the most general setting in [90].

This latter paper is important as it unifies the theory of zeta- and L-functions satisfying functional equations with multiple gamma factors in such a way that by a proper choice of the extraneous function $\Phi(z)$, approximate functional equations with uniform estimate for the error terms are obtained, that it can contain the case of periodic zeta-function including that of an algebraic function field of one variable for which uniqueness is proved, and that from Theorem A one can deduce other known important functional relation that played a brilliant role in the development of number theory in the 20th century (also it has a good list of references).

Regarding the last point, the class described by Lavrik's Theorem A includes not only those functional relations due to Hardy and Littlewood and Linnik [92, 93] but also those papers of Siegel, Bochner and Chandrasekharan [27] and Chandrasekharan and Mandelbrojt [32] which form the bases of the proof of determination of the zeta-function through its functional equation, the first instance being given by Hamburger [62] (cf. references of Knopp [82]), as well as those for Heilbronn's zeta-function and Schmidt's zeta-function.

5.2   The converse Hecke theory or the determination of the zeta-, L-function through their functional equation

Following the last passage in §5.2, we shall say a few words about this important problem, part of which was already expounded in Remark 2.5. As in Remark 2.5, Knopp's survey [82] is very readable and we shall state only those points which are not referred to there fully.

After Hecke's most prominent theory [64, 65, 66], what are most important are probably the works of Siegel [111] which contains all promising seeds of ingredients used in subsequent research (and discussed in Knopp), of Weil ([125], cf. also [99]) which introduces the twists by Dirichlet characters (a continuation exits of this work which was done by Asai-Tanigawa, unpublished) and of Maass who introduced the non-analytic automorphic forms which has made his name immortal. The remarkable feature of Maass' works lies in the fact that the core function $f(x_1, x_2)$ is determined not from the functional equation nor from the modular relation for the associated theta-function, but from the uniqueness of the solution of the partial differential equation satisfied by $f(x_1, x_2)$. Here as well as in Koshlyakov [83] we may see that the more extensive use of differential equation may prevail in number theory.

Speaking of the differential equation we should refer to the recent intriguing paper of Chan [29] where two identities are shown to be equivalent under the modular relation, the first connected with partition mod 5 and the other with the differential equation for the Rogers-Ramanujan continued fraction starting from Raghavan [100]. This paper is unique not only from the fact that it connects two very remote-looking objects, but it opens up possibility for working with the other world of q-series etc.

We wish to turn to this direction of research subsequently. Chan communicated some relation among theta-functions connected with the number of representations as 24 squares etc. which seems interesting but difficult (cf. Berndt's survey, [25]).

5.3   Berndt's series of papers

B. C. Berndt during 6 years published in a series of 7 papers related to "Identities involving the coefficients of a class of Dirichlet series" and unified many of the existing results under this paradigm.

In spite of their importance and thorough nature of his treatment (as with his series of books "Ramanujan's notebooks I–V"), these papers did not seem to receive due attention, and we shall now briefly review the contents which are relevant to our present view point.

[12] treats the functional equation

$$\Gamma^m(s)\varphi(s) = \Gamma^m(\delta - s)\psi(\delta - s) \tag{5.44}$$

for m a positive integer extending Hecke's functional equation and gives the equivalence of (5.44) to the modular relation (Theorem 1),

$$\sum a_n E_m(\lambda_n x) = x^{-\delta} \sum b_n E_m(\mu_n/x) + P(x), \tag{5.45}$$

where $\varphi(s) = \sum_{n=1}^{\infty} a_n \lambda_n^{-s}$, $\psi(s) = \sum_{n=1}^{\infty} b_n \mu_n^{-s}$ and $E_m(x)$ denotes Voronoï's function

$$E_m(x) = \frac{1}{2\pi i} \int_{(c)} \Gamma^m(s) x^{-s} ds \quad (x > 0), \tag{5.46}$$

and to the Riesz sum $\frac{1}{\Gamma(\rho+1)} \sum'_{\lambda_n \le x} a_n (x - \lambda_n)^\rho$ involving the generalized Bessel function $K_\nu(x; \mu; m)$ (Definition 4).

[13] treats the logarithmic Riesz sum

$$\frac{1}{\rho!} \sum_{\lambda_n \le x}{}' a_n \log^\rho \left(\frac{x}{\lambda_n}\right),$$

complementing [12]

[14] deals with the special cases where the gamma factors have some given form of two factors and obtains identities for both Riesz sums and logarithmic Riesz sums referred to above. The most relevant to our work are first, a parenthetical remark on p.324 to the effect that the functional equation

$$\Gamma\left(s + \frac{\nu}{2}\right)\Gamma\left(s - \frac{\nu}{2}\right)\varphi(s) = \Gamma\left(\delta - s + \frac{\nu}{2}\right)\Gamma\left(\delta - s - \frac{\nu}{2}\right)\psi(\delta - s) \tag{5.47}$$

is equivalent to the (explicit) modular relation (see [77]).

$$2\sum a_n K_\nu(2\sqrt{\lambda_n x}) = 2x^{-\delta} \sum b_n K_\nu\left(2\sqrt{\frac{\mu_n}{x}}\right) + P(x), \tag{5.48}$$

secondly an identify involving exponential integral (§7) as a prototype of the incomplete gamma (function coefficients) series (expression), thirdly general modular relation involving Laguerre polynomials (§10), and lastly the most significant contribution of generalized Dirichlet series in §8, where Berndt derives the Bessel series expression for the generalized Dirichlet series (as remarked in §5.1)

$$\varphi(s, a) = \sum a_n (\lambda_n + a)^{-s} \quad (a > 0, \sigma \gg 0), \tag{5.49}$$

which was first introduced and studied in [11]. In the latter paper Berndt uses the Mellin-Barnes integral in an efficient way, which again is used in [15] to prove a general form of his Theorem 8.1 [14] for the case where the gamma factor is of general form. (The proof of Theorem 8.1 [14] is somewhat more structural and is different from other proofs). As remarked in §5.1, he also obtains an Abel summation result extending Atkinson's result [6].

[16] is concerned with the Voronoï summation formula which contains, as a special cases, the modular relation (cf. Example 1).

[17] develops the theory of Epstein zeta-functions by using his theory of generalized Dirichlet series which in turn depends on the Mellin-Barnes integrals or the modular relation. This paper, pushing the study of Epstein zeta-functions forward after the fundamental work of Epstein [41], but probably because of its title, not catch due attention, but undoubtedly, has later been extended by Terras [115].

5.4 Epstein zeta-function, Dedekind eta-function and Dedekind sums

In another series of papers [18, 19, 21], Berndt develops an extensive theory of generalized Lambert series and their transformations which are naturally accompanied by generalizations of the Dedekind eta-function and their transformations. These transformation formulas for generalized Dedekind eta-function includes most of the hitherto published results on the same topic. Associated with them are reciprocity laws for (generalized) Dedekind sums, for which we refer to a rather thorough survey of Hiramatsu, Mimura and Takeda [67].

We will return to the study of these transformation formulas from the point of view of modular relation elsewhere.

Regarding the Epstein zeta-function, it is Terras [115, 121] who, after Epstein [41], Bateman-Grosswald [9], Stark [114], Berndt [17], made a rather thorough study from two different angles, which is culminated in her two books [122].

The point of departure of Terras for Fourier expansion of the Epstein zeta-function was the Chowla-Selberg formula ([35], [105], also Kurokawa [84]) which gives a Fourier expansion of the Epstein zeta-function for a positive definite binary quadratic form, in terms of the modified K-Bessel function. Noticing that there appear the terms $\zeta(s)$ and $\zeta(s-1)$ in the expansion, she generalizes the Chowla-Selberg formula to n dimensions in $n-1$ different ways according to the choice of n_1, n_2 in $n = n_1 + n_2$, by viewing it as a non-holomorphic Eisenstein series for $GL(n, \mathbb{R})$, thus generalizing Berndt's result (l.c. §5.3).

Then putting $n = 2$, thus $n_1 = n_2 = 1$ and eliminating the Epstein zeta-function, she deduced the K-Bessel expansion relation between $\zeta(s)$ and $\zeta(s-1)$. In [116], a result is obtained connecting the special values $\zeta(2m)$ and $\zeta(2m-1)$, which is interpreted as the once differentiated form of Ramanujan's formula. Pushing forward the same reasoning, she was successful in deriving a K-Bessel series relation for $\zeta_K(s)$ and $\zeta_K(s-1)$ for an arbitrary algebraic number field. It appears that the obtained K-Bessel expansion is the expansion for the Epstein zeta-function (non-holomorphic Eisenstein series) used as an expansion for two adjacent values of the relevant zeta-function. We hope to return to a closer study of the relation between Terras's situation and ours at a later occasion.

On the other hand, Terras [121] turns her attention to the incomplete gamma expansion for the Epstein zeta-function, which she further continues in her papers [117, 119, 120] and books [122]. As we mentioned, her method is a direct descendent of that of Riemann and is a special case $w = 1$ of our case (and Bochner-Lavrik). In view of the presence of the parameter w, we are in a more lucid and easily tractable situation where we appeal to the modular relation. In the literature there are other papers on incomplete gamma series [49], [118] and we hope to return to them at another occasion.

We should add still two papers to the above list which are of the same nature as our consideration in §2. One is Barner [8] who evaluated the values of the ring class zeta- and L-functions of a real quadratic field at positive integral arguments in closed form using Dedekind sums, and the other is Smart [112] who evaluated the values of the Epstein zeta-function (*i.e.* the zeta-function of an imaginary quadratic field, so to say) at positive integral arguments using derivatives of (generalized) Lambert series. The use of (generalized) Lambert series is common in both and the present papers. Barner uses the Fourier expansion (*i.e.* the K-Bessel series) of the Eisenstein series (the sum of which is the Epstein zeta-function), or Maass wave forms, to express the values in question in term of derivatives of Lambert series and then applies Hecke's integral representation. We wish to return to investigation of such a situation at another occasion.

5.5 Numerical calculation and our series of papers

Extensive numerical calculations have been done of special values of various zeta-functions. The methods are either the Fourier expansion, *i.e.* the effective use of the K-Bessel functions (Borwens [28], Chaba-Pathria [31], Glasser-Zucker [48]) or incomplete gamma series (Akiyama-

Tanigawa [1] and Terras [121]) save for the elementary Euler-Maclaurin formula.

We refer to Akiyama-Tanigawa [1] and the papers therein referred to, especially, Fermiger [46], Manin [94, 95, 96], Mazur-Swinnerton-Dyer [97] for zeta-functions of elliptic curves etc. to Borwens [28] where a nice exposition of Madelung constants (*i.e.* constants associated to lattices, or the Epstein zeta-function for which we refer to §5.4), and also to fundamental works of Chaba-Pathria [31] and Glasser-Zucker [48] for Madelung and related constants.

Finally, we shall say a few words about our own series of papers and some related ones.

In [72, 73] we started proving the closed form for the series of the form

$$F(z) = \sum_{m=2}^{\infty} \frac{z^m}{m+\lambda} Z(m),$$

where $Z(m) = \zeta(m)$ and $L(m, \chi)$ in [72] and [73], respectively. The closed form evaluation of the above series was started by Srivastava et al (cf. [113]) and almost completed by Srivastava et al and by M. Katsurada [79] who used the binomial expansion and the Mellin-Barnes integral, respectively. If we view the Mellin-Barnes integral as an analytic form of the binomial expansion, then we may interpret the both of these results as based on the binomial expansion. We note that Srivastava et al's research dates back to 1772 Euler [45] while Katsurada's to Wilton [128]. In a sense Wilton went further as far as the closed form of $F(z)$ is concerned, and both Srivastava et al's and Kantsurada's results are bisection (or trisection) formulas evaluating

$$\frac{1}{2}\left(F(z) + F(-z)\right) \quad \text{or} \quad \frac{1}{2}\left(F(z) - F(-z)\right),$$

(or trisection, etc.). The evaluation of $F(z)$ itself was done by KKY [71, 72] in the case of the Hurwitz zeta-function and in [73] in the case of the Dirichlet L-function.

As stated in Remark 2.2, the book [113] of Srivastava and Choi contains, as one of main results, the acceleration formula for special values of zeta-functions in terms of the closed form of

$$G(z) = \sum_{m=2}^{\infty} \frac{z^m}{m(m+1)\cdots(m+\lambda)} \zeta(m, \alpha).$$

As is proved in [71], the closed form formulas for $F(z)$ and $G(z)$ are equivalent under a combinatorial identity. We have given at least four different proofs for the former, whose prototype is to be found in Ramanujan's notebook [22, p.279], while Srivastava et al gave a proof of the latter.

Then in §2 of [77], [76] we went on incorporating the modular relation for the Riemann zeta-function and the Dirichlet L-function, respectively, as the stronger acceleration formulas. Then, in §3 of [77] we found that it is not merely an acceleration formula but a more important principle which has been perceived fragmentarily in vast amount of literature, *i.e.* the modular relation principle which is partly described in this paper.

As stated in Remark 2.3, our papers [74, 75, 78] are devoted to explicit evaluation of special values of zeta- and L-function at rational points. In [74], after stating the corrected version of D. Klusch [81], we were successful in evaluating the values of the Riemann zeta-function at $1/3$ (and $2/3$). Then in [78], we interpreted the true meaning of Entry 8, Chapter 15 of Ramanujan's second notebook [23] connecting the Lambert series expression for the value $\zeta(\frac{1}{2})$ by extending previous results of Egami [39] and Katsurada [80]. Again Katsurada [80] obtained the thereto most general formula for the value at $1/2$ of the Hurwitz zeta-function. We extended this result of Katsurada to the most general case of arbitrary rational argument $b/2N$. The result of [75] is that one can obtain the corresponding evaluation in the case of the Riemann zeta-function at any rational argument b/N without the restriction to the denominator. In a forthcoming paper, we shall state results for Dirichlet L-functions and make some numerical calculations.

Acknowledgments

The authors would like to express their hearty thanks to Professor Dr. Winfried Kohnen for very enlightening discussion on §§2 and 3 whose remarks enabled them to include aspects of Eichler integrals.

The first author wishes to thank Dr. H. H. Chan especially for enthusiastic discussion at his house, Feb. 2001, on various aspects of our principle.

References

[1] S. Akiyama and Y. Tanigawa, Calculation of values of L-functions associated to elliptic curves, *Math. Comp.* **68** (1999), no. 227, 1201–1231.

[2] T. M. Apostol, Generalized Dedekind sums and transformation formulae of certain Lambert series. *Duke Math. J.* **17** (1950), 147–157.

[3] T. M. Apostol, Theorems on generalized Dedekind sums, *Pacific J. Math.* **2** (1952), 1–9.

[4] T. M. Apostol and A. Sklar, The approximate functional equation of Hecke's Dirichlet series, *Trans. Amer. Math. Soc.* **86** (1957), 446–462.

206

[5] L. Atkin and J. Lehner, Hecke operators on $\Gamma_0(m)$, *Math. Ann.* **185** (1970), 134–160.

[6] F.V. Atkinson, The Riemann zeta-function, *Duke Math. J.* **17** (1950), 63–38.

[7] F.V. Atkinson, The mean value of the Riemann zeta-function, *Acta Math.* **81** (1949), 353–376.

[8] K. Barner, Über die Werte der Ringklassen-L-Funktionen reelle-quadratischer Zahlkorper an naturlichen Argumentstellen, *J. Number Theory* **1** (1969), 28–64.

[9] P. T. Bateman and E. Grosswald, On Epstein's zeta function, *Acta Arith.* **9** (1964), 365–373.

[10] R. Bellman, An analog of an identity due to Wilton, *Duke Math. J.* **16** (1949), 539–545.

[11] B. C. Berndt, Generalized Dirichlet series and Hecke's functional equation, *Proc. Edinburgh Math. Soc.* **15** (1967), 309–313.

[12] B. C. Berndt, Identities involving the coefficients of a class of Dirichlet series I, *Trans. Amer. Math. Soc.* **137** (1969), 345–359.

[13] B. C. Berndt, Identities involving the coefficients of a class of Dirichlet series II, *Trans. Amer. Math. Soc.* **137** (1969), 361–374.

[14] B. C. Berndt, Identities involving the coefficients of a class of Dirichlet series III, *Trans. Amer. Math. Soc.* **146** (1969), 323–348.

[15] B. C. Berndt, Identities involving the coefficients of a class of Dirichlet series IV, *Trans. Amer. Math. Soc.* **149** (1970), 179–185.

[16] B. C. Berndt, Identities involving the coefficients of a class of Dirichlet series V, *Trans. Amer. Math. Soc.* **160** (1971), 139–156.

[17] B. C. Berndt, Identities involving the coefficients of a class of Dirichlet series VI, *Trans. Amer. Math. Soc.* **160** (1971), 157–167.

[18] B. C. Berndt, Generalized Dedekind eta-functions and generalized Dedekind sums, *Trans. Amer. Math. Soc.* **178** (1973), 495–508.

[19] B. C. Berndt, Generalized Eisenstein series and modified Dedekind sums, *J. Reine Angew. Math.* **272** (1975), 182–193.

[20] B. C. Berndt, Modular transformations and generalizations of several formulae of Ramanujan, *Rocky Mountain J. Math.* **7** (1977), 147–189.

[21] B. C. Berndt, Analytic Eisenstein series, theta functions, and series relations in the spirit of Ramanujan, *J. Reine Angew. Math.* **303/304** (1978), 332–365.

[22] B. C. Berndt, *Ramanujan's Notebooks Part I*, Springer Verlag, New York, 1985.

[23] B. C. Berndt, *Ramanujan's Notebooks Part II*, Springer Verlag, New York, 1989.

[24] B. C. Berndt, *Ramanujan's Notebooks Part IV*, Springer Verlag, New York, 1994.

[25] B. C. Berndt and Ae Ja Yee, Ramanujan's contributions to Eisenstein series, especially in his lost notebook, this volume.

[26] S. Bochner, Some properties of modular relations. *Ann. of Math.* (2) **53** (1951), 332–363.

[27] S. Bochner and K. Chandrasekharan, On Riemann's functional equation, *Ann. of Math.* **63** (1956), 336–360.

[28] M. Borwens and B. Borwens, *Pi and the AGM*, John Willey & Sons, Inc., New York, 1987.

[29] H. H. Chan, On the equivalence of Ramanujan's partition identities and a connection with the Rogers-Ramanujan continued fraction, *J. Math. Anal. Appli.* **198** (1996), 111–120.

[30] H. H. Chan and K. S. Chua, Representations of integers as sums of 32 squares, submitted for publication.

[31] A. N. Chaba and R. K. Pathria, Evaluation of a class of lattice sums in arbitrary dimensions, *J. Math. Phys.* **16** (1975), 1457–1460.

[32] K. Chandrasekharan and S. Mandelbrot, On Riemann's functional equation, *Ann. of Math.* **66** (1957), 285–296.

[33] K. Chandrasekharan and Raghavan Narasimhan, Hecke's functional equation and arithmetical identities. *Ann. of Math.* (2) **74** (1961), 1–23.

[34] K. Chandrasekharan and Raghavan Narasimhan, Functional equations with multiple gamma factors and the average order of arithmetical functions. *Ann. of Math.* (2) **76** (1962), 93–136.

[35] S. Chowla and A. Selberg, On Epstein's zeta-function (I), *Proc. Nat. Acad. Sci. U.S.A.* **35** (1949), 371–374 = Collected Papers of Atle Selberg, Vol. I, 367–370, Springer, 1989.

[36] J. B. Conrey and D. W. Farmer, An extension of Hecke's converse theorem, *International Math. Research Notices* **9** (1995), 445–463.

[37] K. Doi and T. Miyake, *Number Theory and Automorphic Forms*, Kinokuniya Pub. Tokyo, (in Japanese) 1976.

[38] H. M. Edwards, *Riemann's Zeta-function*, Academic Press, New York-London, 1974.

[39] S. Egami, A χ-analogue of a formula of Ramanujan for $\zeta(1/2)$, *Acta Arith.* **69** (1995), 189–191.

[40] M. Eichler, Eine Verallgemeinerung der Abelschen Integrale, *Math. Z.* **67** (1957), 267–298.

[41] P. Epstein, Zur Theorie allgemeiner Zetafunktionen, *Math. Ann.* **56** (1903), 615–644=*Arch. Math. u. Phys.* **7** (1902), 614–644 ; II, *ibid.* **63** (1907), 205–216.

[42] A. Erdélyi, *Higher Transcendental Functions I*, McGraw-Hill, New York, 1953.

[43] A. Erdélyi, *Higher Transcendental Functions II*, McGraw-Hill, New York, 1953.

[44] A. Erdélyi, W. Magnus, F. Oberhettinger and F. G. Tricomi, *Tables of integral transforms. Vol. I.* McGraw-Hill Book Company, Inc., New York-Toronto-London, 1954.

[45] L. Euler, Exercitationes analyticae, *Novi Comment. Acad. Sci. Petropol* **17** (1772), 173–204 = Opera omnia, Ser. I, Vol. 15, Leipzig (1927), 131–167.

[46] S. Ferminger, Zéros des fonction L de courbres elliptiques, *Experimental Math.*, **1** (1992), 167–173.

[47] A. O. Gel'fond, Some functional equations implied by equations of Riemann type, *Izv. Akad. Nauk SSSR Ser. Mat.* **24** (1960), 469–474.

[48] M. L. Glasser and I. J. Zucker, Lattice sums, *Theoretical Chemistry: Advances and Perspectives*, Vol. 5, ed. by D. Henderson, Academic Press 1980, 67–139.

[49] D. Goldfeld and C. Viola, Mean values of *L*-functions associated to elliptic, Fermat and other curves at the center of the critical strip, *J. of Number Theory* **11** (1979), 305–320.

[50] L. Goldstein, Zeta functions and Eichler integrals, *Acta Arith.* **36** (1980), 229–256.

[51] L. J. Goldstein and P. de la Torre, On the transformation of $\log \eta(\tau)$, *Duke Math. J.* **41** (1974). 291–297.

[52] I. S. Gradshteyn and I. M. Ryzhik, *Table of integrals, series, and products.* Translated from the fourth Russian edition. Fifth edition. Translation edited and with a preface by Alan Jeffrey. Academic Press, Inc., Boston, MA, 1994.

[53] E. Grosswald, Die Werte der Riemannschen Zetafunktion an ungeraden Argumentstellen, *Nachr. Akad. Wiss. Göttingen Math.-Phys. Kl.* II (1970), 9–13.

[54] E. Grosswald, Comments on some formulae of Ramanujan. *Acta Arith.* **21** (1972), 25–34.

[55] E. Grosswald, Relations between the values at integral arguments of Dirichlet series that satisfy functional equations. *Proc. Sympos. Pure Math.*, Vol. 24, Amer. Math. Soc., Providence, 1973, 111–122.

[56] A. P. Guinand, Functional equations and self-reciprocal functions connected with Lambert series. *Quart. J. Math.* Oxford Ser. **15** (1944), 11–23.

[57] A. P. Guinand, Some rapidly convergent series for the Riemann ξ-function. *Quart. J. Math.* Oxford Ser. (2) **6** (1955), 156–160.

[58] A. Guthman, The Riemann-Siegel integral formula for Dirichlet series associated to cusp forms, in *Analytic and Elementary Number Theory*, Vienna, 1996, 53–69.

[59] A. Guthman, Die Riemann-Siegel-Integralformel für die Mellintransformation von Spitzenformen. (German) *Arch. Math.* (Basel) **69** (1997), 391–402.

[60] A. Guthman, New integral representations for the square of the Riemann zeta-function. *Acta Arith.* **82** (1997), 309–330.

[61] A. Guthman, Asymptotic expansions for Dirichlet series associated to cusp forms. *Publ. Inst. Math.* (Beograd) (N.S.) **65** (79) (1999), 69–96.

[62] H. Hamburger, Über die Riemannsche Funktionalgleichung der ζ-Funktion I,II,III, *Math. Zeit.* **10** (1921), 240–254; **11** (1922), 224–245; **13** (1922) 283–311.

[63] G. H. Hardy and J. E. Littlewood, The zeros of Riemann's zeta-function on the critical line, *Math. Zeit.* **10** (1921), 283–317.

[64] E. Hecke, Über die Bestimmung Dirichletscher Reihen durch ihre Funktionalgleichung, *Math. Ann.* **11** (1936), 664–699.

[65] E. Hecke, *Lectures on Dirichlet Series, Modular Functions and Quadratic Forms*, Edwards, Ann Arbor, 1938.

[66] E. Hecke, Herleitung des Euler-Produkts der Zetafuncktion und einiger *L*-Reihen aus ihr Funktionalgleichung, *Math. Ann.* **119** (1944), 266–287.

[67] T. Hiramatsu, Y. Mimura and I. Takada, Dedekind sum and automorphic forms, *RIMS Kokyuroku* **572** (1985), 151–175 (in Japanese).

[68] J-I. Igusa, *Lectures on forms of higher degree*, Tata Institute of Fundamental Research, Bombay, 1978.

[69] A. Ivić, *The Riemann Zeta-Function*, John Wiley & Sons, New York, 1985.

[70] M. Jutila, On the approximate functional equation for $\zeta^2(s)$ and other Dirichlet series, *Quart. J. Math.* Oxford(2) **37** (1986), 193–209.

[71] S. Kanemitsu, M. Katsurada and M. Yoshimoto, On the Hurwitz-Lerch zeta function, *Aeq. Math.* **59** (2000), 1–19.

[72] S. Kanemitsu, H. Kumagai and M. Yoshimoto, Sums involving the Hurwitz zeta function, *The Ramanujan J.* **5** (2001), 5–19.

[73] S. Kanemitsu, H. Kumagai and M. Yoshimoto, On rapidly convergent series expressions for zeta- and L-values, and log sine integrals, *The Ramanujan J.* **5** (2001), 91–104.

[74] S. Kanemitsu, Y. Tanigawa and M. Yoshimoto, On zeta- and L-function values at special rational arguments via the modular relation, *Proc. Int. Conf. SSFA*, Vol. I (2001), 21–42.

[75] S. Kanemitsu, Y. Tanigawa and M. Yoshimoto, On the values of the Riemann zeta-function at rational arguments, *Hardy Ramanujan J.* **24** (2001), 10–18.

[76] S. Kanemitsu, Y. Tanigawa and M. Yoshimoto, On rapidly convergent series for Dirichlet L-function values via the modular relation, *Proc. of the International Conference on Number Theory and Discrete Mathematics in honour of Srinivasa Ramanujan*, 114–133, Hindustan Book Agency, 2002.

[77] S. Kanemitsu, Y. Tanigawa and M. Yoshimoto, On rapidly convergent series for the Riemann zeta-values via the modular relation, preprint.

[78] S. Kanemitsu, Y. Tanigawa and M. Yoshimoto, On multiple Hurwitz zeta-function values at rational arguments, preprint.

[79] M. Katsurada, Rapidly convergent series representations for $\zeta(2n+1)$ and their χ-analogue, *Acta Arith.* **90** (1999), 79–89.

[80] M. Katsurada, On an asymptotic formula of Ramanujan for a certain theta-type series, *Acta Arith.* **97** (2001), 157–172.

[81] D. Klusch, On Entry 8 of Chapter 15 of Ramanujan's Notebook II, *Acta Arith.* **58** (1991), 59–64.

[82] M. Knopp, Hamberger's theorem on $\zeta(s)$ and the abundance principle for Dirichlet series with functional equations, *Number Theory* (ed. by R. P. Bambah et al.), 201–216, Hindustan Book Agency, New Delhi, 2000.

[83] N. Koshlyakov, Investigation of some questions of analytic theory of the rational and quadratic fields, I-III (Russian), *Izv. Akad. Nauk SSSR, Ser. Mat.* **18** (1954), 113–144, 213–260, 307–326; Errata **19** (1955), 271.

[84] N. Kurokawa, *100 years of zeta regularized product*, *The 39th algebra symposium* (1994), 153–166 (in Japanese).

[85] R. Kuz'min, Contributions to the theory of a class of Dirichlet series (Russian), *Izv. Akad. Nauk SSSR, Ser. Math. Nat. Sci.* **7** (1930), 115–124.

[86] R. Kuz'min, On the roots of Dirichlet series, *Izv. Akad. Nauk SSSR, Ser. Math. Nat. Sci.* **7** (1934), 1471–1491.

[87] S. Lang, *Introduction to Modular Forms*, Springer, Berlin-New York, 1976.

210

[88] A. F. Lavrik, Approximate functional equations of Dirichlet functions (Russian). *Izv. Akad. Nauk SSSR*, Ser. Mat. **32** (1968), 134–185; English translation in *Math. USSR-Izv.* **2** (1968), 129–179.

[89] A. F. Lavrik, An approximate functional equation for the Dirichlet *L*-function, *Trudy Moskov Mat. Obšč.* **18** (1968), 91–104 =Trans. *Moscow Math. Soc.* **18** (1968), 101–115.

[90] A. F. Lavrik, The principle of the theory of nonstandard functional equation for Dirichlet functions, consequences and applications of it, *Trudy Mat. Inst. Steklov* **132** (1973), 70–76= *Proc. Steklov Int. Math.* **132** (1973), 77–85.

[91] A. F. Lavrik, Functional equations with a parameter for zeta-functions (Russian). *Izv. Akad. Nauk SSSR* Ser. Mat. **54** (1990), 501–521; English translation in *Math. USSR-Izv.* **36** (1991), 519–540.

[92] Ju. V. Linnik, An asymptotic formula in an additive problem of Hardy and Littlewood, *Izv. Akad. Nauk SSSR Ser. Mat.* **24** (1960), 629–706; English transl., *Amer. Math. Soc. Transl.* (2) **46** (1965), 65–148.

[93] Ju. V. Linnik, All large numbers are sum of a prime and two squares (A problem of Hardy and Littlewood) II, *Mat. Sb.* **53** (1961), 3–38; English transl., *Amer. Math. Soc. Transl.* (2) **37** (1964), 197–240.

[94] J. Manin, Cyclotomic fields and modular curves, *Russian Math. Surveys* **26** (1971), no. 6, 7–71.

[95] J. Manin, Parabolic points and zeta functions of modular curves, *Math. USSR Izvestia* **6** (1972), 19–64.

[96] J. Manin, Periods of parabolic forms and p-adic Hecke series, *Math. USSR Sbornik* **21** (1973), 371–393.

[97] B. Mazur and H.P.F. Swinnerton-Dyer, Arithmetic of Weil curves, *Invent. Math.* **25** (1974), 1–61.

[98] Y. Motohashi, *Lectures on the Riemann-Siegel Formula*, Ulam Seminar, Dept. of Math., Univ. of Colorado, Boulder 1987.

[99] A. P. Ogg, *Modular Forms and Dirichlet Series*, Benjamin, New York, 1969.

[100] S. Raghavan, On certain identities due to Ramanujan, *Quart. J. Math.* Oxford (2) **37** (1986), 221–229.

[101] R. A. Rankin, *Modular forms and functions*, Cambridge University Press, Cambridge, 1977.

[102] M. J. Razar, Values of Dirichlet series at integers in the critical strip, *Modular Functions of One Variable VI*, Bonn 1976, Lecture Notes in Math. 627, Springer-Verlag, Berlin (1977), 1–10.

[103] B. Riemann, Über die Anzahl der Primzahlen unter einer gegebenen Grösse, *Monatsber. der Berliner Akad.* (1859), 671–680=*Ges. Math. Werke*, 145–153, Dover, New York, 1953.

[104] F. Sato, Searching for the origin of prehomogeneous vector spaces, at annual meeting of the Math. Soc. Japan 1992 (in Japanese).

[105] A. Selberg and S. Chowla, On Epstein's zeta-function, *J. Reine Angew. Math.* **227** (1967) 86–110= Collected Papers of Atle Selberg, Vol. I, 521–545, Springer, 1989.

[106] J. P. Serre, *A course in Arithmetic*, Springer-Verlag, New York, 1973.

[107] G. Shimura, Sur les intégral attachées aux formes automorphes, *J. Math. Soc. Japan* **11** (1959), 291–311.

[108] G. Shimura, *Introduction to the Theory of Automorphic Functions*, Princeton University Press, Princeton, N. J. (1971).

[109] C. L. Siegel, Über Riemanns Nachlaß zur analytischen Zahlentheorie, *Quellen u. Studien zur Geschichte der Math., Astr. Phys.*, **2** (1932),45–80 =*Ges. Abh., I*, 275–310, Springer, Berlin-New York 1966.

[110] C. L. Siegel, Contribution to the theory of the Dirichlet *L*-series and the Epstein zeta-functions, *Ann. of Math.* **44** (1943), 143–172 = *Ges. Abh., II*, 360–389, Springer, Berlin-New York 1966.

[111] C. L. Siegel, A simple proof of $\eta(-1/\tau) = \eta(\tau)\sqrt{\tau/i}$, *Mathematika* **1** (1954), p.4 =*Ges. Abh., III*, 188, Springer, Berlin-New York 1966.

[112] J. R. Smart, On the values of the Epstein zeta function, *Galsgow Math. J.* **14** (1973), 1–12.

[113] H. M. Srivastava and J. Choi, *Series Associated with the Zeta and Related Functions*, Kluwer Academic Publishers, Dordrecht-Boston-London, 2001.

[114] H. M. Stark, *L*-functions and character sums for quadratic forms (I) and (II), *Acta Arith.* **14** (1968), 35–50, and *ibid.* **15** (1969), 307–317.

[115] A. Terras, Bessel series expansion of the Epstein zeta function and the functional equation, *Trans. Amer. Math. Soc.* **183** (1973), 477–486.

[116] A. Terras, Some formulas for the Riemann zeta function at odd integer argument resulting from Fourier expansions of the Epstein zeta function, *Acta Arith.* **29** (1976), 181–189.

[117] A. Terras, The Fourier expansion of Epstein's zeta function for totally real algebraic number fields and some consequences for Dedekind's zeta function, *Acta Arith.* **30** (1976), 187–197.

[118] A. Terras, Applications of special functions for the general linear group to number theory, *Séminaire Delange-Pisot-Poitou*, 18e année, 1976/77, No. 23, 1–16.

[119] A. Terras, The Fourier expansion of Epstein's zeta function over an algebraic number field and its consequences for algebraic number theory. *Acta Arith.* **32** (1977), 37–53.

[120] A. Terras, A relation between $\zeta(s)$ and $\zeta(s-1)$ for any algebraic number field, in *Algebraic Number Fields* A. Frohlich (Ed.), Academic Press, N.Y., 1977, 475–483.

[121] A. Terras, The minima of quadratic forms and the behavior of Epstein and Dedekind zeta functions, *J. Number Theory* **12** (1980), 258–272.

[122] A. Terras, *Harmonic Analysis on Symmetric Spaces and Applications I, II*, Springer Verlag, New York-Berlin-Heidelberg, 1985.

[123] E. C. Titchmarsh, *The Theory of the Riemann Zeta-Function*, (second edition revised by D. R. Heath-Brown), OUP 1986.

[124] M. Toyoizumi, Ramanujan's formulae for certain Dirichlet series. *Comment. Math. Univ. St. Paul.* **30** (1981), 149–173; **31** (1982), 87.

[125] A. Weil, Sur une formule classique, *J. Math. Soc. Japan* **20** (1968), 400-402 = *Coll. Papers, III*, 198–200, Springer, New York, 1980.

212

[126] A. Weil, Über die Bestimmung Dirichletscher Reihen durch Funktionalgleichungen, *Math. Ann.* **168** (1967), 149–156 = *Coll. Papers, III*, 165–172, Springer, New York, 1979.

[127] A. Weil, Remarks on Hecke's lemma and its use, *Algebraic Number Theory*, Intern. Symposium Kyoto 1976, S. Iyanaga (ed.), Jap. Soc. for the Promotion of Science 1977, pp. 267–274=*Coll. Papers III*, 405–412, Springer, New York, 1980.

[128] J. R. Wilton, A proof of Burnside's formula for $\log \Gamma(x + 1)$ and certain allied properties of Riemann's ζ-function. *Messenger Math.* **52** (1922/1923), 90–93.

[129] D. Zagier, Modular forms whose Fourier coefficients involve zeta-functions of quadratic forms, *Modular Functions of One Variable VI*, Bonn 1976, Lecture Notes in Math. 627, Springer-Verlag, Berlin (1977), 105–169.

WALDSPURGER'S FORMULA AND CENTRAL CRITICAL VALUES OF *L*-FUNCTIONS OF NEWFORMS IN WEIGHT ASPECT

Winfried Kohnen

Universität Heidelberg, Mathematisches Institut, INF 288,
D-69120 Heidelberg, Germany

winfried@mathi.uni-heidelberg.de

Jyoti Sengupta

Jyoti Sengupta, T.I.F.R., School of Mathematics, Homi Bhabha Road, 400 005 Bombay,
India

sengupta@math.tifr.res.in

Abstract   We shall give a proof of the Lindelöf hypothesis in weight aspect on the average for central critical values of quadratic character twists of Hecke *L*-functions attached to cuspidal Hecke eigenforms. One of the basic tools will be Waldspurger's results on central critical values of *L*-functions in weight aspect.

Keywords: Waldspurger's formula, Jacobi form, Lindelöf hypothesis, Poincaré series

1.   Introduction

As demonstrated in many places, Waldspurger's formula relating central critical values of quadratic character twists of Hecke *L*-functions of newforms of even integral weight to the squares of Fourier coefficients of modular forms of half-integral weight, is a very useful tool both in the theory of modular forms of half-integral weight as well as in the study of arithmetic properties of special values of *L*-functions.

Here we would like to point out that this formula (and variants of it) can also very easily be used –once the constant of proportionality has been set up explicitly– to study average sums of central values of *L*-

functions attached to newforms of an arbitrary level N in *weight* aspect (in the case $N = 1$, see our previous paper [5]).

2. Statement of result

Fix positive integers N and k and denote by $\mathcal{F}_{2k,N}^{new}$ the set of normalized cuspidal Hecke eigenforms of weight $2k$ and level N with trivial character which are newforms in the sense of Atkin-Lehner.

For D a fundamental discriminant with $(D, N) = 1$, let $L(f, D, s)$ ($s \in \mathbb{C}$) be the Hecke L-function of f twisted with the associated quadratic character. Recall that $L(f, D, s)$ is an entire function and satisfies the functional equation

$$L^*(f, D, s) := (2\pi)^{-s}(ND^2)^{s/2}\Gamma(s)L(f, D, s)$$

$$= (-1)^k(\frac{D}{-N})w_f L^*(f, D, 2k - s)$$

where w_f is the eigenvalue of f under the Fricke involution.

We set

$$\mathcal{F}_{2k,N}^{new,-} := \{f \in \mathcal{F}_{2k,N}^{new} \mid w_f = (-1)^{k+1}\}.$$

Then for $f \in \mathcal{F}_{2k,N}^{new,-}$ and for $D < 0$, D a square modulo $4N$ the L-function $L^*(f, D, s)$ has root number $+1$.

Theorem. *Let $D < 0$ be a fundamental discriminant with $(D, N) = 1$ which is a square modulo $4N$. Then for any $\epsilon > 0$ one has*

$$\sum_{f \in \mathcal{F}_{2k,N}^{new,-}} L(f, D, k) \ll_{\epsilon, D, N} k^{1+\epsilon} \quad (k \to \infty) \tag{1}$$

where the implied constant in $\ll$ only depends on ϵ, D and N, and is effective.

Note that it is widely believed that $L(f, D, k) \ll_{\epsilon, D, N} k^\epsilon$ when f varies in $\mathcal{F}_{2k,N}^{new,-}$ and $k \to \infty$ ("Lindelöf hypothesis in weight aspect"). However, this hypothesis seems far from being proved.

The proof of (1) which will be given in the next section, uses an explicit version of Waldspurger's formula in the context of Jacobi forms, valid for arbitrary level N and for D as above [2]. We shall also make use of the upper bound for the Petersson norm of f given in [3, 4].

3. Proof.

Let $S_{2k}^{new,-}(N)$ be the space of those newforms of weight $2k$ and level N which have eigenvalue $(-1)^{k+1}$ under the Fricke involution.

Denote by $J_{k+1,N}^{cusp}$ the space of Jacobi cusp forms of weight $k+1$ and index N and by $J_{k+1,N}^{cusp,new}$ its subspace of newforms [1].

Then according to [6], the spaces $J_{k+1,N}^{cusp,new}$ and $S_{2k}^{new,-}(N)$ are isomorphic as Hecke modules. Moreover, if $f \in \mathcal{F}_{2k,N}^{new,-}$, then according to [2] there is a non-zero Hecke eigenform $\phi \in J_{k+1,N}^{cusp,new}$ uniquely determined up to multiplication by non-zero scalars, such that

$$\frac{|c(n,r)|^2}{\langle \phi, \phi \rangle} = \frac{(k-1)!}{2^{2k-1}\pi^k N^{k-1}}|D|^{k-1/2}\frac{L(f,D,k)}{\langle f,f \rangle}. \tag{2}$$

Here $c(n,r)$ (with $D = r^2 - 4Nn < 0$, $n \in \mathbf{N}$, $r \in \mathbf{Z}$) is the (n,r)-th Fourier coefficient of ϕ and $\langle , \rangle$ denote properly normalized Petersson scalar products. The left-hand side of (2) does not depend on the chosen r with $D \equiv r^2 \pmod{4N}$.

Let $P_{k+1,N,(n,r)}$ be the (n,r)-th Poincaré series in $J_{k+1,N}^{cusp}$. Thus

$$\langle \psi, P_{k+1,N,(n,r)} \rangle = \alpha_{k+1,N}|D|^{-k+1/2}c_\psi(n,r) \tag{3}$$

for any $\psi \in J_{k+1,N}^{cusp}$ where $c_\psi(n,r)$ is the (n,r)-th Fourier coefficient of ψ and

$$\alpha_{k+1,N} := \frac{N^{k-1}\Gamma(k-\frac{1}{2})}{2\pi^{k-1/2}}.$$

Write

$$\mathcal{F}_{2k,N}^{new,-} = \{f_1, \ldots, f_g\}$$

and for each ν with $1 \leq \nu \leq g$ let $\phi_\nu \in J_{k+1,N}^{cusp,new}$ be a non-zero Hecke eigenform corresponding to f_ν and satisfying (2). We complete $\{\phi_1, \ldots, \phi_g\}$ to an orthogonal basis $\{\phi_1, \ldots, \phi_g, \phi_{g+1}, \ldots, \phi_d\}$ of $J_{k+1,N}^{cusp}$. Then from (3) we conclude the usual Petersson formula

$$c_{k+1,N,(n,r)}(n,r) = \alpha_{k+1,N}|D|^{-k+1/2}\sum_{\nu=1}^{d}\frac{|c_{\psi_\nu}(n,r)|^2}{\langle \psi_\nu, \psi_\nu \rangle} \tag{4}$$

where $c_{k+1,N,(n,r)}(n,r)$ denotes the (n,r)-th Fourier coefficient of $P_{k+1,N,(n,r)}$. From (2) and (4) it follows that

$$\sum_{\nu=1}^{g} \frac{L(f_\nu, D, k)}{\langle f_\nu, f_\nu \rangle} = \frac{2^{2k-1}\pi^k N^{k-1}}{(k-1)! |D|^{k-1/2}} \sum_{\nu=1}^{g} \frac{|c_{\psi_\nu}(n,r)|^2}{\langle \psi_\nu, \psi_\nu \rangle}$$

$$\leq \frac{2^{2k-1}\pi^k N^{k-1}}{(k-1)! |D|^{k-1/2}} \sum_{\nu=1}^{d} \frac{|c_{\psi_\nu}(n,r)|^2}{\langle \psi_\nu, \psi_\nu \rangle}$$

$$= \frac{(2\pi)^{2k}}{\sqrt{\pi}\,\Gamma(k-\frac{1}{2})\Gamma(k)} c_{k+1,N,(n,r)}(n,r). \qquad (5)$$

On the other hand, by [3, 4] for any $\epsilon > 0$ we have

$$\langle f_\nu, f_\nu \rangle \ll_{\epsilon,N} \frac{\Gamma(2k)}{(4\pi)^{2k}} k^\epsilon \quad (\nu = 1, \ldots, g)$$

where the implied constant in $\ll$ depends only on ϵ and N and is effective.

Hence using (5) we find that

$$\sum_{\nu=1}^{g} L(f_\nu, D, k) \ll_{\epsilon,N} \frac{\Gamma(2k)}{\sqrt{\pi}\,\Gamma(k-\frac{1}{2})\Gamma(k)2^{2k}} k^\epsilon \cdot c_{k+1,N,(n,r)}(n,r)$$

$$= \frac{(k-\frac{1}{2})k^\epsilon}{2\pi} \cdot c_{k+1,N,(n,r)}(n,r) \qquad (6)$$

where in the last line we have used Legendre's duplication formula for the gamma function.

The Fourier coefficients of $P_{k+1,N,(n,r)}$ were explicitly computed in [2]. In particular, under the assumption $(D, N) = 1$ one has

$$c_{k+1,N,(n,r)}(n,r) = 1 + i^k \frac{\pi\sqrt{2}}{\sqrt{N}} \sum_{c \geq 1} \left(H_{N,c}(n,r,n,r) \right.$$

$$\left. + (-1)^k H_{N,c}(n,r,n,-r) \right) J_{k-\frac{1}{2}} \left(\frac{\pi|D|}{Nc} \right)$$

where

$$H_{N,c}(n, r, n, \pm r)$$

$$= c^{-3/2} \sum_{\substack{\rho \pmod{c}^* \\ \lambda \pmod{c}}} e_c \left(\left((N\lambda)^2 + r\lambda + n \right) \rho^{-1} + n\rho \pm r\lambda \right) e_{2Nc} \left(\pm r^2 \right),$$

ρ runs over integers modulo c prime to c, we have written $e_c(x) = e^{2\pi i x/c}$ $(x \in \mathbf{R})$, ρ^{-1} denotes an integer with $\rho^{-1}\rho \equiv 1 \pmod{c}$ and $J_{k-\frac{1}{2}}$ is the Bessel function of order $k - \frac{1}{2}$.

Now trivially we have

$$|H_{N,c}(n,r,n,\pm r)| \leq \sqrt{c}.$$

Moreover, as is well-known

$$|J_{k-\frac{1}{2}}(x)| \leq \sqrt{\frac{2}{\pi}} \frac{x^{k-1/2}}{2^k \Gamma(k)} \quad (x > 0).$$

We therefore obtain

$$c_{k+1,N,(n,r)}(n,r) \ll_{D,N} 1 \quad (k \to \infty).$$

Hence using (6) our assertion follows.

References

[1] M. Eichler and D. Zagier, The theory of Jacobi forms, Progress in Math. vol. 55, Birkhäuser: Boston 1985

[2] B. Gross, W. Kohnen and D. Zagier, Heegner points and derivatives of L-series. II, Math. Ann. 278, 497-562 (1987)

[3] H. Iwaniec, Small eigenvalues of Laplacian for $\Gamma_0(N)$, Acta Arith. 56, 65-82 (1990)

[4] H. Iwaniec and P. Michel, The second moment of the symmetric square L-functions, Ann. Acad. Sci. Fennicae 26, 465-482 (2001)

[5] W. Kohnen and J. Sengupta, On quadratic character twists of Hecke L-functions attached to cusp forms of varying weights at the central point, Acta Arith. 99, 61-66 (2001)

[6] N.-P. Skoruppa and D. Zagier, Jacobi forms and a certain space of modular forms, Invent. math. 94, 113-146 (1988)

PRIMITIVE ROOTS: A SURVEY

Shuguang Li

Department of Mathematics, Natural Sciences Division, University of Hawaii–Hilo, 200 W. Kawili Street, Hilo, HI 96720-4091

shuguang@hawaii.edu

Carl Pomerance

Fundamental Mathematics Research, Mathematics Center, Lucent Technologies, Bell Laboratories, 600 Mountain Ave., Murray Hill, NJ 07974-0636

carlp@lucent.com

Abstract   For primes p, the multiplicative group of reduced residues modulo p is cyclic, with cyclic generators being referred to as primitive roots. Here we survey a few results and conjectures on this subject, and we discuss generalizations to arbitrary moduli. A primitive root to a modulus n is a residue coprime to n which generates a cyclic subgroup of maximal order in the group of reduced residues modulo n. [1]

Keywords: primitive root, Artin's conjecture, Generalized Riemann Hypothesis

Introduction

For a prime p, the multiplicative group $(\mathbf{Z}/p\mathbf{Z})^*$ is cyclic. Number theorists refer to any cyclic generator of this group as a primitive root modulo p. There are many attractive theorems and conjectures concerning primitive roots, and we shall survey some of them here. But it is also our intention to broaden the playing field, so to speak, and introduce the concept of a primitive root for a composite modulus n. It is well-known that for most numbers n, the multiplicative group $(\mathbf{Z}/n\mathbf{Z})^*$ is *not* cyclic (namely, $(\mathbf{Z}/n\mathbf{Z})^*$ is not cyclic for any number $n > 4$ that is not of the form p^a, $2p^a$ for p an odd prime). So what then do we mean by a primitive root for n? In any finite group G one may look at

---

[1] This article also appeared in "New Aspects of Analytic Number Theory" (RIMS kokyuroku No. 1274), 2002.

elements whose order is the maximum order over all elements in G. We do precisely this, and say that such elements for the group $G = (\mathbf{Z}/n\mathbf{Z})^*$ are primitive roots modulo n. That is, a primitive root modulo n is an integer coprime to n such that the multiplicative order of this integer modulo n is the maximum over all integers coprime to n. This concept reduces to the usual notion in the case that G is cyclic, so there should be no confusion. We shall see that there are a few surprises in store when we consider primitive roots for composite moduli. For a more traditional survey on primitive roots, see Murty [12].

The number of primitive roots for a given modulus

A basic question that one might ask is a formula for $R(n)$, the number of primitive roots for a given modulus n, and beyond that, a study of the order of magnitude of $R(n)$ as a function of n. For primes, the situation is straightforward. If g is a primitive root modulo p then all of the primitive roots for p are of the form g^a where a is coprime to $p-1$. Thus $R(p) = \varphi(p-1)$ where φ is Euler's function. This fact is well-known, but less well-known is that $\varphi(p-1)/(p-1)$ has a continuous distribution function. That is, let $D(u)$ denote the relative asymptotic density in the set of all primes of the set $\{p \text{ prime} : R(p)/(p-1) \leq u\}$. Then $D(u)$ exists for every real number u, $D(u)$ is a continuous function of u, and $D(u)$ is strictly increasing on $[0, 1/2]$, with $D(0) = 0$, $D(1/2) = 1$. This beautiful result, which echoes Schoenberg's theorem on $\varphi(n)/n$, is due to Kátai [5].

It is not so easy to get a formula for $R(n)$ in general. It may be instructive to first consider the case of a general finite abelian group G. Write G as a product of cyclic groups of prime power order. For each prime p dividing the order of G, let p^{λ_p} be the highest power of p that appears as an order of one of these cyclic factors, and let ν_p be the number of times that this cyclic factor appears. Then the maximal order of an element in G is

$$\prod_{p \mid |G|} p^{\lambda_p},$$

and the number of elements of G with this order is

$$|G| \prod_{p \mid |G|} \left(1 - p^{-\nu_p}\right).$$

To see the latter assertion, note that an element g will have p^{λ_p} dividing its order if and only if at least one of its projections in the ν_p cyclic factors of order p^{λ_p} has order p^{λ_p}. The chance that one particular projection does not have this order, that is, it is killed by the exponent $p^{\lambda_p - 1}$, is

$1/p$. Thus, the fraction of elements g for which each of the ν_p projections is killed by the exponent $p^{\lambda_p - 1}$ is $p^{-\nu_p}$, so the fraction for which at least one projection has order p^{λ_p} is $1 - p^{-\nu_p}$. The assertion follows.

To apply this result to $G = (\mathbf{Z}/n\mathbf{Z})^*$ we must compute the numbers ν_p for this group. By the Chinese remainder theorem, G has a decomposition into the product of the groups $(\mathbf{Z}/q^a\mathbf{Z})^*$, where q is prime and $q^a \| n$. Further, the groups $(\mathbf{Z}/q^a\mathbf{Z})^*$ are themselves cyclic unless $q = 2$ and $a \geq 3$, in which case $(\mathbf{Z}/2^a\mathbf{Z})^*$ is the product of a cyclic group of order 2 and a cyclic group of order 2^{a-2}. It is thus a simple task to further refine the decomposition afforded by the Chinese remainder theorem into a factorization of $(\mathbf{Z}/n\mathbf{Z})^*$ into cyclic groups of prime power order. We thus can work out a formula, albeit not so simple, for $R(n)$. For the sake of completeness, we record this formula: If q^a is a prime power, let $\lambda(q^a)$ be the order of the largest cyclic subgroup of $(\mathbf{Z}/q^a\mathbf{Z})^*$; thus, $\lambda(q^a) = \varphi(q^a)$ if q is odd or if $q = 2$ and $a < 3$, while if $q = 2$ and $a \geq 3$, then $\lambda(2^a) = \frac{1}{2}\varphi(2^a) = 2^{a-2}$. If the prime factorization of n is $\prod_{i=1}^k q_i^{a_i}$, and p is a prime with $p | \varphi(n)$, let λ_p be the largest number such that $p^{\lambda_p} | \lambda(q_i^{a_i})$ for some i. If p is odd, let ν_p be the number of i's with $p^{\lambda_p} | \lambda(q_i^{a_i})$. If $p = 2$ and either $\lambda_2 > 1$ or $n \not\equiv 8 \pmod{16}$, the definition of ν_2 is the same. If $p = 2$, $\lambda_2 = 1$, and $n \equiv 8 \pmod{16}$, then $\nu_2 = k + 1$. Then

$$R(n) = \varphi(n) \prod_{p | \varphi(n)} \left(1 - p^{-\nu_p}\right).$$

In analogy with Kátai's theorem about $R(p)/(p-1)$, one might ask if $R(n)/\varphi(n)$, has a distribution function. That is, for a given real number u does the set

$$\mathcal{R}_u := \{n : R(n)/\varphi(n) \leq u\}$$

have a natural density? Our first surprise is that the answer is no. It is shown by the first author in [7], [8] that there are values of u so that $\mathcal{R}_u$ does not have a natural density. In fact, there is a small positive number δ such that for every $u > 0$, $\mathcal{R}_u$ has upper density at least δ, but the lower density tends to 0 as $u \to 0$.

There are other naturally occurring sets in number theory where there is no natural density. For example, consider the set of integers n with an even number of decimal digits. While the natural density does not exist (the fraction of numbers in the set at 10^{2n} is at least $9/10$ while the fraction in the set at 10^{2n+1} is at most $1/10$), note that this set does have a logarithmic density. That is, the sum of the reciprocals of the numbers in the set that are $\leq x$, when divided by $\ln x$, approaches a limit, namely $1/2$. (It is interesting to note that logarithmic density is equivalent to the concept of Dirichlet density from analytic number theory.) Well, perhaps the set of numbers with an even number of decimal digits is not so natural a concept. But also consider the set of integers n with $\pi(n) > \text{li}(n)$. (Here, $\pi(x)$ is the number of primes in the interval $[1, x]$ and $\text{li}(x) = \int_0^x dt/\ln t$, where the principal value is taken for the singularity at $t = 1$.) It was once thought that there should be no values of n with $\pi(n) > \text{li}(n)$, until Littlewood showed that there are infinitely many with the inequality holding, and also infinitely many with the reverse inequality. It is shown in Rubinstein and Sarnak [13] that assuming reasonable conjectures concerning the zeroes of the Riemann zeta function, the set of integers n with $\pi(n) > \text{li}(n)$ does not have a natural density, but it does have a logarithmic density. Similar results pertain to the set of integers n with $\pi(n, 4, 1) > \pi(n, 4, 3)$, where $\pi(x, k, l)$ denotes the number of primes p in $[1, x]$ that are in the residue class $l \pmod{k}$.

So maybe the sets $\mathcal{R}_u$ have a logarithmic density? Alas, the answer is again no, as is shown in [7]. In fact, the oscillation persists at even the double logarithmic density (where one sums $1/a \ln a$ for members a of the set that are in $[2, x]$ and divides the sum by $\ln \ln x$). Maybe the *triple* logarithmic density exists: In [7] it is shown that at the triple level, $\mathcal{R}_u$ has upper density tending to 0 as $u \to 0$.

The source of the oscillation

Where does this surprising oscillation come from? The answer lies in the numbers ν_p described above. Consider a game played with n coins: We give you n coins, and at the end of the game you will either have given us back all n of the coins, or you will have given us back $n - 1$ coins, keeping one for yourself. Here's how the game is played. You flip the n coins (assume they are all fair coins with a $1/2$ probability of landing heads—the front of the coin—and a $1/2$ probability of landing tails—the back of the coin.), returning to us all of the coins that land tails. If there is more than one coin left, you repeat the process. If at any time you have exactly one coin left, you get to keep it. What is the probability P_n that you win the game by getting to keep a coin? It is not so hard to work out an expression for P_n, it is

$$P_n = \sum_{k=1}^{\infty} n2^{-k} \left(1 - 2^{1-k}\right)^{n-1}.$$

Indeed, if one keeps flipping until no coins are left, and the last coin leaves on round k, with the other $n - 1$ coins leaving on earlier rounds, then the probability of this is $n2^{-k} \left(1 - 2^{1-k}\right)^{n-1}$. (There are n choices for the "last" coin, the probability it falls heads $k - 1$ straight times followed by a tails is 2^{-k}, and the probability that each of the other

$n - 1$ coins has at least one tails in the first $k - 1$ flips is $\left(1 - 2^{1-k}\right)^{n-1}$.)
But more interestingly, one can ask:

$$\text{What is } \lim_{n \to \infty} P_n?$$

It is easy to convince oneself that when n is large, the biggest contribution to the sum for P_n is from the terms k with $2^k \approx n$. Suppose $0 \le \alpha < 1$ and S_α is an infinite set of natural numbers n such that the fractional part of the base-2 logarithm for $n \in S_\alpha$ converges to α modulo 1. For example, if $\alpha = 0$, then we might take S_0 as the set of powers of 2. Or we might also throw in the numbers of the form $2^m - 1$ and numbers of the form $2^m + m^2$. Then

$$\lim_{n \to \infty, n \in S_\alpha} P_n = \sum_{j=-\infty}^{\infty} 2^{-\alpha-j} e^{-2^{1-\alpha-j}}.$$

From this result it surely looks like the limiting value of P_n actually depends on α, the limiting value of the fractional part of the base-2 logarithm of n. That is, it looks like $\lim_{n \to \infty} P_n$ does not exist!

And this is indeed the case, though the oscillation in P_n is very gentle. We have $\limsup P_n \approx 0.72135465$ which is achieved when $\alpha \approx 0.139$, and $\liminf P_n \approx 0.72134039$ which is achieved when $\alpha \approx 0.639$. That is, the oscillation is only in the fifth decimal place! (For more on this kind of oscillation in probability theory, see [1] and the references in the acknowledgment of priority therein, and [6].)

It may be unclear what this game has to do with $R(n)$. Consider the number ν_2: If 2^{λ_2} is the highest power of 2 dividing the order of an element modulo n, then ν_2 is the number of cyclic factors of order 2^{λ_2} in $(\mathbf{Z}/n\mathbf{Z})^*$. We might ask where these ν_2 factors come from. But for a set of numbers n of density 0 we have ν_2 equal to the number of primes $p \mid n$ with $p \equiv 1 \pmod{2^{\lambda_2}}$. Now think of the odd primes dividing n as the coins in the game. Those primes $p \equiv 3 \pmod 4$ are the coins that turn tails on the first round and are returned. Those primes $p \equiv 5 \pmod 8$ are returned on the second round of flips, and so on. The number of primes that are alive in the last round is ν_p, and from our coin experience, we see that there is some oscillation for the probability that $\nu_p = 1$. But what corresponds to the number of coins? This is the number of odd prime factors of n, which is normally about $\ln \ln n$. Thus the limiting probability should depend on the fractional part of $\ln \ln \ln n / \ln 2$. With all of these iterated logarithms, it may begin to be clear why the density oscillation persists at logarithmic and double logarithmic levels.

But we noticed that the oscillation for the coin game is very slight. To see why there are great oscillations in the normal value of $R(n)$, we need to bring the other numbers ν_p into play for higher values of p. This then suggests a game played with unfair coins, where the probability of landing heads is $1/p$. An analysis of this game shows that there is again oscillation for the probability of winning, and as p tends to infinity, the ratio of the limsup of the probability to the liminf of the probability tends to infinity. In particular, if x tends to infinity in such a way that the fractional part of $\ln \ln \ln x / \ln p$ is very nearly 1 for all small primes p, then for most numbers n up to x, the values ν_p will frequently be 1 for these small primes p, so that then $R(n) = o(\varphi(n))$ for most numbers n up to x. But if x tends to infinity in such away that $\ln \ln \ln x / \ln p$ has fractional part about $1/2$ for all small primes p, then the values ν_p will mostly be > 1, so that $R(n)/\varphi(n)$ is bounded away from 0 for most numbers n up to x.

Artin's conjecture

Rather than fixing the modulus and asking for the number of primitive roots, as we have been doing, we may do the reverse: Fix an integer a and ask for the number of primes (or integers) for which a is a primitive root. Artin's famous conjecture deals with primes, and gives a supposedly necessary and sufficient condition on when there are infinitely many primes p with primitive root a. For example, take $a = 10$. (Note that in the special case $a = 10$, Gauss already had conjectured that there are infinitely many primes p with primitive root 10.) Note that 10 is a primitive root for a prime $p \neq 2, 5$ if and only if the length of the period for the decimal for $1/p$ has length $p - 1$. Thus, the Artin–Gauss problem might even be understandable to a school child.

Since primes $p > 2$ are all odd, the groups $(\mathbf{Z}/p\mathbf{Z})^*$ all have even order, so that squares cannot be cyclic generators. Clearly too, the number -1 has order dividing 2 in $(\mathbf{Z}/p\mathbf{Z})^*$, so that -1 cannot be a cyclic generator when $p > 3$. Thus, a necessary condition on a for there to be infinitely many primes p with primitive root a is that a should not be a square and that a should not be -1. Artin's conjecture is that these trivially necessary conditions are also sufficient:

Artin's conjecture. *If the integer a is not a square and not -1, then there are infinitely many primes with primitive root a.*

Artin also formulated a strong form of this conjecture:

Artin's conjecture, strong form. *If the integer a is not a square and not -1, then there is a positive number $A(a)$ such that the number of primes $p \leq x$ with primitive root a is $\sim A(a)\pi(x)$.*

Artin gave a heuristic argument for a formula for the numbers $A(a)$ appearing in the conjecture, but, as reported in [4], after some numerical experiments of the Lehmers which cast some doubt on Artin's formula, Heilbronn revised Artin's heuristic argument and came up with a formula which agreed better with the numerical experiments. Let

$$A = \prod_{q \text{ prime}} \left(1 - \frac{1}{q(q-1)}\right) = 0.3739558136\ldots,$$

the number known as *Artin's constant*. Write a as $a_1 a_2^2$, where a_1 is squarefree. We are assuming that a is not a square, but it might be some other power. Let h be the largest integer for which a is an h-th power, so that necessarily h is odd. In the case that $h = 1$, that is, that a is not any power higher than the first power, the Artin–Heilbronn formula for $A(a)$ is fairly simple; it is

$$A(a) = \begin{cases} A, & \text{if } a_1 \not\equiv 1 \pmod 4 \text{ and } h = 1 \\ A\left(1 - \prod_{q|a_1} \frac{1}{1+q-q^2}\right), & \text{if } a_1 \equiv 1 \pmod 4 \text{ and } h = 1. \end{cases}$$

In particular, if $h = 1$, then $A(a)$ is uniformly bounded away from 0. In the case that $h > 1$, the formula is more complicated:

$$A(a) = \begin{cases} A \prod_{q|h} \dfrac{q^2 - 2q}{q^2 - q - 1}, & \text{if } a_1 \not\equiv 1 \pmod 4 \\ A \prod_{q|h} \dfrac{q^2 - 2q}{q^2 - q - 1} \left(1 - \prod_{q|a_1} \dfrac{1}{1+q-q^2} \prod_{q|(a_1,h)} \dfrac{q^2 - q - 1}{q - 2}\right), & \\ & \text{if } a_1 \equiv 1 \pmod 4. \end{cases}$$

(This more complicated formula reduces to the earlier one in the case that $h = 1$.)

Where do these formulas come from? Understanding at least the appearance of Artin's constant is relatively simple. Assume that $h = 1$, that is, assume that a is not a nontrivial power. For a to be a primitive root modulo a prime p, it must be the case that for each prime q that divides $p - 1$ (namely, the order of the group $(\mathbf{Z}/p\mathbf{Z})^*$), a is not a q-th power modulo p. These conditions are not only necessary, they are sufficient. Say that p "passes the q-test" if either q does not divide $p-1$ or $q|p-1$ and a is not a q-th power modulo p. (Passing the q-test for a prime p is equivalent to q not dividing the index of the subgroup generated by

a in $(\mathbf{Z}/p\mathbf{Z})^*$). By the Chebotarev density theorem, the proportion of primes p with $p \equiv 1 \pmod{q}$ and a is a q-th power modulo p is $1/q(q-1)$. Thus, the proportion of primes p that pass the q-test is $1 - 1/q(q-1)$. Assuming "independence", the product of these expressions, which is Artin's constant, should then give the density of primes p for which a is a primitive root.

But are the events independent? In fact, if a_1, the squarefree part of a, is not 1 (mod 4), then it can be shown by the Chebotarev theorem that for fixed primes q, the q-tests are independent. And in the general case, the correct joint densities may be computed.

So why then is the strong form of Artin's conjecture not a theorem? The answer lies in the tail of the inclusion-exclusion. One can prove rigorously that if $\psi(x)$ tends to infinity very slowly with x, then the proportion of primes p for which the index of the subgroup generated by a in $(\mathbf{Z}/p\mathbf{Z})^*$ is not divisible by any prime $q \leq \psi(x)$ is indeed asymptotically $A(a)$. To complete the proof one needs to exclude those primes p which fail the q-test for some prime $q > \psi(x)$. We would only need a crude upper bound for these counts, such as $\leq c/q^2$ of all primes, or even $\leq c/q \ln q$ of all primes. However, we have nothing better than $\leq c/q$ afforded by the Brun–Titchmarsh inequality. And so, the strong form of Artin's conjecture remains just that, a conjecture.

Hooley [4] however, has made the above heuristic into a rigorous proof under the assumption of the Generalized Riemann Hypothesis. This hypothesis allows a stronger form of the Chebotarev theorem which gives an estimate of $O(\pi(x)/q^2 + x^{1/2} \ln x)$ primes up to x which fail the q-test, uniformly for $q \leq x^{1/2}/\ln^2 x$. Larger primes q may then be handled by an elementary argument that does not involve the GRH.

A parallel with another problem may be instructive here. Let $S(x)$ be the number of primes $p \leq x$ with $p - 1$ squarefree. Here the "q-test" is that we should not have $q^2 | p - 1$. The proportion of primes p which pass this q-test is then $1 - 1/q(q-1)$, by the prime number theorem for arithmetic progressions. The Chinese remainder theorem implies we have independence, without any exceptional cases, so that we may conjecture that $S(x) \sim A\pi(x)$. However, in this case, the heuristic may be turned into a rigorous and unconditional proof, since Brun–Titchmarsh allows a good uniform upper estimate for the distribution of primes p which fail the q-test for the primes $q < x^\epsilon$, and a trivial argument can be used for larger primes q. (Actually, we can use the Page–Siegel–Walfisz theorem instead of Brun–Titchmarsh.) The difference here is that we have tools for handling large primes q that are not readily available in the Artin context.

It is interesting that not only do we not have a proof of the strong form of Artin's conjecture, we do not have a proof of the weak form either, not for any single number a. However, if several numbers a are thrown in together, there are theorems. The most intriguing perhaps is the result of Heath-Brown [3] (based on earlier work of Gupta and Murty [2]) that there are at most two *prime* values of a for which the weak form of Artin's conjecture is false. Nevertheless, we repeat, we do not know a single value of a for which the conjecture is true.

Allowing more values of a, we can even show the strong form of Artin's conjecture unconditionally on average. Let $P_a(x)$ denote the number of primes $p \leq x$ which have a as a primitive root. It is relatively easy to estimate $\frac{1}{x \ln x} \sum_{1 \leq a \leq x \ln x} P_a(x)$, showing it to be $\sim A\pi(x)$. (Note that the average of the numbers $A(a)$ is $\sim A$, so that this result for $P_a(x)$ on average is consistent with the strong form of Artin's conjecture.) The sum of $P_a(x)$ may be thought of as the number of pairs a, p with $1 \leq a \leq x \ln x$, p a prime with $p \leq x$, and a is a primitive root modulo p. Thus, the sum may be reorganized as a sum over primes p, and then we may use the trivial result that there are $\varphi(p-1)$ primitive roots modulo p in every interval of p consecutive integers. Far less trivial is to get an average estimate with a running over an interval of the shape $[1, x^\epsilon]$. The champion theorem here is due to Stephens [14] (improving on earlier work of Goldfeld), and ϵ may be taken as $4(\ln \ln x / \ln x)^{1/2}$.

Artin's conjecture for composite moduli

We saw that it makes perfectly good sense to consider primitive roots for composite moduli, namely, a is a primitive root for n if the order of a in $(\mathbf{Z}/n\mathbf{Z})^*$ is as large as possible. Let $N_a(x)$ denote the number of integers n in $[1, x]$ with primitive root a. In analogy with Artin's conjecture for primes, it is tempting to conjecture that if a does not lie in some exceptional set, yet to be determined, then there is a positive constant $B(a)$ with $N_a(x) \sim B(a)x$. However, the experience above with the normal value of $R(n)$, the number of primitive roots modulo n, shows that we might be wary of such a conjecture.

To gain some further insight, we might begin by first considering the kind of result on average that was relatively easy in the case of primes. Namely, what can be said about $\frac{1}{x} \sum_{1 \leq a \leq x} N_a(x)$? Now the problem is not as easy as before, but the same sort of trick works, namely reorganizing the sum, so that now we are summing over integers $n \leq x$, and for each n we would like to know how many primitive roots it has in $[1, x]$. This estimate was worked out by the first author in [10], and sure enough, there is an oscillation. It is shown that

$$\liminf_{x\to\infty} \frac{1}{x^2} \sum_{1\le a\le x} N_a(x) = 0, \quad \limsup_{x\to\infty} \frac{1}{x^2} \sum_{1\le a\le x} N_a(x) > 0.$$

(The reason for the extra factor of $1/x$ is that it is natural to begin with the assumption that each term $N_a(x)$ is of order of magnitude x.)

Thus, while this result is not inconsistent with the assertion that $N_a(x) \sim B(a)x$, it certainly causes some serious doubt. In addition, the first author of this survey believes he may be able to generalize the Goldfeld–Stephens argument and achieve results like the ones above, but with the average taken over an interval of a-values of the form $[1, x^\epsilon]$.

Before proceeding, we note that there are certain numbers a for which we always have $N_a(x) = o(x)$. Namely if a is a nontrivial power, or if a is a square times -1 or a square times ± 2, then $N_a(x) = o(x)$. To get the idea of this, consider for example the case of $a = 2$, which is not exceptional at all when one considers prime moduli, but is exceptional for composite moduli. For all odd numbers n but for a set of density 0, the highest power of 2 which divides an order of an element in $(\mathbf{Z}/n\mathbf{Z})^*$, as before call it 2^{λ_2}, has $\lambda_2 \ge 3$. (That is, almost all numbers are divisible by a prime that is 1 (mod 8).) If $p|n$ where $p \equiv 1$ (mod 2^{λ_2}) (at least one such prime must divide n), then necessarily, since $p \equiv 1$ (mod 8), we have that 2 is a quadratic residue modulo p. Thus, 2^{λ_2} cannot divide the order of 2 in $(\mathbf{Z}/n\mathbf{Z})^*$ and so 2 cannot be a primitive root modulo n. The number of exceptional numbers $n \le x$ where this argument is not valid is $O(x/(\ln x)^{1/4})$, which is $o(x)$ as claimed.

Let $\mathcal{E}$ denote the set of integers a such that either a is a nontrivial power, or a is a square times -1 or a square times ± 2. Thus, if $a \in \mathcal{E}$, then $N_a(x) = o(x)$. The set $\mathcal{E}$ should then stand as a candidate for the exceptional set in a generalization of Artin's conjecture for composite moduli.

But beyond this exceptional set, the first author in [9] was able to show that for *any* integer a, we have

$$\liminf_{x\to\infty} \frac{1}{x} N_a(x) = 0. \qquad (1)$$

Moreover, this result was obtained on a set of real numbers x that is independent of the choice of a, in some sense. That is, there is an unbounded set S of positive reals such that for every integer a,

$$\lim_{x\to\infty,\, x\in S} \frac{1}{x} N_a(x) = 0.$$

So, we definitely do *not* have $N_a(x) \sim B(a)x$ for a positive number $B(a)$, not for any integer a.

With these thoughts in place, the first author in [7] made the conjecture that if a is a fixed integer not in $\mathcal{E}$, then there is a positive number $B(a)$ with

$$\limsup_{x\to\infty} \frac{1}{x} N_a(x) = B(a).$$

Recently, see [11], we have been able to prove this conjecture, under assumption of the GRH. In fact, we have been able to show that there is an unbounded set S' of positive reals and a positive constant c such that for each integer $a \notin \mathcal{E}$,

$$\limsup_{x\to\infty,\, x\in S'} \frac{1}{x} N_a(x) \geq c\frac{\varphi(|a|)}{|a|}. \tag{2}$$

One might ask about the weak Artin conjecture for composite moduli. Actually on this question, it is indeed possible to unconditionally prove that there are infinitely many n with primitive root a for many values of a. For example, take $a = 2$. We have that 2 is a primitive root for all of the numbers 3^j. In general, if a is a primitive root for p^2, where p is an odd prime, then a is a primitive root for p^j for every j. Other examples: Any number $a \equiv \pm 3 \pmod 8$ is a primitive root for all of the numbers 2^j. What is still unsolved, and may be tractable without the GRH: Given an integer a that is not a square nor -1, are there infinitely many squarefree integers n with primitive root a?

Local densities and a stronger conjecture

Let us first consider an easier question. Given a fixed prime q and a fixed integer $a \notin \mathcal{E}$, what is the distribution of the set of natural numbers n coprime to a such that the power of q in the order of a in the group $(\mathbf{Z}/n\mathbf{Z})^*$ is as large as possible over all elements in the group? Say the number of such integers $n \leq x$ is $N_a^q(x)$. This problem, see [11], can be analyzed unconditionally, giving

$$N_a^q(x) = (1 + o(1))\frac{\varphi(|a|)}{|a|} x \left(1 - F_q(x)\right),$$

where

$$F_q(x) = \sum_{j=0}^{\infty} \left(\exp\left(-\left(\frac{1}{\varphi(q^j)} - \frac{1}{q^{j+1}}\right)\ln\ln x\right) - \exp\left(-\frac{1}{\varphi(q^j)}\ln\ln x\right)\right).$$

As with the coin-flip problem, the density $1 - F_q(x)$ does not tend to a limit as $x \to \infty$. It is possible to show that

$$\liminf_{x\to\infty} F_q(x) \sim \frac{\ln q}{q^2}, \quad \limsup_{x\to\infty} F_q(x) \sim \frac{1}{eq},$$

as $q \to \infty$. By choosing a sequence of x-values where $F_q(x) \geq c/q$ occurs for many small primes q, it is possible to prove (1). It is also possible to choose a sequence of x-values where $F_q(x) = O(1/q \ln q)$ for most small primes q, but this is not sufficient for (2), since larger primes q can spoil the result. To show that larger primes usually do not pose too great an influence, the GRH comes into play.

Let

$$F_q = \liminf_{x \to \infty} F_q(x) = \inf_{t>0} \sum_{j=-\infty}^{\infty} \frac{\exp(t/q^{j+1}) - 1}{\exp(t/(q^j - q^{j-1}))}. \tag{3}$$

(Notice that the function of t is invariant under $t \mapsto tq$.) It seems reasonable to conjecture that the upper density $B(a)$ may be taken as $\alpha \varphi(|a|)/|a|$, where

$$\alpha := \prod_q (1 - F_q) \approx 0.326,$$

a conjecture made in [11]. That is, it is conjectured that for every integer a not in $\mathcal{E}$,

$$\limsup_{x \to \infty} \frac{1}{x} N_a(x) = \alpha \frac{\varphi(|a|)}{|a|}, \tag{4}$$

and that this limsup is attained on a set of positive reals independent of the choice of a. Note that for the number c in (2), we do have $c \leq \alpha$; in fact this is unconditional. That is, for every integer $a \neq 0$,

$$\limsup_{x \to \infty} \frac{1}{x} N_a(x) \leq \alpha \frac{\varphi(|a|)}{|a|}.$$

Let t_q be a value of t in $[1, q)$ where the infimum in (3) occurs. Then $t_q = \ln q + \ln \ln q + o(1)$ as $q \to \infty$. If $x \to \infty$ in such a way that the fractional part of $\ln \ln \ln x / \ln q$ tends to the fractional part of $\ln t_q / \ln q$, then $F_q(x) \to F_q$. Part of the problem in showing the conjecture (4) is to show that there is an unbounded sequence of values of x such that simultaneously, for all small primes q, the fractional part of $\ln \ln \ln x / \ln q$ approaches the fractional part of $\ln t_q / \ln q$. That such a sequence of x-values exists follows from Schanuel's conjecture in transcendental number theory. Indeed, from this conjecture, it follows that if $q_1, \ldots, q_k$ are distinct primes, then the real numbers $\ln q_1, \ldots, \ln q_k$ are algebraically independent. It would follow that the real numbers $1/\ln q_1, \ldots, 1/\ln q_k$ are linearly independent over the rationals, allowing simultaneous diophantine approximation of the quantities $\ln \ln \ln x / \ln q_1, \ldots, \ln \ln \ln x / \ln q_k$ modulo 1. However, even with Schanuel's conjecture and the GRH, there still seems to be some difficulties with the stronger conjecture.

Perhaps somewhat more tractable may be the conjecture from [9] that for a fixed integer a_0 not in $\mathcal{E}$, the individual count $N_{a_0}(x)$ is asymptotically equal to the average count over all integers a in $[1, x]$. That is, as $x \to \infty$,

$$N_{a_0}(x) = (1 + o(1)) \frac{1}{x} \sum_{1 \leq a \leq x} N_a(x).$$

We close with another conjecture that is perhaps tractable:

$$\limsup_{x \to \infty} \frac{1}{x^2} \sum_{1 \leq a \leq x} N_a(x) = \frac{6\alpha}{\pi^2}.$$

References

[1] J. S. Athreya and L. M. Fidkowski, *Number theory, balls in boxes, and the asymptotic uniqueness of maximal discrete order statistics*, Integers—The Electronic Journal of Combinatorial Number Theory (http://www.integers-ejcnt.org) **0** (2000), article A3.

[2] R. Gupta and M. R. Murty, *A remark on Artin's conjecture*, Invent. Math. **78** (1984), 127–130.

[3] D. R. Heath-Brown, *Artin's conjecture for primitive roots*, Quart. J. Math. Oxford Ser. (2) **37** (1986), 27–38.

[4] C. Hooley, *On Artin's conjecture*, J. reine angew. Math. **225** (1967), 209–220.

[5] I. Kátai, *On distribution of arithmetical functions on the set prime plus one*, Compositio Math. **19** (1968), 278–289.

[6] P. Kirschenhofer and H. Prodinger, *The number of winners in a discrete geometrically distributed random sample*, Ann. Appl. Probab. **6** (1996), 687–694; *Addendum*, ibid. **8** (1998), 647.

[7] S. Li, *On Artin's conjecture for composite moduli*, Ph. D. dissertation, University of Georgia, 1998.

[8] S. Li, *On the number of elements with maximal order in the multiplicative group modulo n*, Acta Arith. **86** (1998), 113–132.

[9] S. Li, *On extending Artin's conjecture to composite moduli*, Mathematika **46** (1999), 373–390.

[10] S. Li, *Artin's conjecture on average for composite moduli*, J. Number Theory **84** (2000), 93–118.

[11] S. Li and C. Pomerance, *On generalizing Artin's conjecture on primitive roots to composite moduli*, J. reine angew. Math., to appear.

[12] M. R. Murty, *Artin's conjecture for primitive roots*, Math. Intelligencer **10** (1988), no. 4, 59–67.

[13] M. Rubinstein and P. Sarnak, *Chebyshev's bias*, Experiment. Math. **3** (1994), 173–197.

[14] P. J. Stephens, *An average result for Artin's conjecture*, Mathematika **16** (1969), 178–188.

ZETA-FUNCTIONS DEFINED BY TWO POLYNOMIALS

Kohji Matsumoto
Graduate School of Mathematics, Nagoya University, Chikusa-ku, Nagoya 464-8602, Japan

kohjimat@math.nagoya-u.ac.jp

Lin Weng
Graduate School of Mathematics, Kyushu University, Fukuoka 812-8581, Japan

wen@math.kyushu-u.ac.jp

(former address)

Graduate School of Mathematics, Nagoya University, Chikusa-ku, Nagoya 464-8602, Japan

weng@math.nagoya-u.ac.jp

Abstract   The analytic continuation of certain multiple zeta-functions is shown. In particular, the analytic continuation of the zeta-function $\zeta(s; P, Q)$, defined by two polynomials P and Q, follows. Then the holomorphy of $\zeta(s; P, Q)$ at non-positive integers is proved, and explicit formulas for the values $\zeta(0; P, Q)$ and $\zeta'(0; P, Q)$ are given. The latter formula gives a generalization of an explicit formula for the regularized determinant of the Laplacian on the high-dimensional sphere.

Keywords: zeta-function, multiple zeta-function, Mellin-Barnes integral, special values, Laplacian

Introduction

Let $P(x), Q(x)$ be two non-zero polynomials with complex coefficients. Define the associated Dirichlet series $\zeta(s; P, Q)$ by

$$\zeta(s; P, Q) := \sum_{n=1}^{\infty} \frac{P(n)}{Q(n)^s}. \tag{1}$$

Clearly, $\zeta(s; P, Q)$ is well-defined when Re s is sufficiently large, provided that $Q(n) \neq 0$ for all $n = 1, 2, 3, \ldots$, which in this paper we always assume.

It is the principal motivation of the present paper to study the analytic properties and the special values of $\zeta(s; P, Q)$. However, we actually introduce the multi-variable series

$$\zeta_r((s_1,...,s_r);(\alpha_1,...,\alpha_r)) := \sum_{n=1}^{\infty} (n+\alpha_1)^{-s_1}(n+\alpha_2)^{-s_2} \cdots (n+\alpha_r)^{-s_r},$$

$$(2)$$

where $s_1, \ldots, s_r$ are complex variables and $\alpha_1, \ldots, \alpha_r \in \mathbf{C}\backslash\{-1, -2, \ldots\}$, and study its analytic behavior. Note that $(n+\alpha_j)^{-s_j} = \exp(-s_j \log(n + \alpha_j))$, where the branch of logarithm is fixed as $-\pi < \arg(n + \alpha_j) \leq \pi$. The properties of (1) can be easily deduced from those of (2), as we shall explain later.

Our interests in $\zeta(s; P, Q)$ date back to the beginning of 90's. In order to evaluate precisely Ray-Singer analytic torsions for certain special symmetric spaces, during that time, the second author introduced $\zeta(s; P, Q)$ and calculated the special values $\zeta(0; P, Q)$ and $\zeta'(0; P, Q)$. However, the series (1) is indeed a classical object: Various properties of the more general Dirichlet series

$$\sum_{n_1=1}^{\infty} \cdots \sum_{n_k=1}^{\infty} \frac{P(n_1, \ldots, n_k)}{Q(n_1, \ldots, n_k)^s},$$

$$(3)$$

where $P(x_1, \ldots, x_k)$, $Q(x_1, \ldots, x_k)$ are two polynomials of k indeterminants with complex coefficients, have been studied by many mathematicians. Under certain assumptions on the properties of P and Q, Mellin [20, 21] and Mahler [17] established the meromorphic continuation of (3) to the whole complex plane $\mathbf{C}$, and studied the location of poles. This direction of research was revived in 80's. Many new results were obtained by, for instance, Cassou-Noguès [1, 2, 3], Sargos [24, 25], Lichtin [10, 11, 12], Eie [4, 5] and Peter [22]. These authors were mainly concerned with the single variable series (3), though the included polynomials are of several indeterminants. However, Lichtin [11] proposed the problem of investigating the multi-variable series

$$\sum_{n_1=1}^{\infty} \cdots \sum_{n_k=1}^{\infty} \frac{P(n_1, \ldots, n_k)}{Q_1(n_1, \ldots, n_k)^{s_1} \cdots Q_r(n_1, \ldots, n_k)^{s_r}},$$

$$(4)$$

where $P, Q_1, \ldots, Q_r$ are polynomials of k indeterminants with complex coefficients. Lichtin indeed carried out such studies in his papers [13, 14, 15, 16] when $Q_1, \ldots, Q_r$ are hypoelliptic. Under the assumption of the hypoellipticity, he proved in [13] the meromorphic continuation of (4) to the whole $\mathbf{C}^r$. This result especially implies the meromorphic continuation of (2) and hence of (1). But Lichtin's method, based on the theory of $\mathcal{D}$-modules, is rather sophisticated and it is not clear how to deduce further explicit information from Lichtin's results.

In the first section of the present paper we prove the meromorphic continuation of (2) by a quite different method (Theorem A(i)(ii)), which obviously implies the meromorphic continuation of (1) (Theorem B). Moreover we prove that (1) is holomorphic at any non-positive integers (Theorem C). The starting point of our method is the Mellin-Barnes integral formula ((4) in Section 1). The prototype of the method can already be found in Mellin [20] (see also Cassou-Noguès [1, 2]). Katsurada [7, 8] discovered that the Mellin-Barnes formula is useful to study the analytic behavior of double zeta sums. The first author [18, 19] generalized Katsurada's idea to obtain a proof of meromorphic continuation of Euler-Zagier multiple zeta sums. The method in this paper is a modification of the argument developed in [18, 19].

An advantage of our method is that it gives explicit information of the behavior of the multiple zeta functions $\zeta_r((s_1,...,s_r);(\alpha_1,...,\alpha_r))$. For instance, we will prove certain convergent infinite series expansions (Theorem A(iii) and Remark 2). These results show an interesting new feature of our present situation, because in [18, 19] we have obtained similar asymptotic expansions for multiple zeta sums but they are not convergent (Remark 3).

In the second section, we evaluate the special values of our zeta functions and the associated derivatives at $s = 0$ in terms of the special values of Hurwitz zeta functions (Theorems D,E) after establishing a certain combinatorial identity. We here only give the formulas for $\zeta(0; P, Q)$ and $\zeta'(0; P, Q)$, but by extending our method, it is possible to show similar formulas for the values at $s = 0$ of higher derivatives, and also corresponding formulas at any negative integers.

The evaluation of $\zeta'(0; P, Q)$ for some special cases has been studied extensively in connection with the regularized determinant of the Laplacian [27], [28], [23], [9]. The regularized determinant of the Laplacian on the (C^∞-functions over) g-dimensional sphere is given by $\exp(-Z_g'(0))$, where

$$Z_g(s) = \sum_{n=1}^{\infty} \left(\binom{n+g}{g} - \binom{n+g-2}{g} \right) (n(n+g-1))^{-s}.$$

Hence the evaluation of the regularized determinant in this case is reduced to that for $H_{g,d}'(0)$, where

$$H_{g,d}(s) = \sum_{n=1}^{\infty} \frac{n^d}{(n(n+g))^s}. \tag{5}$$

Based on [27], an explicit formula for $H'_{g,d}(0)$ may be given. See e.g., [9]. In the last part of the present paper we show that our Theorem E includes such a result as a special case.

The authors express their gratitude to Professor Yoshio Tanigawa and the referee for useful comments.

1. Analytic Properties

1.1 Meromorphic Extension

Let $P(x)$, $Q(x)$ be two non-zero polynomials with complex coefficients. The associated Dirichlet series

$$\zeta(s; P, Q) := \sum_{n=1}^{\infty} \frac{P(n)}{Q(n)^s} \qquad (1)$$

is well-defined for $\mathrm{Re}\,(s)$ sufficiently large, provided that $Q(n) \neq 0$ for all $n = 1, 2, 3, \ldots$ which in this paper we always assume.

In this section, we prove $\zeta(s; P, Q)$ admits the analytic continuation to the whole complex plane. For this, we study the following more general multi-variable zeta functions.

Let $\alpha_1, \ldots, \alpha_r \in \mathbf{C} \backslash \{-1, -2, \ldots\}$. Define

$$\zeta_r((s_1,\ldots, s_r); (\alpha_1,\ldots, \alpha_r)) := \sum_{n=1}^{\infty} (n + \alpha_1)^{-s_1} (n + \alpha_2)^{-s_2} \cdots (n + \alpha_r)^{-s_r}.$$

$$(2)$$

This series is well-defined and is clearly convergent absolutely if $\mathrm{Re}\,(s_1 + s_2 + \cdots + s_r) > 1$. For our later convenience, we further assume that $\mathrm{Re}\,s_j > 1, j = 1, \ldots, r$. (So $\mathrm{Re}\,(s_1 + \cdots + s_r) > r$ indeed.)

Now to write $\zeta(s; P, Q)$ in terms of ζ_r, first factor $Q(x)$ as $Q(x) = b \prod_{j=1}^{m} (x + \beta_j)$, then expand $P(x)$ in terms of $x + \beta_1$ to get $P(x) =: \sum_{i=0}^{n} \tilde{a}_i (x + \beta_1)^i$ where $\tilde{a}_n \neq 0$. Clearly

$$\zeta(s; P, Q) = b^{-s} \sum_{i=0}^{n} \tilde{a}_i \zeta_m((s - i, s,\ldots, s); (\beta_1, \beta_2,\ldots, \beta_m)). \qquad (3)$$

Hence our problem is reduced to the analytic continuation of ζ_r.

If $r = 1$, nothing should be added here. So we may assume that $r \geq 2$. The key is the following classical Mellin-Barnes integral formula

$$\Gamma(s)(1 + \lambda)^{-s} = \frac{1}{2\pi i} \int_{(C)} \Gamma(s + z)\Gamma(-z)\lambda^z dz, \qquad (4)$$

where $\Gamma(s)$ denotes the standard gamma function, $s, \lambda \in \mathbf{C}$, $\mathrm{Re}\,s > 0$, $\lambda \neq 0$, $|\arg \lambda| < \pi$, $-\mathrm{Re}\,s < C < 0$, and the path is the vertical line $\mathrm{Re}\,z = C$.

Indeed, if $\alpha_r \notin \mathbf{R}_{\leq 0}$, from (4) with $s = s_r$ and $\lambda = \alpha_r/n$, we get

$$(n + \alpha_1)^{-s_1} \cdots (n + \alpha_r)^{-s_r}$$
$$= \frac{1}{2\pi i} \int_{(C)} \frac{\Gamma(s_r + z)\Gamma(-z)}{\Gamma(s_r)}$$
$$\times (n + \alpha_1)^{-s_1} \cdots (n + \alpha_{r-1})^{-s_{r-1}} \cdot n^{-s_r - z} \alpha_r^z dz, \qquad (5)$$

where $-\operatorname{Re} s_r < C < 0$. Note that

$$\operatorname{Re}(s_1 + \cdots + s_{r-1} + s_r + z) > (r - 1) + \operatorname{Re}(s_r) + C > r - 1 \geq 1.$$

Hence

$$\zeta_r((s_1, ..., s_r); (\alpha_1, ..., \alpha_r))$$
$$= \frac{1}{2\pi i} \int_{(C)} \frac{\Gamma(s_r + z)\Gamma(-z)}{\Gamma(s_r)}$$
$$\times \zeta_r((s_1, ..., s_{r-1}, s_r + z); (\alpha_1, ..., \alpha_{r-1}, 0)) \alpha_r^z dz. \qquad (6)$$

Similarly, from (4) (this time with $s = s_{r-1}$ and $\lambda = \alpha_{r-1}/n$ assuming $\alpha_{r-1} \notin \mathbf{R}_{\leq 0}$), we obtain the following relation

$$\zeta_r((s_1, ..., s_{r-1}, s_r); (\alpha_1, ..., \alpha_{r-1}, 0))$$
$$= \frac{1}{2\pi i} \int_{(C_{r-1})} \frac{\Gamma(s_{r-1} + z_{r-1})\Gamma(-z_{r-1})}{\Gamma(s_{r-1})}$$
$$\times \zeta_{r-1}((s_1, ..., s_{r-2}, s_{r-1} + s_r + z_{r-1}); (\alpha_1, ..., \alpha_{r-2}, 0))$$
$$\times \alpha_{r-1}^{z_{r-1}} dz_{r-1}, \qquad (7)$$

where $-\operatorname{Re} s_{r-1} < C_{r-1} - 1 < C_{r-1} < 0$ and $\operatorname{Re} s_j > 1, j = 1, ..., r$. This (7) is our basis of the induction procedure.

Remark 1. If $\alpha_j \in \mathbf{R}_{\leq 0} \backslash \{-1, -2, ...\}$, there exists a positive integer n_0 such that $n_0 + \alpha_j > 0$. Hence (2) becomes

$$\sum_{n=1}^{n_0} (n + \alpha_1)^{-s_1} \cdots (n + \alpha_r)^{-s_r} + \sum_{n=n_0+1}^{\infty} (n + \alpha_1)^{-s_1} \cdots (n + \alpha_r)^{-s_r},$$

and the holomorphy of the first sum with respect to $s_1, \ldots, s_r$ is obvious. Therefore we can reduce the problem to the case $\alpha_j \notin \mathbf{R}_{\leq 0}$.

Proposition 1. *For any $r \geq 1$,*
(i) The r-ple zeta function ζ_r defined by (2) with $\alpha_r = 0$ can be continued meromorphically with respect to $s_1, \ldots, s_r$ to the whole $\mathbf{C}^r$-space, and is holomorphic in $\alpha_1, \ldots, \alpha_{r-1}$ if $\alpha_j \in \mathbf{C} \backslash \mathbf{R}_{\leq(-1)} (1 \leq j \leq r - 1)$;

(ii) *The possible singularities of ζ_r are only located on $s_1 + s_2 + \cdots + s_r = 1 - k$, $(k \in \mathbf{N}_0 := \mathbf{Z}_{\geq 0})$;*

(iii) *The order estimate*

$$\zeta_r \ll \mathcal{F}(t_1, ..., t_r) \cdot e^{\rho_1 |t_1| + \rho_2 |t_2| + \cdots + \rho_{r-1} |t_{r-1}|}$$

holds, where $t_j = \mathrm{Im} s_j$ and $\rho_j = |\arg \alpha_j|$. Here, and in what follows, $\mathcal{F}(\cdots)$ denotes a quantity, not necessarily the same at each occurrence, which is of polynomial order in the indicated variables.

Proof. If $r = 1$, ζ_r is simply the Riemann zeta function. So we are done.

Now assume $r \geq 2$. First consider the case $\alpha_j \notin \mathbf{R}_{\leq 0}$ $(1 \leq j \leq r - 1)$. Then we may apply (7). Hereafter we assume the validity of Proposition for ζ_{r-1} and prove the case for ζ_r. We write $s_j = \sigma_j + it_j$ and $z_j = x_j + iy_j$ $(1 \leq j \leq r)$.

By using the order assumption (iii) for ζ_{r-1}, we see that the integrand on the right-hand side of (7) is

$$\ll e^{-\frac{\pi}{2}|t_{r-1} + y_{r-1}|}(|t_{r-1} + y_{r-1}| + 1)^{\sigma_{r-1} + x_{r-1} - \frac{1}{2}}$$
$$\times e^{-\frac{\pi}{2}|y_{r-1}|}(|y_{r-1}| + 1)^{-x_{r-1} - \frac{1}{2}} e^{\frac{\pi}{2}|t_{r-1}|}(|t_{r-1}| + 1)^{\frac{1}{2} - \sigma_{r-1}}$$
$$\times \mathcal{F}(t_1, ..., t_{r-2}, t_{r-1} + t_r + y_{r-1})$$
$$\times e^{\rho_1 |t_1| + \cdots + \rho_{r-2} |t_{r-2}|} |\alpha_{r-1}|^{x_{r-1}} e^{\rho_{r-1} |y_{r-1}|}. \tag{8}$$

Since we may assume $\rho_{r-1} < \pi$, we find that the above tends to 0 when $|y_{r-1}| \to \infty$. Hence we may shift the path of integration of (7) to $\mathrm{Re}\, z_{r-1} = M - \varepsilon$, where M is a positive integer. The poles of the integrand are as follows:

(I) poles $-s_{r-1} - k$ $(k \in \mathbf{N}_0)$, coming from the factor $\Gamma(s_{r-1} + z_{r-1})$,

(II) poles k $(k \in \mathbf{N}_0)$, coming from the factor $\Gamma(-z_{r-1})$, and

(III) poles coming from the factor ζ_{r-1}.

By the assumption (ii) for ζ_{r-1}, the poles (III) are

$$s_1 + \cdots + s_{r-2} + (s_{r-1} + s_r + z_{r-1}) = 1 - k \qquad (k \in \mathbf{N}_0),$$

that is,

$$z_{r-1} = 1 - k - (s_1 + \cdots + s_r). \tag{9}$$

Since

$$\mathrm{Re}\, z_{r-1} \leq 1 - \mathrm{Re}\,(s_1 + \cdots + s_r)$$
$$= -\mathrm{Re}\,(s_1 + \cdots + s_{r-2}) - \mathrm{Re}\, s_{r-1} + (1 - \mathrm{Re}\, s_r)$$
$$< 0 + C_{r-1} + 0 = C_{r-1},$$

we see that the poles (9) are all located to the left of the line $\operatorname{Re} z_{r-1} = C_{r-1}$. The poles (I) are clearly to the left of the same line, because $-\operatorname{Re} s_{r-1} < C_{r-1}$. Hence the only relevant poles are type (II). Thus, we obtain

$$\zeta_r((s_1,...,s_r);(\alpha_1,...,\alpha_{r-1},0))$$

$$= \sum_{j=0}^{M-1} \binom{-s_{r-1}}{j} \zeta_{r-1}((s_1,...,s_{r-2},s_{r-1}+s_r+j);(\alpha_1,...,\alpha_{r-2},0))\alpha_{r-1}^j$$

$$+ \frac{1}{2\pi i} \int_{(M-\varepsilon)} \frac{\Gamma(s_{r-1}+z_{r-1})\Gamma(-z_{r-1})}{\Gamma(s_{r-1})}$$

$$\times \zeta_{r-1}((s_1,...,s_{r-2},s_{r-1}+s_r+z_{r-1});(\alpha_1,...,\alpha_{r-2},0))\alpha_{r-1}^{z_{r-1}} dz_{r-1}. \tag{10}$$

Obviously, the poles of the integrand of the above integral, which are (I), (II) and (III) listed above, do not lie on the path of integration if

$$-\operatorname{Re} s_{r-1} < M - \varepsilon \quad \text{and} \quad 1 - \operatorname{Re}(s_1 + \cdots + s_r) < M - \varepsilon,$$

that is,

$$\operatorname{Re} s_{r-1} > -M + \varepsilon \quad \text{and} \quad \operatorname{Re}(s_1 + \cdots + s_r) > 1 - M + \varepsilon.$$

Therefore, since M is arbitrary, (10) gives the continuation of ζ_r to the whole $\mathbf{C}^r$-space. The holomorphy with respect to $\alpha_1, \ldots, \alpha_{r-1}$ is clear from (10) in case they are $\notin \mathbf{R}_{\leq 0}$. Even if $-1 < \alpha_j \leq 0$ for some j, we can show the meromorphy with respect to s_j and the holomorphy with respect to α_j by using Remark 1, because $n + \alpha_j > 0$ for any positive integer n. (If $\alpha_j < -1$, we encounter the problem of multi-valuedness of $\log(n+\alpha_j)$.) Moreover, the singularities are only coming from the factor

$$\zeta_{r-1}((s_1,...,s_{r-2},s_{r-1}+s_r+j);(\alpha_1+\cdots+\alpha_{r-2},0))$$

in the first term on the right-hand side of (10). Those singularities are, by the assumption (ii) for ζ_{r-1},

$$s_1 + \cdots + s_{r-2} + (s_{r-1}+s_r+j) = 1 - k \qquad (k \in \mathbf{N}_0),$$

that is,

$$s_1 + \cdots + s_r = 1 - j - k \qquad (j, k \in \mathbf{N}_0).$$

Therefore we now obtain the assertions (i) and (ii) of Proposition 1 for ζ_r.

Next we prove (iii). By using (8), we find that the integral on the right-hand side of (10) is

$$\ll \int_{-\infty}^{\infty} \exp\left(-\frac{\pi}{2}|t_{r-1} + y_{r-1}| - \frac{\pi}{2}|y_{r-1}| + \frac{\pi}{2}|t_{r-1}|\right)$$
$$\times \exp\left(\rho_1|t_1| + \cdots + \rho_{r-2}|t_{r-2}| + \rho_{r-1}|y_{r-1}|\right)$$
$$\times \mathcal{F}(t_1,\ldots,t_{r-2}, t_{r-1}, t_r, y_{r-1}) dy_{r-1}$$
$$= \exp\left(\rho_1|t_1| + \cdots + \rho_{r-2}|t_{r-2}| + \frac{\pi}{2}|t_{r-1}|\right)$$
$$\times \int_{-\infty}^{\infty} \mathcal{F}(t_1,\ldots,t_r, y_{r-1})$$
$$\times \exp\left(-\frac{\pi}{2}|t_{r-1} + y_{r-1}| + (\rho_{r-1} - \frac{\pi}{2})|y_{r-1}|\right) dy_{r-1}.$$

Lemma 1 ([19, §5, Lemma 3]). *Let* $p, A, B, \alpha, \beta \in \mathbf{R}$, $A + B < 0$. *Then*

$$\int_{-\infty}^{\infty} (|y| + 1)^p \exp\left(A|y + \alpha| + B|y + \beta|\right) dy$$
$$= O\left((|\alpha| + 1)^{p+1} e^{B|\alpha - \beta|} + (|\beta| + 1)^{p+1} e^{A|\alpha - \beta|}\right),$$

and the O-constant depends only on p, A *and* B.

By this lemma, the integral of the right hand side of (10) is

$$\ll \exp\left(\rho_1|t_1| + \cdots + \rho_{r-2}|t_{r-2}| + \frac{\pi}{2}|t_{r-1}|\right)$$
$$\cdot \mathcal{F}(t_1,\ldots,t_r) \cdot \left(e^{-\frac{\pi}{2}|t_{r-1}|} + e^{(\rho_{r-1} - \frac{\pi}{2})|t_{r-1}|}\right)$$
$$\ll \mathcal{F}(t_1,\ldots,t_r) \cdot \exp\left(\rho_1|t_1| + \cdots + \rho_{r-2}|t_{r-2}| + \rho_{r-1}|t_{r-1}|\right).$$

On the other hand, by using the assumption (iii) for ζ_{r-1} again, we find that the first term on the right-hand side of (10) is

$$\mathcal{F}(t_1,\ldots,t_r) \cdot e^{\rho_1|t_1| + \cdots + \rho_{r-2}|t_{r-2}|}.$$

Therefore, we obtain

$$\zeta_r((s_1,\ldots,s_r); (\alpha_1,\ldots,\alpha_{r-1}, 0))$$
$$\ll \mathcal{F}(t_1,\ldots,t_r) \cdot \exp\left(\rho_1|t_1| + \cdots + \rho_{r-2}|t_{r-2}| + \rho_{r-1}|t_{r-1}|\right),$$

which is the assertion (iii) of Proposition 1 for ζ_r (with $\alpha_r = 0$). The proof of Proposition 1 is now complete. $\square$

From (10) we obtain the asymptotic expansion

$$\zeta_r((s_1,...,s_r);(\alpha_1,...,\alpha_{r-1},0))$$

$$= \sum_{j=0}^{M-1} \binom{-s_r-1}{j} \zeta_{r-1}((s_1,...,s_{r-2},s_{r-1}+s_r+j);(\alpha_1,...,\alpha_{r-2},0))\alpha_{r-1}^j$$

$$+ O(|\alpha_{r-1}|^{M-\varepsilon}) \tag{11}$$

with respect to $|\alpha_{r-1}|$ when $|\alpha_{r-1}| \to 0$. Similarly, we may deduce the asymptotic expansion with respect to $|\alpha_{r-1}|$ when $|\alpha_{r-1}| \to \infty$ by shifting the path of integration to the left.

Proposition 1 gives sufficient analytic information on the function $\zeta_r((s_1,...,s_r);(\alpha_1,...,\alpha_{r-1},0))$. Now finally, combining Proposition 1 with the formula (6), we deduce analytic information on the general zeta function $\zeta_r((s_1,...,s_r);(\alpha_1,...,\alpha_r))$.

We shift the path of integration on the right-hand side of (6) to $\mathrm{Re}z = M - \varepsilon$. By using the estimate Proposition 1(iii), we find that the integrand of (6) is

$$\ll e^{-\frac{\pi}{2}|t_r+y|}(|t_r+y|+1)^{\sigma_r+x-\frac{1}{2}} e^{-\frac{\pi}{2}|y|}(|y|+1)^{-x-\frac{1}{2}} e^{\frac{\pi}{2}|t_r|}(|t_r|+1)^{\frac{1}{2}-\sigma_r}$$

$$\cdot \mathcal{F}(t_1,...,t_{r-1},t_r+y) \cdot \exp\left(\rho_1|t_1|+\cdots+\rho_{r-1}|t_{r-1}|\right)|\alpha_r|^x e^{\rho_r|y|}$$

($\rho_j = |\arg\alpha_j|$, $1 \le j \le r$) which tends to zero when $|y| \to \infty$ if $\rho_r < \pi$, and we have already known (Remark 1) that we may assume this inequality $\rho_r < \pi$. Hence the shifting (indicated above) is possible. From Proposition 1(ii) we see that the poles of $\zeta_r((s_1,...,s_{r-1},s_r+z);(\alpha_1,...,\alpha_{r-1},0))$ are

$$s_1 + \cdots + s_{r-1} + (s_r + z) = 1 - k,$$

that is

$$z = 1 - (s_1 + \cdots + s_r) - k \qquad (k \in \mathbf{N}_0).$$

When $\mathrm{Re}\, s_j > 1$ $(1 \le j \le r)$, we have

$$\mathrm{Re}\, z \le 1 - \mathrm{Re}\,(s_1 + \cdots + s_r) = 1 - \mathrm{Re}(s_1 + \cdots + s_{r-1}) - \mathrm{Re}\, s_r$$

$$< -\mathrm{Re}\, s_r \qquad (\text{if } r \ge 2)$$

$$< C,$$

that is, these poles are on the left of the original path $\mathrm{Re}\, z = C$. Thus the only relevant poles are $z = 0, 1, 2, \ldots, M - 1$. Counting the residues of those poles, we obtain

$$\zeta_r((s_1,...,s_r);(\alpha_1,...,\alpha_r))$$

$$= \sum_{k=0}^{M-1} \binom{-s_r}{k} \zeta_r((s_1,...,s_{r-1},s_r+k);(\alpha_1,...,\alpha_{r-1},0))\alpha_r^k$$

$$+ \frac{1}{2\pi i} \int_{(M-\varepsilon)} \frac{\Gamma(s_r+z)\Gamma(-z)}{\Gamma(s_r)}$$

$$\times \zeta_r((s_1,...,s_{r-1},s_r+z);(\alpha_1,...,\alpha_{r-1},0))\alpha_r^z dz. \qquad (12)$$

The poles of the integrand are

$$z = -s_r - k, \quad z = k, \quad z = 1 - (s_1 + \cdots + s_r) - k \qquad (k \in \mathbf{N}_0),$$

which do not lie on the path $\operatorname{Re} z = M - \varepsilon$ if

$$-\operatorname{Re} s_r < M - \varepsilon \qquad \text{and} \qquad 1 - \operatorname{Re}(s_1 + \cdots + s_r) < M - \varepsilon.$$

Therefore the integral on the right-hand side of (12) is holomorphic (as a function in $s_1,...,s_r$) in the region

$$\operatorname{Re} s_r > -M + \varepsilon \qquad \text{and} \qquad \operatorname{Re}(s_1 + \cdots + s_r) > 1 - M + \varepsilon.$$

Since M is arbitrary, now (12) implies the meromorphic continuation of $\zeta_r((s_1,...,s_r);(\alpha_1,...,\alpha_r))$ to the whole $\mathbf{C}^r$-space. All singularities are coming from the first term

$$\sum_{k=0}^{M-1} \binom{-s_r}{k} \zeta_r((s_1,...,s_{r-1},s_r+k);(\alpha_1,...,\alpha_{r-1},0))\alpha_r^k.$$

Hence, by using Proposition 1(ii), we find that $\zeta_r((s_1,...,s_r);(\alpha_1,...,\alpha_r))$ is holomorphic except for the possible singularities at

$$s_1 + \cdots + s_r = 1 - k \qquad (k \in \mathbf{N}_0).$$

Therefore we now obtain the first and the second assertions of the following

Theorem A. *For any $r \geq 1$,*
(i) The r-ple zeta function ζ_r defined by (2) can be continued meromorphically with respect to $s_1, \ldots, s_r$ to the whole $\mathbf{C}^r$-space, and is holomorphic in $\alpha_1, \ldots, \alpha_r$ if $\alpha_j \in \mathbf{C} \backslash \mathbf{R}_{\leq(-1)}$ ($1 \leq j \leq r$);
(ii) The possible singularities of ζ_r are only located at $s_1 + s_2 + \cdots + s_r = 1 - k$ ($k \in \mathbf{N}_0 := \mathbf{Z}_{\geq 0}$);

(iii) *If $|\alpha_r| < 1$, then*

$$\zeta_r((s_1,...,s_r);(\alpha_1,...,\alpha_r))$$

$$= \sum_{k=0}^{\infty} \binom{-s_r}{k} \zeta_r((s_1,...,s_{r-1},s_r+k);(\alpha_1,...,\alpha_{r-1},0))\alpha_r^k.$$

Proof of Theorem A(iii). This is just the Taylor expansion of ζ_r with respect to α_r. The radius of convergence is 1, because the assertion (i) implies that the singularity of α_r nearest to the origin is $\alpha_r = -1$. $\quad\square$

Obviously, from Theorem A, we obtain the following

Theorem B. *Let $P(x)$, $Q(x)$ be polynomials with complex coefficients. Assume that all roots $-\beta_1,\ldots,-\beta_m$ of $Q(x)$ are not in $\mathbf{R}_{\geq 1}$. Define the Dirichlet series $\zeta(s;P,Q)$ associated to P and Q by*

$$\zeta(s;P,Q) := \sum_{k=1}^{\infty} \frac{P(k)}{Q(k)^s}.$$

Then, $\zeta(s;P,Q)$ can be meromorphically extended to the whole complex s-plane, and is holomorphic in $\beta_1,\ldots,\beta_m$.

Now set

$$I_M = \frac{1}{2\pi i} \int_{(M-\varepsilon)} \frac{\Gamma(s_r+z)\Gamma(-z)}{\Gamma(s_r)}$$

$$\times \zeta_r((s_1,...,s_{r-1},s_r+z);(\alpha_1,...,\alpha_{r-1},0))\alpha_r^z dz.$$

From (12) and Theorem A(iii), we obtain

Lemma 2. *With the same notation as above, $\lim_{M\to\infty} I_M = 0$ if $|\alpha_r| < 1$.*

In view of Remark 3 below, it is of interest to give a direct proof of this fact. The following argument is valid if $|\arg\alpha_r| < \pi$.

Using

$$\Gamma(-z) = -\frac{\pi}{\Gamma(1+z)\sin(\pi z)}$$

and Stirling's formula

$$\Gamma(z) = \sqrt{2\pi}e^{-z}z^{z-\frac{1}{2}}\left(1+O\left(\frac{1}{|z|}\right)\right),$$

we have

$$I_M = \frac{ie}{2e^{s_r}\Gamma(s_r)} \int_{(M-\varepsilon)} \frac{(z(1+\frac{s_r}{z}))^{s_r+z-\frac{1}{2}}}{(z(1+\frac{1}{z}))^{z+\frac{1}{2}}\sin(\pi z)}\left(1+O\left(\frac{1}{|z|}\right)\right)$$

$$\times \zeta_r((s_1,...,s_{r-1},s_r+z);(\alpha_1,...,\alpha_{r-1},0))\alpha_r^z dz.$$

Clearly,

$$\left(1 + \frac{s_r}{z}\right)^{s_r + z - \frac{1}{2}} = \exp\left(s_r + O\left(\frac{1}{|z|}\right)\right) = e^{s_r}\left(1 + O\left(\frac{1}{|z|}\right)\right),$$

and

$$\left(1 + \frac{1}{z}\right)^{z + \frac{1}{2}} = e \cdot \left(1 + O\left(\frac{1}{|z|}\right)\right).$$

Hence,

$$I_M = \frac{ie}{2e^{s_r}\Gamma(s_r)}e^{s_r - 1}\int_{(M-\varepsilon)} \frac{z^{s_r - 1}}{\sin(\pi z)}\left(1 + O\left(\frac{1}{|z|}\right)\right)$$
$$\times \zeta_r((s_1, \ldots, s_{r-1}, s_r + z); (\alpha_1, \ldots, \alpha_{r-1}, 0))\alpha_r^z dz.$$

Now write $\mu = M - \varepsilon$, $z = \mu + iy$. Note that, since $\mu \notin \mathbf{Z}$, $|e^{\pi i(\mu + iy)} - e^{-\pi i(\mu + iy)}| \gg e^{|y|}$ for any $y \in \mathbf{R}$. Thus, writing $\sin(\pi z) = \frac{1}{2i}(e^{\pi iz} - e^{-\pi iz})$, we have

$$I_M \ll \frac{1}{|\Gamma(s_r)|}\int_{-\infty}^{\infty} \frac{|(\mu + iy)^{s_r - 1}|}{e^{\pi|y|}}\left(1 + O\left(\frac{1}{1 + |y|}\right)\right)$$
$$\times |\zeta_r((s_1, \ldots, s_{r-1}, s_r + z); (\alpha_1, \ldots, \alpha_{r-1}, 0))| \cdot |\alpha_r|^{\mu}e^{\rho_r|y|}dy.$$

Write $s_r = \sigma_r + it_r$. Then

$$|(\mu + iy)^{s_r - 1}| \le |\mu + iy|^{\sigma_r - 1}e^{\frac{\pi}{2}|t_r|}.$$

Thus, from $1 + O(\frac{1}{1 + |y|}) \ll 1$, we have

$$I_M \ll \frac{e^{\frac{\pi}{2}|t_r|}}{|\Gamma(s_r)|}\int_{-\infty}^{\infty} \frac{|(\mu + iy)^{\sigma_r - 1}|}{e^{\pi|y|}}$$
$$\times |\zeta_r((s_1, \ldots, s_{r-1}, s_r + z); (\alpha_1, \ldots, \alpha_{r-1}, 0))| \cdot |\alpha_r|^{\mu}e^{\rho_r|y|}dy.$$

Thus, by the fact that $\zeta_r((s_1, \ldots, s_{r-1}, s_r + z); (\alpha_1, \ldots, \alpha_{r-1}, 0))$ is absolutely convergent if $\mathrm{Re}\,(z) = \mu = M - \varepsilon$ is sufficiently large, we get

$$I_M \ll \int_{-\infty}^{\infty} \frac{|(\mu + iy)^{\sigma_r - 1}|}{e^{\pi|y|}} \cdot |\alpha_r|^{\mu}e^{\rho_r|y|}dy.$$

But if $|y| \le \mu$ (resp. $|y| > \mu$), then $|\mu + iy| \sim \mu$ (resp. $|\mu + iy| \sim |y|$), hence we have

$$I_M \ll \int_{-\mu}^{\mu} \frac{\mu^{\sigma_r - 1}}{e^{\pi|y|}} \cdot |\alpha_r|^{\mu}e^{\rho_r|y|}dy + \int_{|y| > \mu} \frac{|y|^{\sigma_r - 1}}{e^{\pi|y|}} \cdot |\alpha_r|^{\mu}e^{\rho_r|y|}dy$$

$$\le \mu^{\sigma_r - 1}|\alpha_r|^{\mu}\int_{-\infty}^{\infty} \frac{dy}{e^{(\pi - \rho_r)|y|}} + |\alpha_r|^{\mu}\int_{|y| > \mu} \frac{|y|^{\sigma_r - 1}}{e^{(\pi - \rho_r)|y|}}dy.$$

If $\rho_r = |\arg \alpha_r| < \pi$, we have

$$I_M \ll \mu^{\sigma_r - 1}|\alpha_r|^\mu + |\alpha_r|^\mu$$

where $\ll$ depends on σ_r. Therefore $\lim_{M \to \infty} I_M = 0$ if $|\alpha_r| < 1$.

Remark 2. Similarly to Theorem A(iii), we can prove that, if $|\alpha_{r-1}| < 1$,

$$\zeta_r((s_1,...,s_r); (\alpha_1,...,\alpha_{r-1}, 0))$$

$$= \sum_{j=0}^{\infty} \binom{-s_{r-1}}{j} \zeta_{r-1}((s_1,...,s_{r-2}, s_{r-1} + s_r + j); (\alpha_1,...,\alpha_{r-2}, 0))\alpha_{r-1}^j.$$

Remark 3. In [18, 19], we have encountered asymptotic expansions which are similar to (11), (12) but each term in the expansions includes an additional factor like $\zeta(-k, b)$, where $\zeta(s, b)$ stands for the Hurwitz zeta function. By using the formula (2.17.3) of [26] it is easily seen that $|\zeta(-k, b)| \sim (2e\pi)^{-k}k^{k+\frac{1}{2}}$, which implies that the expansions in [18, 19] are not convergent.

1.2   Regularity at non-positive integers

Now we consider the situation at non-positive integers of $\zeta_r((s - k, s,..., s); (\alpha_1,..., \alpha_r))$. By (12), it is enough to consider the case $\alpha_r = 0$, that is, $\zeta_r((s - k, s,..., s); (\alpha_1,..., \alpha_{r-1}, 0))$. Let $n \in \mathbf{N}$ and $h \in \mathbf{N}_0$. We study the following somewhat more general form: $\zeta_r((s - k, s,..., s, ns + h); (\alpha_1,..., \alpha_{r-1}, 0))$.

First we discuss the case $r = 2$. By (10) we have

$$\zeta_2((s - k, ns + h); (\alpha_1, 0))$$

$$= \sum_{j=0}^{M-1} \binom{-s + k}{j} \zeta((n + 1)s - k + h + j)\alpha_1^j + \text{(integral term)}. \quad (13)$$

Since the integral term is holomorphic, the poles are coming only from the factors $\zeta((n + 1)s - k + h + j)$. Hence

Lemma 3. *All poles of* $\zeta_2((s - k, ns + h); (\alpha_1, 0))$ *are at most of order* 1.

The poles of $\zeta((n + 1)s - k + h + j)$ is at $(n + 1)s - k + h + j = 1$, that is

$$s = \frac{k - h - j + 1}{n + 1}.$$

On the other hand, we have

$$\binom{-s + k}{j} = \begin{cases} 1 & \text{if } j = 0, \\ \frac{1}{j!}(-s + k)(-s + k - 1)\cdots(-s + k - j + 1) & \text{if } j \geq 1, \end{cases}$$

and the zeros of the latter are at

$$s = k, k - 1, \ldots, k - j + 1. \tag{14}$$

When $j = 0$, the pole $s = \frac{k-h+1}{n+1}$ is really a pole of $\zeta_2((s - k, ns + h); (\alpha_1, 0))$. When $j \geq 1$, the pole $\frac{k-h-j+1}{n+1}$ is cancelled by a zero in the list of (14) if and only if there exists an

$$l \in \mathbf{N}_0, \ 0 \leq l \leq j - 1, \ \frac{k - h - j + 1}{n + 1} = k - l. \tag{15}$$

These conditions are valid if and only if $\frac{k-h-j+1}{n+1} \in \mathbf{Z}$ and $j \geq k+1+\frac{h}{n}$. That is,

$$\frac{k - h - j + 1}{n + 1} \in \mathbf{Z}$$

and

$$\frac{k - h - j + 1}{n + 1} \leq \frac{k - h - (k + 1 + \frac{h}{n}) + 1}{n + 1} = -\frac{h}{n}.$$

Summarizing the above argument, we now obtain

Lemma 4. *The set of all poles of $\zeta_2((s - k, ns + h); (\alpha_1, 0))$ is*

$$\left\{ \frac{k - h - j + 1}{n + 1}; \ j \geq 0 \right\} \setminus \left\{ \text{integers} \leq -\frac{h}{n} \right\},$$

and poles are all simple.

Note that, from (13), the residue of $\zeta_2((s - k, ns + h); (\alpha, 0))$ at $s = \frac{k-h-j+1}{n+1}$ is

$$= \frac{\alpha_1^j}{n + 1} \binom{k - \frac{k-h-j+1}{n+1}}{j}. \tag{16}$$

Next we consider the case $r = 3$. By (10), we have

$$\zeta_3((s - k, s, ns + h); (\alpha_1, \alpha_2, 0))$$
$$= \sum_{j=0}^{M-1} \binom{-s}{j} \zeta_2((s - k, (n + 1)s + h + j); (\alpha_1, 0))\alpha_2^j + (\text{integral term}).$$

Using Lemma 4, we have that $\zeta_2((s - k, (n + 1)s + h + j); (\alpha_1, 0))$ has a pole at $s = \frac{k-(h+j)-j'+1}{(n+1)+1}, j' \geq 0$ and the residue there is by (16)

$$\frac{\alpha_1^{j'}}{(n + 1) + 1} \binom{k - \frac{k-(h+j)-j'+1}{(n+1)+1}}{j'}.$$

Let $j + j' = l$. The same $s = \frac{k-h-l+1}{n+2}, l \geq 0$ appears from the pairs $(j, j') = (0, l), (1, l-1), \ldots, (l, 0)$. (We assume M is sufficiently large so $M - 1 \geq l$.) Therefore, the residue at $s = \frac{k-h-l+1}{n+2}$ of $\zeta_3((s - k, s, ns + h); (\alpha_1, \alpha_2, 0))$ is equal to

$$\sum_{j=0}^{l} \binom{-\frac{k-h-l+1}{n+2}}{j} \cdot \frac{\alpha_1^{j'}}{(n+1)+1} \binom{k - \frac{k-h-l+1}{(n+1)+1}}{j'} \alpha_2^j$$

$$= \frac{1}{n+2} \sum_{j=0}^{l} \binom{-\frac{k-h-l+1}{n+2}}{j} \binom{k - \frac{k-h-l+1}{n+2}}{l-j} \alpha_1^{l-j} \alpha_2^j. \tag{17}$$

If

$$\frac{k - h - l + 1}{n + 2} = -m \in \mathbf{Z}, m \geq 0, \tag{18}$$

then $\binom{-\frac{k-h-l+1}{n+2}}{j} = 0$ for $j \geq m+1$ and $\binom{k-\frac{k-h-l+1}{n+2}}{l-j} = 0$ for $j \leq l - k - m - 1$. Therefore we can conclude that if

$$m \leq l - k - m - 1, \tag{19}$$

then the right hand side of (17) is zero, so actually $\zeta_3((s - k, s, ns + h); (\alpha_1, \alpha_2, 0))$ is holomorphic at $s = -m$. The condition (19) is equivalent to $l \geq k + 2m + 1$, hence from (18) we have

$$-m(n + 2) = k - h - l + 1 \leq k - h - (k + 2m + 1) + 1,$$

i.e., $m \geq \frac{h}{n}$. This argument implies that the negative integers $-m$ with $m \geq \frac{h}{n}$ is not singular.

Now we study the general case. The above results suggest that the function

$$\zeta_r((s - k, s, \ldots, s, ns + h); (\alpha_1, \ldots, \alpha_{r-1}, 0))$$

is holomorphic at $s = -m$, which is an integer and $\leq -\frac{h}{n}$. We prove this fact by induction. Assume that the above claim is true for ζ_{r-1}. Recall that we have

$$\zeta_r((s - k, s, \ldots, s, ns + h); (\alpha_1, \ldots, \alpha_{r-1}, 0))$$

$$= \sum_{j=0}^{M-1} \binom{-s}{j} \zeta_{r-1}((s - k, s, \ldots, s, (n+1)s + h + j); (\alpha_1, \ldots, \alpha_{r-2}, 0)) \alpha_{r-1}^j$$

$$+ (\text{holomorphic term}) \tag{20}$$

where in particular the binomial coefficients are $\binom{-s}{j}$ instead of $\binom{-s+k}{j}$ as $r \geq 3$. First, it is easy to see that

Lemma 5. *All poles of* $\zeta_r((s - k, s,..., s, ns + h); (\alpha_1,..., \alpha_{r-1}, 0))$ $(n \in \mathbf{N}, h \in \mathbf{N}_0)$ *are at most of order 1.*

Proof. We already know this fact for $r = 2$. (See Lemma 3.) Hence, by using (20), the general case immediately follows by induction. $\qquad \square$

The induction hypothesis says that $\zeta_{r-1}((s - k, s,..., s, (n + 1)s + h + j); (\alpha_1,..., \alpha_{r-2}, 0))$ is holomorphic at $s = -m, m \in \mathbf{Z}, m \geq \frac{h+j}{n+1}$. On the other hand, by Lemma 5 we see that

$$\binom{-s}{j} \zeta_{r-1}((s - k, s,..., s, (n + 1)s + h + j); (\alpha_1,..., \alpha_{r-2}, 0)) \qquad (21)$$

is not singular at $s = 0, -1, -2, -3, \ldots, -(j - 1)$. Hence we find that if

$$\frac{h + j}{n + 1} \leq j, \qquad (22)$$

then the term (21) is non-singular for any non-positive integer. The condition (22) is equivalent to $h + j \leq (n + 1)j$, hence $j \geq \frac{h}{n}$. This implies that the terms on the right-hand side of (20) with $j \geq \frac{h}{n}$ are non-singular at non-positive integers.

Consider the terms (21) with $0 \leq j < \frac{h}{n}$. We have already shown that these terms are holomorphic at $s = -m \in \mathbf{Z}, m \geq \frac{h+j}{n+1}$. Hence if $s = -m, m \in \mathbf{Z}_{\geq 0}$ is singular, then $0 \leq m < \frac{h+j}{n+1}$. Since $j < \frac{h}{n}$ we find

$$m < \frac{h + \frac{h}{n}}{n + 1} = \frac{h}{n}.$$

Therefore if $s = -m$, $m \in \mathbf{Z}_{\geq 0}$, is a pole of the right hand side of (20), it is necessary that $m < \frac{h}{n}$. In other words, $s = -m$, $m \in \mathbf{Z}_{\geq 0}$, $m \geq \frac{h}{n}$ is not a singular point of (20) as desired. We have proved the following

Proposition 2. *The function* $\zeta_r((s - k, s,..., s, ns + h); (\alpha_1,..., \alpha_{r-1}, 0))$ *is holomorphic at any integer* $s = -m \leq -\frac{h}{n}$. *In particular,* $\zeta_r((s - k, s,..., s); (\alpha_1,..., \alpha_{r-1}, 0))$ *is holomorphic at any non-positive integers.*

The same conclusion holds for $\zeta_r((s-k, s,..., s, ns+h); (\alpha_1,..., \alpha_{r-1}, \alpha_r))$, which can be easily checked by using (12). As a direct consequence, we obtain the following

Theorem C. *For any two polynomials* $P(x)$ *and* $Q(x)$, *the associated zeta function* $\zeta(s; P, Q)$ *is holomorphic at any non-positive integers.*

2.   Special Values

2.1   The Value $\zeta(0; P, Q)$

Let $P(x)$, $Q(x)$, $\zeta(s; P, Q)$ be as in the previous section. Denote $n :=$ deg$P$, $m := $ deg$Q$ and in this section we write $P(x) = \sum_{i=0}^{n} a_i x^i$ and $Q(x) = b \prod_{j=1}^{m} (x + \beta_j)$. Assume that $\beta_j \notin \mathbf{R}_{\leq(-1)}, 1 \leq j \leq m$. Then by the results in Section 1, $\zeta(s; P, Q)$ is holomorphic at $s = 0$. Hence it makes sense to talk about $\zeta(0; P, Q)$ and $\zeta'(0; P, Q)$. In this section, we evaluate those values.

Lemma 6. *Assume* $|\beta_j| < 1$ $(1 \leq j \leq m)$. *Then we have*

$$\zeta(0; P, Q) = \sum_{i=0}^{n} a_i \zeta(-i) + \frac{1}{m} \sum_{j=1}^{m} \sum_{i=0}^{n} \frac{a_i}{i+1} (-\beta_j)^{i+1}.$$

Proof. By a direct calculation using

$$\frac{1}{(k + \beta_j)^s} = \frac{1}{k^s} \sum_{l_j=0}^{\infty} \binom{-s}{l_j} \left(\frac{\beta_j}{k} \right)^{l_j},$$

we have

$$\zeta(s; P, Q) = \sum_{k=1}^{\infty} \frac{P(k)}{b^s (k + \beta_1)^s \cdots (k + \beta_m)^s}$$

$$= \sum_{l_1, \ldots, l_m = 0}^{\infty} b^{-s} (\beta_1)^{l_1} \cdots (\beta_m)^{l_m} \binom{-s}{l_1} \cdots \binom{-s}{l_m}$$

$$\times \sum_{k=1}^{\infty} \sum_{i=0}^{n} a_i k^i k^{-ms - l_1 - \cdots - l_m}$$

$$= \sum_{i=0}^{n} a_i \sum_{l_1, \ldots, l_m = 0}^{\infty} b^{-s} (\beta_1)^{l_1} \cdots (\beta_m)^{l_m} \binom{-s}{l_1} \cdots \binom{-s}{l_m}$$

$$\times \zeta(ms + l_1 + \cdots + l_m - i)$$

$$= \sum_{i=0}^{n} a_i b^{-s} \left[\zeta(ms - i) + \sum_{j=1}^{m} \sum_{l_j=1}^{\infty} (\beta_j)^{l_j} \binom{-s}{l_j} \zeta(ms + l_j - i) \right.$$

$$+ \sum_{\#\{j; l_j \neq 0\} \geq 2} (\beta_1)^{l_1} \cdots (\beta_m)^{l_m} \binom{-s}{l_1} \cdots \binom{-s}{l_m}$$

$$\left. \times \zeta(ms + l_1 + \cdots + l_m - i) \right].$$

Therefore, noting

$$\zeta(s) = \frac{1}{s-1} + c_0 + c_1(s-1) + \cdots,$$

and $\binom{0}{l} = 0$ if $l \neq 0$, we get

$$\zeta(0; P, Q) = \sum_{i=0}^{n} a_i \left[\zeta(-i) + \sum_{j=1}^{m} \sum_{l_j=1}^{\infty} (\beta_j)^{l_j} \binom{-s}{l_j} \zeta(ms + l_j - i)|_{s=0} \right]$$

$$= \sum_{i=0}^{n} a_i \left[\zeta(-i) + \sum_{j=1}^{m} (\beta_j)^{i+1} \binom{-s}{i+1} \zeta(ms+1)|_{s=0} \right]$$

$$= \sum_{i=0}^{n} a_i \zeta(-i) + \sum_{i=0}^{n} a_i \sum_{j=1}^{m} (\beta_j)^{i+1}$$

$$\times \frac{(-s)(-s-1)\cdots(-s-(i+1)+1)}{(i+1)!} \zeta(ms+1) \Big|_{s=0}$$

$$= \sum_{i=0}^{n} a_i \zeta(-i) + \sum_{i=0}^{n} a_i \sum_{j=1}^{m} (\beta_j)^{i+1} \frac{(-1)^{i+1}}{i+1} \cdot \frac{1}{m}$$

$$= \sum_{i=0}^{n} a_i \zeta(-i) + \frac{1}{m} \sum_{j=1}^{m} \sum_{i=0}^{n} \frac{a_i}{i+1} (-\beta_j)^{i+1}.$$

This completes the proof of Lemma 6. $\qquad\square$

In view of Theorem B, the left-hand side of the formula in the statement of Lemma 6 is holomorphic in $\beta_j \in \mathbf{C} \backslash \mathbf{R}_{\leq(-1)}$ $(1 \leq j \leq m)$. Hence by analytic continuation with respect to $\beta_1, \ldots, \beta_m$, we have the following

Theorem D. *Let $P(x), Q(x)$ be as above. Assume that $\beta_j \notin \mathbf{R}_{\leq(-1)}, 1 \leq j \leq m$ and define the Dirichlet series $\zeta(s; P, Q) := \sum_{n=1}^{\infty} \frac{P(n)}{Q(n)^s}$. Then,*

$$\zeta(0; P, Q) = \sum_{i=0}^{n} a_i \zeta(-i) + \frac{1}{m} \sum_{j=1}^{m} \sum_{i=0}^{n} \frac{a_i}{i+1} (-\beta_j)^{i+1}.$$

2.2 The Value $\zeta'(0; P, Q)$

In this subsection, we evaluate $\zeta'(0; P, Q)$. We first introduce the following auxiliary functions.

$$A(s) := \sum_{j=1}^{m} \sum_{k=1}^{\infty} \frac{P(k)}{Q(k)^s} \log(k + \beta_j),$$

$$B(s) := b^{-s} \sum_{i=0}^{n} a_i \sum_{j=1}^{m} \sum_{u=0}^{i} \binom{i}{u} (-\beta_j)^{i-u}$$

$$\times \sum_{l_1,\ldots,\hat{l}_j,\ldots,l_m=0}^{\infty} (\beta_1 - \beta_j)^{l_1} \cdots (\beta_m - \beta_j)^{l_m} \binom{-s}{l_1} \cdots \binom{-s}{l_m}$$

$$\times \sum_{k=1}^{\infty} (k + \beta_j)^{-ms - l_1 - \cdots - l_m + u} \log(k + \beta_j).$$

Here $\hat{l}_j$ means that l_j is omitted from the summation.

Lemma 7. *With the same notation as above, under the assumption $|\beta_j| < 1/3$ $(1 \le j \le m)$, we have,*
(i) $\zeta'(0; P, Q) = -\zeta(0; P, Q) \log b - A(0)$;
(ii) $A(0) = B(0)$ and hence $\zeta'(0; P, Q) = -\zeta(0; P, Q) \log b - B(0)$.

Proof. The proof of (i) is given by a direct calculation. As for (ii), we first note that the assumption $|\beta_j| < 1/3$ $(1 \le j \le m)$ implies

$$\left| \frac{\beta_p - \beta_j}{k + \beta_j} \right| < 1 \qquad (1 \le j \le m, \ 1 \le p \le m)$$

for any positive integer k. Therefore

$$(k + \beta_p)^{-s} = (k + \beta_j)^{-s} \left(1 + \frac{\beta_p - \beta_j}{k + \beta_j} \right)^{-s} = (k + \beta_j)^{-s} \sum_{l_p=0}^{\infty} \binom{-s}{l_p} \left(\frac{\beta_p - \beta_j}{k + \beta_j} \right)^{l_p}$$

and similarly

$$k^i = (k + \beta_j)^i \sum_{u=0}^{i} \binom{i}{u} \left(\frac{-\beta_j}{k + \beta_j} \right)^{i-u}.$$

Using these formulas we obtain $A(0) = B(0)$, hence (ii) follows.

Next we consider $B(0)$. For this, we again introduce the following auxiliary functions. First, for β_j, define

$$\zeta_j(s) = \sum_{k=1}^{\infty} \frac{1}{(k + \beta_j)^s},$$

and denote its derivatives by ζ_j'. Note that $\zeta_j(s) = \zeta(s, 1 + \beta_j)$, where the right-hand side denotes the standard Hurwitz zeta-function. In the sequel we sometimes write $\zeta_j'(ms+c)$, which means $(d/dw)\zeta_j(w)|_{w=ms+c}$. We set

$$B^1(s) := b^{-s} \sum_{i=0}^{n} a_i \sum_{j=1}^{m} \sum_{u=0}^{i} \binom{i}{u} (-\beta_j)^{i-u} \zeta_j'(ms - u),$$

$$B^2(s) := b^{-s} \sum_{i=0}^{n} a_i \sum_{j=1}^{m} \sum_{u=0}^{i} \binom{i}{u} (-\beta_j)^{i-u}$$

$$\times \sum_{\substack{1 \le p \le m \\ p \ne j}} \sum_{l_p=1}^{\infty} (\beta_p - \beta_j)^{l_p} \binom{-s}{l_p} \zeta_j'(ms + l_p - u),$$

$$B^3(s) = b^{-s} \sum_{i=0}^{n} a_i \sum_{j=1}^{m} \sum_{u=0}^{i} \binom{i}{u} (-\beta_j)^{i-u}$$

$$\times \sum_{\substack{1 \le p, q \le m \\ j \ne p, j \ne q, p \ne q}} \sum_{l_p, l_q=1}^{\infty} (\beta_p - \beta_j)^{l_p} (\beta_q - \beta_j)^{l_q}$$

$$\times \binom{-s}{l_p} \binom{-s}{l_q} \zeta_j'(ms + l_p + l_q - u),$$

$$B^4(s) := b^{-s} \sum_{i=0}^{n} a_i \sum_{j=1}^{m} \sum_{u=0}^{i} \binom{i}{u} (-\beta_j)^{i-u}$$

$$\times \sum_{\#\{p \ne j; l_p \ne 0\} \ge 3} \sum_{l_1, \ldots, \hat{l}_j, \ldots, l_m=0}^{\infty} (\beta_1 - \beta_j)^{l_1} \cdots (\beta_m - \beta_j)^{l_m}$$

$$\times \binom{-s}{l_1} \cdots \binom{-s}{l_m} \zeta_j'(ms + l_1 + \cdots + \hat{l}_j + \cdots + l_m - u).$$

Since the innermost sum in the definition of $B(s)$ is $-\zeta_j'(ms + l_1 + \cdots + \hat{l}_j + \cdots + l_m - u)$, by a direct calculation, we have the following

Lemma 8. *With the same notation as above, under the assumption* $|\beta_j| < 1/3$ $(1 \le j \le m)$, *we have,*

$$B(0) = -B^1(0) - B^2(0) - B^3(0) - B^4(0).$$

Therefore,

$$\zeta'(0; P, Q) = -\zeta(0; P, Q) \log b + B^1(0) + B^2(0) + B^3(0) + B^4(0).$$

Lemma 9. *With the same notation as above, under the assumption* $|\beta_j| < 1/3$ $(1 \le j \le m)$, *we have,*

(i) $B^1(0) = \sum_{j=1}^{m} \sum_{i=0}^{n} a_i \sum_{u=0}^{i} \binom{i}{u} (-\beta_j)^{i-u} \zeta_j'(-u);$

(ii) $B^2(0) = -\dfrac{1}{m^2} \sum_{i=0}^{n} a_i \sum_{j=1}^{m} \sum_{u=0}^{i} \binom{i}{u} (-\beta_j)^{i-u} \sum_{p=1}^{m} (\beta_j - \beta_p)^{u+1} \dfrac{1}{u+1} \sum_{a=1}^{u} \dfrac{1}{a};$

(iii) $B^3(0) = -\dfrac{1}{m^2} \sum_{j=1}^{m} \sum_{i=0}^{n} a_i \sum_{u=0}^{i} \binom{i}{u} (-\beta_j)^{i-u} \sum_{\substack{1 \le p,q \le m \\ p \neq q}}$

$$\times \sum_{l_p, l_q \ge 1, l_p + l_q = 1+u} \frac{(-\beta_p + \beta_j)^{l_p}}{l_p} \frac{(-\beta_q + \beta_j)^{l_q}}{l_q};$$

(iv) $B^4(0) = 0.$

Proof. First, from the definition, (i) follows immediately.

Note that $\zeta_j(s) = \frac{1}{s-1} + c_0 + c_1(s-1) + \cdots$ (see e.g., [6]). Thus by $\binom{0}{l} = 0$ if $l \neq 0$, from the definition, $B^4(0) = 0$, which gives (iv).

As for (iii), we have

$$B^3(0)$$

$$= b^{-s} \sum_{i=0}^{n} a_i \sum_{j=1}^{m} \sum_{u=0}^{i} \binom{i}{u} (-\beta_j)^{i-u}$$

$$\times \sum_{\substack{1 \le p,q \le m \\ j \neq p, j \neq q, p \neq q \\ l_p + l_q - u = 1}} (\beta_p - \beta_j)^{l_p} (\beta_q - \beta_j)^{l_q}$$

$$\times \frac{(-s)(-s-1)\cdots(-s-l_p+1)}{l_p!} \frac{(-s)(-s-1)\cdots(-s-l_q+1)}{l_q!} \frac{-1}{m^2 s^2} \Big|_{s=0}$$

$$= \sum_{i=0}^{n} a_i \sum_{j=1}^{m} \sum_{u=0}^{i} \binom{i}{u} (-\beta_j)^{i-u} \sum_{\substack{1 \le p,q \le m \\ j \neq p, j \neq q, p \neq q}}$$

$$\times \sum_{l_p, l_q \ge 1, l_p + l_q = 1+u} (\beta_p - \beta_j)^{l_p} (\beta_q - \beta_j)^{l_q} \frac{(-1)^{l_p-1}}{l_p} \cdot \frac{(-1)^{l_q-1}}{l_q} \cdot \frac{-1}{m^2}$$

$$= -\frac{1}{m^2} \sum_{j=1}^{m} \sum_{i=0}^{n} a_i \sum_{u=0}^{i} \binom{i}{u} (-\beta_j)^{i-u} \sum_{\substack{1 \le p,q \le m \\ p \neq q}}$$

$$\times \sum_{l_p, l_q \geq 1, l_p+l_q=1+u} \frac{(-\beta_p + \beta_j)^{l_p}}{l_p} \frac{(-\beta_q + \beta_j)^{l_q}}{l_q}.$$

This gives (iii). □

Finally, let us consider $B^2(s)$. For this, we need the following

Sublemma. *With the same notation as above, we have*

$$\sum_{j=1}^{m} \sum_{p=1}^{m} \left(\sum_{u=0}^{i} \binom{i}{u} \frac{1}{u+1} (-\beta_j)^{i-u} (\beta_j - \beta_p)^{u+1} \right) = 0.$$

Proof. Clearly,

$$\sum_{j=1}^{m} \sum_{p=1}^{m} \left(\sum_{u=0}^{i} \binom{i}{u} \frac{1}{u+1} (-\beta_j)^{i-u} (\beta_j - \beta_p)^{u+1} \right)$$

$$= \frac{1}{i+1} \sum_{j,p=1}^{m} \sum_{u=0}^{i} \binom{i+1}{u+1} (-\beta_j)^{(i+1)-(u+1)} (\beta_j - \beta_p)^{u+1}$$

$$= \frac{1}{i+1} \sum_{j,p=1}^{m} \sum_{k=1}^{i+1} \binom{i+1}{k} (-\beta_j)^{(i+1)-k} (\beta_j - \beta_p)^k$$

$$= \frac{1}{i+1} \sum_{j,p=1}^{m} \left(\sum_{k=0}^{i+1} \binom{i+1}{k} (-\beta_j)^{(i+1)-k} (\beta_j - \beta_p)^k - (-\beta_j)^{i+1} \right)$$

$$= \frac{1}{i+1} \sum_{j,p=1}^{m} \left((-\beta_j + \beta_j - \beta_p)^{i+1} - (-\beta_j)^{i+1} \right)$$

$$= 0.$$

This completes the proof of the sublemma. □

Now let us come back to the proof of Lemma 9(ii). By the definition,

$$B^2(0) = b^{-s} \sum_{i=0}^{n} a_i \sum_{j=1}^{m} \sum_{u=0}^{i} \binom{i}{u} (-\beta_j)^{i-u}$$

$$\times \sum_{p=1}^{m} \sum_{l_p=1}^{\infty} (\beta_p - \beta_j)^{l_p} \binom{-s}{l_p} \zeta_j'(ms + l_p - u)|_{s=0}$$

$$= b^{-s} \sum_{i=0}^{n} a_i \sum_{j=1}^{m} \sum_{u=0}^{i} \binom{i}{u} (-\beta_j)^{i-u}$$

$$\times \sum_{p=1}^{m} (\beta_p - \beta_j)^{u+1} \binom{-s}{u+1} \zeta_j'(ms+1)\big|_{s=0}$$

$$= b^{-s} \sum_{i=0}^{n} a_i \sum_{j=1}^{m} \sum_{u=0}^{i} \binom{i}{u} (-\beta_j)^{i-u} \sum_{p=1}^{m} (\beta_p - \beta_j)^{u+1}$$

$$\times \frac{(-s)(-s-1)\cdots(-s-(u+1)+1)}{(u+1)!}$$

$$\times \left(-\frac{1}{m^2} \cdot \frac{1}{s^2} + \text{holomorphic terms} \right)\Bigg|_{s=0}$$

$$= -\frac{1}{m^2} b^{-s} \sum_{i=0}^{n} a_i \sum_{j=1}^{m} \sum_{u=0}^{i} \binom{i}{u} (-\beta_j)^{i-u}$$

$$\times \sum_{p=1}^{m} (\beta_j - \beta_p)^{u+1} \frac{(s+1)\cdots(s+u)}{(u+1)!} \left(\frac{1}{s}\right)\Big|_{s=0}$$

$$= -\frac{1}{m^2} b^{-s} \sum_{i=0}^{n} a_i \sum_{j=1}^{m} \sum_{u=0}^{i} \binom{i}{u} (-\beta_j)^{i-u}$$

$$\times \sum_{p=1}^{m} (\beta_j - \beta_p)^{u+1} \frac{u! + \sum_{a=1}^{u} u!/a \cdot s}{(u+1)!} \left(\frac{1}{s}\right)\Bigg|_{s=0}$$

$$= -\frac{1}{m^2} b^{-s} \sum_{i=0}^{n} a_i \sum_{j=1}^{m} \sum_{u=0}^{i} \binom{i}{u} (-\beta_j)^{i-u}$$

$$\times \sum_{p=1}^{m} (\beta_j - \beta_p)^{u+1} \left(\frac{1}{u+1} + \frac{\sum_{a=1}^{u} u!/a \cdot s}{(u+1)!} \right) \left(\frac{1}{s}\right)\Bigg|_{s=0}$$

$$= -\frac{1}{m^2} \sum_{i=0}^{n} a_i \sum_{j=1}^{m} \sum_{u=0}^{i} \binom{i}{u} (-\beta_j)^{i-u} \sum_{p=1}^{m} (\beta_j - \beta_p)^{u+1} \frac{1}{u+1} \sum_{a=1}^{u} \frac{1}{a}$$

provided that

$$\sum_{i=0}^{n} a_i \sum_{j=1}^{m} \sum_{u=0}^{i} \binom{i}{u} (-\beta_j)^{i-u} \sum_{p=1}^{m} (\beta_j - \beta_p)^{u+1} \frac{1}{u+1} = 0.$$

But by the sublemma above,

$$\sum_{u=0}^{i} \binom{i}{u} \frac{1}{u+1} \sum_{j=1}^{m} \sum_{p=1}^{m} (-\beta_j)^{i-u} (\beta_j - \beta_p)^{u+1} = 0.$$

This then completes the proof of (ii) and hence of Lemma 9.

By analytic continuation, we may remove the assumption $|\beta_j| < 1/3$ $(1 \leq j \leq m)$. Therefore, we have the following

Theorem E. Let $P(x) = \sum_{i=0}^{n} a_i x^i$, $Q(x) = b \prod_{j=1}^{m}(x + \beta_j)$ be polynomials with complex coefficients. Assume that $\beta_j \notin \mathbf{R}_{\leq(-1)}, 1 \leq j \leq m$. Define the Dirichlet series $\zeta(s; P, Q) := \sum_{n=1}^{\infty} \frac{P(n)}{Q(n)^s}$. Then,

$$\zeta'(0; P, Q) = -\zeta(0; P, Q)\log b + \sum_{j=1}^{m}\sum_{i=0}^{n} a_i \sum_{u=0}^{i}\binom{i}{u}(-\beta_j)^{i-u}\zeta_j'(-u)$$

$$-\frac{1}{m^2}\sum_{j=1}^{m}\sum_{i=0}^{n} a_i \sum_{u=0}^{i}\binom{i}{u}(-\beta_j)^{i-u}\sum_{\substack{1\leq p,q\leq m \\ p\neq q}}$$

$$\times \sum_{l_p,l_q\geq 1, l_p+l_q=1+u}\frac{(-\beta_p+\beta_j)^{l_p}}{l_p}\frac{(-\beta_q+\beta_j)^{l_q}}{l_q}$$

$$-\frac{1}{m^2}\sum_{j=1}^{m}\sum_{i=0}^{n} a_i \sum_{u=0}^{i}\binom{i}{u}(-\beta_j)^{i-u}$$

$$\times \sum_{p=1}^{m}(\beta_j - \beta_p)^{u+1}\frac{1}{u+1}\sum_{a=1}^{u}\frac{1}{a}.$$

Here $\zeta_j(s) := \sum_{k=1}^{\infty}\frac{1}{(k+\beta_j)^s} = \zeta(s, 1+\beta_j)$ denotes the standard Hurwitz zeta function associated to β_j.

2.3   Determinant of the Laplacian

Concerning the zeta-function $H_{g,d}(s)$ defined by (5) in the Introduction, Kumagai [9, Lemma 3] proves that

$$H_{g,d}'(0) = \sum_{k=1}^{g}(k-g)^d \log k$$

$$-\frac{1}{2^d}\cdot\frac{(-g)^{d+1}}{d+1}\sum_{1\leq l\leq d; l:\text{odd}}\binom{d+1}{l+1}\sum_{1\leq j\leq l, j:\text{odd}}\frac{1}{j}$$

$$+\zeta'(-d) + (-g)^d\sum_{r=0}^{d}\binom{d}{r}\frac{\zeta'(-r)}{(-g)^r}. \tag{1}$$

On the other hand, Theorem E with $P(x) = x^d$, $Q(x) = x(x + g)$ (and hence $\beta_1 = 0$, $\beta_2 = g$) implies

$$
\begin{aligned}
H'_{g,d}(0) = &\zeta'(-d) + \sum_{u=0}^{d} \binom{d}{u} (-g)^{d-u} \zeta'(-u, 1 + g) \\
&- \frac{1}{4} \cdot (-g)^{d+1} \left(\frac{1}{d+1} \sum_{j=1}^{d} \frac{1}{j} + \sum_{u=1}^{d} \binom{d}{u} \frac{(-1)^{u+1}}{u+1} \sum_{j=1}^{u} \frac{1}{j} \right). \quad (2)
\end{aligned}
$$

Substituting the standard formula

$$
\zeta'(-u, 1 + g) = \zeta'(-u) + \sum_{k=2}^{g} k^u \log k,
$$

we find that the right-hand side of (2) is

$$
\begin{aligned}
=&\zeta'(-d) + (-g)^d \sum_{u=0}^{d} \binom{d}{u} (-g)^{-u} \zeta'(-u) + \sum_{u=0}^{d} \binom{d}{u} (-g)^{d-u} \sum_{k=2}^{g} k^u \log k \\
&- \frac{1}{4} \cdot \frac{(-g)^{d+1}}{d+1} \left(\sum_{j=1}^{d} \frac{1}{j} + \sum_{u=1}^{d} \binom{d+1}{u+1} (-1)^{u+1} \sum_{j=1}^{u} \frac{1}{j} \right).
\end{aligned}
$$

We will show that the above coincides with the right-hand side of (1). Since the third term of the above is equal to $\sum_{k=2}^{g} (k - g)^d \log k$, what we have to check is that

$$
\sum_{1 \le l \le d, l:\text{odd}} \binom{d+1}{l+1} \sum_{1 \le j \le l, j:\text{odd}} \frac{1}{j}
$$

$$
= 2^{d-2} \left(\sum_{j=1}^{d} \frac{1}{j} + \sum_{u=1}^{d} (-1)^{u+1} \binom{d+1}{u+1} \sum_{j=1}^{u} \frac{1}{j} \right). \quad (3)
$$

Moreover, noting

$$
\binom{d+1}{u+1} = \binom{d}{u+1} + \binom{d}{u} \qquad \text{for } u < d, \quad (4)
$$

we have

$$\sum_{u=1}^{d}(-1)^{u+1}\binom{d+1}{u+1}\sum_{j=1}^{u}\frac{1}{j}$$

$$=\sum_{u=1}^{d-1}(-1)^{u+1}\binom{d}{u+1}\sum_{j=1}^{u}\frac{1}{j}-\sum_{u=1}^{d-1}(-1)^{u}\binom{d}{u}\sum_{j=1}^{u}\frac{1}{j}+(-1)^{d+1}\sum_{j=1}^{d}\frac{1}{j}$$

$$=\sum_{u=2}^{d}(-1)^{u}\binom{d}{u}\sum_{j=1}^{u-1}\frac{1}{j}-\sum_{u=1}^{d-1}(-1)^{u}\binom{d}{u}\left(\sum_{j=1}^{u-1}\frac{1}{j}+\frac{1}{u}\right)+(-1)^{d+1}\sum_{j=1}^{d}\frac{1}{j}$$

$$=d-\sum_{j=2}^{d}(-1)^{j}\binom{d}{j}\frac{1}{j}.$$

Therefore, it suffices to prove the following

Lemma 10. *The following identity holds:*

$$\sum_{1\le l\le d,l:\mathrm{odd}}\binom{d+1}{l+1}\sum_{1\le j\le l,j:\mathrm{odd}}\frac{1}{j}$$

$$=2^{d-2}\left(\sum_{j=1}^{d}\frac{1}{j}+d-\sum_{j=2}^{d}(-1)^{j}\binom{d}{j}\frac{1}{j}\right). \tag{5}$$

Before proving this Lemma, let us consider the left-hand side first. For this, let

$$S(l):=\sum_{1\le j\le l,j:\mathrm{odd}}\frac{1}{j},$$

$$I(d):=\sum_{1\le l\le d,l:\mathrm{odd}}\binom{d+1}{l+1}S(l),$$

$$J(d):=\sum_{0\le l\le d-1,l:\mathrm{even}}\binom{d+1}{l+1}S(l+1).$$

Then the left-hand side of (5) is just $I(d)$, and we have

$$I(d)=I(d-1)+J(d-1)+\delta(d)S(d), \tag{6}$$

$$J(d)=J(d-1)+I(d-1)+(2\delta(d)-1)S(d)+\frac{2^{d}}{d+1}-\frac{1-\delta(d)}{d+1}, \tag{7}$$

where $\delta(d) = 1$ or 0 according as d is odd or even. In fact, (6) comes directly from (4). As for (7), by definition and (4), we have

$$J(d) = J(d-1) + \delta(d)S(d) + \sum_{0 \le l \le d-1, l:\text{even}} \binom{d}{l} S(l+1)$$

$$= J(d-1) + \delta(d)S(d) + 1 + \sum_{0 \le k \le d-2, k:\text{odd}} \binom{d}{k+1} S(k+2)$$

$$= (J(d-1) + \delta(d)S(d) + 1) + I(d-1) - (1 - \delta(d))S(d-1)$$

$$+ \sum_{1 \le k \le d-2, k:\text{odd}} \binom{d}{k+1} \frac{1}{k+2}.$$

Here in the last step we use the fact that $S(k+2) = S(k) + \frac{1}{k+2}$ for k odd. But $\binom{d}{k+1}\frac{1}{k+2} = \frac{1}{d+1}\binom{d+1}{k+2}$. Hence to complete the proof of (7), it suffices to show that

$$\sum_{1 \le k \le d-2, k:\text{odd}} \binom{d+1}{k+2} = 2^d - (d+1) - (1 - \delta(d)),$$

which may be deduced from

$$\sum_{1 \le k \le d-2} \binom{d+1}{k+2} = \sum_{h=3}^{d} \binom{d+1}{h} = 2^{d+1} - d - 3 - \frac{1}{2}(d+1)(d+2)$$

and the similar identity

$$\sum_{1 \le k \le d-2} (-1)^k \binom{d+1}{k+2} = d - \frac{1}{2}(d+1)(d+2) - (-1)^{d+1}.$$

Now, from (6) and (7), we obtain the following recursion formula

$$I(d+1) - I(d) = I(d) + \delta(d+1)S(d+1) + (\delta(d)-1)S(d) + \frac{2^d}{d+1} - \frac{1 - \delta(d)}{d+1}.$$

Easily,

$$\delta(d+1)S(d+1) + (\delta(d)-1)S(d) - \frac{1 - \delta(d)}{d+1} = 0.$$

Hence, we arrive at the relation

$$I(d+1) = 2I(d) + \frac{2^d}{d+1}. \tag{8}$$

Proof of Lemma 10. Induction on d. Clearly if $d = 1, 2, 3$, (5) holds. Assume now that (5) is valid for some $d \geq 3$. Then from the induction hypothesis and (8), we have

$$I(d+1) = 2^{d-1}\Big(\sum_{j=1}^{d+1} \frac{1}{j} + \frac{1}{d+1} + d - \sum_{j=2}^{d}(-1)^j \binom{d}{j}\frac{1}{j}\Big).$$

Thus it suffices to show that

$$\frac{1}{d+1} - \sum_{j=2}^{d}(-1)^j \binom{d}{j}\frac{1}{j} = 1 - \sum_{j=2}^{d+1}(-1)^j \binom{d+1}{j}\frac{1}{j}. \tag{9}$$

Noting $\binom{d}{j-1}\frac{1}{j} = \frac{1}{d+1}\binom{d+1}{j}$ as before, we have

$$\sum_{j=2}^{d+1}(-1)^j \binom{d+1}{j}\frac{1}{j}$$

$$= \sum_{j=2}^{d}(-1)^j \Big(\binom{d}{j} + \binom{d}{j-1}\Big)\frac{1}{j} + (-1)^{d+1}\frac{1}{d+1}$$

$$= \sum_{j=2}^{d}(-1)^j \binom{d}{j}\frac{1}{j} + \frac{1}{d+1}\sum_{j=2}^{d}(-1)^j\binom{d+1}{j} + (-1)^{d+1}\frac{1}{d+1}$$

$$= \sum_{j=2}^{d}(-1)^j \binom{d}{j}\frac{1}{j} + \frac{1}{d+1}\big(d - (-1)^{d+1}\big) + (-1)^{d+1}\frac{1}{d+1}$$

$$= \sum_{j=2}^{d}(-1)^j \binom{d}{j}\frac{1}{j} + 1 - \frac{1}{d+1}$$

which is equivalent to (9). This completes the proof of Lemma 10, and hence establishes the equivalence of (1) and (2). $\qquad\square$

References

[1] P. Cassou-Noguès, Prolongement de certaines séries de Dirichlet, *Amer. J. Math.* **105** (1983), 13–58.

[2] P. Cassou-Noguès, Valeurs aux entiers négatifs des series de Dirichlet associées a un polynôme I, II, III, *J. Number Theory* **14** (1982), 32–64; *Amer. J. Math.* **106** (1984), 255–299; ibid. **109** (1987), 71–89.

[3] P. Cassou-Noguès, Séries de Dirichlet et intégrales associées à un polynôme à deux indèterminèes, *J. Number Theory* **23** (1986), 1–54.

[4] M. Eie, On a Dirichlet series associated with a polynomial, *Proc. Amer. Math. Soc.* **110** (1990), 583–590.

[5] M. Eie, The special values at negative integers of Dirichlet series associated with polynomials of several variables, *Proc. Amer. Math. Soc.* **119** (1993), 51–61.

[6] A. Ivić, *The Riemann zeta-function*, Wiley (1985).

[7] M. Katsurada, An application of Mellin-Barnes' type integrals to the mean square of Lerch zeta-functions, *Collect. Math.* **48** (1997), 137–153.

[8] M. Katsurada, An application of Mellin-Barnes type integrals to the mean square of *L*-functions, *Liet. Mat. Rink.* **38** (1998), 98–112.

[9] H. Kumagai, The determinant of the Laplacian on the *n*-sphere, *Acta Arith.* **91**(1999), 199–208.

[10] B. Lichtin, Generalized Dirichlet series and *b*-functions, *Compositio Math.* **65** (1988), 81–120. (Erratum: ibid. **72** (1989), 237–239.)

[11] B. Lichtin, Poles of Dirichlet series and *D*-modules, in "Théorie des Nombres/Number Theory" (Proc. Intern. Number Theory Conf., Laval, 1987), J.-M. De Coninck and C.Levesque (eds.), Walter de Gruyter, 1989, pp.579–594.

[12] B. Lichtin, On the moderate growth of generalized Dirichlet series for hypoelliptic polynomials, *Compositio Math.* **80** (1991), 337–354.

[13] B. Lichtin, The asymptotics of a lattice point problem associated to a finite number of polynomials I, *Duke Math. J.* **63** (1991), 139–192.

[14] B. Lichtin, Volumes and lattice points – proof of a conjecture of L.Ehrenpreis, in "Singularities" (Lille, 1991), J.-P. Brasselet (ed.), London Math. Soc. Lecture Note Ser. Vol. 201, Cambridge Univ. Press, 1994, pp.211–250.

[15] B. Lichtin, The asymptotics of a lattice point problem associated to a finite number of polynomials II, *Duke Math. J.* **77** (1995), 699–751.

[16] B. Lichtin, ine Asymptotics determined by pairs of additive polynomials, *Compositio Math.* **107** (1997), 233–267.

[17] K. Mahler, Über einen Satz von Mellin, *Math. Ann.* **100** (1928), 384–398.

[18] K. Matsumoto, Asymptotic expansions of double zeta-functions of Barnes, of Shintani, and Eisenstein series, preprint.

[19] K. Matsumoto, The analytic continuation and the asymptotic behavior of multiple zeta-functions I, preprint.

[20] H. Mellin, E Formel für den Logarithmus transcendenter Funktionen von endlichem Geschlecht, *Acta Soc. Sci. Fenn.* **29**, no.4 (1900).

[21] H. Mellin, Die Dirichlet'schen Reihen, die zahlentheoretischen Funktionen und die unendlichen Produkte von endlichem Geschlecht, *Acta Math.* **28** (1904), 37–64.

[22] M. Peter, Dirichlet series associated with polynomials, *Acta Arith.* **84** (1998), 245–278.

[23] J.R. Quine and J. Choi, Zeta regularized products and functional determinants on spheres, *Rocky Mountain J. Math.* **26**(1996) 719–729.

[24] P. Sargos, Prolongement méromorphe des séries de Dirichlet associées à des fractions rationnelles de plusieurs variables, *Ann. Inst. Fourier* **33** (1984), 83–123.

[25] P. Sargos, Croissance de certaines séries de Dirichlet et applications, *J. Reine Angew. Math.* **367** (1986), 139–154.

262

[26] E.C. Titchmarsh, *The theory of the Riemann zeta-function*, Oxford Sci. Pub., (1986).

[27] I. Vardi, Determinants of Laplacians and multiple gamma functions, *SIAM J. Math. Anal.* **19**(1988), 493–507.

[28] L. Weng and Y. You, Analytic torsions of spheres, *Intern. J. Math.* **7** (1996), 109–125.

SOME ASPECTS OF INTERACTIONS BETWEEN ALGEBRAIC NUMBER THEORY AND ANALYTIC NUMBER THEORY

Katsuya MIYAKE*
Department of Mathematics
Tokyo Metropolitan University
miyakek@comp.metro-u.ac.jp

Abstract   This work is a historical exposition of mathematical ideas, methods and research programs which supported the birth and growth of modern Algebraic Number Theory. The mathematicians picked up here are Cardano, Fermat, Euler, Lagrange, Legendre, Gauss, Abel, Dirichlet, Kummer, Kronecker, Dedekind, Weber and Zolotareff.

Keywords: Historical approach to Algebraic Number Theory; The Birth of Analytic Number Theory; Divisor Theories of Kronecker, Dedekind and Zolotareff; Kronecker's Dream in his Youth; Complex Multiplication and Weber's Congruence Ideal Classes

1.   Prehistory of Algebraic Number Theory

1.1   P. de Fermat (1601-1665)

Pierre de Fermat is the great grand father of modern number theory. He himself published only a few of his findings and proofs which he suggested to have obtained. However, what he found on numbers succeeded in attracting Euler. Even with his gifts for mathematics it was not an easy task for Euler to reconstruct what was in the wide view of Fermat. Spending much of his life time, however, he was finally able to give proofs and a few disproofs to all but 'the last theorem', which Fermat stated on numbers. Then Lagrange and Legendre followed; and

*The author was partly supported by the Grant-in-Aid for Scientific Research (B) No. 11440013, Japan Society for the Promotion of Science while he prepared this work.

Gauss founded the basic frame work of a modern science on numbers on the fertile ground. Hence it may be allowed to call Fermat the great grand father of the modern number theory and Euler its grand father.

We point out here just one of Fermat's theories which we may clearly understand as arithmetic of the quadratic field $\mathbb{Q}(\sqrt{-1})$. (For more and detailed information, see Weil [We-1984] and Cox [Co-1989].)

$\Diamond$ Determination of the numbers of the form $a^2 + b^2$, a, $b \in \mathbb{Z}$.

In the days of Fermat it was well recognized that the numbers of the form $a^2 + b^2$, a, $b \in \mathbb{Z}$, are closed under multiplication; we may even say that Fermat and some of his contemporaries must have been familiar with the formula

$$(a^2 + Nb^2)(c^2 + Nd^2) = (ac \pm Nbd)^2 + N(ad \mp bc)^2$$

for integers N with small absolute values.

Fermat found all the 'generators' (= 'atoms') of sums of two squares under the multiplicative structure ([Fe-1891].II.213-214,221-222);

a prime p divides a primitive $a^2 + b^2$, a, $b \in \mathbb{Z}$
$$\Longleftrightarrow p = a^2 + b^2, \ a, \ b \in \mathbb{Z}$$
$$\Longleftrightarrow p \equiv 1 \mod 4$$
$$\Longleftrightarrow p \in N_{\mathbb{Q}(\sqrt{-1})/\mathbb{Q}}(\mathbb{Z}[\sqrt{-1}]).$$

$\Diamond$ He also handled binary quadratic forms $x^2 + 2y^2, x^2 + 3y^2$ ([Fe-1891].II. 313,403,431-436) and $x^2 - 2y^2$ ([Fe-1891].II.221,224-226,434,441). As for the last case, he recognized the importance of the solution $(1,1)$ of the equation $x^2 - 2y^2 = -1$ which represents the fundamental unit $\varepsilon = 1 + \sqrt{2}$ of the real quadratic field $\mathbb{Q}(\sqrt{2})$ though, of course, he did not directly work with these irrational numbers; he used $(3,2)$ and $(3,-2)$, i.e. $\varepsilon^2 = 3 + 2\sqrt{2}$ and $\varepsilon^{-2} = 3 - 2\sqrt{2}$, when he needed a unit e with $N_{\mathbb{Q}(\sqrt{2})/\mathbb{Q}}(e) = 1$.

$\Diamond$ He had a firm belief that he found a *new science* on numbers in his various findings ([We-1984].118-119).

1.2 L. Euler (1707-1783)

$\Diamond$ Euler investigated in binary quadratic forms $x^2 + Ny^2$ for $N = $ 2, 3, 5, 6, 7, 14 and also 17 ([Eu-1911].I-2.6-17,196-199,556-575,I-3.218-239,273-275); he added an important remark that a solution (a,b) of $x^2 - Ny^2 = 1$ provides a good rational approximation a/b for $\sqrt{N}$.

As for quadratic irrational numbers, he finally introduced them into the theory of binary quadratic forms in his book *Algebra* ([Eu-1911].I-1.1-498).

◇ He was aware of the quadratic reciprocity law ([We-1984].209,218-219).

1.3 J. L. Lagrange (1736-1813)

◇ Lagrange started to handle all of the binary quadratic forms with a fixed discriminant D simultaneously, and introduced the equivalence classes,

$$\{\text{binary quadratic forms with a fixed discriminant } D\}/GL_2(\mathbb{Z})$$

in his *Recherches d'Arithmetique* ([La-1867].III.697-758,759-795). This set of classes corresponds to the ideal class group of the order $\mathbb{Z} + \mathbb{Z}\sqrt{D}$ in the quadratic field $\mathbb{Q}(\sqrt{D})$.

2. Prelude to the Birth of Analytic Number Theory

2.1 L. Euler (1707-1783)

Euler found and proved important results on the infinite series

$$\zeta(s) = 1 + \frac{1}{2^s} + \frac{1}{3^s} + \frac{1}{4^s} + \frac{1}{5^s} \cdots$$

which we now call Riemann's zeta function, and on some other related series (cf. e.g. *Introductio in Analysin Infinitorum*, [Eu-1911].I-8). One of them is the Euler product formula:

$$1 + \frac{1}{2^s} + \frac{1}{3^s} + \frac{1}{4^s} + \frac{1}{5^s} + etc. = \frac{1}{(1 - \frac{1}{2^s})(1 - \frac{1}{3^s})(1 - \frac{1}{5^s})(1 - \frac{1}{7^s})\, etc.}$$

where the product on the right hand side is taken over all prime numbers.

◇ In the case of $s = 1$, we have

$$1 + \frac{1}{2} + \frac{1}{3} + \frac{1}{4} + \frac{1}{5} + etc. = \frac{1}{(1 - \frac{1}{2})(1 - \frac{1}{3})(1 - \frac{1}{5})(1 - \frac{1}{7})\, etc.}.$$

The left hand side is the harmonic series whose n-th partial sum is as large as $\log n$ as n tends to infinity. (Euler denoted it by $\log \infty$.) Hence, first of all, the Euler product shows that there exist infinitely many prime numbers. By taking the logarithm of both sides Euler pointed out, for example, that the sum of all reciprocals of primes

$$1 + \frac{1}{2} + \frac{1}{3} + \frac{1}{5} + \frac{1}{7} + \ etc.$$

has infinite magnitude as large as $\log \log \infty$ ([Eu-1911].I-14.87-100).

$\Diamond$ He was also able to determine the values of $\zeta(s)$ at positive even integers;

$$\zeta(2) = \frac{\pi^2}{6}, \quad \zeta(2n) = \frac{1}{2}(-1)^{n-1}\frac{b_{2n}}{2n!}(2\pi)^{2n}$$

where b_{2n} is the nth Bernoulli number ([Eu-1911].I-14.434-439).

2.2　　A.-M. Legendre (1752-1833)

$\Diamond$ In his book *Recherches d'Analyse Indéterminée* ([Le-1785]), Legendre clearly stated the quadratic reciprocity law and tried to prove it. In the course of his trial, he used the Prime Number Theorem in Arithmetic Progressions. He had a strong belief in it. He himself, however, could not find any effective ways to go. And he could not complete his proof of the reciprocity law either. It was Gauss who first gave a full proof. He published two different proofs at the beginning of the new century; he put them in his epoch-making book *Disquisitiones Arithmeticae* ([G-1801]). A little later in 1837, Dirichlet was to give a genuine proof to the prime number theorem in arithmetic progressions in [Di-1837b]; it should be regarded as the year of the birth of analytic number theory.

$\Diamond$ Legendre is the first author who used the terms 'theory of numbers' instead of 'arithmetic'. The book was published in 1798 with the title *Essai sur la théorie des nombres* ([Le-1798]). The third edition of the book appeared in 1830 in two big volumes with the simplified title *Théorie des Nombres* ([Le-1830]).

In the book, he introduced the counting function of prime numbers

$$\pi(x) = \text{the number of primes not exceeding } x,$$

and stated that $\pi(x)$ is approximately equal to $x/(\log x - 1.08366)$. (The symbol '$\pi(x)$' for the function was introduced later by N. Nielsen [Ni-1906].) He was unable to prove this. After some contributions ([Tc-1848,-52]) of P. L. Tchebychef (1821-1894), Jacques Hadamard (1865-1963) and Charles Jean de la Vallée Poussin (1866-1962) independently proved the Prime Number Theorem later at the end of the 19th century ([Ha-1896] and [VP-1896]); it states

$$\lim_{x \to \infty} \frac{\pi(x)}{\frac{x}{\log x}} = 1.$$

In a letter with the date $^3\sqrt{6064321219}$ (August 24, 1823) to his friend B. Holmboe, N. H. Abel wrote about the statement of Legendre on $\pi(x)$; he picked up no other than this from the Essai because he thought it the most remarkable result in mathematics ([Ab-1902], Correspondence 5).

In 1863, the second volume of Gauss' Work [G-1863] was published. It contains a letter of Gauss to Encke dated December 24,1849 (pp.444–447). According to it, he obtained Lambert's table of logarithms with a table of prime numbers as a supplement in 1792 or 1793 and was aware that the integral $\int \frac{dn}{\log n}$ numerically approximates $\pi(n)$ very well.

3.   Algebraic Equations and the Fundamental Theorem of Algebra

3.1   Ars Magna

In the 16th century a monumental step toward a flourishment of algebra was taken in Italy. Scipione del Ferro (1465-1526), Professor of University of Bologna, found a formula for a root of the cubic equation $x^3 + ax = b$, $a > 0$, $b > 0$. At the time the common way to write and handle algebraic equations was the geometric one. Neither the concept nor the symbol of the 'number 0' were introduced in Europe yet. Then cubic equations were classified into several types. Scipione del Ferro succeeded in solving one of them. Challenged by Niccolò Tartaglia (1500?-1557), he selected one of his disciples, Antonio Maria Fiore, for the mathematical contest. The challenger Tartaglia worked hard for it and succeeded in finding a formula for a root of another type of cubic equations besides del Ferro's by the day of the contest, and won an overwhelming victory over Fiore who armed only with the formula of his teacher.

After a while, Girolamo Cardano (1501-1576) learned the formula for the above equation of del Ferro from Tartaglia after eager and insistent requests. Then he succeeded in solving all types of cubic equations. Moreover, one of his disciples, Lodovico Ferrari (1522-1565), was able to solve biquadratic equations. He reduced it to cubic and quadratic equations. Cardano published all of these results in *Artis Magnæ Sive de Regulis Algebraicis* [Ca-1545] (cf. [Ca-1968]).

◇ A root of the cubic equation $x^3 + ax + b = 0$ is given by the formula

$$\sqrt[3]{\frac{-b}{2} + \sqrt{(\frac{-b}{2})^2 + (\frac{a}{3})^3}} + \sqrt[3]{\frac{-b}{2} - \sqrt{(\frac{-b}{2})^2 + (\frac{a}{3})^3}}$$

if we choose the two cubic roots properly. Cardano and his disciples have already well understood that the cubic equation has three real roots if and only if the square root $\sqrt{(\frac{-b}{2})^2 + (\frac{a}{3})^3}$ is imaginary, that is, $(\frac{-b}{2})^2 + (\frac{a}{3})^3 < 0$. Then they became well acquainted with imaginary numbers, perhaps to be ready for mathematical contests. Rafael Bombelli (1526-1572) wrote up a perfect treatment of complex numbers in his book *Algebra* ([Bo-1572]).

The Ars Magna contains many cubic equations with three real roots. Cardano, however, did not use imaginary numbers nor the formula to obtain these roots. Here he handled only those equations for which one can find a real root almost at once. Then he factored out the linear term to get quadratic equations.

He introduced imaginary numbers in Chapter XXXVII *On the Rule for Postulating a Negative* with the following problem ([Ca-1968], p.219):

Divide 10 into two parts the product of which is 30 or 40.

He gave the answer $5 + \sqrt{-15}$ and $5 - \sqrt{-15}$ in the case of 40.

It should be noted that imaginary numbers could not have been numbers in reality for them at their time. These were not mathematical reality but something belonging to arts with which they tacitly handled cubic equations and other surely existing objects of algebra.

3.2   The Fundamental Theorem of Algebra

D. S. Smith says that Peter Roth (1580-1617) was the first author who openly stated the Fundamental Theorem of Algebra in 1608 ([Ro-1608]); cf. [Sm-1925].II.474. Then Albert Girard (1595-1632) put it forth in his book *Invention nouvelle en l'algébre* [Gi-1637]. An important step was made by J. le Rond d'Alembert (1717-1783) in his memoirs [dA-1746]. He tried hard to show that a (non-constant) polynomial with real coefficients has a root of the form $a + b\sqrt{-1}$, a, $b \in \mathbb{R}$, if it does not have any real roots; then it was not very hard for him to show by this that a polynomial has the same number of roots as its degree.

Lagrange opened the gate toward investigations in the mechanism hidden behind the relations of roots and coefficients of a polynomial in [La-1770]; there he introduced 'Lagrangian resolvents'.

C. F. Gauss (1777-1855) published his first proof to the fundamental theorem of algebra in [G-1799]. (He implicitly used the completeness of the field of real numbers.) By this, anyway, he provided not only algebra but also analysis with a rigid universal domain, the field of complex numbers.

4. The 19th Century begins

4.1 C. F. Gauss (1777-1855)

◇ The number theory of the 19th century began with the celebrated book of Gauss, *Disquisitiones Arithmeticae* ([G-1801]). He began it with the concept of the congruence relation of integers, and introduced the term 'modulus' and the symbol ≡ with numerical examples, $-16 \equiv 9 \pmod 5$ and $-7 \equiv 15 \pmod{11}$. This book contains two complete proofs to the quadratic reciprocity law, and a modern theory of cyclotomy in the last *Section Seven*. He published his papers toward biquadratic reciprocity law [G-1828] and [G-1932] 27 years later. In 1801, however, he had already prepared *Section Eight* ([G-1801*]) of *Disquisitiones Arithmeticae* which was posthumously published by Dedekind in 1863. Günther Frei recently pointed out in [Fre-2001] that *"Section Seven on Cyclotomy served only as a preparation for Section Eight which was to contain a Third Proof of Quadratic Reciprocity Law, a proof Gauss planned to generalize to Higher Reciprocity"*.

Anyway Gauss introduced the ring $\mathbb{Z}[\sqrt{-1}]$ for the biquadratic case in 1828, and proposed a research problem of establishing Higher Power Reciprocity Laws though he might not have done it explicitly. Hereafter through the century or more, this became a principal motivation for developing algebraic number theory. First in 1844 Gotthold Eisenstein (1823-1852) made important contributions for the cubic case in a series of papers [Ei-1844a-e]; he had naturally to develop arithmetic in the ring $\mathbb{Z}[\frac{-1+\sqrt{-3}}{2}]$ of the cyclotomic field of cubic roots of unity. Then Kummer, Kronecker, Dedekind, and so on, followed.

4.2 N. H. Abel (1802-1829)

We have to pick up N. H. Abel who did not leave any distinguished works on number theory but supplied important sources of ideas for the development of algebraic number theory due particularly to Kronecker, and indirectly to Zolotareff. One of his works we mention here is on algebraic equations, and the other is on elliptic functions.

Before we go into these main topics, however, we should also pay attention to his work [Ab-1826] on elliptic and hyperelliptic integrals; this indirectly motivated Zolotareff for his divisor theory in algebraic number fields (cf. Section 7.4 below). Abel studied there those hyperelliptic differential forms $\frac{\rho dx}{\sqrt{R}}$ with polynomials ρ and R in x whose integrals are given as logarithm functions of the form $\log \frac{y+\sqrt{R}}{y-\sqrt{R}}$ with a polynomial y, and characterized them in terms of the continued fraction expansion of the square root $\sqrt{R}$. His criterion is the expansion to be periodic and of a certain special form. In the last section Abel deals with elliptic integrals, that is, the case where R is a monic polynomial of degree 4. He explicitly stated that the integral

$$\int \frac{\left(x + \frac{\sqrt{5}+1}{14}\right) dx}{\sqrt{\left(x^2 + \frac{\sqrt{5}-1}{2}\right)^2 + \left(\sqrt{5} - 1\right)^2 x}}$$

could be expressed by logarithms.

It is apparent that the work of Legendre [Le-1811] or its revised version [Le-1825] on elliptic integrals was in the background of this work of Abel.

◇ Abelian Equations

As is well known, Abel succeeded in proving that there cannot exist any algebraic formulas for roots of general polynomials of degree equal to 5 ([Ab-1824]). This is an important and essential development on algebraic equations after Lagrange opened the door by his work [La-1770] as we pointed it out above in Section 3.2. Abel left the full-scale theory of algebraic equations to Évariste Galois. His interest tended to characterization of solvable equations, and found the *Abelian criterion* because of which we now have the names *Abelian groups* and *Abelian equations*. It is Kronecker who introduced the word 'Abelian equations'. He first used it in [Kr-1853] to mean cyclic polynomials, that is, polynomials with cyclic Galois groups. Then he enlarged its use to mean polynomials with Abelian Galois groups (cf. [Kr-1857a,-1877]).

◇ Elliptic Functions

Abel developed a beautiful theory of elliptic functions in [Ab-1827], and found quite new Abelian polynomials in the work. Legendre used the word 'elliptic functions' before Abel; however, all he worked on were elliptic integrals with analysis in the real number field. In [Ab-1827] Abel started his investigation by considering inverse functions of elliptic integrals by utilizing complex analysis. Since then we have customarily been using the word 'elliptic functions' in Abel's sense. His arithmetical instinct did not miss smelling out the importance of elliptic functions with complex multiplication; and he attracted Kronecker especially with a few explicit numerical examples. The terminology 'complex multiplication' was also introduced by Kronecker who fostered 'Kronecker's dream in his youth' (see below Section 7.1 and [Kr-1857a,-1880b]). The theory of complex multiplication must also have supplied sources for Dedekind to formulate the concepts of 'modules', 'orders' and 'ideals' of an algebraic number field in [De-1871,1877a,-1879,-1893]. (See below Sections 7.2 and 7.3.)

5.   Birth of Analytic Number Theory

As Euler is the father of modern number theory (cf. 1-1), then so is P. G. Lejeune Dirichlet (1805-1859) of Analytic Number Theory. It was born in his paper [Di-1837b] where he proved the Prime Number Theorem in Arithmetic Progressions conjectured by Legendre (cf. 2-2). His strategy was to follow Euler's idea of utilizing Euler product formulas. However, none of the modified harmonic series for arithmetic progressions have necessary product formulas as they are. To overcome the difficulty he brought out a brilliant idea; he utilized all of the modified harmonic series with initial terms which are relatively prime to the fixed common difference, and put them together by Dirichlet characters defined for the common difference. Then the orthogonal relations of the characters lead us to the desired end. Thus Dirichlet's L-series were sent out into the world. Let us see them more closely:

Let d be the fixed common difference; for simplicity, Dirichlet restricted himself to the case where d is an odd prime number. For each a, $0 \leq a \leq d - 1$, put

$$C_a = \{a + dn \mid n = 0,\ 1,\ 2, \ldots\},$$

$$\zeta(s; C_a) = \sum_{n=0}^{\infty} \frac{1}{(a + dn)^s}.$$

Let χ be a character of the Abelian group $(\mathbb{Z}/d\mathbb{Z})^{\times}$, that is, a homomorphism of the group to $\mathbb{C}^{\times}$. The values are of finite order and hence roots of unity ($\varphi(d)$-th roots of 1 where φ is the Euler function). We naturally regard χ as a map from $\mathbb{Z}$ to $\mathbb{C}$ with $\chi(m) = 0$ if m is not relatively prime to d, i.e. $(m, d) \neq 1$; thus we get Dirichlet characters modulo d which still remain multiplicative. Define Dirichlet's L-functions by

$$L(s; \chi) = \sum_{a=0}^{d-1} \chi(a) \zeta(s; C_a) = \sum_{n=1}^{\infty} \frac{\chi(n)}{n^s}.$$

Then we have product formulas

$$L(s; \chi) = \prod_{p,\ prime} \frac{1}{1 - \chi(p)p^{-s}}.$$

Note that $L(s; \chi)$ is a multiple of Riemann's zeta function by a finite product $\prod_{p \mid d}(1 - p^{-s})$ for the trivial character $\chi = 1$. On one hand, therefore, we have $L(1; 1) = +\infty$ at once. Dirichlet, on the other hand, could show that $L(s; \chi)$ converges and is not equal to 0 at $s = 1$ for every non-trivial χ. Hence the value of $\log L(s; \chi)$ at $s = 1$ is $+\infty$ if $\chi = 1$ and finite if $\chi \neq 1$. Now let us consider the series

$$\lambda(s; \chi) := \sum_{p,\ prime} \frac{\chi(p)}{p^s}.$$

Then we see from the values of $\log L(s; \chi)$ at $s = 1$ that $\lambda(1; 1) = +\infty$ and $\lambda(1; \chi)$ is a finite value for each non-trivial χ. It follows from the orthogonal relations of characters that

$$\sum_{p,\ prime \in C_a} \frac{1}{p^s} = \varphi(d)^{-1} \sum_{\chi} \chi(a)^{-1} \lambda(s; \chi)$$

for every $a \in (\mathbb{Z}/d\mathbb{Z})^{\times}$. Therefore the values of the right hand side at $s = 1$ shows

$$\sum_{p,\ prime \in C_a} \frac{1}{p} = +\infty.$$

Hence we conclude that the set C_a of integers in an arithmetic progression contains infinitely many primes if the initial term a is relatively prime to the common difference d.

◇ Dirichlet's Class Number Formula

Dirichlet expanded his analytic method to investigate binary quadratic forms. He had already concerned the works of Fermat, Euler and Lagrange on the subject in [Di-1833,-1834]. In [Di-1834] he saw the basic structure of the solutions of a Fermat equation $x^2 - Dy^2 = 1$ with $D > 0$, i.e., that of the units of the real quadratic field $\mathbb{Q}(\sqrt{D})$.

For the first time in [Di-1838] he proved his class number formula of binary quadratic forms with a *negative* prime discriminant; he introduced Dirichlet series of the form $\sum_{n=1}^{\infty} (\frac{n}{q}) \frac{1}{n^s}$ and $\sum_{n,odd} (\frac{n}{p}) \frac{(-1)^{(n-1/2)}}{n^s}$, and also $\sum \frac{1}{(ax^2 + 2bxy + cy^2)^s}$ where q and p are prime numbers of the form $4\nu + 3$ and $4\nu + 1$, respectively, $(\frac{n}{q})$ and $(\frac{n}{p})$ are Legendre symbols for quadratic residues, and $ax^2 + 2bxy + cy^2$ is a quadratic form with discriminant $-q$ or $-p$. This time he observed and compared asymptotic behaviors of these series as s tends to 1 from the right. Then in [Di-1839] he successfully handled general binary quadratic forms in both cases with positive and negative discriminant D.

As soon as we translate his works in arithmetic of the quadratic field $\mathbb{Q}(\sqrt{D})$, we find the zeta function of the field and its decomposition into a product of Riemann's zeta function and an L-function. It would have then become a central motivation of Dedekind in number theory to seek similar results for pure cubic fields, that is, cubic fields of the form $\mathbb{Q}(\sqrt[3]{D})$.

◇ Dirichlet's Unit Theorem

This may be a suitable place to give a remark on Dirichlet's Unit Theorem. As we pointed out above, he actually found the structure of the unit group of a real quadratic field in [Di-1834] though he did not openly handle any irrational quadratic numbers. In [Di-1841] he introduced irrational numbers and norm forms of algebraic number fields as an important, interesting class of homogeneous forms of cubic and higher degree. More precisely, let $P(X)$ be a monic polynomial with rational integer coefficients:

$$P(X) = X^n + a_1 X^{n-1} + a_2 X^{n-2} + \ldots + a_{n-1} X + a_n, \ a_1, a_2, \ldots, a_n \in \mathbb{Z};$$

suppose that $P(X)$ is irreducible over $\mathbb{Q}$, and let $\alpha = \alpha_1, \ \alpha_2, \cdots, \ \alpha_n$ be all of the roots of $P(X) = 0$ in $\mathbb{C}$. Let $X_1, \ X_2, \ldots, \ X_n$ be a set of independent variables and put

$$F(X_1, X_2, \ldots, X_n) = \prod_{j=1}^{n} (X_1 \alpha_j^{n-1} + X_2 \alpha_j^{n-2} + \ldots + X_{n-1} \alpha_j + X_n).$$

This is a homogeneous polynomial of degree n in $X_1, \ X_2, \ldots, \ X_n$ with coefficients in $\mathbb{Z}$. The favorite example of Dirichlet must be the quadratic form $X^2 - DY^2 = (X - \sqrt{D}Y)(X + \sqrt{D}Y)$ of a Fermat equation. As was classically well noticed, the solutions of the Fermat equation $X^2 - DY^2 = 1$, $D > 0$, play an important role also in solutions of $X^2 - DY^2 = m$ for an integer $m \geq 1$. Suppose, in general, that we have a solution $T_1, \ T_2, \ldots, T_n \in \mathbb{Z}$ for $F(X_1, X_2, \ldots, \ X_n) = m$ and $U_1, \ U_2, \ldots, U_n \in \mathbb{Z}$ for $F(X_1, \ X_2, \ldots, \ X_n) = 1$. Then we can construct many other solutions of the former equation by *complex numbers*

$$(T_1 \alpha^{n-1} + T_2 \alpha^{n-2} + \ldots + T_{n-1} \alpha + T_n)(U_1 \alpha^{n-1} + U_2 \alpha^{n-2} + \ldots + U_{n-1} \alpha + U_n)^N$$

$(N \in \mathbb{Z})$ in the field $\mathbb{Q}(\alpha)$ if we express them in the form

$$S_1 \alpha^{n-1} + S_2 \alpha^{n-2} + \ldots + S_{n-1} \alpha + S_n, \ S_1, S_2, \ldots, S_n \in \mathbb{Z}.$$

Finally in 1846, Dirichlet stated his unit theorem in the paper *Zur Theorie der Complexen Einheiten* [Di-1846] which clarifies the structure of the solutions $U_1, \ U_2, \ldots, \ U_n \in \mathbb{Z}$ for $F(X_1, X_2, \ldots, X_n) = 1$ through units

$$U_1 \alpha^{n-1} + U_2 \alpha^{n-2} + \ldots + U_{n-1} \alpha + U_n, \ U_1, U_2, \ldots, Un \in \mathbb{Z},$$

in the field $\mathbb{Q}(\alpha)$.

E. E. Kummer started his research in cyclotomic integers in 1844 ([Ku-1844]) and published his first paper on '*ideale complexe Zahlen*' in 1846 ([Ku-1846b,-1847b]). It was eventually almost in the middle of the 19th century that algebraic numbers became fully recognized as proper arithmetic objects.

6. Cyclotomic Fields

In the 1840's, the German number theorists, Gauss, Jacobi, Eisenstein, Dirichlet, Kummer, and so on, had already started their studies of algebraic numbers, mainly of cyclotomic integers, and compiled some amount of knowledges on them in their pockets. The principal motivation of them was to obtain higher power residue reciprocity laws.

E. Kummer (1810-1893) may be the first person who dared to move openly with a big stride.

$\Diamond$ Divisor theory and arithmetic in cyclotomic number fields

In 1844 Kummer submitted a paper *Über die complexen Primfactoren der Zahlen, und deren Anwendung in der Kreisteilung* [Ku-1844a] to the Berlin Academy of Science. This was not published because of Kummer's request of withdrawal. The fully revised and enlarged version [Ku-1844b] was written in Latin and published in a few months; this contains a big table of data on decomposition of prime numbers up to 1000 in the cyclotomic field of the l-th root of unity for a prime l up to 23. In the first paper he erroneously stated that a prime p would be fully decomposed into a product of prime elements in $\mathbb{Q}(\zeta_l)$ if p is congruent to 1 modulo l, where ζ_l is a primitve l-th root of unity. In the table attached to the revised paper contains the smallest counter example, $l = 23$ and $p = 47$. It was Jacobi who pointed out the error in the first paper with a counter example for $l = 23$.

Kummer was, however, confident that *prime decomposition* should be uniquely done in $\mathbb{Q}(\zeta_l)$ even if it would not contain sufficiently many *prime elements*. He published an outline of his epoch-making theory in [Ku-1846b] which was republished in Journal für reine und angew. Math. as *Zur Theorie der complexen Zahlen* [Ku-1847b] together with a full scale paper [Ku-1847c] (see [We-1975, p.4, footnote]). His terminology of '*ideale complexe Zahlen*' may be misleading. What he did was to develop a divisor theory in $\mathbb{Q}(\zeta_l)$. What he needed was congruence relations modulo '*eine complex ideale Modul*'. There was a serious gap in this paper which was not realized for a while. It was finally filled almost ten years later in 1856 by the short paper [Ku-1857a] which was written up on June 5, 1856. With this indispensable result his theory was completed in the paper *Theorie der idealen Primfactoren der complexen Zahlen, welche aus den Wurzeln der Gleichung $\omega^n = 1$ gebildet sind, wenn n eine zusammengesetzte Zahl ist* ([Ku-1856]).

◇ Fermat's Last Theorem and Bernoulli Numbers

Now Kummer picked up Fermat's Last Theorem to demonstrate the effectiveness of his divisor theory; the reason may be partly because it would not be so easy for him to get any substantial results on higher power reciprocity laws at once. Perhaps he did not think the Last Theorem itself a very serious problem in number theory if we take his words in the first letter to Dirichlet in [Ku-1847a]; he even called it 'ein Curiosum' (p.139). (This paper consists of his two letters to Dirichlet with a comment of Dirichlet on the first one.) He thought the arithmetic in cyclotomic fields he developed was much more important than the Last Theorem. In his second letter of [Ku-1847a] he analyses the ideal class number h of the cyclotomic field $\mathbb{Q}(\zeta_l)$ for an odd prime l and decomposes it into a product of the first and the second factors h_1 and h_2, respectively, where h_2 is the class number of the maximal real subfield $\mathbb{Q}(\zeta_l + \zeta_l^{-1})$. He believed that he could have showed Fermat's Last Theorem for those odd primes l which did not divide h. He also declared that l divides h if and only if l divides h_1. He states his class number formula for the first factor h_1 and from it he gives a criterion for l not to divide h_1 in terms of Bernoulli numbers B_2, $B_4, \ldots$, B_{l-3}; he believed that infinitely many prime numbers l satisfy this criterion. We now call an odd prime *regular* if it divides none of these $(l-3)/2$ Bernoulli numbers. Kummer actually showed that Fermat's Last Theorem holds for every regular prime. It is not, however, proved yet that there exist infinitely many regular primes. (K. L. Jensen [Je-1915] could prove that there exist infinitely many *irregular* primes of form $4n + 3$; at the time he was a student, and then did not seem to pursue any mathematical career.) Kummer gave a precise proof for his class number formulas in [Ku-1850a] and for the criterion of the regularity in [Ku-1850b]. A full account of his results on Fermat's Last Theorem for regular primes is demonstrated in the paper [Ku-1850c]. Later in his paper [Ku-1857b] he could prove the Last Theorem also for some irregular primes. He showed in [Ku-1851b] that the prime numbers 37, 59 and 67 are the only irregular primes up to 100. Later in [Ku-1874] he showed as results of laborious calculations that 101, 103, 131, 149, 157 are consecutive irregular primes after the first three.

He also gave the following conjecture in the paper: let $\alpha(x)$ and $\beta(x)$ be the numbers of irregular primes and regular ones less than or equal to x (≥ 3), respectively; then the ratio $\nu(x) = \alpha(x)/\beta(x)$ would tend to

1/2 as x tends to $+\infty$. Siegel [Si-1964] proposed another value in place of 1/2 in the conjecture so that we have

Conjecture of Siegel: $\nu(x)$ would tend to $e^{1/2} - 1 = 0.648\ldots$ as x tends to $+\infty$.

◇ Higher Power Residue Reciprocity Law

In 1850 Kummer announced his prospect on the higher power residue reciprocity law for an odd prime in [Ku-1850d]; this is a letter to Dirichlet. Then a big scale paper [Ku-1852] appears; here he made an extensive study on *cyclotomic units*. The final version was published in 1859 ([Ku-1859a,b]). In 1855 he introduced a new tool, *Lagrangian Resolvents for cyclotomy*, in [Ku-1855]. In [Ku-1859b] we see his investigations in *Kummer extensions* fully demonstrated.

As we have pointed out above, he proved Fermat's Last Theorem for some irregular primes in [Ku-1857b]. He handled there such an irregular prime l for which the class number of $\mathbb{Q}(\zeta_l)$ is divisible by l but not by l^2. Therefore there exists just one unramified cyclic extension of degree l over $\mathbb{Q}(\zeta_l)$ by class field theory. It is realized as a Kummer extension $\mathbb{Q}(\zeta_l, {}^l\sqrt{e})$ with some unit e in $\mathbb{Q}(\zeta_l)$ as Weil pointed out in [We-1975]. This is the background where Kronecker started his career as a number theorist.

7.   Algebraic Numbers — From Divisor Theories to Class Field Theory

In this chapter we pick up mainly four mathematicians, L. Kronecker, R. Dedekind, H. Weber and E. I. Zolotareff. For a historical study on the process of the establishment of the Takagi-Artin Class Field Theory interested readers are suggested to see [Mi-1994] for example.

7.1   L. Kronecker (1823-1891)

The mathematical style of Kronecker seems very singular. It is true that he discovered at least a few profound arithmetic phenomena and could bravely formulate big research projects from them. T. Takagi once called Kronecker a prophet ([Ta-1948], footnote, p.261); he made a comment related to Tschebotareff's Density Theorem, "His speculation has turned out well here again.", and chose the terminology 'Kronecker density'.

◇ Abelian polynomials over $\mathbb{Q}$

In 1853 Kronecker stated the following proposition in [Kr-1853]:

Kronecker-Weber Theorem: Roots of every Abelian polynomial with rational integer coefficients are expressed as a rational function of a root of unity.

Here he means by an Abelian polynomial the one with a cyclic Galois group. Later in [Kr-1877] he extends it to mean the one with an Abelian Galois group. As for the proof of the theorem, H. Weber made a certain contribution in [Wb-1886]. The basic tool of both authors for their trials to prove it was Lagrangian resolvents. Both of them, however, could not give a complete proof. An error in Weber's paper did not seem to be realized for a while. He could finally give a complete proof of his own in [Wb-1909].(II). The first complete proof of the theorem was given by D. Hilbert in [Hi-1896]. Olaf Neumann gave a clear and detailed explanation on the errors in the proofs based on Lagrangian resolvents in [Ne-1981].

It must be noted that Kronecker mentions a generalization of the theorem at the end of [Kr-1853]; namely, roots of an Abelian polynomial with coefficients in the Gaussian ring $\mathbb{Z}[\sqrt{-1}]$ may be similarly treated with the division of the lemniscate. He also indicates further generalization. It is probable that, at the time, he had already studied elliptic functions with complex multiplication to some extent through Abel's works. Kronecker's dream in his youth must have appeared in these days. His words, "... um meinen liebsten Jugendtraum ...", appear in his letter [Kr-1880b] to Dedekind written in 1880.

◇ Elliptic functions with complex multiplication

In 1857 he published the paper *Über die elliptischen Functionen, für welche complexe Multiplication stattfindet* [Kr-1857a] on arithmetic of elliptic functions with complex multiplication. He also wrote a letter [Kr-1857b] to Dirichlet about his findings. Though the statements of the letter are not mathematically exact, we can vividly look over what he had found:

Let $\mathbb{Q}(\sqrt{-D})$, $D > 0$, be an imaginary quadratic number field. Let k be the singular modulus (in the sense of Kronecker) of an elliptic function which has complex multiplication by the ring of integers of $\mathbb{Q}(\sqrt{-D})$, and H be the class number of binary quadratic forms over $\mathbb{Z}$ with discriminant $-D$ (the class number of $\mathbb{Q}(\sqrt{-D})$). His findings are as follows:

(1) the singular modulus k is a root of a polynomial of degree H (in [Kr-1857a] he gives the correct value $6H$) over $\mathbb{Q}(\sqrt{-D})$ which is algebraically solvable;

(2) the polynomial has the property which Abel treated: namely, when any one of the roots is chosen, every other root can be expressed as a rational function of it, and the Galois group is commutative;

(3) H values of singular moduli k respectively correspond to the H equivalence classes of the binary quadratic forms with discriminant $-D$;

(4) a certain rational function of the irrational number k may be regarded as the ideal complex number corresponding to each class of quadratic forms; etc. (For these statements we should take j-invariants instead of k.)

In the paper [Kr-1862] he investigated the different of the Abelian polynomial. We know that the Abelian extension $\mathbb{Q}(\sqrt{-D}, j)/\mathbb{Q}(\sqrt{-D})$ is unramified, and, corresponding to (4), we have

The Principal Ideal Theorem: Every ideal of an algebraic number field of finite degree is realized by an irrational number (i.e. becomes a principal ideal) in the maximal unramified Abelian extension field.

Kronecker formulated this theorem as 'die Frage der zu associirenden Gattungen' in [Kr-1882a]. After the Takagi-Artin class field theory was established, it was finally proved in 1930 by Ph. Furtwängler [Fw-1930] by making use of the general reciprocity law of E. Artin [Ar-1927,-1930]. (See [Mi-1988] for further developments on the subject including a historical overview.)

We state here Kronecker's dream in his youth. This is not mathematically precise. For a detailed mathematical discussion about it, we refer to Zusatz 35 of [Kr-1895, pp.510-515] written by H. Hasse.

Kronecker's Dream in his Youth: All Abelian extensions of an imaginary quadratic field $\mathbb{Q}(\sqrt{-D})$ are obtained by adjoining the j-invariant of an elliptic function with complex multiplication by the ring of integers of $\mathbb{Q}(\sqrt{-D})$ and the values of the elliptic function at division points of the periods.

This research problem seems the principal motivation of not only H. Weber but also T. Takagi for their works on class fields. Takagi first proved Kronecker's Dream for the Gaussian field $\mathbb{Q}(\sqrt{-1})$ in his doctoral thesis [Ta-1903]. Then he finally gave a complete proof in [Ta-1920] with his class field theory after an important contribution by R. Fueter [Ft-1914]. (For a historical study of class field theory, see [Mi-1994].)

◇ Divisor theory for algebraic number fields

Kronecker had a strong opinion on Kummer's theory of 'ideale complexe Zahlen'; he thought that an important concept like 'ideale complexe Zahlen' must be given a clear mathematical description. He seemed to have his divisor theory for algebraic number fields of finite degree around 1857 if we adopt Kummer's testimony in [Ku-1859b], p.57. He published it in [Kr-1882a] later in 1882. In this paper he also formulated the Principal Ideal Theorem as we mentioned above. He uses (indefinitely) many independent variables for his theory.

Let $\tilde{\mathbb{Q}}$ be the rational function field $\mathbb{Q}(X, Y, Z, \ldots)$ in independent variables X, Y, Z, ... with coefficients in $\mathbb{Q}$. We call a polynomial in $\tilde{\mathbb{Q}}$ an integral element if its coefficients are integers, and a primitive one if the g.c.d. of the coefficients is equal to 1. An element f of $\tilde{\mathbb{Q}}$ is expressed in the form

$$f = r \cdot (E_1(X, Y, Z, \ldots)/E_2(X, Y, Z, \ldots))$$

with $r \in \mathbb{Q}$, $r > 0$, and primitive integral elements E_1 and $E_2 \in \tilde{\mathbb{Q}}$; it is clear that the rational number r is uniquely determined by f and called the number factor of f. We also call f an *integral element* if $r \in \mathbb{Z}$ to widen the concept. Then we can introduce division among integral elements in a natural way. A divisor of 1 is a unit. An element of $\tilde{\mathbb{Q}}$ is a unit if and only if its number factor is equal to 1; hence there are many units, indeed.

Now let K be an algebraic number field of finite degree and $\tilde{K}$ be the rational function field $K(X, Y, Z, \ldots)$. An element of $\tilde{K}$ is an integral element if it is a root of a monic polynomial whose coefficients are integral elements of $\tilde{\mathbb{Q}}$. Then we are able to define a unit, a prime element, etc., of $\tilde{K}$ in a natural manner. Note that there are plenty of units in $\tilde{K}$. We can also define a g.c.d. of a finite number of integral elements of $\tilde{K}$ though it is only determined up to units. Then we have

Theorem: For a finite number of integral elements of $\tilde{K}$, there exists a g.c.d. of them in $\tilde{K}$.

An algebraic integer in K is an integral element of $\tilde{K}$. It is, therefore, decomposed into a product of prime elements of $\tilde{K}$ uniquely up to units.

As for the relation with the ideal theory of Dedekind, let $\alpha_1, \alpha_2, \ldots, \alpha_n$ be algebraic integers in K, and $X_1, X_2, \ldots, X_n$ be independent variables in $\tilde{K}$. Then

$$\varphi := \alpha_1 X_1 + \alpha_2 X_2 + \ldots + \alpha_n X_n \in \tilde{K}$$

is a g.c.d. of $\alpha_1, \alpha_2, \ldots, \alpha_n$ in $\tilde{K}$. Hence the ideal (in the sense of Dedekind) of K

$$(\alpha_1, \ \alpha_2, \ldots, \ \alpha_n)$$

which is generated by $\alpha_1, \alpha_2, \ldots, \alpha_n$ corresponds to φ as a g.c.d. of $\alpha_1, \alpha_2, \ldots, \alpha_n$.

$\Diamond$ Kronecker's density of primes

In 1880 Kronecker [Kr-1880a] introduced a kind of density of a set of (rational) prime numbers in connection with a polynomial.

Let $F(x)$ be a polynomial with integer coefficients. For a prime p, let ν_p be the number of roots of the equation $F(x) \equiv 0$ mod p in the finite field $\mathbb{Z}/p\mathbb{Z}$; here we count the multiplicity, of course. Then Kronecker states

Theorem: The notation being as above, the limit of the value of the series

$$\sum_{p,\ prime} \nu_p \cdot p^{-1-w}$$

as the positive w tends to 0 is proportional to the value of $\log(1/w)$ and coincides with $\log(1/w)$ multiplied by the number of irreducible factors of $F(x)$.

For each integer k, $0 \le k \le n := \deg F$, let us denote those primes p for which the equation $F(x) \equiv 0$ mod p has just k roots by p_k. Then the series become

$$\sum_{k=0}^{n} k \cdot \sum p_k^{-1-w}.$$

Assume now that the limit

$$D_k := \lim_{w \to 0+} \frac{\sum p_k^{-1-w}}{\log(1/w)} = \lim_{w \to 0+} \frac{\sum p_k^{-1-w}}{\sum p^{-1-w}}$$

exists. Then by the theorem we have

$$\sum_{k=0}^{n} k \cdot D_k = 1$$

if $F(x)$ is irreducible. This formula was so attractive that G. Frobenius finally formulated a conjecture out of it in [Fr-1896a] by means of the Galois group of the polynomial $F(x)$ and Frobenius automorphisms. N. Tschebotareff [Ts-1926] proved it in 1926; hence it is now called Tschebotareff's Density Theorem. His proof was well analyzed by O. Schreier [Sc-1927] and supplied an essential method for E. Artin to prove his General Reciprocity Law in 1927 ([Ar-1927]).

7.2   R. Dedekind (1831-1916)

There are two well known contributions of Dedekind to algebraic number theory, Theory of Ideals and Dedekind's Zeta Function. It is, however, probable that his direct and indirect influences on Frobenius, H. Weber and E. Artin do not seem to be well recognized. If we closely

study his works on number theory, we find a typical model of interactions between algebraic number theory and analytic number theory.

◇ Basis of Algebraic Number Theory

His theory of ideals first appears in 1871 as Supplement X *Über die
Komposition der binären quadratischen Formen* to the second edition of
Dirichlet's book of number theory, *Vorlesungen über Zahlentheorie* ([De-
1871]). Then his theory of algebraic numbers grew well step by step in his
three works [De-1877a,-1879,-1893], and became a basic standard. His
terminology and even some from his notation are still commonly used.
On the contrary, for example, we now see few of Kronecker's (except
'Abelian polynomials' or 'Abelian extensions' and 'complex multiplication').

In 1900 he published a paper [De-1900] on pure cubic fields; its title
is *Über die Anzahl der Ideal-Klassen in reinen kubischen Zahlkörpern*.
In the introduction he says that this is a revised version of what he
prepared in nearly two years, 1871 and 1872. It seems apparent to
me that his primary motivation toward algebraic number theory is to
generalize Dirichlet's class number formula for a quadratic number field
to that for a pure cubic field.

The first thing he had to do for this purpose is to develop a good
divisor theory at least for a pure cubic field. Since it is not contained in
any cyclotomic field, he could not directly use Kummer's theory of 'ideale
complexe Zahlen', and so gently modified Kummer's 'ideale complexe
Modul'. He did not need any ideal or imaginary objects because he was
ready to introduce infinite sets as concrete mathematical objects. His
Supplement X [De-1871] consists of five sections:

§159. Endlich Körper,

§160. Ganze Algebraische Zahlen,

§161. Theorie der Moduln,

§162. Ganze Zahlen eines endlichen Körpers,

§163. Theorie der Ideale eines endlichen Körpers.

After he introduces an algebraic number field K of finite degree and
algebraic integers in the first two sections, he presents a 'Modul' to
support congruence relation in K as an additive subgroup of K. The
word 'Modul' must have been taken on account of 'modulus' of Gauss
and 'ideale complexe Modul' of Kummer. For a finitely generated $\mathbb{Z}$-
submodule M of K of the maximal rank, an order $\mathfrak{o}_M$ is defined as

$$\mathfrak{o}_M = \{a \in K \mid a \cdot M \subset M\}.$$

Since M is finitely generated over $\mathbb{Z}$, every element of $\mathfrak{o}_M$ is an algebraic
integer. Hence $\mathfrak{o}_M$ is contained in *the maximal order* $\mathfrak{o}$ which is the

ring of all the integers in K. Dedekind could observe examples of such structures in imaginary quadratic fields through complex multiplication of elliptic functions (see Section 7.3 below). In the final section he develops his divisor theory with those modules whose orders coincide with the maximal $\mathfrak{o}$.

◊ From Dedekind's Zeta Functions to Artin's L-functions

His next target was to define a zeta function $\zeta_K(s)$ for an algebraic number field K, and calculate its residue at $s = 1$ or, more precisely, the value $S = \lim_{s \to 1+}(s - 1)\zeta_K(s)$. This was done in [De-1877b]. He could express the limit value by the class number, the discriminant and the regulator. Here he also handled the class numbers of non-maximal orders.

To obtain a class number formula like Dirichlet's one, he had to decompose $\zeta_K(s)$ as a product of Riemann's zeta function and some suitable L-functions. To define suitable L-functions, he needs 'characters' with orthogonal relations. This was not an easy task for him. He was, however, lucky enough to find Frobenius. Encouraged by Dedekind, Frobenius succeeded in establishing the desired theory of group characters for finite groups in [Fr-1896b]. (Cf. also Th. Hawkins [Hk-1970,-1974], and [Mi-1989].) However, neither Dedekind nor Frobenius defined L-functions with the group characters even though the latter formulated Frobenius' conjecture in [Fr-1896a]. The task was left to Artin [Ar-1923,-1924b].

In 1923 E. Artin (1898-1962) published his paper [Ar-1923] under the influence of Dedekind [De-1900]. The theme was to express the quotient $\zeta_K(s)/\zeta_k(s)$ of Dedekind's zeta functions for a finite meta-cyclic extension K/k of algebraic number fields in terms of L-functions. He must have been much encouraged by Takagi [Ta-1920]. In case of an Abelian extension, K is characterized as a congruence class field of k by Takagi's class field theory. Hence (modified) Weber's L-functions with characters of the corresponding congruence ideal class group give a perfect answer. (Weber did not consider congruence relations by archimedian primes. They were first introduced by Hilbert in his theory of relative quadratic extensions [Hi-1899]; the main theme of the paper is to show the quadratic reciprocity law in an arbitrary algebraic number field.) In 1924 Artin gave his L-functions for an arbitrary Galois extension with his conjectural general reciprocity law in [Ar-1924b]. Then he could give its proof in [Ar-1927] to complete the Takagi-Artin class field theory as we mentioned above at the end of Section 7.1.

◊ Rational Function Fields over finite fields

It may be of some interest to note Dedekind's influence on 'analytic theory' of arithmetic in function fields of one variable over a finite field.

Dedekind published a paper on the polynomial ring over a finite field in 1857 ([De-1857]). He presented here a divisor theory in the ring. It is apparent that Gauss [G-1801*] gave a direct motivation to him. The article of Gauss was posthumously published in his collected works [G-1863].II in 1863 as was mentioned in Section 4.1 above. Dedekind attached a note to it which he referred in a footnote of [De-1857]. In the article Gauss had already discussed roots of a polynomial over a finite prime field $\mathbb{Z}/p\mathbb{Z}$ where p is a prime number, by utilizing the p-th power map. (It was Galois who first (posthumously) published a paper on finite algebraic extensions of a finite prime field $\mathbb{Z}/p\mathbb{Z}$ together with the p-th power map in [Gal-1846]. He could not see Gauss [G-1801*] which was not yet published at the time when he prepared his paper.) Meanwhile Dedekind treated the polynomial ring *analogously* with the ring of rational integers as he clearly stated. He obtains a finite extension of the prime field by a congruence relation modulo a prime polynomial (eine (irreductibel Function order) Primfunction) and generalize Fermat's theorem. He shows, moreover, 'the quadratic reciprocity law' for the ring by means of quadratic extensions.

After 62 years later H. Kornblum [Ko-1919] picked up the polynomial ring and proved an analogue of Dirichlet's Prime Number Theorem in Arithmetic Progressions.

Then in his thesis [Ar-1924a] Artin investigated the arithmetic of quadratic extensions of the rational function field of one variable over a finite field, and for the congruence zeta functions showed the analogue of the functional equation of Riemann's zeta function, and introduced the analogue of Riemann's hypothesis.

7.3   H. Weber (1842-1913)

At the international conference on class field theory[1] held at Tokyo in 1998, P. Roquette stated that Germany was the father of class field theory and Japan the mother. H. Weber and D. Hilbert may be most paternal because both of them independently began to use the word 'class field' in different contexts ([Wb-1891] and [Hi-1897]).

Hilbert introduced the word in relation to his Theorem 94 which he thought as the first step toward the Principal Ideal Theorem.

---

[1] The Proceedings: Class Field Theory – its Centenary and Prospect, ed. K. Miyake, Advanced Studies in Pure Math. 30, Math. Soc. Japan, Tokyo, 2001. This contains a paper of H. Suzuki which gives a most general form of the Principal Ideal Theorem.

Weber did it in his investigation on 'Kronecker's dream in his youth' which was his principal interest. He extended his concept of class fields finally to cover those Abelian extensions of an imaginary quadratic field which are constructed by the singular moduli and the values at division points of the periods of elliptic functions with complex multiplication by the base quadratic field; in [Wb-1908] he defined a class field of an imaginary quadratic field as an Abelian extension which 'canonically' corresponds to a congruence ideal class group of the base field. He was eager to determine Abelian extensions of imaginary quadratic fields generated by the special values, but not so much to see class field theory in general algebraic number fields even though he introduced congruence ideal class groups and his L-functions to show one of the two fundamental inequalities in general algebraic number fields ([Wb-1897]).

◇ Weber and Number Theory

When he started his career as a mathematician Weber worked in some area of analysis related to partial differential equations and mathematical physics. He was also interested in Abelian functions and Abelian integrals from an analytical point of view associated, e.g., with the Dirichlet Principle. Meanwhile these Abelian functions gave him a chance to get an acquaintance with Dedekind. The latter was trying to publish the works on Abelian functions of the late colleague B. Riemann at Göttingen; Riemann died in 1866. He asked Clebsch to help him who, however, died soon in 1872. Then he decided to invite Weber for the help. Their cooperation produced the first edition of Riemann's Collected Works in 1876. (Cf. Aurel Voss [Vo-1914] and Frei [Fre-1989].) Then in 1882 they published the big work [DW-1882] on the theory of algebraic functions. Here we see clearly an analogy between algebraic number fields and algebraic function fields. For example, 'ideale Theiler' and 'Modul' were introduced in the theory. In 1882 Weber also published his first paper of number theory, *Beweis des Satzes, dass jede eigentlich primitive quadratische Form unendlich viele Primzahlen darstellen* ([Wb-1882]); as the title shows, this is a quadratic version of Dirichlet's Prime Number Theorem. In the introduction Weber also mentioned Kronecker's papers [Kr-1857a] and [Kr-1880a]; the former is on elliptic functions with complex multiplication and the latter on the densities of sets of primes determined by congruence properties of a polynomial which also attracted Frobenius (see Section 7.1).

◇ Kronecker-Weber Theorem

In 1886 Weber published a colossal work [Wb-1886] in two parts to prove the Kronecker-Weber Theorem. As it was noted in Section 7.1, this contains a gap though it did not seem to be immediately realized. The theorem itself was, however, soon given a new proof by Hilbert [Hi-1896]; his approach, based on Minkowski's result in the Geometry of Numbers ([M-1896]), was quite new with his theory of ramification of ideals ([Hi-1894]). Weber published another paper [Wb-1909].(I) on the theorem later in 1909. Perhaps, he was stimulated by Mertens [Me-1906]. Both of them, however, contain errors. As for Weber's paper, Frobenius pointed out two errors. Then in 1911 Weber published his corrected and first perfect proof of the Kronecker-Weber Theorem in [Wb-1909].(II). For the details on the history of the theorem and a complete proof based on Lagrangian Resolvents the reader should see the interesting paper [Ne-1981] of Olaf Neumann.

◇ Congruence ideal class groups and Weber's *L*-functions

As we have pointed out a few times, Weber eagerly investigated extension fields of an imaginary quadratic field constructed with the singular moduli and the values at division points of an elliptic function which has complex multiplication with numbers of the base field. Through them he exstracted the concept of congruence ideal class groups and developed his analytic theory with his *L*-functions in an arbitrary algebraic number field in his paper [Wb-1897].

Here we explain how a congruence ideal class group comes out from division points of periods of such an elliptic function. For the sake of simplicity, we do not adhere to historical context.

Let $\varphi = \varphi(z)$ be an elliptic function; it is a meromorphic function on the complex plane $\mathbb{C}$ with two independent periods ω_1, ω_2 over $\mathbb{R}$; $\varphi(z) = \varphi(z + \omega_1) = \varphi(z + \omega_2)$, ω_1, $\omega_2 \in \mathbb{C}^\times$, $\omega_2 \notin \mathbb{R}\omega_1$. Then we have $\varphi(z) = \varphi(z + \omega)$ for each element ω of

$$\mathbb{Z}\omega_1 + \mathbb{Z}\omega_2 = \{m\omega_1 + n\omega_2 \mid m,\ n \in \mathbb{Z}\}.$$

We may assume that the imaginary part $\mathrm{Im}(\tau)$ is positive for $\tau = \omega_1/\omega_2$ by changing ω_1 and ω_2 if necessary. If an elliptic function is not a constant, all of its periods are given as a $\mathbb{Z}$-module $\mathbb{Z}\omega_1 + \mathbb{Z}\omega_2$ with a suitable pair ω_1 and ω_2.

Conversely, for such a pair ω_1 and ω_2, there exist those elliptic functions the set of periods of which coincides with the module $\Omega := \mathbb{Z}\omega_1 + \mathbb{Z}\omega_2$; for example, the Weierstrass $\wp$-function $\wp(z) = \wp(\omega_1, \omega_2; z)$ and its derivative $\wp'(z)$ are such.

The set of all elliptic funcions that admit Ω as their periods (including constant functions) form an algebraic function field $\mathfrak{R}_\Omega$; if we take $x := \wp(z)$ and $y := \wp'(z)$, then we have $\mathfrak{R}_\Omega = \mathbb{C}(x, y)$ with the relation $y^2 = 4x^3 - g_2 x - g_3$. The discriminant $g_2^3 - 27g_3^2$ of the cubic polynomial is not equal to 0. Put $j = g_2^3/(g_2^3 - 27g_3^2)$. Then both of the numerator and the denominator are homogeneous functions of ω_1 and ω_2 of the same degree.

Therefore, this may be considered as a function of $\tau = \omega_1/\omega_2$: $j = j(\tau)$, $\mathrm{Im}(\tau) > 0$. For another $\Omega' = \mathbb{Z}\omega_1' + \mathbb{Z}\omega_2'$, $\tau' = \omega_1'/\omega_2'$, two function fields $\mathfrak{R}_\Omega$ and $\mathfrak{R}_{\Omega'}$ are isomorphic over $\mathbb{C}$ if and only if $j(\tau) = j(\tau')$. The fields $\mathfrak{R}_{\Omega'}$ and $\mathfrak{R}_\Omega$ are naturally identified with the fields of all meromorphic functions on the complex tori $\mathbb{C}/\Omega'$ and $\mathbb{C}/\Omega$, respectively. Therefore, an isomorphism of the two fields corresponds to an isomorphism of these two complex tori $f : \mathbb{C}/\Omega \to \mathbb{C}/\Omega'$. Combining f with the translation on $\mathbb{C}/\Omega'$ by $-f(0)$, we may assume $f(0) = 0$. Then, as is well known, this isomorphism f is induced from the multiplication by $\alpha \in \mathbb{C}$ on $\mathbb{C}$; α must satisfy the condition $\alpha\Omega = \Omega'$. In other words, an isomorphism of the algebraic function fields $\mathfrak{R}_{\Omega'}$ and $\mathfrak{R}_\Omega$ is essentially obtained by the variable change from z to αz. If we take $\alpha = \omega_2^{-1}$ then the periods ω_1 and ω_2 are changed to 1 and τ respectively.

Let us take $\Omega = \mathbb{Z} + \mathbb{Z}\tau$. For $A = \begin{pmatrix} a & b \\ c & d \end{pmatrix} \in \mathrm{SL}_2(\mathbb{Z})$, put $A(\tau) = (a\tau + b)/(c\tau + d)$. Since A induces just a basis change on Ω, we have $j(A(\tau)) = j(\tau)$ by the isomorphism invariant property of j explained above.

If Ω' is a submodule of Ω, then the complex torus $\mathbb{C}/\Omega'$ is a Galois covering of $\mathbb{C}/\Omega$ with the covering group Ω/Ω'; and hence $\mathfrak{R}_{\Omega'}/\mathfrak{R}_\Omega$ is an algebraic extension of finite degree.

Suppose now that an elliptic function $\varphi(z)$ with the periods Ω has complex multiplication by $\mu \in \mathbb{C} - \mathbb{R}$. This means that two elliptic functions $\varphi(z)$ and $\varphi(\mu z)$ have an algebraic relation; in other words, we may say that the periods Ω and $\mu^{-1}\Omega$ are commensurable, i.e. that the intersection of the two modules has finite indices in both of Ω and $\mu^{-1}\Omega$. Hence we can find a positive integer N such as $N\Omega \subset \mu^{-1}\Omega$, and then we have $(\mu N)\Omega \subset \Omega$. For simplicity we may consider the case of $\mu\Omega \subset \Omega$. Then there is $A = \begin{pmatrix} a & b \\ c & d \end{pmatrix} \in \mathrm{GL}_2(\mathbb{Q})$ with a, b, c, $d \in \mathbb{Z}$ so that we have $\mu \begin{pmatrix} \tau \\ 1 \end{pmatrix} = A \begin{pmatrix} \tau \\ 1 \end{pmatrix}$. Hence μ is an eigenvalue of A, and is an algebraic integer in the quadratic field $\mathbb{Q}(\mu)$. We easily see $\mathbb{Q}(\mu) = \mathbb{Q}(\tau)$, and that it is an imaginary quadratic field. Thus we have a 'Modul' Ω in $k := \mathbb{Q}(\mu) = \mathbb{Q}(\tau)$ and a fractional ideal Ω of the order $\mathfrak{o}_\Omega$. It is now clear that two fractional ideals Ω and Ω' of the same order $\mathfrak{o}' := \mathfrak{o}_\Omega = \mathfrak{o}_{\Omega'}$ give isomorphic elliptic function fields $\mathfrak{R}_\Omega$ and $\mathfrak{R}_{\Omega'}$ if and only if there exists an element $\alpha \in k$ with the property $\alpha\Omega = \Omega'$. Here we have the ideal class group of an oder $\mathfrak{o}'$ of the imaginary quadratic field k. For the maximal oder $\mathfrak{o}$ of k we have the following theorem:

Theorem: Let k be an imaginary quadratic field and $Cl(k)$ be the (absolute) ideal class group of k. Each class of $Cl(k)$ corresponds to an isomorphism class of elliptic function fields among whose modules of periods we can choose an ideal $\mathfrak{w} = \mathbb{Z} + \mathbb{Z}\tau$, $\text{Im}(\tau) > 0$, from the ideal class as is explained above. Then the extension $k(j(\tau))/k$ is Abelian and unramified, and the Galois group is isomorphic to $Cl(k)$. Every ideal of k becomes a principal one if it is lifted to the ideal of $k(j(\tau))$.

All of the contents of the theorem were finally proved by Weber in [Wb-1908].

Weber further considered the field obtained by adjoining the values of elliptic functions at division points of the periods. Let d be a positive integer. Then the d-th division points of the periods are given by the set $d^{-1}\mathfrak{w}/\mathfrak{w}$ on the complex torus $\mathbb{C}/\mathfrak{w}$. Let $\varphi(z)$ be an elliptic function whose period module is $\mathfrak{w}$, and α an element of $k^{\times}$. We have $\varphi(\alpha d^{-1}\omega) = \varphi(d^{-1}\omega)$ for every $\omega \in \mathfrak{w}$ if and only if $\alpha d^{-1}\omega \equiv d^{-1}\omega \mod \mathfrak{w}$ for every $\omega \in \mathfrak{w}$; that is, α acts trivially on $d^{-1}\mathfrak{w}/\mathfrak{w}$. In other words, we have

$$(\alpha - 1)d^{-1}\omega \equiv 0 \mod \mathfrak{w} \quad \text{for every } \omega \in \mathfrak{w}$$

and hence $(\alpha - 1)d^{-1}\mathfrak{w} \subset \mathfrak{w}$. Multiplying both sides by the ideal $d\mathfrak{w}^{-1}$, we finally have the condition

$$\alpha \equiv 1 \mod d\mathfrak{o}.$$

If we replace d by an integral ideal $\mathfrak{m}$ and consider $\mathfrak{m}$-th division points $\mathfrak{m}^{-1}\mathfrak{w}/\mathfrak{w}$, then we have a condition

$$\alpha \equiv 1 \mod \mathfrak{m}$$

on $\alpha \in k^{\times}$. The multiplicative subgroup $\{\alpha \in k^{\times} \mid \alpha \equiv 1 \mod \mathfrak{m}\}$ of $k^{\times}$ is the Strahl or the ray (in English) modulo $\mathfrak{m}$. Thus we have the *congruence ideal class group* $A(\mathfrak{m})/S(\mathfrak{m})$ modulo $\mathfrak{m}$ where $A(\mathfrak{m})$ is the multiplicative group of those ideals which are relatively prime to $\mathfrak{m}$ and $S(\mathfrak{m})$ is the group of principal ideals coming from the Strahl:

$$S(\mathfrak{m}) = \{(\alpha) \mid \alpha \in k^{\times}, \ \alpha \equiv 1 \mod \mathfrak{m}\}.$$

In [Wb-1897] Weber introduced congruence ideal class groups of the form $A(\mathfrak{m})/H(\mathfrak{m})$ where $H(\mathfrak{m})$ is an intermediate group of $A(\mathfrak{m}) \supset S(\mathfrak{m})$ in an arbitrary algebraic number field K of finite degree as well as his L-functions

$$L(s; \chi) = \sum_{C \in A(\mathfrak{m})/H(\mathfrak{m})} \chi(C)\zeta(s; C) = \sum_{\mathfrak{a} \in A(\mathfrak{m})}{}' \frac{\chi(\mathfrak{a})}{N_{K/\mathbb{Q}}(\mathfrak{a})^s}$$

through the partial zeta-functions

$$\zeta(s; C) = \sum_{\mathfrak{a} \in C}' \frac{1}{N_{K/\mathbb{Q}}(\mathfrak{a})^s}$$

where C is a class in $A(\mathfrak{m})/H(\mathfrak{m})$, $\sum'$ is the summation over integral ideals and χ is a character of the abelian group $A(\mathfrak{m})/H(\mathfrak{m})$.

By the time of his article [Wb-1900], he had proved that the extension

$$k(j(\tau), \varphi(u) \mid u \in \mathfrak{m}^{-1}\mathfrak{w})/k(j(\tau))$$

is Abelian with the Galois group isomorphic to $\mathfrak{m}^{-1}\mathfrak{w}/\mathfrak{w}$. In [Wb-1908] he was able to show finally that the field $k(j(\tau), \varphi(u) \mid u \in \mathfrak{m}^{-1}\mathfrak{w})$ is Abelian over the base quadratic field k with the Galois group isomorphic to $A(\mathfrak{m})/S(\mathfrak{m})$. Then he decisively called all of these fields *class fields* of k.

Hilbert introduced sign distribution at Archimedian primes in his paper [Hi-1899] to handle quadratic extensions over an arbitrary algebraic number field and show the quadratic reciprocity law in the most general framework. Then Takagi used congruence ideal class groups for modulus including Archimedian primes to establish his class field theory in [Ta-1915,-1920].

Once the existence of the class field M/K for each congruence ideal class group $A(\mathfrak{m})/H(\mathfrak{m})$ is assured by Takagi, then Weber's L-functions supply a natural decomposition of the quotient of the Dedekind zeta functions,

$$\zeta_M(s)/\zeta_K(s) = \prod_{\chi} L(s; \chi),$$

where $\prod_\chi$ is the product over all non-trivial characters of the Abelian group $A(\mathfrak{m})/H(\mathfrak{m})$. These examples stimulated Artin to define his L-functions in [Ar-1924b] as was noted in the previous Section.

7.4   E. I. Zolotareff (1847-1878)

In this final section of the present article, we see the motivation of Zolotareff who also developed a divisor theory in an algebraic number field. Although his work is not directly related to our main purpose of this article, it may be of some interest to look into the St. Petersburg school of number theory where Tschebotareff soon came to make a big contribution toward Artin's proof of his general reciprocity law.

P. L. Tchebychef was probably the most influential Russian mathematician who raised the modern school of number theory at St. Petersburg. We have mentioned his works [Tc-1849,-1852] on distribution of prime numbers in Section 2.2. Beside them he published several papers on continued fraction expansions. And he came across Abel's paper [Ab-1826]; here Abel gave a criterion by which one can determine whether hyperelliptic integrals be expressed by logarithm functions and which depends on periodicity of the continued fraction expansions of the square roots in the integrals (cf. Section 4.2). In his paper [Tc-1861] he dealt with the elliptic case $\frac{x+A}{\sqrt{x^4+\alpha x^3+\beta x^2+\gamma x+\delta}}dx$ with *rational numbers* α, β, γ, δ, and gave an *effective* criterion for the case. Since Abel's criterion says that the expansion is to be periodic in a specific form, it so happened that there cannot exist any effective bounds of the periods for all of hyperelliptic integrals of the kind. Here is the point of Tchebychef's criterion; and he had to restrict himself to rational coefficients. In this case one can assume without losing generality that the coefficients are rational integers. The basic tool is Jacobi transformations. Then the integrability by logarithms is reduced to a kind of Diophantine problems which have only a finite number of solutions. In the process one cannot dispense with the Fundamental Theorem of Arithmetic. He states as one of examples that the integral function of $\frac{x+A}{\sqrt{x^4+5x^3+3x^2-x}}dx$ cannot be expressed by logarithm functions for any values of A. He did not, however, give any proofs nor brief explanations in the paper. Then Zolotareff published a detailed proof of Tchebychef's criterion in [Zo-1874].

In his paper [Zo-1880] of his divisor theory in algebraic number fields (which was posthumously published), Zolotareff says that he tried to generalize Kummer's theory to extend Tchebychef's method for any real coefficients (des valeurs réelles quelconques). He even states, after referring to Selling [Se-1865] and Dedekind [De-1871], that there have not been any theories yet which matches Kummer's. (E. Selling [Se-1865] contains a serious error.) It appears that he faithfully followed Kummer's way. To utilize it to his concrete problems, he found it best. Here he means real *algebraic* numbers by 'des valeurs réelles quelconques'. We may safely suppose that he would have liked to handle much wider integrals at least including Abel's example

$$\int \frac{\left(x + \frac{\sqrt{5}+1}{14}\right)dx}{\sqrt{\left(x^2 + \frac{\sqrt{5}-1}{2}\right)^2 + \left(\sqrt{5}-1\right)^2 x}}$$

with Tchebychef's method.

Zolotareff, as a young hope of Tchebychef's school at the time, obtained his degree in 1874 and was selected to be a member of St. Petersburg Academy of Science in 1876. It is probable that he was the first number theorist in the school who worked on algebraic numbers. It is, however, regrettable to say that he passed away at the age of 31 in 1878 because of blood poisoning caused by a car accident; cf. A. N. Kolmogorov and A. P. Yushkevich [KY-1992].

References

[Ab-1824] N. H. Abel. Mémoire sur les équations algébriques oú l'on démontre l'impossibilité de la résolution de l'équation générale du cinquième degré, Brochure imprimée chez Grondahl, Christiania, 1824; Œuvres complètes I, 28–33.

[Ab-1826] N. H. Abel. Sur l'intégration de la formule différentielle $\frac{\rho dx}{\sqrt{R}}$, R et ρ étant des fonctions entières, Jour. reine angew. Math. 1 (1826), 185–221; Œuvres complètes I, 104–144.

[Ab-1827] N. H. Abel. Recherches sur les fonctions elliptiques, Jour. reine angew. Math. 2 (1827), 101–181, 3 (1828), 160–190; Œuvres complètes I, 263–388.

[Ab-1881] N. H. Abel. Œuvres complètes I, II; Christiania, 1881; Reprint from Johnson Reprint, New York, 1965.

[Ab-1902] N. H. Abel. Mémorial publié à l'occasion du centenaire de sa naissance, Christiania, 1902.

[dA-1746] J. le Rond d'Alembert. Réflexions sur la cause générale des vents, Mémoires de l'Académie des Sciences et Belles Lettres de Berlin, 1747.

[Ar-1923] E. Artin. Über die Zetafunktionen gewisser algebraischer Zahlkörper, Math. Ann. 89 (1923), 147-156; Collected Papers, 95–104.

[Ar-1924a] E. Artin. Quadratische Körper im Gebiete der höheren Kongruenzen I, II, Math. Zeitschrift 19 (1924), 153–246; Collected Papers, 1–94.

[Ar-1924b] E. Artin. Über eine neue Art von L-Reihen, Abh. Math. Sem. Univ. Hamburg, 3 (1924), 89–108; Collected Papers, 105–124.

[Ar-1927] E. Artin. Beweis des allgemeinen Reziprozitätsgesetzes, Abh. Math. Sem. Univ. Hamburg, 5 (1927), 353–363; Collected Papers, 131–141.

[Ar-1930] E. Artin. Idealklassen in Oberkörpern und allgemeines Reziprozitäts-gesetz, Abh. Math. Sem. Univ. Hamburg, 7 (1930), 46–51; Collected Papers, 159–164.

[Ar-1931] E. Artin. Zur Theorie der L-Reihen mit allgemeinen Gruppencharak-teren, Abh. Math. Sem. Univ. Hamburg, 8 (1931), 292–306; Collected Papers, 165–179.

[Ar-1965] E. Artin. The Collected Papers, Addison Wesley, 1965.

[Bo-1572] Rafael Bombelli. L'Algebra, Opera di Rafael Bombelli da Bolognia, divisa in tre libri, per Giovanni Rossi, Bologna, 1572.

[Ca-1545] Girolamo Cardano. Artis Magnæ Sive de Regulis Algebraicis. Lib. unus. Qui & totius operis de Arithmetica, quod OPUS PERFECTUM inscrip-sit, est in ordine Decimus, 1545.

[Ca-1968] Girolamo Cardano. The Great Art or The Rules of Algebra, trasl. and ed. by T. Richard Witmer, The M.I.T. Press, Cambridge, Massachusetts and London, England, 1968.

[Co-1989]   David A. Cox. Primes of the Form $x^2 + ny^2$ – Fermat, Class Field Theory, and Complex Multiplication, John Wiley & Sons, Inc., New York, 1989.

[De-1857]   R. Dedekind. Abriß einer Theorie der höheren Kongruenzen in bezug auf einen reellen Primzahl-Modulus, Jour. reine angew. Math., Bd.54 (1857), 1–26; Werke I, 40–66.

[De-1871]   R. Dedekind. Supplement X. Über die Komposition der binären quadratischen Formen, von Vorlesungen über Zahlentheorie von P. G. Lejeune Dirichlet (2. Auflage), 423–462, (1871); Werke III, 223–261.

[De-1877a] R. Dedekind. Sur la théorie des nombres entiers algébriques, Paris, Gauthier-Villars, 1877, 1–121, Bulletin des Sci. math. astron., 1er série, t. XI, 2e série, t. I, 1876, 1877; Werke III, 262–296.

[De-1877b] R. Dedekind. Über die Anzahl der Ideal-Klasse in den verschiedenen Ordnungen eines endlichen Körpers, Festschrift Techn. Hochschule Braunschweig zur Säkularfeier des Geburtstages von C. F. Gauß, Braunschweig, 1877, pp.1–55; Werke I, 105–157.

[De-1878]   R. Dedekind. Über den Zusammenhang zwischen der Theorie der Ideale und der Theorie der höheren Kongruenzen, Abhandl. kgl. Ges. Wiss. Göttingen, Bd.23 (1878), 1–23; Werke I, 202–230.

[De-1879]   R. Dedekind. Supplement XI. Über die Theorie der ganzen algebraischen Zahlen, von Vorlesungen über Zahlentheorie von P. G. Lejeune Dirichlet (3. Auflage), 515–530, (1879); Werke III, 297–313.

[De-1882]   R. Dedekind. Über die Discriminanten endlicher Körper, Abhandl. kgl. Ges. Wiss. Göttingen Bd.29 (1882), 1–56; Werke I, 351–396.

[De-1882*] R. Dedekind. Aus Briefen an Frobenius, Werke II, 414–442.

[De-1892]   R. Dedekind. Erläuterungen zu zwei Fragmenten von Riemann, B. Riemanns gesammelte math. Werke und wissenschaftlicher Nachlass, 2. Auflage, 1892, pp.466–478; Werke I, 159-172.

[De-1893]   R. Dedekind. Supplement XI. Über die Theorie der ganzen algebraischen Zahlen, von Vorlesungen über Zahlentheorie von P. G. Lejeune Dirichlet (4. Auflage), 434–657, (1893). Reprint from Chelsea, New York, 1968.

[De-1894]   R. Dedekind. Zur Theorie der Ideale, Nachr. kgl. Ges. Wiss. Göttingen, Math.-phys. Klasse, 1894, 272–277; Werke II, 43–48.

[De-1900]   R. Dedekind. Über die Anzahl der Ideal-Klassen in reinen kubischen Zahlkörpern, Jour. reine angew. Math. 121 (1900), 40–123; Werke II, 148-233.

[De-1930]   R. Dedekind. Gesammelte Mathematische Werke I, II, III, Braunschweig, 1930–1932; Reprint, Chelsea, New York, 1969.

[D&W-1882] R. Dedekind and H. Weber. Theorie der algebraischen Functionen einer Veränderlichen, Jour. reine angew. Math. 92 (1882), 181–290.

[Di-1833]   P. G. Lejeune Dirichlet. Untersuchungen über die Theorie der Quadratischen Formen, Abhandl. König. Preuss. Akad. Wiss. Jahrg. 1833, 101–121; Werke I, 195–218.

292

[Di-1834]   P. G. Lejeune Dirichlet. Einige Neue Sätze über Unbestimmte Gleichungen, Abhandl. König. Preuss. Akad. Wiss. Jahrg. 1834, 649–664; Werke I, 219–236.

[Di-1837a]   P. G. Lejeune Dirichlet. Beweis eines Satzes über die arithmetische Progression, Bericht Verhandl. König. Preuss. Akad. Wiss. Jahrg. 1837, 108–110; Werke I, 307–312.

[Di-1837b]   P. G. Lejeune Dirichlet. Beweis des Satzes, dass jede unbegrenzte arithmetische Progression, deren erstes Glied und Differenz ganze Zahlen ohne gemeinschaftlichen Factor sind, unendlich viele Primzahlen enthält, Abhandl. König. Preuss. Akad. Wiss. Jahrg. 1837, 45–81; Werke I, 313–342.

[Di-1838]   P. G. Lejeune Dirichlet. Sur l'usage des séries infinies dans la théorie des nombres, Jour. reine angew. Math. 18 (1838), 259–274; Werke I, 357–374.

[Di-1839]   P. G. Lejeune Dirichlet. Recherches sur diverses applications de l'analyse infinitésimale á la théorie des nombres, Jour. reine angew. Math. 19 (1839), 324–369, und 21 (1840), 1–12 und 134–155; Werke I, 411–496.

[Di-1841]   P. G. Lejeune Dirichlet. Einige Resultate von Untersuchungen über eine Classe homogener Functionen des dritten und der höheren Grade, Sitzung der physik.-math. Classe der Acad. Wissens. Jahrg. (1841), 280–285.

[Di-1842]   P. G. Lejeune Dirichlet. Recherches sur les formes quadratiques à coefficients et à indéterminées complexes, Jour. reine angew. Math. 24 (1842), 291–371; Werke I, 533–618.

[Di-1846]   P. G. Lejeune Dirichlet. Zur Theorie der complexen Einheiten, Bericht König. Preuss. Akad. Wiss. Jahrg. 1846, 103–107; Werke I, 639–644.

[Di-1889]   P. G. Lejeune Dirichlet. G. Lejeune Dirichlet's Werke I, II, Berlin, 1889–1897; Reprint, Chelsea, New York, 1969.

[Ed-1975]   H. M. Edwards. The Background of Kummer's Proof of Fermat's Last Theorem for Regular Primes, Arch. Hist. Exact Sci. 14 (1975), 219–236.

[Ed-1977]   H. M. Edwards. Postscript to "The Background of Kummer's Proof of Fermat's Last Theorem for Regular Primes", Arch. Hist. Exact Sci. 17 (1977), 381–394.

[Ei-1844a]   Gotthold Eisenstein. Beweis des Reciprocitätssatzes für die cubischen Restes in der Theorie der aus dritten Wurzeln der Einheit zusammengesetzten complex Zahlen, Jour. reine angew. Math. 27 (1844), 289–310; Mathematische Werke, I, 59–80.

[Ei-1844b]   Gotthold Eisenstein. Nachtrag zum cubischen Reciprocitätssatze für die aus dritten Wurzeln der Einheit zusammengesetzten complex Zahlen. Criterien des cubischen Characters der Zahl 3 und ihrer Theiler, Jour. reine angew. Math. 28 (1844), 28–35; Mathematische Werke, I, 81–88.

[Ei-1844c]   Gotthold Eisenstein. Lois de réciprocité, Jour. reine angew. Math. 28 (1844), 53–67; Mathematische Werke, I, 126–140.

[Ei-1844d]   Gotthold Eisenstein. Einfach Beweis und Verallgemeinerung des Fundamentaltheorems für die Biquadratischen Reste, Jour. reine angew. Math. 28 (1844), 223–245; Mathematische Werke, I, 141–163.

[Ei-1844e]   Gotthold Eisenstein. Allgemeine Untersuchungen über die Formen dritten Grades mit drei Variabeln, welche der Kreistheilung ihre Entstehung verdanken, Jour. reine angew. Math. 28 (1844), 289–374 und 29 (1845), 19–53; Mathematische Werke, I, 167–286.

[Ei-1975]   Gotthold Eisenstein. MATHEMATSCHE WERKE I, II, Chelsea Publ. Comp., New York, 1975.

[Eu-1911]   Leonhardi Euleri Opera omnia, sub ausp. Soc. scient. Nat. Helv. Seres I-IV A, 1911.

[Eu-1862]   Leonhardi Euleri Opera Postuma Mathematica et Phisica I, Anno 1844 detecta, edd.P.-H. Fuss et Nic. Fuss, Petropoli 1862 (Reprint, Kraus 1969).

[Fe-1891]   Pierre de Fermat. Œuvres de Fermat, pub. par Paul Tannery et Charles Henry, Paris, 4 vol., 1891-1912 (& Supplément, pub. par M.C. de Waard, 1 vol., 1922).

[Fe-1679]   Pierre de Fermat. Varia Opera Mathematica D·Petri de Fermat, Senatoris Tolosani, Tolosae, Apud Joannem Pech, juxta Collegium PP. Societatis Jesu M.DC.LXXIX.

[Fre-1989]   Günther Frei. Heinrich Weber and the Emergence of Class Field Theory, in The History of Modern Mathematics, ed. D. E. Rowe and J. McCleary, Acad. Press, 1989, pp.425–450.

[Fre-2001]   Günther Frei. Gauss' unpublished Section Eight of the Disquisitiones arithmeticae: The Beginning of the Theory of Function Fields over a Finite Field, Lecture held in Oberwolfach on June 21, 2001; Preprint, 24th of June, 2001.

[Fr-1887a]   G. Frobenius. Neuer Beweis des Sylowschen Satzes, Jour. reine angew. Math. 100 (1887), 179–181; Ges. Abh. II, 301–303.

[Fr-1887b]   G. Frobenius. Über die Congruenz nach einem aus zwei endlichen Gruppen gebildeten Doppelmodul, Jour. reine angew. Math. 101 (1887), 273–299; Ges. Abh. II, 304–330.

[Fr-1896a]   G. Frobenius. Über Beziehungen zwischen den Primidealen eines algebraischen Körpers und den Substitutionen seiner Gruppe, Sitzungsber. König. Preuss. Akad. Wiss. Berlin (1896), 689–703; Ges. Abh. II, 719–733.

[Fr-1896b]   G. Frobenius. Über Gruppencharaktere, Sitzungsber. König. Preuss. Akad. Wiss. Berlin (1896), 985–1021; Ges. Abh. III, 1–37.

[Fr-1968]   G. Frobenius. Gesammelte Abhandlungen I, II, III, Springer-Verlag, 1968.

[Ft-1914]   R. Fueter. Abelsche Gleichungen in quadratisch-imaginären Zahlkörpern, Math. Ann. 75 (1914), 177–255.

[Fw-1930]   Ph. Furtwängler. Beweis des Hauptidealsatzes für Klassenkörper algebraischer Zahlkörper, Abh. Math. Sem. Univ. Hamburg 7 (1930), 14–36.

[Gal-1846]   E. Galois. Sur la théorie des nombres, Jour. Math. pures appl. 11 (1846), 398–407; Écrite et Mémoires Mathématiques d'Évariste Galois, par R. Bourgne et J.-P. Azra, Gauthier-Villars, Paris, 1962, 113–127.

[G-1799]   C. F. Gauss. Demonstratis Nova Theorematis Omnem Functionem Algebraicam Rationalem Integram Unius Variabilis in Factores Reales

294

Primi vel Secondi Gradus Resolvi Posse (1799), Gauss Werke III, Göttingen, 1866, 1-30.

[G-1801] C. F. Gauss. Disquisitiones Arithmeticae, Leibzig, 1801; Gauss Werke I, Göttingen, 1863.

[G-1801*] C. F. Gauss. Disquisitiones Generales de Congruentiis. Analysis Residuorum Caput Octavum, Gauss Werke II, Göttingen, 1863, 212-242.

[G-1828] C. F. Gauss. Theoria Residuorum Biquadraticorum, Commentatio Prima, Comment. soc. reg. sci. Gottingen 1828; Werke II, 65-92.

[G-1832] C. F. Gauss. Theoria Residuorum Biquadraticorum, Commentatio Secunda, Comment. soc. reg. sci. Gottingen 1828; Werke II, 93-148.

[G-1863] C. F. Gauss. CARL FRIEDRICH GAUSS WERKE Bd.I - Bd.XII, König. Gesel. Wiss. Göttingen, 1863 - 1929; Reprint, Georg Olms Verlag, Hildesheim·New York, 1973.

[Gi-1637] Albert Girard. Invention nouvelle en l'algèbre, Amsterdam, 1629 (Reprint by Bierens de Haan, Leyde, 1884).

[Ha-1896] Jacques Hadamard. Sur la distribution des zéros de la fonction $\zeta(s)$ et ses conséquences arithmétiques, Bull. Soc. Math. France, 24 (1896), 199-228.

[Hk-1970] Th. Hawkins. The Origin of the Theory of Group Characters, Arch. History Exact Sci.7 (1970/71), 142-170.

[Hk-1974] Th. Hawkins. New Light on Frobenius' Creation of the Theory of Group Characters, Arch. History Exact Sci. 12 (1974), 217-243.

[Hi-1896] D. Hilbert. Ein neuer Beweis des Kroneckerschen Fundamentalsatzes über Abelsche Zahlkörper, Nachr. Akad. Wiss. Göttingen (1896), 29-39; Ges.Abh.I, 53-62.

[Hi-1897] D. Hilbert. Bericht: Die Theorie der algebraischen Zahlkörper, Jber. Deutschen Math.-Ver.4 (1897), 175-546; Ges.Abh.I, 63-363.

[Je-1915] Kaj Loechte Jensen. Om talteoretiske Egenskaber ved de Bernoulliske Ta (Danish; On number theoretic properties of the Bernoulli numbers), Nyt Tidsskrift for Matematik, Afd. B, vol. 26 (1915), pp. 73-83.

[KY-1992] A. N. Kolmogorov and A. P. Yushkevich (Edit.). Mathematics of the 19th Century, Mathematical Logic, Algebra, Number Theory, Probability Theory, Birkhäuser, Basel·Boston·Berlin, 1992.

[Kr-1853] L. Kronecker. Über die algebraisch auflösbaren Gleichungen, Monatsber. König. Preuss. Acad. Wiss. Berlin (1853), 365-374; Werke IV, 1-11.

[Kr-1857a] L. Kronecker. Über die elliptischen Functionen, für welche complexe Multiplication stattfindet, Monatsber. König. Preuss. Acad. Wiss. Berlin (1857), 455-460; Werke IV, 177-183.

[Kr-1857b] L. Kronecker. Brief an G. L. Dirichlet vom 17 Mai 1857, Nachr. König. Akad. Wiss. Göttingen (1885), 253-297; Werke V, 418-421.

[Kr-1862] L. Kronecker. Über die complexe Multiplication der elliptischen Functionen, Monatsber. König. Preuss. Acad. Wiss. Berlin (1862), 363-372; Werke IV, 207-217.

[Kr-1874] L. Kronecker. Über die congruenten Transformation der biliniear Formen, Monatsber. König. Preuss. Acad. Wiss. Berlin (1874), 397-447; Werke I, 423-483.

[Kr-1877]   L. Kronecker. Über Abelsche Gleichung (Auszug aus der am 16. April 1877 gelesenen Abhandlung), Monatsber. König. Preuss. Acad. Wiss. Berlin (1877), 845–851; Werke IV, 63–72.

[Kr-1880a]   L. Kronecker. Über die Irreductibilität von Gleichungen, Monatsber. König. Preuss. Acad. Wiss. Berlin (1880), 155–162; Werke II, 85–93.

[Kr-1880b]   L. Kronecker. Auszug aus einem Briefe von L. Kronecker an R. Dedekind, 15. März, 1880, Werke V, 455–457.

[Kr-1882a]   L. Kronecker. Grundzüge einer arithmetschen Theorie der algebraischen Grössen, Jour. reine angew. Math. 92 (1882), 1–122; Werke II, 237–388.

[Kr-1882b]   L. Kronecker. Die Composition Abelscher Gleichungen, Sitzungsber. Acad. Wiss. Berlin (1882), 1059–1064; Werke IV, 115–121.

[Kr-1895]   L. Kronecker. Mathematische Werke I-V, Leipzig, 1895-1930; Reprint, Chelsea, New York, 1968.

[Ku-1844a]   Ernst Eduard Kummer. Über die complexen Primfactoren der Zahlen, und deren Anwendung in der Kreisteilung, April, 1844, Appendix 1 to [Ed-1977], pp.388–393.

[Ku-1844b]   Ernst Eduard Kummer. De numeris complexis, qui radicibus unitatis et numeris integris realibus constant, In: Gratulationsschrift d. Univ. Breslau zur Jubelfeier d. Univ. Königisberg 1844, pp.28; Jour. pures et appl. 12 (1847), 185–212; Collected Papers I, 165–192.

[Ku-1846a]   Ernst Eduard Kummer. Über die Divisoren gewisser Formen der Zahlen, welche aus der Theorie der Kreistheilung entstehen, Jour. reine angew. Math. 30 (1846), 107–116; Collected Papers I, 193–202.

[Ku-1846b]   Ernst Eduard Kummer. Vervollständigung der Theorie der complexen Zahlen, Monatsber. König. Preuss. Wiss. Berlin (1846), 87–96.

[Ku-1847a]   Ernst Eduard Kummer. Über die Zerlegung der aus Wurzeln der Einheit gebildeten complexen Zahlen in ihre Primfactoren, Jour. reine angew. Math. 35 (1847), 327–367; Collected Papers I, 211–251.

[Ku-1847b]   Ernst Eduard Kummer. Zur Theorie der complexen Zahlen (Republ. of [Ku-1846a]), Jour. reine angew. Math. 35 (1847), 319–326; Collected Papers I, 203–210.

[Ku-1847c]   Ernst Eduard Kummer. Beweis des Fermat'schen Satzes der Unmöklivhkeit von $x^\lambda + y^\lambda = z^\lambda$ für eine unendliche Anzahl Primzahlen λ, Monatsber. König. Preuss. Akad. Wiss. Berlin, 1847, 132–141, 305–319; Collected Pares I, 274–283, 283–297.

[Ku-1850a]   Ernst Eduard Kummer. Bestimmung der Anzahl nicht äquivalenter Classen für die aus λten Wurzel der Einheit gebildeten complexen Zahlen und die idealen Factoren derselben, Jour. reine angew. Math. 40 (1850), 93–116; Collected Pares I, 299–322.

[Ku-1850b]   Ernst Eduard Kummer. Zwei besondere Untersuchngen über die Classen-Anzahl und über die Einheiten der aus λten Wurzel der Einheit gebildeten complexen Zahlen, Jour. reine angew. Math. 40 (1850), 117–129; Collected Papers I, 323–335.

[Ku-1850c]   Ernst Eduard Kummer. Allgemeiner Beweis des Fermat'schen Satzes, dass die Gleichung $x^\lambda + y^\lambda = z^\lambda$ durch ganze Zahlen unlösbar ist, für alle diejenigen Potenz-Exponenten λ, welche ungerade Primzahlen sind und in den Zahlen der ersten $(\lambda - 3)/2$ Bernoulli'schen Zahlen als Factoren nicht vorkommen, Jour. reine angew. Math. 40 (1850), 368–372; Collected Papers I, 336–344.

[Ku-1850d] Ernst Eduard Kummer. Allgemeine Reciprocitätsgesetze für beliebig hohe Potenzreste, Monatsber. König. Preuss. Akad. Wiss. Berlin, 1850, 154–165; Collected Papers I, 345–357.

[Ku-1851a] Ernst Eduard Kummer. Über eine allgemeine Eigenschaft der rationalen Entwiklungscoefficienten einer bestimmten Gattung analytischer Funktionen, Jour. reine angew. Math. 41 (1851), 130–138; Collected Papers I, 358–362.

[Ku-1851b] Ernst Eduard Kummer. Mémoire sur la théorie des nombres complexes composés de raciness de unité et de nombres entiers, Jour. math. pures et appl. 16 (1851), 377–498; Collected Papers I, 363–484.

[Ku-1852] Ernst Eduard Kummer. Über die Ergänzungssätze zu den allgemeinen Reciprocitätsgesetzen, Jour. reine angew. Math. 44 (1852), 93–146; Collected Papers I, 485–538.

[Ku-1855] Ernst Eduard Kummer. Über eine besondere Art, aus complexen Einheiten gebildeter Ausdrücke, Jour. reine angew. Math. 50 (1855), 212–232; Collected Papers I, 552–572.

[Ku-1856] Ernst Eduard Kummer. Theorie der idealen Primfactoren der complexen Zahlen, welche aus den Wurzeln der Gleichung $\omega^n = 1$ gebildet sind, wenn n eine Zusammengesetzte Zahl ist, Math. Abhandl. König. Preuss. Akad. Wiss. Berlin (1856), 1–47; Collected Papers I, 583–629.

[Ku-1857a] Ernst Eduard Kummer. Über die den Gaussischen Perioden der Kreistheilung entsprechenden Congruenzwurzeln, Jour. reine angew. Math. 53 (1857), 142–148; Collected Papers I, 573–580.

[Ku-1857b] Ernst Eduard Kummer. Einige Sätze über die aus den Wurzeln der Gleichung $\alpha^\lambda = 1$ gebildete complexen Zahlen, für den Fall, dass die Klassenzahl durch λ teilbar ist, nebst Anwendung derselben auf einen weiteren Beweis des letzten Fermat'schen Lehrsatzes, Monatsber. König. Preuss. Akad. Wiss. Berlin, 1857, 275–282; Collected Papers I, 631–638.

[Ku-1857c] Ernst Eduard Kummer. Einige Sätze über die aus den Wurzeln der Gleichung $\alpha^\lambda = 1$ gebildete complexen Zahlen, für den Fall, dass die Klassenzahl durch λ teilbar ist, nebst Anwendung derselben auf einen weiteren Beweis des letzten Fermat'schen Lehrsatzes, Math. Abhandl. König. Preuss. Akad. Wiss. Berlin, 1857, 41–74; Collected Papers I, 639–672.

[Ku-1858] Ernst Eduard Kummer. Über die allgemeinen Reciprocitätsgesetze, Monatsber. König. Preuss. Akad. Wiss. Berlin, 1858, 158–171; Collected Papers I, 673–687.

[Ku-1859a] Ernst Eduard Kummer. Über die Ergänzungssätze zu den allgemeinen Reciprocitätsgesetzen, Jour. reine angew. Math. 56 (1859), 270–279; Collected Papers I, 688–697.

[Ku-1859b] Ernst Eduard Kummer. Über die allgemeinen Reciprocitätsgesetze unter den Resten und Nichtresten der Potenzen, deren Grad eine

Primzahl ist, Math. Abhandl. König. Preuss. Akad. Wiss. Berlin (1859), 19–159; Collected Papers I, 699–839.

[Ku-1874] Ernst Eduard Kummer. Über diejenigen Primzahlen λ, für welche die Klassenzahl der aus λten Einheitswurzeln gebildeten complexen Zahlen durch λ teilbar ist, Monatsber. König. Preuss. Akad. Wiss. Berlin, 1874, 239–248; Collected Papers I, 945–954.

[Ku-1887] Ernst Eduard Kummer. Zwei neue Beweise der allgemeinen Reciprocitätsgesetze unter den Resten und Nichtresten der Potenzen, deren Grad eine Primzahl ist, Jour. reine angew. Math. 100 (1887), 10–50; Collected Papers I, 842–882.

[Ku-1975] Ernst Eduard Kummer. Collected Papers I, II; Springer-Verlag, 1975.

[La-1770] J. L. Lagrange. Réflexions sur la Résolution Algébraique des Équations, Nouveaux Mémoires de l'Académie des Sciences et Belles Lettres de Berlin, Année 1770 et 1771; Œuvres de Lagrange III, 203–421.

[La-1867] J. L. Lagrange. Œuvres de Lagrange, pub. par M. J.-A. Serret et M. Gaston Darboux, Paris, 14 vol., 1867–1892.

[Le-1785] A.-M. Legendre. Recherches d'Analyse Indéterminée, Histoire de l'Acad. Royale des Sci., Paris, 1785; Mémoires de Mathématique et de Physique de l'Imprimerie Royale, Paris, 1788, 465–559.

[Le-1792] A.-M. Legendre. Mémoire sur les Transcendantes elliptiques, Académie Sciences 1792, Paris, l'an deuxième de la Répablique.

[Le-1798] A.-M. Legendre. Essai sur la Théorie des Nombres, Chez Duprat, Libraire pour Mathématiques, quai des Augustins, An VI, Paris, 1798; 2nd ed., Courcier, Paris, 1808.

[Le-1811] A.-M. Legendre. Exercices de calcul intégral, 3 vol., Paris, 1811-17.

[Le-1825] A.-M. Legendre. Traité des fonctions elliptiques, 3 vol., Paris, 1825-1828.

[Le-1830] A.-M. Legendre. Théorie des Nombres, troisième édition, 2 vol., Firmin-Didot, 1830.

[Mi-1988] Katsuya Miyake. The capitulation problem, Sugaku Expositions Vol.1 (1988), No.2, 175–194, Amer. Math. Soc.

[Mi-1989] Katsuya Miyake. A Note on the Arithmetic Background to Frobenius' Theory of Group Characters, Expo. Math. 7 (1989), 347–358.

[Mi-1994] Katsuya Miyake. The establishment of the Takagi-Artin class field theory, in The Intersection of History and Mathematics (ed. J.W.Dauben et al), Birkhäuser Verlag, Basel·Boston·Berlin, 1994, pp.109–128.

[Na-1990] Władysław Narkiewicz. Elementary and Analytic Theory of Algebraic Numbers, Second Edit., Springer-Verlag, Berlin ·Heidelberg·New York·London·Paris ·Tokyo·Hong Kong, 1990.

[Ne-1981] Olaf Neumann. Two proofs of the Kronecker-Weber theorem "according to Kronecker, and Weber", Jour. reine angew. Math. 323 (1981), 105–126.

[Ni-1906] N. Nielsen. Theorie des Integrallogarithmus und verwandter Transzendenten, Taubner, 1906; Chelsea reprint, 1965.

298

[Ro-1608]   Peter Roth. Arithmetica philosophica, order schöne, neue, wohlgegründete, überaus künstliche Rechnung der Coss oder Algebra, Nürnberg, 1608.

[Sc-1927]   O. Schreier. Über eine Arbeit von Herrn Tschebotareff, Abh. Math. Sem. Univ. Hamburg, 5 (1927), 1–6.

[Se-1865]   E. Selling. Über die idealen Primfactoren der complexen Zahlen, welche aus den wurzeln einer beliebigen irreductibeln Gleichung rational gebildet sind, Schlömilch's Zeitschr. für Math. u. Phys. 10 (1865), 17–47.

[Si-1964]   C. L. Siegel. Zu zwei Bemerkungen Kummers, Nachr. Akad. Wiss. Göttingen, Math.-Phys. Klasse, 1964, Nr.6, 51–57; Gesammelte Abhandlungen III, 436–442.

[Ta-1903]   Teiji Takagi. Über die im Bereiche der rationalen complexen Zahlen Abel'scher Zahlkörper, J. Coll. Sci. Tokyo 19 (1903), 1–42; Collected Papers, 13–39.

[Ta-1915]   Teiji Takagi. Zur Theorie der relativ-Abel'schen Zahlkörper I, Proc. Phys.-Math. Soc. Japan, Ser.II, 8 (1915), 154–162; II, 243–254; Collected Papers, 43–50, 51–60.

[Ta-1920]   Teiji Takagi. Ueber eine Theorie des relativ Abel'schen Zahlkörpers, J. Coll. Sci. Tokyo 41 (1920), 1–133; Collected Papers, 73–167.

[Ta-1948]   Teiji Takagi. Algebraic Number Theory (in Japanese), Iwanami Shoten, Tokyo, 1948.

[Ta-1973]   Teiji Takagi. The Collected Papers, Iwanami Shoten, 1973; the 2nd edit., Springer-Verlag, 1990.

[Tc-1849]   P. L. Tchebychef. Sur la fonction qui détermine la totalité des nombres premiers inférieures à une limite donnée, Jour. math. pures et appl. I sérei, 17 (1852), 341–365; Œuvres I, 27–48.

[Tc-1852]   P. L. Tchebychef. Mémoire sur les nombres premiers, Jour. math. pures et appl. I sérei, 17 (1852), 366–390; Œuvres I, 49–70.

[Tc-1861]   P. L. Tchebychef. Sur l'intégration de la différentielle $\frac{x+A}{\sqrt{x^4+\alpha x^3+\beta x^2+\gamma x+\delta}}$; Bull. l'Acad. Impér. sci. St.-Pétersbourg, T. III (1861), 1–12 = Œuvres I, 517–530.

[Tc-18**]   P. L. Tchebychef. Œuvres I, II; Reprint, Chelsea, New York, 1962.

[Ts-1926]   N. Tschebotareff. Die Bestimmung der Dichtigkeit einer Menge von Primzahlen, welche zu einer gegebenen substitutionsklasse gehören, Math. Ann. 95 (1926), 191–228.

[VP-1896]   C. de la Vallée Poussin. Recherches analytiques sur la théorie des nombres, Annal. de la Soc. Bruxells, 20 (1896), 183–256, 281–397.

[Vo-1914]   Aurel Voss. Heinrich Weber, Jahresbericht der DMV 23 (1914), 431–444.

[Wb-1882]   H. Weber. Beweis des Satzes, dass jede eigentlich primitive quadratische Form unendlich viele Primzahlen darstellen fähig ist, Math. Ann. 20 (1882), 301–329.

[Wb-1884]   H. Weber. Ueber die Galois'sche Gruppe der Gleichung 28ten Grades, von welcher die Doppeltangenten einer Curve vierter Ordnung abhängen, Math. Ann. 23 (1884), 489–503.

[Wb-1886]   H. Weber. Theorie der Abelschen Zahlkörper I, Acta Math. Stokh. 8 (1886), 193–263; II, 9 (1887), 105–130.

[Wb-1891]   H. Weber. Elliptischen Functionen und algebraischen Zahlen, Braunschweig, 1891.

[Wb-1894]   H. Weber. Lehrbuch der Algebra, Vol. I - II, Braunschweig, 1894, 1896.

[Wb-1897]   H. Weber. Ueber Zahlengruppen in algebraischen Körpern I, Math. Ann. 48 (1897), 433–473; II, 49 (1897), 83–100; III, 50 (1898), 1–26.

[Wb-1900]   H. Weber. Komplexe Multiplikation, in: Encyklopödie der Mathematischen Wissenschaften, Leibzig, 1900-1904, I.2.C. Zahlentheorie, 6, pp.718–732.

[Wb-1908]   H. Weber. Lehrbuch der Algebra, Vol. III, Braunschweig, 1908; (the second edition of [W-1891]).

[Wb-1909]   H. Weber. Zur Theorie der zyklischen Zahlkörper (I), Math. Ann. 67 (1909), 32–60; (II), 70 (1911), 459–470.

[We-1975]   André Weil. Introduction of "Ernst Eduard Kummer Collected Papers I", Springer-Verlag, 1975.

[We-1984]   André Weil. Number Theory, An approach through history, From Hammurapi to Legendre, Birkhäuser, Boston·Basel·Stuttgart, 1984.

[Zo-1874]   E. I. Zolotareff. Sur la méthode d'intégration de M. Tchebychef, Jour. de Math. pures appl. 2e série, 16 (1874), 161–188.

[Zo-1880]   E. I. Zolotareff. Sur la théorie des nombres complexes, Jour. de Math. pures appl. 3e série, 6 (1880), 51–84, 129–166.

ON G-FUNCTIONS AND PADÉ APPROXIMATIONS

Makoto Nagata
Research Institute for Mathematical Sciences, Kyoto University
Oiwake-cho Kitashirakawa Sakyo-ku Kyoto, 606-8502, Japan
mnagata@kurims.kyoto-u.ac.jp

Abstract   We intend to introduce some state-of-the-art results on G-functions, which take algebraic values at some algebraic points, seen from the viewpoint of Padé approximations.

Keywords: G-function, G-operator, Padé approximation.

By a Padé approximation we mean an approximation of a set of formal power series by a set of polynomials such that a linear combinations of these formal power series with these polynomials as coefficients vanishes at the origin with high order.

More specifically, let K be a field and let $f_1, \ldots, f_n \in K[[x]]$ be formal power series with coefficients in K. Then for given $0 < \delta < n$, $N \geq 1$, we call the linear combination of $f_1, \ldots, f_n$ with polynomials $P_1, \ldots, P_n \in K[x]$ a *Padé approximation* if

$$\operatorname{ord}_{x=0} \sum_{i=1}^{n} P_i f_i \geq (n - \delta) N$$

with

$$\max_{i} \deg P_i(x) < N.$$

Putting off the precise definition of a G-function somewhat later, we shall first give a rough description of what it is like to give the reader a comprehensive impression.

A G-function is, first of all, a solution of a linear differential equation with polynomial coefficients, satisfying some additional conditions, such that it can be regarded as an extension of an algebraic function. Although it can be so regarded, it seems difficult to define a G-function in the global sense. E.g. immediately ensure questions as to how we can regard a G-function as an extension of a (projective) algebraic curve, or how we can describe them in terms of the monodromy or some related groups.

This problem of defining G-function in the global sense is one of the major problems in the theory of G-functions. The author knows neither the answer to this problem nor which methods are efficacious in it.

There is, however, an approach, of arithmetical nature, to G-functions in the local sense, which is the Padé approximation, and is one of a few potential approaches to G-functions. This method has been successfully applied to some problems involving G-functions, and we are led to suspect that there may be something profound larking in Padé approximations for G-functions. In this paper we shall present some topics on G-functions with this intuition in mind.

First we shall give a precise definition of G-functions and go on to state some related known results.

The notion of G-functions, sometimes called G-series, was introduced by Siegel [13] through the following.

Definition. We call y a G-*function* if it satisfies the following conditions.

(1) y is a solution of the linear differential equation with coefficients in $K[x]$

$$a_0 y^{(n)} + a_1 y^{(n-1)} + \cdots + a_{n-1} y' + a_n y = 0 \qquad \text{(eq)}$$

given by a formal power series with coefficients $\alpha_i \in K$:

$$y = \sum_{i=0}^{\infty} \alpha_i x^i \quad (\in K[[x]]),$$

where α_i's are subject to conditions:

(2) the maximum of absolute values of the conjugates $\alpha_i^{(j)}$ of α_i grows at most geometrically in i: there exists a constant $C > 0$ such that

$$\max_j |\alpha_i^{(j)}| < C^i \quad \text{for } i = 0, 1, \ldots$$

(where in general $\alpha^{(j)}$ denotes a conjugate of α; no confusion with the symbol of derivative should arise).

(3) the positive l.c.m. of the common denominator of $\alpha_0, \ldots, \alpha_i$ grows at most geometrically in i: there exists a sequence $\{d_i\} \subset \mathbb{N}$ with $d_i \leq C^i$ such that $d_i \alpha_0, \ldots, d_i \alpha_i \in \mathcal{O}_K$, the ring of integers in K.

We may extend the above definition to introduce the notion of vectorial *G*-functions.

We recall the (eq) may be expressed in the form

$$\frac{d}{dx}\bar{y} = A\bar{y}, \tag{EQ}$$

where $\bar{y} = {}^t(y, y', \ldots, y^{(n-1)})$ and $A \in M_n(K(x))$.

Now we think the system (EQ) of differential equations is given, and look at its (vectorial) solution $\bar{y} = {}^t(y_1, \ldots, y_n)$. We call $\bar{y}$ again a (vectorial) *G*-function if each of its components is a *G*-function in the sense of the above definition, i.e. a solution $\bar{y}$ of (EQ) is a (vectorial) *G*-function if and only if each of its components is a formal power series in $K[[x]]$ whose coefficients satisfy growth conditions (2) and (3).

Examples of *G*-functions.

(1) Algebraic functions over $\mathbb{Q}$ are *G*-functions.

(2) The logarithm and more generally, polylogarithms

$$L_k(x) = \sum_{i=1}^{\infty} \frac{x^i}{i^k}, \quad k \in \mathbb{N}$$

are *G*-functions.

(3) Gauss' hypergeometric series:

$$_2F_1(\alpha, \beta, \gamma; x) \quad \text{with } \alpha, \beta, \gamma \in \mathbb{Q}$$

are *G*-functions.

E.g.

$$_2F_1(1/2, 1/2, 1; x) = \frac{1}{\pi} \int_0^1 \frac{dt}{\sqrt{t(1-t)(1-xt)}}$$

is a *G*-function.

(4) The derivatives and integrals of *G*-functions are again *G*-functions.

We note that some of the above examples are proved by using some important facts from number theory, to establish condition (3) in Definition above. Indeed, Example (2) is proved by the prime number theorem, while the rationality condition of parameters α, β, γ in Example (3) is associated with Tchebotarev's density theorem.

We see that although the definition of a *G*-function looks simple, it is not so easy to determine whether a solution of a given (eq) is a *G*-function.

Now we state two results on the values of G-functions obtained by Padé approximations. There are interesting applications of these results, but we shall not state them here.

For simplicity, we always assume $K = \mathbb{Q}$ hereafter.

Proposition 1. ([5], [6], etc) *Suppose $\overline{y}$ is a solution of (EQ) which is a (vectorial) G-function and whose components $y_1, \ldots, y_n \in \mathbb{Q}[[x]]$ are linearly independent over $\mathbb{C}(x)$, the quotient field of $\mathbb{C}[x]$. Then for given small $\epsilon > 0$, there exist constants $c_1, c_2, c_3 > 0$ such that for $p, q \in \mathbb{Z}$ and $H_1, \ldots, H_n \in \mathbb{Z}$ with $H := \max_i |H_i|$, the inequality*

$$|H_1 y_1(p/q) + \cdot + H_n y(p/q)| > H^{-n+1-\epsilon}$$

holds provided that $|q| > c_1$, $|q|^{\epsilon c_2} > |p| \geq 1$, and $H > c_3$.

Proposition 2. ([8]) *Under the assumptions of Proposition 1, let $R > 0$ be the radius of convergence of $\overline{y}$ at $x = 0$. For $q \in \mathbb{N}$ and $r > 0$, put*

$$U_r(q) = \{\zeta \in \mathbb{Q} \mid |\zeta| \leq r, \text{ the denominator of } \zeta = q\}$$

and for $r < \min(R, 1)$, suppose $n \geq 6$ and that every components of A in (EQ) has no poles in $\{\zeta \in \mathbb{C} \mid |z| < r\}$. Then the following (non-trivial) estimate holds:

$$\overline{\lim_{q \to \infty}} \frac{\log \#\{\zeta \in U_r(q) \mid \overline{y}(\zeta) \in \mathbb{P}_{\mathbb{Q}}^{n-1}\}}{\log q} \leq \frac{35}{6n} \ (\leq 1).$$

Remark. After the symposium at University of Kinki, Kyushu School of Engineering, we obtained the following improvement on Proposition 2.

Let $H(\zeta)$ be the absolute Height of $\zeta \in K$. Then under the assumptions of Proposition 2, the following holds:

$$\overline{\lim_{B \to \infty}} \frac{\log \#\{\zeta \in \mathbb{Q} \mid \overline{y}(\zeta) \in \mathbb{P}_{\mathbb{Q}}^{n-1}, |\zeta| \leq r, H(\zeta) \leq B\}}{\log B} \leq \frac{4}{n} \ (\leq 2)$$

for $r < 1/2$ and for $n \geq 3$ instead of $n \geq 6$ in Proposition 2.

Moreover, if $y_1, \ldots, y_n$ are homogeneously algebraically independent over $\mathbb{C}(x)$, then the left side $\overline{\lim} \cdots$ of the above inequality is 0.

We note that the trivial estimate of the left side is 2, and that in the case of the number of rational points of algebraic functions, the similar estimate of the left side is $\leq 2/n$. See [12].

Unfortunately, from the point of view of transcendental number theory, it is hard to say that these are very powerful results, in the sense that they cannot yield any transcendency or irrationality results of any interesting concrete special values of G-functions. E.g. we cannot deduce irrationality of $\zeta(3)$, the special value of the trilogarithm $L_3(x)$ at

$x = 1$, the reason of which is that $x = 1$ is on the boundary of convergence of $L_3(x)$. Likewise we cannot deduce transcendency of values of the logarithm at algebraic numbers.

However, it is interesting to notice that Proposition 2 or its Remark can deduce, without the use of the exponential function, that the number of rational values of the logarithm function at rational numbers is small.

We hope to discover some remarkable transcendency results on G-functions, like those on E-functions, for which we need to find appropriate conditions.

However, it is not easy to find out what candidates are for these conditions, for as is known, non-algebraic G-functions (e.g. some concrete hypergeometric functions) take algebraic values at some algebraic numbers (See [3], [14]).

Now we state a result on a relation between G-functions and differential equations. To this end we introduce the notion of G-operator.

Definition. We call $(\frac{d}{dx} - A)$ (or simply (EQ)) a *G-operator* if

$$\varlimsup_{m \to \infty} \sum_{p:\text{ prime}} \frac{1}{m} \max_{i \leq m} \log^+ \left| \frac{1}{i!} \, {}^t\!\left(\frac{d}{dx} + {}^t\!A \right)^i I \right|_p < \infty,$$

where I is the identity matrix, ${}^t(\frac{d}{dx} + {}^t\!A)\,{}^t\!B$ means ${}^t(\frac{d}{dx}B + B\,{}^t\!A)$ for $B \in M_n(K(x))$ and $|\cdots|_p$ mean the so-called Gauss norm.

The following proposition is also obtained by using Padé approximations. We are interested in the fact that Padé approximations appear in the proof of a result related to differential equations.

Proposition 3. ([5]) *Under the assumptions of Proposition 1, (EQ) is a G-operator.*

We note again that these results are obtained by Padé approximations.

Actually, the existence of the Padé approximations in these proofs is derived by the box principle (i.e. the so-called Siegel's lemma). And one may then suspect that the Padé approximations (obtained by the box principle) would be poor. With these in mind, we are led to ask a question.

Question. *Are the Padé approximations (obtained by the box principle) mere auxiliary tools for G-functions or do they contain anything essential?*

In the following we shall give some evidence for believing the Padé approximations to contain some essence of G-functions, or at least to be a nice approach to them.

306

In order to state the following propositions we introduce some new notations.

Let p be a prime. Given a differential equation (eq) over $\mathbb{Z}[x]$, we associate to it the differential equation (eq_p) over $\mathbb{F}_p[x]$:

$$\overline{a_0}y^{(n)} + \overline{a_1}y^{(n-1)} + \cdots + \overline{a_{n-1}}y' + \overline{a_n}y = 0, \qquad (\text{eq}_p)$$

where $\overline{a_i}$ denotes a polynomial in $\mathbb{F}_p[x]$ obtained from a_i in $\mathbb{Q}[x]$ by replacing the coefficients of a_i by their reduction mod. p, where we assume that $\overline{a_0} \neq 0$.

Let (EQ_p) denote the matrix form of (eq_p) with the coefficient matrix A.

Definition. We say that (EQ_p) has a *nilpotent p-curvature* if the operator ${}^t(\frac{d}{dx} + {}^tA)^p$ on $M_n(\mathbb{F}_p[x])$ is nilpotent for A, and that the *p-curvature is 0* if the operator ${}^t(\frac{d}{dx} + {}^tA)^p$ is 0.

Especially, if (EQ_p) has a nilpotent p-curvature, ${}^t(\frac{d}{dx} + {}^tA)^{p^n}I$ ($\in M_n(\mathbb{F}_p[x])$) is 0.

Proposition 4. (See [2]) *If* (EQ) *is a G-operator, then there exists a set T of prime numbers such that*

$$\sum_{p \notin T} 1/p < \infty, \quad (\sum_{p \in T} 1/p = \infty,)$$

(EQ_q) *has a nilpotent p-curvature for all $p \in T$.*

We digress slightly and state two conjectures which say that the G-functions are like extensions of algebraic functions in the following sense (cf. Proposition 6 below in this regard).

Conjectures. (See [2])

(0) (Grothendieck) The n solutions of (eq) which are linearly independent over the constant field are algebraic functions $\overset{?}{\Longleftrightarrow}$ the p-curvature of (EQ_p) is 0 for almost all primes p.

(1) (EQ) is a G-operator $\overset{?}{\Longleftrightarrow}$ (EQ_p) has a nilpotent p-curvature for almost every prime p, where we mean by *almost every prime p* all primes in a set T of primes whose Dirichlet density is 1.

We now return to the main stream and state two propositions.

Proposition 5. ([7], (ch $= p$ case)) *For any prime $p \geq n$, (EQ_p) has a nilpotent p-curvature if and only if (EQ_p) has n (vectorial) solutions at $x = 0$ as "polynomials in $\log x$ with polynomial coefficients" $(\mathbb{F}_p[x][\log x])^n$ which are linearly independent over the constant field.*

Proposition 6. ([1], [10], (ch = 0 case)) *Under some conditions, (EQ) is a G-operator if and only if (EQ) has n (vectorial) solutions at x = 0 as "polynomials in* log x *with G-function coefficients" which are linearly independent over the constant field.*

Of course, the constant fields in Propositions 5 and 6 are completely different from each other, and so the polynomials in $\log x$ with polynomial (and with G-function) coefficients in these propositions have little in common. However, their similarity is encouraging the apprehension as to disbelieve that Padé approximations are mere auxiliary tools.

The next proposition is remarkable in that it reminds us of the reason (= the box principle) why the index "2" is Roth's theorem is the best possible. The details are complicated, and we state it in concise form (See [4] for details)

Proposition 7. ([4]) *Except for certain cases, algebraic functions defined over K with genus ≥ 1 do not have substantially better Padé approximations than those obtained by Siegel's lemma (i.e. the box principle).*

It is reasonable to think that Padé approximation for functions which take algebraic values at all algebraic numbers are not so good, because the converse is against our experiences on transcendence theory that good Padé approximation should bring out good transcendence properties. In this regard, the fact in Proposition 7 that lower bounds for the Height of polynomials in Padé approximations are nearly the same as that obtained by the box principle is far beyond our imagination. This assertion is to the effect that there are no better Padé approximations, and in a sense give a bound of the method, which may lead one to think Proposition 7 is not supportive of our view.

However, we may interpret it as supporting our view in the following respect. Fix the δ prescribed in the definition of Padé approximations[1], and compare the following two points: The Height of the coefficients of power series for a G-function grows at most geometrically, and the Height of the coefficients of Padé approximations, obtained by Siegel's lemma, for G-functions at most grows geometrically in N. We note that the Height of the coefficients of Padé approximations for algebraic functions *at least* grows geometrically in N, supplied by Proposition 7.

This point common in both suggests that assertions for algebraic functions, typical examples of G-functions, Proposition 7 retains G-functionological information and that one may even conclude that the

Padé approximations obtained by the box principle are not poor in this special case.

Lastly, we state a proposition which gives a lower bound for the height of polynomials in Padé approximations by G-functions. Although the result is weaker than that in Proposition 7, it holds for G-functions in general.

Proposition 8. *Let $\overline{y} = {}^t(y_1, \ldots, y_n)$ be an analytic vectorial solution for (EQ) on $D = \{z \in \mathbb{C} \mid |z| < 1/2\}$.*

Let l be a positive integer, $\zeta_0, \ldots, \zeta_l$ be in $D \cap K$, and let non-negative real numbers $a_0, \ldots, a_l$ with $a_0 + \cdots + a_l = n - \delta$ be fixed.

For any large $N \in \mathbb{N}$, we consider any Padé approximations such that

$$\mathrm{ord}_{x=\zeta_t} \sum_{i=1}^{n} P_i(x) y_i(x) \geq a_t N \quad \text{for} \quad t = 0, \ldots, l$$

with $\max_i \deg P_i(x) < N$.

Suppose that $\overline{y}(\zeta_0), \ldots, \overline{y}(\zeta_l) \in \mathbb{P}_K^{n-1}$, and that there are no singularities of (EQ) on D.

Then under the assumptions of Proposition 1, there exist positive constants C_1, C_2 (depending only on A) such that

$$\frac{\max_i h(P_i)}{N} \geq \frac{1}{[K : \mathbb{Q}]} \sum_{t=1}^{l} a_t \log \frac{1}{|\zeta_t - \zeta_0|}$$
$$- ((1 + \epsilon + \delta)h(\zeta_0) + (\epsilon + \delta)C_1 + C_2),$$

where $h(\,\cdot\,)$ means the absolute logarithmic heights in suitable senses.

Lines of proof. Use Jensen's formula and Theorem IV (ii) in [11].

We make some remarks on Proposition 8. It assumes the existence of at least two ζ's satisfying $\overline{y}(\zeta) \in \mathbb{P}_K^{n-1}$, and in order to deduce non-trivial lower bounds from the above inequality we need to assume some appropriate conditions on such ζ's.

In conclusion, in view of the fact that the existence of the constant C_1 is implied by the definition of G-functions, this lower bound, if it is non-trivial, suggests that it does not lose G-functionological information, and further asserts that Padé approximations by the box principle are not poor.

The above propositions are supporters of our opinion on Padé approximations. They may not give enough evidence to make us believe that the Padé approximations for *G*-functions contain something important, yet they give good evidence to believe the Padé approximations to be useful tools for *G*-functions. To say the least of it, Padé approximations are not mere auxiliary tools, and those by the box principle are significant for *G*-functions.

Finally, we make a general remark. In the function-theoretic sense we consider the existence of Padé approximations as an analogy with Dirichlet's inequality in Diophantine approximations, and indeed, in case the constant field is a number field, Siegel's lemma enters into the stage, which has some analogy with Dirichlet's inequality. In this way, the geometry of numbers sets a restriction to Padé approximations from one direction.

Furthermore, if one assumes some arithmetic conditions (like those in Proposition 8), then these conditions imply another restriction to them from another direction, and these arithmetic conditions might be *G*-functions, *G*-operators, nilpotent *p*-curvatures, and so on.

Acknowledgment. The author would like to express his gratitude to Professor Shigeru Kanemitsu for his warm encouragement.

Notes

1. Of course, one of the most important assertions of Proposition 7 is that it allows the case of $\delta \sim 1/N$. But, from application-oriented point of view, such a critical case is not very common, and in this paper, we pay attention to the case of fixed δ.

References

[1] Y. André, *G-functions and Geometry*, Max-Planck-Institut. (1989)

[2] F. Baldassari, *a lecture note, at Chiba University*, (1996)

[3] F. Beukers, *Algebraic values of G-functions*, J. reine angew. Math. **434**, pp. 45–64 (1993)

[4] E. Bombieri, P. B. Cohen, *Siegel's lemma, Padé approximations and Jacobians*, Ann. Scuola Norm. Sup. Pisa Cl. Sci. **XXV**, pp. 155–178 (1997)

[5] D. V. Chudnovsky, G. V. Chudnovsky, *Applications of Padé approximations to Diophantine inequalities in values of G-functions*, Lect. Notes in Math. **1135**, pp. 9–51 Springer-Verlag (1985)

[6] A. I. Galočhkin, *Estimates from below of polymomials in the values of analytic functions of a certain class*, Math. USSR Sbornik **24**, pp. 385–407 (1974)

[7] T. Honda, *Algebraic differential equations*, Symposia Mathematica **XXIV**, pp. 169–204 (1981)

[8] M. Nagata, *An estimation on the number of rational values related values of G-functions*, preprint (RIMS-1231)

310

[9] ———, *Sequences of differential systems*, Proc. Amer. Math. Soc. **124**, pp. 21–25 (1996)

[10] ———, *A generalization of the sizes of differential equations and its applications to G-functions*, Ann. Scoula Norm. Sup. Pisa Cl. Sci (4) **XXX**, pp. 465–497 (2001)

[11] C. F. Osgood, *Nearly Perfect Systems and Effective Generalizations of Shidlovski's Theorem*, J. Number Theory **13**, pp. 515–540 (1981)

[12] J. P. Serre, *Lectures on the Mordell-Weil Theorem (3rd ed.)*, Vieweg (1997)

[13] C. L. Siegel, *Über einige Anwendungen diophantischer Approximationen*, Abh. Preuss. Akad. Wiss., Phys. Math. Kl. nr.1, (1929)

[14] J. Wolfart, *Werte hypergeometrischer Funktionen* , Inv. Math. **92**, pp. 187–216 (1988)

A PENULTIMATE STEP TOWARD CUBIC THETA-WEYL SUMS

Yoshinobu Nakai

Department of Mathematics, Faculty of Education, Yamanashi University, Kofu, Yamanashi, 400-8510, Japan

nakai@edu.yamanashi.ac.jp

Abstract   The attempt at interpreting the Weyl sums as finite theta series (theta-Weyl sums) has been successful only in the case of quadratic polynomials. In this paper we shall present basic ingredients for interpreting cubic Weyl sums as finite theta series, i.e. the cubic continued fraction expansion, the van der Corput reciprocal function, cubic reciprocal and parabolic transformations.

Keywords: Weyl sum, exponential sum, cubic continued fraction expansion, Diophantine approximation, van der Corput reciprocal function

1.   Introduction

The name "theta-Weyl" sums in the title needs an explanation as it is the author's own terminology which he first adopted in his paper [7-1]. The first "theta" suggests the theta series which are the most famous special functions and have been studied extensively in the hitherto published literature, while "Weyl" sums mean special type of exponential sums – these are the sums of the form

$$\sum_{X<n\leq X+N} e^{2\pi i F(n)},$$

where $F(x)$ denotes a real-valued function of wide choice – e.g. with $F(x)$ a polynomial of the form $\alpha f(x)$, where $\alpha \in \mathbb{R}$, $f(x) \in \mathbb{Z}[x]$ (a polynomial with integer coefficients). Multiple Weyl (exponential) sums are also considered with great interest, but we shall be concerned with single sums in this paper.

Needles to say, then, Weyl sums are special (finite) Fourier series of the form

$$\sum_{m\in\mathbb{Z}} a_m e^{2\pi i \alpha m},$$

311

where

$$a_m = \# \left\{ n, X < n \leq X + N \big| f(n) = m \right\},$$

but, since their introduction into number theory by H.Weyl in 1916 in connection with the uniform distribution of sequences mod 1 and subsequent successful application to the 1921 estimate $|\zeta(1 + it)|$, [12], where $\zeta(s)$ denotes the Riemann zeta-function, Weyl sums have been one of the major tools as well as the central themes in analytic number theory. There have appeared numerous papers dealing with or using them, and some of the results are presented in the books by A. Walfisz [12].

There are three major methods known for treating Weyl sums, each of which has its own features (cf. Appendix at the end of the paper):

(i) The Weyl-Hardy-Littlewood method,
(ii) The Vinogradov method,
and
(iii) The van der Corput method.

Most of the researches have been centered around the estimate of Weyl sums, however, and very little effort has been made toward interpreting them in the light of theta-transformation under the action of the (projective) modular group $PSL(2, \mathbb{Z})$.

Supposing the incorporation of automorphic devices possible, what should be expected traditionally would be the following

Conjecture. Let

$$f(x) = x^h + a_1 x^{h-1} + \cdots + a_{h-1} x \in \mathbb{Z}[x]$$

and suppose a given $\alpha \in \mathbb{R}$ is well approximated, for given $N \gg 1$, by an irreducible fraction $\dfrac{a}{q}$ as $|\alpha - \dfrac{a}{q}| < \dfrac{1}{qN^{h-1}}$ for $1 \leq q \leq N^{h-1}$. Then as $N \to \infty$

$$\sum_{X < n \leq X+N} e^{2\pi i \alpha f(n)} = \frac{1}{q} \left(\sum_{1 \leq n \leq q} e^{2\pi i \frac{a}{q} f(n)} \right) \int_X^{X+N} e^{2\pi i \left(\alpha - \frac{a}{q} \right) f(x)} \, dx$$

$$+ O_h \left(N^{1 - \frac{1}{h}} \right) \quad \text{for } 1 \leq q \leq N \qquad (1.1)$$

and

$$\sum_{X < n \leq X+N} e^{2\pi i \alpha f(n)} = O_h \left(N^{1 - \frac{1}{h}} \right) \qquad \text{for } N < q \leq N^{h-1}. \qquad (1.2)$$

If we can deduce satisfactory explicit estimates for Weyl sums such as (1.1) and (1.2) above, by extracting, expectedly, built-in automorphic properties from them, then we call these Weyl sums "theta-Weyl sums".

So far we have succeeded only in the case $h = 2$, i.e. in the case of quadratic Weyl sums, so that we have "quadratic theta-Weyl sums" (cf. [7-1], also [7-3], Chap II).

The aim of this paper is to state those results which the author has obtained up to now, corresponding to the case of quadratic theta-Weyl sums as penultimate steps toward the cubic theta-Weyl sums, a milestone to the eventual interpretation of higher degree Weyl sums, the realization of the author's "Jugendtraum" since [7-1].

To group the necessary ingredients in the cubic case, we review essential elements of the quadratic case, which has been and will be the guiding line.

Suppose α is an irrational number and define the sequences $\{\beta_k\}$ and $\{a_k\}$ (with $[\xi]$ denoting the integral part of ξ) by

$$
\begin{aligned}
a_0 &= [\alpha], & \alpha &= a_0 + \beta_1^{-1}, \\
a_1 &= [\beta_1], & \beta_1 &= a_1 + \beta_2^{-1}, \ldots \\
a_k &= [\beta_k], & \beta_k &= a_k + \beta_{k+1}^{-1}, \ldots
\end{aligned}
$$

Then this gives the regular continued fraction expansion

$$
[a_0; a_1, a_2, \ldots] \quad (a_0 \in \mathbb{Z}, a_1, a_2, \cdots \in \mathbb{N})
$$

of α from which we form the k-th convergent

$$
A_k B_k^{-1} = [a_0; a_1, \ldots, a_k].
$$

For real $\beta \neq 0$ and t, call

$$
\theta(\beta^{-1}, t; X, X + N) = \sum_{X < n \leq X+N} e\left(\frac{1}{2}\beta^{-1}(n + t)^2\right)
$$

the quadratic theta-Weyl sum, where

$$
e(\xi) = e^{2\pi i \xi}.
$$

Then the transition from $\theta(\beta_k, *)$ to $\theta(\beta_{k+1}^{-1}, *)$, where $*$ means an appropriate value of t varying with k, corresponds to the transformation under a parabolic element $\begin{pmatrix} 1 & -a_k \\ 0 & 1 \end{pmatrix}$ and the transition from $\theta(\beta_{k+1}^{-1}, *)$ to $\theta(\beta_{k+1}, *)$ corresponds the one under the elliptic element $\begin{pmatrix} 0 & -1 \\ 1 & 0 \end{pmatrix}$, or what amounts to the same thing, to the most famous

θ-transformation formula for the θ-series defined by

$$\theta(z;t) = \sum_{n=-\infty}^{\infty} e\left(\frac{1}{2}z(n+t)^2\right).$$

The θ-transformation like behavior of the core polynomial $f(x)$ naturally occurs in iterates of van der Corput's inversion (cf. Appendix), the reciprocal function $g(y)$ being defined by

$$g(y) = f(x_y) - y \cdot x_y,$$

where x_y is defined through $f'(x_y) = y$.

Note that if

$$f(x) = \frac{h-1}{h}\alpha x^h,$$

then

$$g(y) = -\frac{h-1}{h}\frac{y^{\frac{h}{h-1}}}{((h-1)\alpha)^{\frac{1}{h-1}}},$$

and the inversion function is of nice form only for $h = 2$, i.e. for quadratic polynomials.

This probably explains why all preceding endeavours at higher degree (finite) θ-series were not in great success, and one should aim at an approximate transformation formula with an admissible error term.

Thus we see that there are three main ingredients to be developed for implementing cubic θ-Weyl sums:

(1) Cubic continued fraction expansion. (Viewing the regular continued fraction expansion given above as one corresponding to the θ-series, or quadratic polynomials, we call the new continued fraction expansion in §2 "cubic"),

(2) Cubic van der Corput's reciprocal function (corresponding to the elliptic element),

and

(3) The formula exhibiting the behavior of the cubic Weyl sum under the action of parabolic elements.

Regarding (1) the cubic continued fraction expansion, the final form being uniquely determined by the van der Corput reciprocal function, we were successful in deriving the final form in [7-4, Cubic II] refining the preceding paper [7-4, Cubic I], which we will present in §2.

The van der Corput reciprocal function, which the author found on September 8, 1988, is presented in §3.

After giving preparatory reductions in §4, we shall state in §5 the complete form of cubic reciprocal (elliptic) transformation incorporating the van der Corput reciprocal function developed in §3. In §6 we shall state a portion of preliminary calculations concerning the determination of the action of parabolic elements. The author found the present form of the expression more than 10 years ago, but was unable to decide if this would be the most proper form (among other possible forms), but is now certain that it is. Finally, in §7 we shall propose a candidate for cubic theta-Weyl sums, i.e.an expected interpretation of cubic Weyl sums as cubic theta-Weyl sums.

We shall also add Appendix which gives a brief survey on three methods for treating Weyl sums.

We remark that for (infinite) theta series, there have been occasional efforts for finding cubic (or higher degree) theta series (cf. [3]–[6], [7-5], [8], [9]. Also [11-1] is worth mentioning), among which T. Kubota [4] found his higher theta series having power-residue symbols as its (automorphic) multipliers. There has, however, been an underground suspicion that only functions that are "invariant" under Fourier transform would be essentially $e^{-\pi t x^2}$ $(t > 0)$, and so there would be no higher theta series in its naive sense (cf. [3], p.138, 1.2 ↑–1.1 ↑).

Thus, our approach, which seems to be the only plausible one, hopefully may shed some light on formation of higher (infinite) theta series.

2.   Cubic continued fraction expansions

As remarked in introduction and will be elucidated in later sections, no other expansions can be expected to be suitable for our cubic theta-Weyl sums.

We begin with the notation. Starting from a real number α_0, $\frac{1}{2} > \alpha_0 > 0$, we form the sequences $\{a_k\}$, $\{\alpha_{k+1}\}$ and $\{\varepsilon_{k+1}\}$ inductively by

$$\sqrt{2\alpha_k}^{-1} = a_k + \varepsilon_{k+1}\alpha_{k+1},$$

where $a_k \in \mathbb{N}$, $\frac{1}{2} > \alpha_{k+1} > 0$, $\varepsilon_{k+1} = \pm 1$ $(k = 0, 1, 2, \ldots)$, in which we exclude the case where the expansion stops at k steps with $(2\alpha_k)^{-\frac{1}{2}} \in \mathbb{N}$.

In the special case where $a_k = 1$, we have $\varepsilon_{k+1} = 1$.

Hence α_0 has the expansion

$$\alpha_0 = \cfrac{1}{2\left(a_0 + \cfrac{\varepsilon_1}{2\left(a_1 + \cfrac{\varepsilon_2}{2\left(a_2 + \cfrac{\ddots}{\quad + \cfrac{\varepsilon_{k-1}}{2\left(a_{k-1} + \frac{\varepsilon_k}{2(a_k + \varepsilon_{k+1}\alpha_{k+1})^2}\right)^2}}\right)^2}\right)^2}\right)^2}$$

$$= \cfrac{1}{2\left(\varepsilon_1 a_0 + \cfrac{1}{2\left(\varepsilon_2 a_1 + \cfrac{1}{2\left(\varepsilon_3 a_2 + \cfrac{\ddots}{\quad + \cfrac{1}{2\left(\varepsilon_k a_{k-1} + \frac{1}{2(\varepsilon_{k+1}a_k + \alpha_{k+1})^2}\right)^2}}\right)^2}\right)^2}\right)^2}$$

For convenience of notation, we further suppose $\varepsilon_0 = 1$.

We introduce polynomials $p_k(x_0, x_1, \ldots, x_k)$, respectively $q_k(x_0, x_1, \ldots, x_k)$, as the numerator, respectively the denominator, of the fraction

$$\cfrac{1}{2\left(x_0 + \cfrac{1}{2\left(x_1 + \cfrac{1}{2\left(x_2 + \cfrac{\ddots}{\quad + \cfrac{1}{2\left(x_{k-1} + \frac{1}{2x_k^2}\right)^2}}\right)^2}\right)^2}\right)^2}$$

(after rewriting it in the form of the fraction, which we abbreviate as $\frac{p_k}{q_k}(x_0, x_1, \ldots, x_k)$), i.e.,

$$p_{-1} = 0, \qquad q_{-1} = 1,$$

$$p_0(x_0) = 1, \qquad q_0(x_0) = 2x_0^2,$$

and

$$p_k(x_o, x_1, \ldots, x_k) = q_{k-1}^2(x_1, \ldots, x_k),$$

$$q_k(x_o, x_1, \ldots, x_k) = 2(x_o q_{k-1}(x_1, \ldots, x_k) + q_{k-2}^2(x_2, \ldots, x_k))^2,$$
$$(k = 1, 2, 3, \ldots).$$

We have, then,

$$q_k(x_0, x_1, \ldots, x_{k-1}, x_k) = (2x_k^2)^{2^k} q_{k-1}\left(x_0, x_1, \ldots, x_{k-2}, x_{k-1} + \frac{1}{2x_k^2}\right),$$
$$(k = 0, 1, 2, 3, \ldots),$$

and in abbreviated notation above,

$$\alpha_0 = \frac{p_k}{q_k}(\varepsilon_1 a_0, \varepsilon_2 a_1, \ldots, \varepsilon_k a_{k-1}, \varepsilon_{k+1} a_k + \alpha_{k+1})$$
$$= \frac{p_{k+1}}{q_{k+1}}(\varepsilon_1 a_0, \varepsilon_2 a_1, \ldots, \varepsilon_k a_{k-1}, \varepsilon_{k+1} a_k, (2\alpha_{k+1})^{-\frac{1}{2}}).$$

For simplicity of notation we further put

$$p_{j,k} = p_{k-j}(\varepsilon_{j+1} a_j, \ldots, \varepsilon_{k+1} a_k), \quad q_{k-j} = q_{j,k}(\varepsilon_{j+1} a_j, \ldots, \varepsilon_{k+1} a_k),$$
$$0 \le j \le k$$

$$p_{k+1,k} = 0, \qquad q_{k+1,k} = 1,$$

$$\hat{p}_{j,k+1} = p_{j,k+1}\left(\varepsilon_{j+1} a_j, \ldots, \varepsilon_{k+1} a_k, (2\alpha_{k+1})^{-\frac{1}{2}}\right),$$

$$\hat{q}_{j,k+1} = q_{j,k+1}\left(\varepsilon_{j+1} a_j, \ldots, \varepsilon_{k+1} a_k, (2\alpha_{k+1})^{-\frac{1}{2}}\right),$$

$$\alpha_{j,k} = p_{j,k} q_{j,k}^{-1}, \qquad 0 \le j \le k$$

and

$$\hat{\alpha}_{j,k+1} = \hat{p}_{j,k+1} \hat{q}_{j,k+1}^{-1}$$

which is, in fact, α_j itself.

Then we have, in the first place, the following consequences.

$$\sqrt{2\alpha_j}\sqrt{2\alpha_{j+1}} < \frac{2}{3} \qquad (j = 0, 1, 2, \ldots),$$

$$q_j \to \infty \qquad \text{as} \qquad j \to \infty,$$

$$2q_{0,k}^{-1} \prod_{j=1}^{k}(2q_{j,k}) = \prod_{j=0}^{k}(2\alpha_{j,k}),$$

$$2\hat{q}_{0,k+1}^{-1} \prod_{j=1}^{k+1} (2\hat{q}_{j,k+1}) = \prod_{j=0}^{k+1} (2\alpha_j).$$

In the second place, we have the following Lemma from which we may prove, by induction, the succeeding proposition.

Lemma 1. *For $k \geq 1$*

$$\alpha_0 - p_{0,k}q_{0,k}^{-1} = (-\varepsilon_1)\left((\alpha_1 - p_{1,k}q_{1,k}^{-1})(2\alpha_0 2p_{0,k}q_{0,k}^{-1})\right)^{\frac{3}{4}}$$

$$\times \sqrt{1 + \frac{1}{4}\sqrt{2\alpha_0 \cdot 2p_{0,k}q_{0,k}^{-1}}(\alpha_1 - p_{1,k}q_{1,k}^{-1})^2}.$$

Proposition 1. *With notation above we have*

$$\alpha_0 - p_{0,k}q_{0,k}^{-1} = (-1)^{k+1}\varepsilon_1 \cdots \varepsilon_{k+1}\left(\frac{1}{8}\alpha_{k+1}\right)^{\frac{1}{4}}$$

$$\times \left(2q_{0,k}^{-1}\prod_{j=1}^{k}(2q_{j,k})\right)^{\frac{3}{4}} \left(2\hat{q}_{0,k+1}^{-1}\prod_{j=1}^{k+1}(2\hat{q}_{j,k+1})\right)^{\frac{3}{4}}$$

$$\times \prod_{j=0}^{k}\sqrt{1 + \frac{1}{4}\sqrt{(2p_{j,k}q_{j,k}^{-1})(2\alpha_j)} \cdot \left(\alpha_{j+1} - p_{j+1,k}q_{j+1,k}^{-1}\right)^2}.$$

Proof of Lemma 1. Recalling the definition of convergents $p_{0,k}q_{0,k}^{-1}$ and $p_{1,k}q_{1,k}^{-1}$, we obtain

$$\alpha_0 - p_{0,k}q_{0,k}^{-1} = \alpha_0 \cdot p_{0,k}q_{0,k}^{-1} \cdot (-\varepsilon_1)(\alpha_1 - p_{1,k}q_{1,k}^{-1})$$

$$\cdot \left((2\alpha_0)^{-\frac{1}{2}} + (2p_{0,k}q_{0,k}^{-1})^{-\frac{1}{2}}\right)$$

$$= (-\varepsilon_1)(\alpha_1 - p_{1,k}q_{1,k}^{-1})(2\alpha_0 \cdot 2p_{0,k}q_{0,k}^{-1})^{\frac{3}{4}}$$

$$\cdot \left(\frac{1}{2}A^{\frac{1}{4}} + \frac{1}{2}A^{-\frac{1}{4}}\right),$$

with $A = (2\alpha_0)^{-1} \cdot (2p_{0,k}q_{0,k}^{-1})$.

By definition, we have

$$A^{-\frac{1}{2}} = \left(\frac{1}{(\varepsilon_1 a_0 + \alpha_1)^2} \right)^{\frac{1}{2}} \left(\frac{1}{(\varepsilon_1 a_0 + p_{1,k} q_{1,k})^2} \right)^{-\frac{1}{2}}$$

$$= \frac{\varepsilon_1 a_0 + p_{1,k} q_{1,k}}{\varepsilon_1 a_0 + \alpha_1}$$

$$= 1 - \frac{\alpha_1 - p_{1,k} q_{1,k}^{-1}}{\varepsilon_1 a_0 + \alpha_1}$$

$$= 1 - \varepsilon_1 \sqrt{2\alpha_0} \left(\alpha_1 - p_{1,k} q_{1,k}^{-1} \right).$$

Substituting this in the identity

$$\frac{1}{2} A^{\frac{1}{4}} + \frac{1}{2} A^{-\frac{1}{4}} = \sqrt{1 + \frac{1}{4} A^{\frac{1}{2}} (1 - A^{-\frac{1}{2}})^2}$$

we complete the proof. $\qquad\square$

As a corollary to Proposition 1, we get the following estimate, which corresponds to the estimate of approximation by convergents in the case of regular (i.e. quadratic) continued fraction expansions.

$$\alpha_0 - p_{0,k} q_{0,k}^{-1} = (-1)^{k+1} \mathrm{sgn}(\varepsilon_1 \dots \varepsilon_{k+1})$$

$$\cdot \left(\frac{\alpha_{k+1}}{8} \right)^{\frac{1}{4}} \left(2 q_{0,k}^{-1} \prod_{j=1}^{k} (2 q_{j,k}) \cdot 2 \hat{q}_{0,k+1}^{-1} \prod_{j=1}^{k+1} (2 \hat{q}_{j,k+1}) \right)^{\frac{3}{4}}$$

$$\cdot \left(1 + O(\alpha_{k+1})^2 \right).$$

The author has not yet tried to prove other results corresponding to those in Lemmas 7, 8 and 10 of [7-1] (with Errata).

3. The reciprocal function of van der Corput type for cubic theta-Weyl sums

In considering general Weyl sums

$$\sum_{X < n \leq X+N} e^{2\pi i \alpha f(n)}, \qquad 1 < N \leq X$$

we impose on the real-valued function $f(n)$ the condition that it has a continuous second derivative which is positive, in the range that contains the range of summation $[X, X + N]$, so that $f'(n)$ is strictly increasing. As in Introduction, we define n_y by $f'(n_y) = y$ for a given y and put

$$g(y) = f(n_y) - y n_y,$$

which is the reciprocal function of $f(n)$ of van der Corput type.

We note that

$$\frac{d}{dy}g(y) = -n_y = -(f')^{-1}(y)$$

and recall that for

$$f(n) = \frac{h-1}{h}\alpha n^h \quad (h > 1)$$

$$g(y) = -\frac{h-1}{h}((h-1)\alpha)^{-\frac{1}{h-1}}y^{\frac{h}{h-1}},$$

and $f(n)$ is self-reciprocal only when $h = 2$, i.e.

$$f(n) = \frac{1}{2}\alpha n^2.$$

On the other hand, the interval of summation $[X, X+N]$ in n is carried over, under the above reciprocation, to $[f'(X), f'(X+N)]$ in y, which has length $\asymp \alpha X^{h-2}N$, longer than the original length N if $h > 2$.

These two phenomena made the straits so narrow that no wonder all efforts to get good "(finite) theta series" strack reefs on their way.

As announced in Introduction, we shall find out the cubic reciprocal function of van der Corput type as the most essential ingredient in cubic reciprocal (elliptic) transformation developed in §5.

Let k be a non-negative integer which we will use in the context of the k-th (iterate of) reciprocation. Let, as before, $(X, X+N]$ $(1 < N \ll X, X \to \infty)$ be the interval of summation for n.

By abuse of language we shall also regard n as a real variable and introduce the second variable $m \in \mathbb{Z}$ such that

$$\left(n - \frac{1}{2}\right)^2 < n^2 + m \leq \left(n + \frac{1}{2}\right)^2 \quad \text{or} \quad -n < m \leq n.$$

We use the expression $x = n^2 + m$, in which we regard n as the global variable and m the local variable.

We consider a function F in $k + 1$ (real) variables $\Xi_0, \Xi_1, \ldots, \Xi_k$, in vector notation $(\Xi) = (\Xi_0, \Xi_1, \ldots, \Xi_k)$:

$$F((\Xi)) = F(\Xi_0, \Xi_1, \ldots, \Xi_k) \tag{3.1}$$

which we suppose has continuous 4-th derivative for $|\Xi_0|, |\Xi_1|, \ldots, |\Xi_k| \ll X^{-1}$ and $F(0, 0, \ldots, 0) = 1$, and we introduce along with the associated function

$$G((\Xi)) = G(\Xi_0, \Xi_1, \ldots, \Xi_k) = G(F)$$

$$= F((\Xi)) - \frac{2}{3}\sum_{j=0}^{k}\Xi_j\left(\frac{\partial}{\partial\Xi_j}F\right). \tag{3.2}$$

Now, with a real variable (of integration) α and with the expression $x = n^2 + m$, we put

$$f(n) = \frac{2}{3}\alpha \cdot x^{\frac{3}{2}} F\left(\frac{\xi_0}{x}, \frac{\xi_1}{x}, \ldots, \frac{\xi_k}{x}\right), \tag{3.3}$$

where real parameters ξ_j satisfy $|\xi_j| \ll X$ $(j = 0, 1, \ldots, k)$.

Lemma 2. *Viewing n as a real variable, we have*

$$\frac{\partial}{\partial n}f(n) = 2\alpha x(1 - Q)^{\frac{1}{2}}G(\Xi), \tag{3.4}$$

where $Q = x^{-1}m$, $x = n^2 + m$ and $\Xi_j = x^{-1}\xi_j$ $(j = 0, 1, \ldots, k)$.
Following the aforementioned prescription, we put, for given y,

$$\frac{\partial}{\partial n}f(n)\Big|_{n=n_y} = y \tag{3.5}$$

and

$$x_y = n_y^2 + m.$$

Now we have the following fundamental result.

Theorem 1. *Under the above notation, we have*

$$\frac{2}{3}\alpha x_y^{\frac{2}{3}} F\left(x_y^{-1}(\xi)\right) - yn_y$$

$$= -\frac{2}{3}\frac{1}{\sqrt{2\alpha}}y^{\frac{3}{2}}\left(\mathrm{tr}\frac{1+\sqrt{-1}}{2}\right.$$

$$\times\left(1 + \sqrt{-1}\left\{y^{-1}\alpha m + \frac{1}{2}\left(G^2\left(y^{-1}\cdot 2\alpha(\xi)\right) - 1\right)\right\}\right)^{\frac{3}{2}}$$

$$\left.-\frac{1}{2}(F - G^3)(y^{-1}\cdot 2\alpha(\xi))\right) + O_{F,k,\alpha}\left(X^{-1}\right)$$

$$= -\frac{2}{3}\frac{1}{\sqrt{2\alpha}}y^{\frac{3}{2}}\left(\begin{array}{c}\mathrm{tr}\dfrac{1+\sqrt{-1}}{2}(\cdots)^{\frac{3}{2}}\\-\frac{1}{2}(F - G^3)\left(y^{-1}\cdot 2\alpha(\xi)\right)\end{array}\right) + O_{F,k,\alpha}(X^{-1}). \tag{3.6}$$

Here for real variables U and V with $|U| \ll X^{-1}$, $|V| \ll X^{-1}$,

$$\mathrm{tr}\left[\frac{1+\sqrt{-1}}{2}\left(1 + U + \sqrt{-1}V\right)^{\frac{3}{2}}\right]$$

$$= \frac{1+\sqrt{-1}}{2}\left(1 + U + \sqrt{-1}V\right)^{\frac{3}{2}} + \frac{1-\sqrt{-1}}{2}\left(1 + U - \sqrt{-1}V\right)^{\frac{3}{2}}.$$

In what follows, the following corollary will play a vital role.

Corollary 1. *The reciprocal function of van der Corput type of*

$$\frac{2}{3}\alpha(n^2+m)^{\frac{3}{2}}F\left((n^2+m)^{-1}\cdot(\xi_0,\xi_1,\ldots,\xi_k)\right)$$

with respect to n (other variables are considered to be fixed) is

$$-\frac{2}{3}\frac{1}{\sqrt{2\alpha}}y^{\frac{3}{2}}\left(\text{tr}\frac{1+\sqrt{-1}}{2}\left(1+\sqrt{-1}\left\{y^{-1}\alpha m\right.\right.\right. \tag{3.7}$$

$$\left.\left.\left.+\frac{1}{2}\left(G^2\left(y^{-1}2\alpha(\xi)\right)-1\right)\right\}\right)^{\frac{3}{2}}-\frac{1}{2}\left(F-G^3\right)\left(y^{-1}2\alpha(\xi)\right)\right).$$

Here, for applications, y is to be decomposed as $y = u^2 + v$ $(u \in \mathbb{N}, v \in \mathbb{Z}, -u < v \le u)$.

Proof of Theorem 1. We shall prove that (the main parts of) both side of (3.6) coincide with the error $O(X^{-1})$. We first note that the partial derivative (in y) of the left-hand side, LH, say, $f(n_y) - yn_y$, of (3.6) is $-n_y$, i.e.,

$$-\frac{\partial}{\partial y}(LH) = n_y.$$

Hence we may appeal to Lemma 2 with x_y for x: the right-hand side of (3.4) is then

$$2\alpha x_y\left(1-\frac{m}{x_y}\right)^{\frac{1}{2}}G\left(\frac{(\xi)}{x_y}\right). \tag{3.8}$$

Substituting $(-1)\frac{\partial}{\partial y}(RH)$ for n_y in (3.8) and noting (3.5), we see that it suffices to show that the right-hand side of (3.4) is $y + O(X^{-2})$, which we do by applying the Taylor expansion of G and formulas

$$(1+u)^{\frac{3}{2}} = 1 + \frac{3}{2}u + \frac{3}{8}u^2 - \frac{1}{16}u^3 + O(u^4)$$

and

$$\text{tr}\frac{1+\sqrt{-1}}{2}(1+\sqrt{-1}v)^{\frac{3}{2}} = 1 - \frac{3}{2}v - \frac{3}{8}v^2 - \frac{1}{16}v^3 + O(v^4)$$

valid for $u \doteqdot 0$ and $v \doteqdot 0$. $\qquad\square$

Remark 1. (i) In the special case where $F = 1$, i.e. $f(n) = \frac{2}{3}\alpha(n^2+m)^{\frac{3}{2}}$, putting $y^{-1}\alpha m = \tan\theta$, we may reduce the proof of Theorem 1 to the triplication formula in trigonometry (thus, suggestive of a Chebyshev's polynomial) or of the identities

$$\sqrt{-3\tilde{Q}' + \tilde{Q}'^3 + (\sqrt{1 + \tilde{Q}'^2})^3} = \operatorname{tr}\frac{1 + \sqrt{-1}}{2}\left(1 + \sqrt{-1}\tilde{Q}'\right)^{\frac{3}{2}},$$

and

$$\sqrt{\mp\tilde{Q}' + \sqrt{1 + \tilde{Q}'^2}} = \operatorname{tr}\frac{1 + \sqrt{-1}}{2}\left(1 + \sqrt{-1}(\pm\tilde{Q}')\right)^{\frac{1}{2}}$$

with $\tilde{Q}' = y^{-1}\alpha m$, and we have then the true identity without error term:

$$f(n_y) - yn_y = -\frac{2}{3}\frac{1}{\sqrt{2\alpha}}y^{\frac{3}{2}}\operatorname{tr}\frac{1 + \sqrt{-1}}{2}\left(1 + \sqrt{-1}\frac{\alpha m}{y}\right)^{\frac{3}{2}}.$$

(ii) The error term in Theorem 1 does not vanish if $m \neq 0$ and the third derivative of $\frac{1}{2}(G^2 - 1) - \sum_j \Xi_j\frac{\partial}{\partial\Xi_j}\left(\frac{1}{2}(G^2 - 1)\right)$ does not vanish.

(iii) The Taylor expansion of $F - G^3$ in Theorem 1 starts from the degree 2 term. This follows from the observation:

If $F = \operatorname{tr}\frac{1 + \sqrt{-1}}{2}\left(1 + U + \sqrt{-1}V\right)^{\frac{3}{2}}$, and if $|U| \ll X^{-1}, |V| \ll X^{-1}$, then

$$F - G^3 = -\frac{3}{2}V^2(1 - \frac{1}{6}V - \frac{1}{2}U) + O(X^{-4}),$$

which follows from

$$\operatorname{tr}\frac{1 + \sqrt{-1}}{2}(1 + u + \sqrt{-1}v)^{\frac{3}{2}} = 1 + \frac{3}{2}(u - v) + \frac{3}{8}(u^2 - 2uv - v^2)$$

$$- \frac{1}{16}(u^3 - 3u^2v - 3uv^2 + v^3) + O(X^{-4}), \tag{3.9}$$

for $u = O(X^{-1})$, $v = O(X^{-1})$.

4.   Preliminary reductions

In §§5-6, we are going to consider transformation formulas for the Weyl sums (cf. §1)

$$\sum_{X < n \leq X+N} e\left(f(n)\right),$$

where $1 \leq N \ll X$, $X \to \infty$ and

$$f(n) = \frac{2}{3}\alpha n^3 + \gamma n \tag{4.1}$$

with $\alpha \in \mathbb{R}$, $2\alpha \notin \mathbb{Z}$, and $\gamma \in \mathbb{R}$.

To this end we shall show in this section that $f(n)$ is of the form

$$\frac{2}{3}\alpha(n^2)^{\frac{3}{2}}F$$

with $F = \text{tr}\frac{1+\sqrt{-1}}{2}\left(1+\sqrt{-1}\Xi_0\right)^{\frac{3}{2}}$ as in Theorem 1 with an admissible error term $O(X^{-1})$.

We may also treat the case of a more general $f(n)$ of the form

$$\frac{2}{3}\alpha n^3 + \beta n^2 + \gamma n$$

(or equivalently, of the form $\frac{2}{3}\alpha(n+\beta')^3 + \gamma'(n+\beta')$) in which case we are to consider

$$\frac{2}{3}\alpha\left(x^{\frac{3}{2}}F_1(x^{-1}(\xi)) + cxF_2(x^{-1}(\xi))\right)$$

with F_i as in §3 and c a constant. We can also obtain the corresponding result in this case, but we shall restrict ourselves to the case $\beta = 0$ given by (4.1).

Here we confirm the above remark that the error term $O(X^{-1})$ is admissible.

Suppose the core function is decomposed as $f(n;m) = \check{f}(n;m) + O(X^{-1})$, then the sum $\sum_{n,m} e(f(n;m)))$, which we are eventually to treat, can be written as

$$\sum_{n,m} e(\check{f}(n;m)) + \sum_{n,m} e(\check{f}(n;m))\left(e(O(X^{-1})) - 1\right)$$

whose second term is tractable by a well-known procedure such as the partial summation.

We shall now show, as a first reduction step of the problem, that in (4.1) we may suppose that

$$0 < \alpha < \frac{1}{2} \quad \text{and} \quad -\frac{1}{2} \le \gamma \le \frac{1}{2}. \tag{4.2}$$

For considering congruences mod 1, we may suppose in the first instance that

$$-\frac{3}{2} < \alpha < \frac{3}{2} \quad \text{and} \quad -\frac{1}{2} \le \gamma \le \frac{1}{2},$$

then taking the complex conjugate ($\bar{z}$ denoting the one for z) $\overline{\sum e(f(n))}$, which is $\sum e(-f(n))$, we may also suppose that $0 < \alpha < \frac{3}{2}$.

If $\frac{1}{2} < \alpha < \frac{3}{2}$, then $\alpha' = 1 - \alpha$ lies in $-\frac{1}{2} < \alpha' < \frac{1}{2}$, and

$$\frac{2}{3}\alpha n^3 = \frac{2}{3}(1-\alpha')n^3 \equiv -\left(\frac{2}{3}\alpha'n^3 + \frac{1}{3}n\right) \mod 1$$

for $n \in \mathbb{Z}$, which gives (-1) times (4.1) when substituted in (4.1). We can incorporate this $-$ sign by complex conjugation and thus we arrive at (4.2).

In what follows we exclude the case where α reduces to 0, and, as remarked at the beginning of §2, the case when its cubic continued fraction expansion terminates in finite steps.

Then (4.1) becomes in the notation of §3 (cf. (3.3))

$$f(n) = \frac{2}{3}\alpha x^{\frac{3}{2}} F + O(X^{-1}), \tag{4.3}$$

where

$$x = n^2 + 0 \ (\text{i.e. } m = 0),$$

$$F = \operatorname{tr}\frac{1+\sqrt{-1}}{2}(1+\sqrt{-1}\Xi)^{\frac{3}{2}},$$

and

$$\Xi_0 = x^{-1}(-1)\alpha^{-1}\gamma.$$

We put (4.3) on record as the initial (0-th) state of the induction process by rewriting it as

$$f_0(n_0) = \frac{2}{3}\alpha_0 x_0^{\frac{3}{2}} \operatorname{tr}\frac{1+\sqrt{-1}}{2}(1+\sqrt{-1}x_0^{-1}\kappa_0)^{\frac{3}{2}}, \tag{4.4}$$

where $0 < \alpha_0 < \frac{1}{2}$, $x_0 = n_0^2 + 0$ (i.e. $m_0 = 0$; m_0 corresponds to $m_0^{(0)}$ in §7), $X_0 < n_0 \leq X_0 + N_0$, and κ_0 is a real constant.

5. The cubic reciprocal transformation (after k times iteration)

In this section we shall work out the transfer principle (as Theorem 2) for repeated applications of the reciprocal transformation.

Recalling Corollary 1, we suppose that $f(n)$ is given (after reciprocal transformation and reduction mod 1) by

$$f(n) = \frac{2}{3}\alpha x^{\frac{3}{2}}\left(\operatorname{tr}\frac{1+\sqrt{-1}}{2}\left(1+\sqrt{-1}x^{-1}\{K((\xi))+\kappa\}\right)^{\frac{3}{2}} \right.$$

$$+ \sum_j H_j L\left(1 + x^{-1}\left(L_j((\xi)) + \lambda_j\right)\right.$$

$$\left. +\sqrt{-1}x^{-1}\left(M_j((\xi)) + \mu_j\right)\right), \tag{5.1}$$

where $\sum_j$ is a finite sum, κ, λ_j, μ_j, and H_j are constants, $K((\xi))$, $L_j((\xi))$, and $M_j((\xi))$ are linear forms in $(\xi) = (\xi_0, \xi_1, \ldots, \xi_k)$ with $\xi_i \in \mathbb{R}$ and $|\xi_i| \ll X$, and where

$$L(1 + U + \sqrt{-1}V) = \mathrm{tr}\frac{1 + \sqrt{-1}}{2}(1 + U + \sqrt{-1}V)^{\frac{3}{2}}$$

$$- \left(\mathrm{tr}\frac{1 + \sqrt{-1}}{2}(1 + U + \sqrt{-1}V)^{\frac{1}{2}} \right)^3 \qquad (5.2)$$

with real U and V satisfying $|U| \ll X^{-1}$ and $|V| \ll X^{-1}$.

Using Remark 1, (iii) in §3, we have

Lemma 3. *For $|U| \ll X^{-1}$ and $|V| \ll X^{-1}$ we have*

$$L(1 + U + \sqrt{-1}V) = -\frac{3}{2}(V^2 - \frac{1}{2}UV^2 - \frac{1}{6}V^3) + O(X^{-4})$$

and a fortiori

$$L(1 + U + \sqrt{-1}V) = O(X^{-2}).$$

In view of (3.3), we have

$$F = \mathrm{tr}\frac{1 + \sqrt{-1}}{2}\left(1 + \sqrt{-1}\,(K((\Xi)) + \dot{\kappa})\right)^{\frac{3}{2}}$$

$$+ \sum_j H_j \cdot L\left(1 + \left(L_j((\Xi)) + \dot{\lambda}_j\right) + \sqrt{-1}\,(M_j((\Xi)) + \dot{\mu}_j)\right), \qquad (5.3)$$

where $\Xi_i = x^{-1}\xi_i$, $\dot{\kappa} = x^{-1}\kappa$, $\dot{\lambda}_j = x^{-1}\lambda_j$ and $\dot{\mu}_j = x^{-1}\mu_j$, and where κ, λ_j and μ_j are constants, so that $|\Xi_i| \ll X^{-1}$ and $|\dot{\kappa}|$, $|\dot{\lambda}_j|$, $|\dot{\mu}_j| \ll X^{-2}$.

To treat the part $L(\cdot)$ in (5.3), we prepare

Lemma 4. *Recalling the dependence of G on F in (3.2), we have*

$$G(L(1 + (L + \dot{\lambda}) + \sqrt{-1}(M + \dot{\mu}))$$

$$= \frac{1}{2}((M + \dot{\mu})(1 + (L + \dot{\lambda}))^{-1})^2$$

$$\cdot \mathrm{tr}\frac{1 + \sqrt{-1}}{2}(1 + (L + \dot{\lambda}) + \sqrt{-1}(M + \dot{\mu}))^{\frac{1}{2}} + O(X^{-4}),$$

where L and M are linear forms in $\Xi = (\Xi_0, \Xi_1, \ldots, \Xi_k)$ with $|\Xi_i| \ll X^{-1}$, and where $\dot{\lambda} = x^{-1}\lambda$ and $\dot{\mu} = x^{-1}\mu$ with constants λ and μ.

Using Lemma 4 etc., we are going to calculate the terms $\tilde{K}$, $\tilde{G}^2 - 1$ and $\tilde{F} - \tilde{G}^3$ corresponding to those in (3.7) (with descriptions made below).

All calculations will be carried out with the admissible error term (cf. §4) $+O(X^{-4})$.

First we obtain from (5.3) and Lemma 4,

$$
G = \text{tr}\frac{1+\sqrt{-1}}{2}\left(1+\sqrt{-1}\,(K+\dot{\kappa})\right)^{\frac{1}{2}}
$$
$$
+\sum_j \left(\frac{1}{2}H_j\right)\left((M_j+\dot{\mu}_j)\left(1+\left(L_j+\dot{\lambda}_j\right)\right)^{-1}\right)^2
$$
$$
\cdot\text{tr}\frac{1+\sqrt{-1}}{2}\left(1+\left(L_j+\dot{\lambda}_j\right)+\sqrt{-1}\,(M_j+\dot{\mu}_j)\right)^{\frac{1}{2}}+O\left(X^{-4}\right),
$$

$$(5.4)$$

whence we deduce that

$$
\frac{1}{2}(G^2-1)=-\frac{1}{2}\,(K+\dot{\kappa})+\Gamma+O\left(X^{-4}\right). \tag{5.5}
$$

Here

$$
\Gamma = \frac{1}{2}\left(\sqrt{1+(K+\dot{\kappa})^2}-1\right)
$$
$$
+\sum_j\left(\frac{1}{2}H_j\right)\left((M_j+\dot{\mu}_j)\left(1+\left(L_j+\dot{\lambda}_j\right)\right)^{-1}\right)^2
$$
$$
\cdot\text{tr}\frac{1+\sqrt{-1}}{2}\left(1+\left(L_j+\dot{\lambda}_j\right)+\sqrt{-1}\,(M_j+\dot{\mu}_j)\right)^{\frac{1}{2}}
$$
$$
\cdot\text{tr}\frac{1+\sqrt{-1}}{2}\left(1+\sqrt{-1}\,(K+\dot{\kappa})\right)^{\frac{1}{2}}, \tag{5.6}
$$

so that

$$
\Gamma = O\left(X^{-2}\right). \tag{5.7}
$$

Now, with the descriptions $\tilde{Q}'=y^{-1}\alpha m$,

$$
\tilde{G}=G\left(y^{-1}2\alpha(\xi)\right),\ \tilde{K}=K(y^{-1}2\alpha(\xi)),\ \tilde{\kappa}=y^{-1}2\alpha\kappa,\ \text{etc.}, \tag{5.8}
$$

we may calculate the two summands in (3.7) as follows:

$$
\text{tr}\frac{1+\sqrt{-1}}{2}\left(1+\sqrt{-1}\left[\tilde{Q}'+\frac{1}{2}\left(\tilde{G}^2-1\right)\right]\right)^{\frac{3}{2}}
$$
$$
=\text{tr}\frac{1+\sqrt{-1}}{2}\left(1+\sqrt{-1}\left[\tilde{Q}'-\frac{1}{2}\left(\tilde{K}+\tilde{\kappa}\right)\right]\right)^{\frac{3}{2}}
$$
$$
+\frac{3}{2}\text{tr}\frac{1+\sqrt{-1}}{2}\left(1+\sqrt{-1}\left[\tilde{Q}'-\frac{1}{2}\left(\tilde{K}+\tilde{\kappa}\right)\right]\right)^{\frac{1}{2}}\sqrt{-1}\Gamma+O\left(X^{-4}\right),
$$

$$(5.9)$$

and

$$-\frac{1}{2}(\tilde{F} - \tilde{G}^3) = -\frac{1}{2}\text{tr}\frac{1+\sqrt{-1}}{2}(1+\sqrt{-1}(\tilde{K}+\tilde{\kappa}))^{\frac{3}{2}}$$

$$-\sum_j \frac{1}{2}H_jL\left(1+(\tilde{L}_j+\tilde{\lambda}_j)+\sqrt{-1}(\tilde{M}_j+\tilde{\mu}_j)\right)$$

$$+\frac{1}{2}\left(\text{tr}\frac{1+\sqrt{-1}}{2}(1+\sqrt{-1}(\tilde{K}+\tilde{\kappa}))^{\frac{1}{2}}\right)^3$$

$$+\frac{3}{2}\left(\text{tr}\frac{1+\sqrt{-1}}{2}(1+\sqrt{-1}(\tilde{K}+\tilde{\kappa}))^{\frac{1}{2}}\right)^2$$

$$\cdot\sum_j \frac{1}{2}H_j\left(\left(\tilde{M}_j+\tilde{\mu}_j\right)\left(1+\left(\tilde{L}_j+\tilde{\lambda}_j\right)\right)^{-1}\right)^2$$

$$\cdot\text{tr}\frac{1+\sqrt{-1}}{2}\left(1+(\tilde{L}+\tilde{\lambda}_j)+\sqrt{-1}(\tilde{M}_j+\tilde{\mu}_j)\right)^{\frac{1}{2}}+O(X^{-4}), \quad (5.10)$$

where we used (5.3) and (5.4).

Substituting (5.9) and (5.10) in (3.7) and expressing it in the form of (5.1), we deduce, after simplification,

Theorem 2. *The reciprocal function of $f(n)$ given by (5.1) is*

$$-\frac{2}{3}\frac{1}{\sqrt{2\alpha}}y^{\frac{3}{2}}\left(\text{tr}\frac{1+\sqrt{-1}}{2}\left(1+\sqrt{-1}\left\{\tilde{Q}'-\frac{1}{2}(\tilde{K}+\tilde{\kappa})\right\}\right)\right)^{\frac{3}{2}}$$

$$-L(1+\tilde{Q}'+\frac{1}{2}\sqrt{-1}(\tilde{K}+\tilde{\kappa})) \qquad\qquad (5.11)$$

$$-\sum_j \frac{1}{2}H_jL(1+\tilde{Q}'+\frac{1}{2}(\tilde{K}+\tilde{\lambda})+(\tilde{L}+\tilde{\lambda})+\sqrt{-1}(\tilde{M}+\tilde{\mu}))\Bigg) + O(X^{-1}),$$

or

$$-\frac{2}{3}\frac{1}{\sqrt{2\alpha}}\left(\text{tr}\frac{1+\sqrt{-1}}{2}(y+\sqrt{-1}\alpha\,[m-(K((\xi))+\kappa)])^{\frac{3}{2}}\right.$$

$$-L\left(y+\alpha m+\sqrt{-1}\alpha\{K((\xi))+\kappa\}\right)$$

$$-\sum_j \frac{1}{2}H_jL(y+\alpha\,[m+\{K((\xi))+\kappa\}]+2\alpha\{L_j((\xi))+\lambda_j\}$$

$$\left.+\sqrt{-1}\cdot 2\alpha\{M_j((\xi))+\mu_j\})\right) + O(X^{-1}), \qquad (5.12)$$

where in the passage from (5.11) to (5.12) we used (5.8) and the interpretation

$$L(y + yU + \sqrt{-1}yV) = y^{\frac{3}{2}} L(1 + U + \sqrt{-1}V), \qquad (5.13)$$

which we will use repeatedly without further remark.

Here as in Corollary 1, y is to be decomposed as $y = u^2 + v$, where now y will be in the range $2\alpha X^2 < y \leq 2\alpha(X + N)^2$ in case $\sqrt{2\alpha}X \gg 1$, and u, the global variable, and v, the local variable, range respectively over $\sqrt{2\alpha}X < u \leq \sqrt{2\alpha}(X + N)$ and $-u < v \leq u$.

6. Cubic parabolic transformations

As announced in Introduction, we shall describe in this section, the action of a parabolic element $\begin{pmatrix} 1 & -a_k \\ 0 & 1 \end{pmatrix}$ on the function $g(y)$, say, given by (5.12), through the cubic continued fraction expansion algorithm developed in §2.

Considering $g(y) = -\dfrac{2}{3}\dfrac{1}{\sqrt{2\alpha}}y^{\frac{3}{2}}(\cdots)$ as the k-th iterate $g_k(y) = -\dfrac{2}{3}\dfrac{1}{\sqrt{2\alpha_k}}y^{\frac{3}{2}}(\cdots)$, we need to find out the effect of the linear fractional transformation $\begin{pmatrix} 1 & -a_k \\ 0 & 1 \end{pmatrix}(\sqrt{2\alpha_k}^{-1})$, or what amounts to the same, that of the algorithm

$$(2\alpha_k)^{-\frac{1}{2}} = a_k + \varepsilon_{k+1}\alpha_{k+1}, \qquad (6.1)$$

where $a_k \in \mathbb{N}$, $\varepsilon_{k+1} = \pm 1$, $0 < \alpha_{k+1} < \frac{1}{2}$ (cf. §§2 and 4).

Since the terms $-\dfrac{2}{3}\dfrac{1}{\sqrt{2\alpha_k}}L_m(\cdots)$ of $g_k(y)$ simply change into the same form

$$-\frac{2}{3}(\varepsilon_{k+1}\alpha_{k+1})\left((\varepsilon_{k+1}\alpha_{k+1}\sqrt{2\alpha_k})^{-1}L(\cdots)\right),$$

it suffices to study the first term only.

Suppose we are at the k-th state of our process (of repeated application of modular transformations). For simplicity of notation, we use $\alpha, a, a', \varepsilon_1$ and α_1 for $\alpha_k, a_k, a'_k, \varepsilon_{k+1}$ and α_{k+1} in (6.1), where a'_k is the absolutely least residue mod 3 of a_k : $a'_k \equiv a_k \bmod 3$, $a'_k = 0, \pm 1$.

We denote the first term of $g(y)$ by $\tilde{g}(y)$:

$$\tilde{g}(y) = \frac{2}{3}(2\alpha)^{-\frac{1}{2}} \operatorname{tr} \frac{1 + \sqrt{-1}}{2}(y + \sqrt{-1}K')^{\frac{2}{3}}, \qquad (6.2)$$

where

$$K' = \alpha(m - \{K((\xi)) + \kappa\}) = O(X). \tag{6.3}$$

Using the decomposition $y = u^2 + v$ prescribed in Corollary 1, where now $u \asymp X$, we obtain

$$\tilde{g}(y) = \frac{2}{3}(2\alpha)^{-\frac{1}{2}} u^3 \mathrm{tr} \frac{1 + \sqrt{-1}}{2} (1 + u^{-2}v + \sqrt{-1}u^{-2}K')^{\frac{3}{2}} \tag{6.4}$$

by clipping out the factor u^3.

We may use (3.9) to expand the factor $\mathrm{tr} \frac{1 + \sqrt{-1}}{2} (\cdots)^{\frac{3}{2}}$. Hence using (6.1) (without suffices), we deduce that

$$\begin{aligned}
\tilde{g}(y) = \frac{2}{3}(a + \varepsilon_1 \alpha_1) & \left(u^3 + \frac{3}{2}u(v - K') \right. \\
& \left. + \frac{3}{8}u^{-1}(v^2 - 2vK' - K'^2) - \frac{1}{16}u^{-3}(v^3 - 3v^2K' - 3vK'^2 + K'^3) \right) \\
& + O(X^{-1}).
\end{aligned} \tag{6.5}$$

Noting that u, v, a are integers and

$$\frac{2}{3}au^3 \equiv -\frac{1}{3}a'u \bmod 1,$$

we further reduce (6.5) mod 1 to

$$\begin{aligned}
\tilde{g}(y) \equiv & \frac{2}{3}\varepsilon_1\alpha_1 u^3 - \frac{1}{3}a'u + \varepsilon_1\alpha_1 uv - (2\alpha)^{-\frac{1}{2}}uK' \\
& + \frac{2}{3}(2\alpha)^{-\frac{1}{2}} \left(\frac{3}{8}u^{-1}(\cdots) - \frac{1}{16}u^{-3}(\cdots) \right) + O(X^{-1}),
\end{aligned} \tag{6.6}$$

where the penultimate term comes from the corresponding terms on the right side of (6.5).

In anticipation of expressing the first four terms on the right-side of (6.6), we put them into

$$\frac{2}{3}\varepsilon_1\alpha_1 u^3 + \frac{2}{3}\varepsilon_1\alpha_1 \cdot \frac{3}{2}u(v - K'') = \frac{2}{3}\varepsilon_1\alpha_1 (u^3 + \frac{3}{2}u(v - K'')),$$

where

$$K'' = (\varepsilon_1\alpha_1)^{-1} \left(\frac{1}{3}a' \right) + (\varepsilon_1\alpha_1\sqrt{2\alpha})^{-1}K'. \tag{6.7}$$

Hence

$$
\tilde{g}(y) \equiv \frac{2}{3}\varepsilon_1\alpha_1 u^3 \mathrm{tr}\frac{1+\sqrt{-1}}{2}(1+u^{-2}v+\sqrt{-1}u^{-2}K'')^{\frac{3}{2}}
$$
$$
-\frac{2}{3}\varepsilon_1\alpha_1\left(\frac{3}{8}u^{-1}(v^2-2vK''-(K'')^2)\right.
$$
$$
-\frac{1}{16}u^{-3}(v^3-3v^2K''-3v(K'')^2+(K'')^3)\Big)
$$
$$
+\frac{2}{3}(2\alpha)^{-\frac{1}{2}}\left(\frac{3}{8}u^{-1}(v^2-2vK'-K'^2)\right.
$$
$$
-\frac{1}{16}u^{-3}(v^3-3v^2K'-3vK'^2+K'^3)\Big)+O(X^{-1}). \qquad (6.8)
$$

Substituting (6.7) in (6.8), we see that the terms containing K' cancel, and after regrouping terms, taking the equality $(2\alpha)^{-\frac{1}{2}}-\varepsilon_1\alpha_1=a$ into account, we arrive at

$$
\tilde{g}(y)=\frac{2}{3}\varepsilon_1\alpha_1\cdot\mathrm{tr}\frac{1+\sqrt{-1}}{2}(y+\sqrt{-1}K'')^{\frac{3}{2}}
$$
$$
-\frac{2}{3}\cdot\frac{1}{4}a\cdot\left(-\frac{3}{2}\right)(u^{-1}(v+\frac{1}{3}a^{-1}a')^2-\frac{1}{6}u^{-3}(v+\frac{1}{3}a^{-1}a')^3)
$$
$$
-\frac{2}{3}\cdot\frac{1}{4}\varepsilon_1\alpha_1\cdot\left(-\frac{3}{2}\right)(u^{-1}(-K'')^2-\frac{1}{6}u^{-3}(3v(-K'')^2+(-K'')^3))
$$
$$
+\frac{2}{3}\cdot\frac{1}{4}(2\alpha)^{-\frac{1}{2}}\left(-\frac{3}{2}\right)(u^{-1}(-K')^2-\frac{1}{6}u^{-3}(3v(-K')^2+(-K')^3))
$$
$$
+O(X^{-1}). \qquad (6.9)
$$

Now appealing to (the first formula in) Lemma 3, we may express the second, third, and fourth term on the right of (6.9), respectively as

$$
-\frac{2}{3}\cdot\frac{1}{4}au^3L(1+\sqrt{-1}u^{-2}(v+\frac{1}{3}a^{-1}a')),
$$
$$
-\frac{2}{3}\cdot\frac{1}{4}\varepsilon_1\alpha_1 u^3L(1+u^{-2}v+\sqrt{-1}u^{-2}(-K'')),
$$

and

$$
\frac{2}{3}\cdot\frac{1}{4}(2\alpha)^{-\frac{1}{2}}u^3L(1+u^{-2}v+\sqrt{-1}u^{-2}(-K')),
$$

whence

$$\tilde{g}(y) = \frac{2}{3}\varepsilon_1\alpha_1 \text{tr}\frac{1+\sqrt{-1}}{2}(y+\sqrt{-1}K'')^{\frac{3}{2}}$$

$$-\frac{2}{3}\cdot\frac{1}{4}aL(y-v+\sqrt{-1}(v+\frac{1}{3}a^{-1}a'))$$

$$-\frac{2}{3}\cdot\frac{1}{4}\varepsilon_1\alpha_1 L(y+\sqrt{-1}(-K''))$$

$$+\frac{2}{3}\cdot\frac{1}{4}(2\alpha)^{-\frac{1}{2}}L(y+\sqrt{-1}(-K'))+O(X^{-1}). \qquad (6.10)$$

Incorporating (6.7) and (6.3) in (6.10) we obtain the following.

Theorem 3 (Cubic parabolic transformation). *Under the general setting in this section, the reciprocal function (5.12) is congruent mod 1 to*

$$-\frac{2}{3}\varepsilon_1\alpha_1 y^{\frac{3}{2}}\left(\text{tr}\frac{1+\sqrt{-1}}{2}\left(1+\sqrt{-1}y^{-1}[(\varepsilon_1\alpha_1\sqrt{2\alpha})^{-1}\alpha(m-(K+\kappa))\right.\right.$$

$$\left.\left.+(\varepsilon_1\alpha_1)^{-1}\left(\frac{1}{3}a'\right)]\right)^{\frac{3}{2}}\right)$$

$$-\frac{1}{4}(\varepsilon_1\alpha_1)^{-1}aL(1-y^{-1}v+\sqrt{-1}y^{-1}(v+\frac{1}{3}a^{-1}a'))$$

$$-\frac{1}{4}L\left(1+\sqrt{-1}y^{-1}[-((\varepsilon_1\alpha_1\sqrt{2\alpha})^{-1}\alpha(m-(K+\kappa))+(\varepsilon_1\alpha_1)^{-1}(\frac{1}{3}a'))]\right)$$

$$+\frac{1}{4}(\varepsilon_1\alpha_1\sqrt{2\alpha})^{-1}L\left(1+\sqrt{-1}y^{-1}[-\alpha(m-(K+\kappa))]\right)$$

$$-(\varepsilon_1\alpha_1\sqrt{2\alpha})^{-1}L\left(1+y^{-1}\alpha m+\sqrt{-1}y^{-1}\alpha(K+\kappa)\right)$$

$$-\sum_j(\varepsilon_1\alpha_1\sqrt{2\alpha})^{-1}\frac{1}{2}H_jL\left(1+y^{-1}\{\alpha(m+(K+\kappa))\right.$$

$$\left.+2\alpha(L_j+\lambda_j)+\sqrt{-1}y^{-1}\{2\alpha(M_j+\mu_j)\}\right)+O(X^{-1}),$$

where K, L_j, and M_j are linear forms in (ξ) and κ, λ_j, μ_j and H_j are constants prescribed in (5.1).

Remark 2. (i) We remark that in the above process, the first $L(\cdot)$-term on the right of (6.10) admits another representation

$$-\frac{2}{3}\cdot\frac{1}{16}aL(y+\sqrt{-1}(-2)(v+\frac{1}{3}a^{-1}a')).$$

(ii) We observe the (seemingly remote) similarly between the quadratic case and the present cubic case under the operations of modular transformations to $f(n)$. In the quadratic case, we are to add $\frac{1}{2}\alpha\gamma^2$ in the transition from $\frac{1}{2}(x+\gamma)^2 - yx$ to $-\frac{1}{2}\alpha^{-1}(y-\alpha^{-1}\gamma)^2$, where x is subject to $\frac{\partial}{\partial x}(\frac{1}{2}\alpha(x+\gamma)^2) = y$. Corresponding to this term $\frac{1}{2}\alpha\gamma^2$, four new $L(\cdot)$-terms appear, each time we operate "cubic" modular transformations $\begin{pmatrix} 0 & -1 \\ 1 & 0 \end{pmatrix}$ and $\begin{pmatrix} 1 & -a \\ 0 & 1 \end{pmatrix}$ $(a \in \mathbb{N})$.

7. Inductive descendents of the clan of cubic theta-Weyl sums

Now that we have Theorems 2 and 3 at hand, we may write down the k-th state of the process inductively as follows. The calculation, however, are not completed as yet in its number-theoretic perspective as in [7-1], and we hope to return to this problem of number-theoretically understanding these inductive descendents elsewhere.

Starting from $f_0(n_0)$ in §4, and applying Theorems 2 and 3, we have, at the k-th step of induction $(k \geq 1)$, the Weyl sums of the form (A_k denoting a constant)

$$A_k \sum_{n_k} n_k^{-\frac{1}{2}k} \left(\sum_{m^{(k)}} e(*) \right) + O_{\alpha,k}(\sqrt{X}), \tag{7.1}$$

where, with the new end point X_k (resp. the length N_k), of the summation interval

$$X_k = \sqrt{(2\alpha_0)(2\alpha_1)\cdots(2\alpha_{k-1})}X_0,$$

$$N_k = \sqrt{(2\alpha_0)(2\alpha_1)\cdots(2\alpha_{k-1})}N_0 \qquad (k = 1, 2, \ldots),$$

the summation variable $n_k \in \mathbb{N}$ ranges over

$$X_k < n_k \leq X_k + N_k$$

and $m^{(k)}$ denotes the (vector) summation variable

$$m^{(k)} = (m_1^{(k)}, \ldots, m_{k-1}^{(k)}, m_k^{(k)})$$

each of whose component $m_j^{(k)} \in \mathbb{Z}$ ranges over the interval

$$-B_j^{(k)} n_k < m_j^{(k)} \leq B_j^{(k)} n_k \qquad (j = 1, \ldots, k-1)$$

$(B_j^{(k)}$ are determined inductively as in the remark that immediately follows Theorem 2), and $m_k^{(k)} = m_k$, where m_k is the local variable in x_k below, ranges over

$$-n_k < m_k^{(k)} \le n_k \qquad (m_0^{(0)} = 0),$$

and where we understand the associated data are implemented:

$$0 < \alpha_0 < \frac{1}{2}, \quad \frac{1}{\sqrt{2\alpha_k}} = a_k + \varepsilon_{k+1}\alpha_{k+1},$$

$$a_k \in \mathbb{N}, \quad \varepsilon_{k+1} = \pm 1, \quad 0 < \alpha_{k+1} < \frac{1}{2},$$

and

$$x_k = n_k^2 + m_k \qquad (m_k = m_k^{(k)}).$$

And, in case $N_k \gg 1$, the core function in $e(*)$ in (7.1) is given by

$$\frac{2}{3}(-1)^k \varepsilon_1 \cdots \varepsilon_k \alpha_k \left(\mathrm{tr} \frac{1 + \sqrt{-1}}{2} \left(x_k + \sqrt{-1}(K_k(\boldsymbol{m}^{(k)}) + \kappa_k) \right)^{\frac{3}{2}} \right.$$

$$+ \sum_{j=1}^{k} \sum_{h=1}^{4} H_j^{(k)} L(x_k + (L_{j,h}^{(k)}(\boldsymbol{m}^{(k)}) + \lambda_{j,h}^{(k)})$$

$$\left. + \sqrt{-1}(M_{j,h}^{(k)}(\boldsymbol{m}^{(k)}) + \mu_{j,h}^{(k)})) \right),$$

where $K_k, L_{j,h}^{(k)}$ and $M_{j,h}^{(k)}$ are linear forms in $\boldsymbol{m}^{(k)}$, and $\kappa_k, \lambda_{j,h}^{(k)}, \mu_{j,h}^{(k)}$ and $H_j^{(k)}$ are constants determined inductively.

If $N_k \ll 1$, then as in Theorem 1 of [7-1], we have to use a multiple integral version of the van der Corput Lemma, and some work is still to be done.

Since the study on the Hessian suggests that the sum in $\boldsymbol{m}^{(k)}$ for a fixed n_k is

$$n_k^{-\frac{1}{2}k} \sum_{\boldsymbol{m}^{(k)}} e(*) = O_{\alpha,k}(1),$$

we speculate that we may regard $\boldsymbol{m}^{(k)}$ as the local variable attached to n_k.

References

[1]   H. Fiedler, W. Jurkat and O. Körner, Asymptotic expansions of finite theta series, *Acta Arith.* **32** (1977), 129–146

[2]   S. W. Graham and G. Kolesnik, *Van der Corput's method of exponential sums*, London Mathematical Society Lecture Note Series 126, Cambridge Univ. Press, London, 1991

[3]   J.-I. Igusa, Lectures on forms of higher degree, Tata Institute of Fundamental Research Lectures on Mathematics and Physics 59, Tata Inst., Bombay, 1978

[4]   T. Kubota, On an analogy to the Poisson summation formula for generalised Fourier transformation, *J. Reine angew. Math.* **268/269** (1974), 180–189

[5]   Y. Kuribayashi, On pseudo-Fourier transform (in Japanese), Sûrikaisekikenkyusho Kôkyûroku **1039** (1998), 152–164

[6]   W. Maier, Transformation der kubischen Thetafunktionen, Math. Ann. **111** (1935), 183–196

[7-1]   Y.-N. Nakai, On a θ-Weyl sum, Nagoya Math. J., **52** (1973), 163–172. Errata, the same J., 60 (1976), 217.

[7-2]   Y.-N. Nakai, Weyl sums and the van der Corput method (in Japanese), Sûrikaisekikenkyusho Kôkyûroku **456** (1982), 2–18

[7-3]   Y.-N. Nakai, On Diophantine inequalities of real indefinite quadratic forms of additive type in four variables, *Advanced Studies in Pure Mathematics* **13**, (1988), Investigations in Number Theory, 25–170

[7-4]   Y.-N. Nakai, Cubic continued fraction expansions I and II (in Japanese), *Mem. Fac. Edu. Yamanashi Univ.* **41-2** (1990), 4-6 (Cubic I) and *ibid.* **44-2** (1993), **1-3** (Cubic II)

[7-5]   Y.-N. Nakai, A strengthening of the conjecture that the functions invariant under Fourier transformation are quadratic polynomials, (in Japanese), *Mem. Fac. Edu. Yamanashi Univ.* **43-2** (1992), 1–3

[7-6]   Y.-N. Nakai, A candidate for cubic theta-Weyl sums (I) (in Japanese), *Mem. Fac. Edu. Yamanashi Univ.* **49-2** (1998), 1–4

[7-7]   Y.-N. Nakai, A candidate for cubic theta-Weyl sums (in Japanese), Sûrikaisekikenkyusho Kôkyûroku **1091** (1999), 298–307

[8]   W. Raab, Kubische und biquadratische Thetafunktionen I und II, Sizungsber. *Österreich. Akad. Wiss. Mat-Natur. Kl.* **188** (1979), 47–77 and 231–246

[9]   T. Suzuki, Weil type representations and automorphic forms, *NagoyaMath. J.,* **77** (1980), 145–166

[10]   E. C. Titchmarsh, *The theory of the Riemann zeta-function*, Oxford Univ. Press, 1951; second ed., rev. by D.R. Heath-Brown, 1996

[11-1] R. C. Vaughan, Some remarks on Weyl sums, *Colloq. Math. Soc. Janos Bolyai,* **34** (II), Topics in Classical Number Theory, Budapest (1987), 1585–1602

[11-2] R. C. Vaughan, The Hardy-Littlewood method, second edition, Cambridge Tracts in Mathematics 125, Cambridge University Press, London 1997.

[12]   A. Walfisz, *Weylsche Exponentialsummen in der neueren Zahlentheorie*, VEB Deutscher Verlag der Wissenschaften, Berlin, 1963

Appendix: Weyl sums and the van der Corput method

Let a polynomial $F(x) = F(x_1, \ldots, x_s)$ in s variables with integer coefficients be given and suppose we wish to study the number $\mathcal{S}_n$ of lattice points $x \in \mathbb{Z}^s$ which lie in a subset $\mathcal{B} \subset \mathbb{R}^s$ and satisfy the equation $F(x) = n$ for a given integer n:

$$\mathcal{S}_n = \# \left\{ x \in \mathbb{Z}^s \cap \mathcal{B} \,\middle|\, F(x) = n \right\}.$$

Exponential sums, to be defined below, are efficient tools for treating this kind of problem mainly in the case where s is relatively large compared to $\deg F$ and $\mathcal{B}$ is of the form on an interval in $\mathbb{R}^s$.

Many of the problems that we commonly encounter reduce to the proof of non-vanishingness of $\mathcal{S}_n$ for each n and given F by choosing $\mathcal{B}$ suitably.

There are two methods known. One is to use Davenport's lemma (cf. [11-2], Chap. 6) or sieve method to show $\mathcal{S}_n \neq 0$ directly, while the other is to use, as a generating function, the power series

$$f(z) = \sum_{x \in \mathbb{Z}^s \cap \mathcal{B}} \rho_x z^{F(x)}$$

with ρ_x a suitable weight and $z \in \mathbb{C}$. If $f(z)$ is regular at the origin, then

$$J_n = \sum_{x \in \mathbb{Z}^s \cap \mathcal{B},\, F(x) = n} \rho_x$$

can be expressed as the Cauchy integral

$$\frac{1}{2\pi i} \int f(z) \frac{dz}{z^{n+1}}.$$

Extracting non-trivial information from this expression is a major problem, and the Hardy-Littlewood circle method is a commonly adopted procedure, which if successful, gives rise to singular series, local data. If Hasse's principle of pasting together these local data applies, then this will give global data.

The set $\mathcal{B}$ is customarily a large one like the set of all natural numbers, and so convergence problems occur. However, in practical applications, it often suffices to consider the case of $\mathcal{B}$ of the form of a finite interval, whence arises the notion of Weyl sums, a finite version of the second method. Namely, if $F(x)$ is a suitable (e.g. of class C^{s+1}) real-valued function in s variables and $\mathcal{B}(\subset \mathbb{R}^s)$ is of the form of an interval, we call the sum

$$\sum_{x \in \mathbb{Z}^s \cap \mathcal{B}} e^{2\pi i F(x)}$$

an *exponential sum*, and its special case where $F(x) \in \mathbb{R}[x_1, \ldots, x_s]$, a polynomial with real coefficients in s variables, a *Weyl sum.*

In analytic number theory, it is often needed to treat Weyl sums with weights, which are done by Abel summation, Vinogradov's method etc.

The forms of the core function $F(x)$ vary from an additive type in which it reduces to the sum of one variable terms to a many variables type whose variables are not separable. But probably what we are to aim at would be those Weyl sums which appear in Waring's problems, the one variable case being stated in §1, there is a mean value type one.

Mean value conjecture. Let $f(x)$ denote a polynomial of degree h with integer coefficients. Then for $\lambda > h$ and any $\varepsilon > 0$,

$$\int_0^1 \left| \sum_{X < n \leq X+N} e^{2\pi i \alpha f(n)} \right|^\lambda d\alpha \ll_{(h,\lambda,\varepsilon)} N^{\lambda - h + \varepsilon}.$$

At present some achievements are attained in the case $\lambda = 2l(l \in \mathbb{N})$ mainly by Vinogradov.

Regarding the van der Corput method, there are four types of devices known:

(i) various estimates based on mean value theorems for integrals

(ii) the so-called van der Corput lemmas

(iii) averaging process based on differencing

(iv) exponent pairs.

Of these we shall state only one typical lemma in category (ii) whose generalization has been our principle motive in this paper.

A van der Corput lemma. Suppose a real-valued function $f(x)$ of class C^3 on $[X, X']$ satisfy the conditions:
$f'(x)$ is monotone decreasing, $\lambda_2 \ll (-f'') \ll \lambda_2$, and $|f^{(3)}| \ll \lambda_3$ for positive constants λ_2, λ_3. Let

$$[Y, Y'] = f'([X, X'])$$

be the corresponding interval. For $y \in [Y, Y']$ define x_y by $f'(x_y) = y$ and put

$$g(y) = f(x_y) - y x_y.$$

Then

$$\sum_{X<x\leq X'} e^{2\pi i f(x)} = e^{-2\pi i \frac{1}{8}} \sum_{Y<y\leq Y'} |f'(x_y)|^{-\frac{1}{2}} e^{2\pi i g(y)}$$

$$+ O(\lambda_2^{-\frac{1}{2}}) + O(\log(2 + (X' - X)\lambda_2))$$

$$+ O\left((X' - X)(\lambda_2\lambda_3)^{\frac{1}{5}}\right).$$

This is Theorem 4.9 of [10] (cf. [2], Lemma 3.6).

SOME RESULTS IN VIEW OF NEVANLINNA THEORY

Junjiro Noguchi*

Graduate School of Mathematical Sciences, University of Tokyo, Komaba, Meguro, Tokyo 153-8914, Japan

noguchi@ms.u-tokyo.ac.jp

Abstract   Here we discuss and survey some results on rational points of algebraic varieties and Nevanlinna theory in relation to Lang's conjectures and Vojta's. We will also announce new results in the case of function fields and also the second main theorem for holomorphic curves into semi-abelian varieties.

Keywords: rational point, Lang's conjecture, Vojta's conjecture, abc-conjecture, Diophantine approximation, Kobayashi hyperbolicity, Nevanlinna theory

1.   Introduction

In this article we deal only with the case of characteristic 0. We recall two Lang's Conjectures:

L.C.1 ([L60], [L74]) Let V be an algebraic variety defined over a number field F. Assume that V with some embedding $F \hookrightarrow \mathbf{C}$ is Kobayashi hyperbolic. Then $|V(F)| < \infty$? Does the analogue over function fields hold, either?

L.C.2 ([L66]) Let $f : \mathbf{C} \to A$ be an analytic 1-parameter subgroup (or holomorphic curve) in an Abelian variety A, and let D be an ample divisor of A. Then $f(\mathbf{C}) \cap D \neq \emptyset$?

For L.C.1 a basic conjectural observation is the correspondence:

1.1.   *non-constant holomorphic*   *an infinite set of*
                                      $\Longleftrightarrow$
            curves $f : \mathbf{C} \to V$ *rational points on* V.

*Research supported in part by Grant-in-Aid for Scientific Research (A)(1), 13304009.

340

Theorem 1.2. (i) *L.C.1 over function fields holds in the case of* dim $V = 1$. (Manin [M63], Grauert [Gra65], Noguchi [No85].)

(ii) *L.C.1 over number fields holds in the case of* dim $V = 1$. (Faltings [Fa83].)

(iii) *L.C.1 over function fields holds in arbitrary* dim $V \geqq 1$. (Noguchi [No85], [No92]; cf. [No81a], too.)

(iv) *For a subvariety V of an Abelian variety L.C.1 holds. Moreover, this holds on semi-Abelian varieties in a generalized sense.* (Faltings [Fa91], Vojta [Vo96].)

L.C.1 over number fields is open for dim $V > 1$. As the Nevanlinna theory is a powerful tool to prove the hyperbolicity of a complex manifold, so is the Diophantine approximation to obtain the finiteness of the set of rational points of an algebraic variety. In this context, Vojta [Vo87] made observation 1.1 deeper to the analogue

1.3. $\qquad$ *Nevanlinna theory $\Longleftrightarrow$ Diophantine approximation.*

2. S.M.T. and integral points

After Vojta's analogue 1.3 Roth-Schmidt's approximation corresponds to the S.M.T. (second main theorem) in Nevanlinna-Cartan theory.

Theorem 2.1. ([Ca33]) *Let $f : \mathbf{C} \to \mathbf{P}^n(\mathbf{C})$ be a linearly non-degenerate holomorphic curve. Let $\{H_j\}_{j=1}^q$ be a finite family of hyperplanes in $\mathbf{P}^n(\mathbf{C})$ in general position. Then*

$$(q - n - 1)T_f(r) < \sum_{j=1}^{q} N_n(r, f^*H_j) + O(\log(rT_f(r))) \ \|.$$

Here $T_f(r)$ is the order function of f with respect the hyperplane bundle $O(1)_{\mathbf{P}^n(\mathbf{C})}$ over $\mathbf{P}^n(\mathbf{C})$, and $N_n(r, f^*H_j)$ the counting function of the divisor f^*H_j, truncated at level n (cf. [Ca33], [Fu93]). The truncation for the counting functions are very important and is related to "*abc*-Conjecture" in number theory. An implication of this theorem is

Corollary 2.2. (Borel's Theorem) *Let $f_1(z), f_2(z), \ldots, f_s(z)$ be entire functions which are units (zero free). Assume the following unit equation.*

$$f_1(z) + f_2(z) + \cdots + f_s(z) = 0.$$

Then there is a partition $\{1, \ldots, s\} = \bigcup I_\lambda$ of the index satisfying the following.

(i) $|I_\lambda| \geqq 2$.

(ii) *For arbitrary $i, j \in I_\lambda$, the function $\dfrac{f_i(z)}{f_j(z)} = c_{ij}$ is constant.*

(iii) *For every λ, $\sum_{i \in I_\lambda} f_i(z) = 0$.*

Theorem 2.3. (Roth-Schmidt) *Let F be a number field, and let S be a finite set of places including all infinite places of F. Let $\{H_j\}_{j=1}^q$ be a finite family of hyperplanes of $\mathbf{P}_F^n$ in general position. Then for an arbitrary $\epsilon > 0$ there is a finite union E of proper linear subspaces such that for $x \in \mathbf{P}^n(F) \setminus E$*

$$(q - n - 1 - \epsilon)\mathrm{Ht}(x) < \sum_{j=1}^q N(S; H_j(x)) + \mathrm{Const}.$$

An immediate consequence analogous to Corollary 2.2 is

Corollary 2.4. *Let $\mathcal{Z}$ be the set of all S-unit solutions of equation*

$$a_1 x_1 + \cdots + a_s x_s = 0 \qquad (s \geqq 2)$$

with $a_j \in F^*$. Then there is a finite decomposition $\mathcal{Z} = \cup_{\mu=1}^{\mu_0} \mathcal{Z}_\mu$ ($\mu_0 < \infty$) such that for every fixed $\mathcal{Z}_\mu$, $1 \leqq \mu \leqq \mu_0$, there is a decomposition of indices $\{1, \ldots, s\} = \bigcup_{l=1}^m I_l$ satisfying the following conditions:

(i) *$|I_l| \geqq 2$ for all l.*

(ii) *If we write $\mathcal{Z}_\mu = \{(x_i(\zeta)); \zeta \in \mathcal{Z}_\mu\}$ and take an arbitrarily fixed I_l, then*

$$\frac{x_j(\zeta)}{x_k(\zeta)} = c_{jk} \in \mathcal{O}_S^*$$

are independent of $\zeta \in \mathcal{Z}_\mu$ for all $j, k \in I_l$.

(iii) *$\sum_{j \in I_l} a_j x_j(\zeta) = 0$ for $\zeta \in \mathcal{Z}_\mu$ and $l = 1, 2, \ldots, m$.*

The ways of the arguments to deduce Corollaries 2.2 and 2.4 from Theorems 2.1 and 2.3 are almost identical by making use of the induction on s ([No97]).

In general, one uses the notation α^+ to denote the positive part $\max\{\alpha, 0\}$ of $\alpha \in \mathbf{R}$.

Theorem 2.5. (M. Ru and P.-M. Wong [RW91]) *Let $\{H_j\}_{j=1}^q$ be hyperplanes in $\mathbf{P}_F^n$ in general position. Let A be a $(\sum_{j=1}^q H_j, S)$-integral point set. Then A is contained in a finite union W of linear subspaces such that*

$$\dim W \leqq (2n + 1 - q)^+.$$

In special, A is finite for $q \geqq 2n + 1$.

In their proof, Nochka's weight was essential. But, in the case of Nevanlinna theory Fujimoto [Fu72] and Green [Gre72] independently obtained an optimal dimension estimate

$$\dim \leqq [n/(q - n)], \qquad q > n,$$

where [∗] stands for Gauss' symbol. In fact, we have the same dimension estimate as above for A in Theorem 2.5 in more general context as follows.

Theorem 2.6. ([NW02]) *Let V be an n-dimensional projective algebraic variety defined over F. Let $\{D_j\}_{j=1}^q$ be a family of effective divisors on V_F in general position.*

(i) *Assume that all D_i are ample and that $q > n(\operatorname{rank}_{\mathbf{Z}} NS(V) + 1)$. Then any $(\sum_{i=1}^l D_i, S)$-integral point set of $V(K)$ is finite.*

(ii) *Let $X \subset \mathbf{P}_F^m$ be an irreducible subvariety, and let $D_j, 1 \leqq j \leqq q$, be distinct hypersurface cuts of X that are in general position as hypersurfaces of X. If $q > 2 \dim X$, then any $(\sum_{j=1}^q D_j, S)$-integral point set of $X(K)$ is finite.*

(iii) *Let $D_j, 1 \leqq j \leqq q$, be ample divisors of V in general position. Let A be a subset of $V(K)$ such that for every D_j, either $A \subset D_j$, or A is a $(\sum_{D_j \not\supset A} D_j, S)$-integral point set. Assume that $q > n$. Then A is contained in an algebraic subvariety W of V such that*

$$\dim W \leqq \left[\frac{n}{q - n} \operatorname{rank}_{\mathbf{Z}} NS(V) \right].$$

In the special case of $V = \mathbf{P}_K^n$ we have

$$\dim W \leqq \left[\frac{n}{q - n} \right].$$

In the proof we use Vojta's result ([Vo96]), which in the Nevanlinna theory is known as log-Bloch-Ochiai's Theorem ([No77], [No81b]).

3.   S.M.T. over function fields

As we saw the force of the estimate of S.M.T., it is interesting to deal with it over function fields. Actually a number of people have obtained such estimates (Mason, Voloch, Brownawell-Masser, J. Wang, Á. Pintér, Noguchi ...). We recall

Theorem 3.1. ([No97]) *Let* $x = [x_0, \ldots, x_n] : R \to \mathbf{P}^n(\mathbf{C})$ *be a morphism from a smooth projective variety (or compact Kähler manifold) R with a Kähler form ω. Let $H_1, \ldots, H_q$ ($q \geqq n+1$) be linear forms on $\mathbf{P}^n(\mathbf{C})$ in general position such that the divisor $(H_j(x))$ is defined for every j. Let r denote the rank of dx at general point, and let l denote the dimension of the smallest linear subspace of $\mathbf{P}^n(\mathbf{C})$ containing $x(R)$. Then*

$$(q - 2n + l - 1)\mathrm{ht}(x; \omega) \leqq \sum_{i=1}^{q} N_{l-r+1}((H_i(x)); \omega)$$

$$+ \left\{ \frac{(l-r+1)(l-r+2)}{2} + r - 1 \right\} \frac{2n - l + 1}{l+1} N(J; \omega).$$

If $\dim R = 1$, $N(J; \omega) = 2g - 2$, *where g denotes the genus of R.*

Here, we set

$$\mathrm{ht}(x; \omega) = \int_R x^* c_1(O(1)_{\mathbf{P}^n(\mathbf{C})}) \wedge \omega^{\dim R - 1},$$

$$N_{l-r+1}((H_i(x)); \omega) = \int_{a \in \{H_j(x) = 0\}} \min\{\mathrm{ord}_a(H_j(x)), l - r + 1\} \, \omega^{\dim R - 1}.$$

J. Wang dealt with the case where the coefficients of $H_i, 1 \leqq i \leqq q$, are not constants, but elements of the function field over R when $\dim R = 1$ (see [W96], [W00]). The case of non-constant coefficients is of interest from the viewpoint of the Diophantine approximation; the variables should belong to the same field as the coefficients'. In the estimate we still need to make effective and clear the following points:

(i) In the proof of her result there was a part to chase the lower-limit

$$\liminf_{r \to \infty} \frac{d(r+1)}{d(r)} = 1,$$

where $d(r)$ denotes the dimension of the vector space generated by the r-th products of coefficients of H_i. This r is involved in the coefficients and the constant terms of the approximation.

(ii) The level of the truncation of counting functions was not dealt with.

If the counting functions are not truncated, then the first and second main theorems coincide with Poincaré's duality, or a special case of complete intersection theory over compact varieties. The truncation in counting functions is as essentially important as in the case of "abc-Conjecture". We will show some S.M.T. where the truncation level is a bit complicated, but computable.

Theorem 3.2. *Let* $\dim R = 1$ *and let the genus of* R *be* g. *Let* L *be a line bundle on* R *whose degree is* $\deg L$. *Let* $H_j, 1 \leq j \leq q$, *be linear forms in general position on* $\mathbf{P}^n$ *with coefficients* $a_{ji} \in H^0(R, L)$. *For an arbitrarily given* $0 < \epsilon < 1$, *let*

$$p_\epsilon = \max \left\{ \left[\frac{2n}{\epsilon} \right] + g, 2g - 2 \right\} + 1.$$

Then for an arbitrary morphism $x : R \to \mathbf{P}^n$, *we have*

$$(q - 2n - \epsilon)\mathrm{ht}(x) \leq \sum_{j=1}^{q} N_{(n+1)h(p_\epsilon+1)-1}(H_j(x)) + C(\epsilon, q),$$

where $h(p_\epsilon + 1) = (p_\epsilon + 1) \deg L - g + 1$ *and*

$$C(\epsilon, q) = q \binom{q}{n+1}(n+1)(p_\epsilon+1) \deg L + 2n((n+1)h(p_\epsilon+1)-1)(2g-2)^+.$$

The proof of Theorem 3.2 is based on Theorem 3.1 and the methods of Steinmetz [St85]-Shirosaki [Sh91] and Wang [W00], combined with the Riemann-Roch: For $\deg L^p > 2g - 2$, where L^p is the pth symmetric tensor power of L,

$$h^0(L^p) = p \deg L - g + 1.$$

Let $\alpha_i, 1 \leq i \leq h(p)$, be bases of $H^0(R, L^p)$. Then $\alpha_i H_j$ are linear forms with coefficients in $H^0(R, L^{p+1})$ of which dimension is $h(p + 1)$. Then one may apply the arguments in Shirosaki [Sh91] and Wang [W00] to deduce Theorem 3.2 [*)].

The next immediately follows from Theorem 3.2

Corollary 3.3. *Let the notation be as in Theorem 3.2. Let* $S \subset R$ *be a finite set, and* $q \geq 2n + 1$. *If* x *is a* $(\sum H_j, S)$-*integral point, i.e., as a mapping* $x^{-1}(\sum H_j) \subset S$, *then*

$$(1 - \epsilon)\mathrm{ht}(x) \leq q((n + 1)h(p_\epsilon + 1) - 1)|S| + C(\epsilon, q), \qquad 0 < \epsilon < 1.$$

For the case of $\dim R \geq 2$, we do not know $h^0(L^p)$ so explicitly, but $h^0(L^p)$ is known to be a polynomial of degree at most $\dim R$ for large p.

[*)] The full proof will appear elsewhere.

Assuming this polynomial, we can work out the above obtained result for R of dim $R \geq 2$. For instance, let dim $R = 2$. Then, by R.-R.

$$h^0(L^p) - h^1(L^p) + h^2(L^p) = \frac{L \cdot L}{2} p^2 - \frac{L \cdot K_R}{2} p + \chi(\mathcal{O}_R).$$

Assume that L is ample. By the vanishing theorem we see that for all large p (geometrically effective), $h^1(L^p) = h^2(L^p) = 0$, so that

$$h^0(L^p) = \frac{L \cdot L}{2} p^2 - \frac{L \cdot K_R}{2} p + \chi(\mathcal{O}_R).$$

Then for a given $\epsilon > 0$ one can determine effectively p; in all dimensions,

$$p \gtrsim 1/\epsilon.$$

Remark. We explain the meaning of the truncation estimate in relation with abc-Conjecture:

$$(1 - \epsilon) \log \max\{|a|, |b|, |c|\} \leq \left(\sum_{\text{prime } p} \min\{1, \text{ord}_p a\} \log p \right.$$

$$+ \sum_{\text{prime } p} \min\{1, \text{ord}_p b\} \log p + \sum_{\text{prime } p} \min\{1, \text{ord}_p c\} \log p \left. \right) + C_\epsilon,$$

where a, b, c are mutually prime integers and $a + b + c = 0$. Here the truncation level is 1. One may relax the claim so that for some $k > 0$

$$(1 - \epsilon) \log \max\{|a|, |b|, |c|\} \leq \left(\sum_{\text{prime } p} \min\{k, \text{ord}_p a\} \log p \right.$$

$$+ \sum_{\text{prime } p} \min\{k, \text{ord}_p b\} \log p + \sum_{\text{prime } p} \min\{k, \text{ord}_p c\} \log p \left. \right) + C_\epsilon.$$

Furthermore, one may allow that k depends on ϵ. Unfortunately, the truncation level of N_* in Theorem 3.2 is depending on ϵ, and it is an open problem if it is taken independently of ϵ.

Taking the analogue to Theorem 3.2 one may pose

Little abc-Conjecture. *For an arbitrary $\epsilon > 0$, there exist positive constants k_ϵ and C_ϵ such that for mutually prime integers a, b, c satisfy-*

ing $a + b + c = 0$

$$(1 - \epsilon) \log \max\{|a|, |b|, |c|\} \leq \left(\sum_{\text{prime } p} \min\{k_\epsilon, \text{ord}_p a\} \log p \right.$$

$$\left. + \sum_{\text{prime } p} \min\{k_\epsilon, \text{ord}_p b\} \log p + \sum_{\text{prime } p} \min\{k_\epsilon, \text{ord}_p c\} \log p \right) + C_\epsilon.$$

4. L.C.2

Answering a question raised by S. Lang [L66], Ax [Ax72] proved

Theorem 4.1. *Let* $f : \mathbf{C} \to A$ *be a non-trivial analytic 1-parameter subgroup of an Abelian variety* A, *and* D *be an ample divisor on* A. *Then*

$$N(r, f^*D) \sim r^2, \qquad r \to \infty.$$

Then Griffiths [Gri74] generalized the question for holomorphic curves which are not necessarily subgroups; then this is a problem of the Nevanlinna theory. Siu-Yeung [SY96] and [No98] proved that non-constant holomorphic f always intersects D.

To deal with more general case we introduce a notion of a semi-torus. Let M be a complex Lie group admitting the exact sequence

$$0 \to (\mathbf{C}^*)^p \to M \xrightarrow{\eta} M_0 \to 0, \tag{4.2}$$

where $\mathbf{C}^*$ is the multiplicative group of non-zero complex numbers, and M_0 is a (compact) complex torus. Such M is called a *complex semi-torus* or a *quasi-torus*. If M_0 is algebraic, that is, an Abelian variety, M is called a *semi-Abelian variety* or a *quasi-Abelian variety*. By making use of the compactification $(\mathbf{C}^*)^p \subset (\mathbf{P}^1(\mathbf{C}))^p$, we get a compactification $\bar{M}$ of M.

Lately we proved

Theorem 4.3. ([NWY00], [NWY02]) *Let* $f : \mathbf{C} \to M$ *be a holomorphic curve into a complex semi-torus* M *such that the image* $f(\mathbf{C})$ *is Zariski-dense in* $\bar{M}$. *Let* D *be an effective divisor on* M *such that the closure* $\bar{D}$ *of* D *in* $\bar{M}$ *is an effective divisor on* $\bar{M}$. *Assume that* D *satisfies a certain boundary condition. Then we have the following.*

(i) *Suppose that* f *is of finite order* ρ_f. *Then there is a positive integer* $k_0 = k_0(\rho_f, D)$ *depending only on* ρ_f *and* D *such that*

$$T_f(r; c_1(\bar{D})) = N_{k_0}(r; f^*D) + O(\log r).$$

(ii) *Suppose that f is of infinite order. Then there is a positive integer $k_0 = k_0(f, D)$ depending on f and D such that*

$$T_f(r; c_1(\bar{D})) = N_{k_0}(r; f^*D) + O(\log(rT_f(r; c_1(\bar{D})))) \; \|_E.$$

Cf. [Kr00], [Mc96] and [SY97] for related results. The following very precise estimate is an immediate consequence of Theorem 4.3 (cf. Theorem 4.1):

Theorem 4.4. *Let $f : \mathbf{C} \to A$ be a 1-parameter analytic subgroup in an Abelian variety A with $a = f'(0) \neq 0$. Let D be an effective divisor on A with the Riemann form $H(\cdot, \cdot)$ such that $D \not\supset f(\mathbf{C})$. Then we have*

$$N(r; f^*D) = H(a, a)\pi r^2 + O(\log r).$$

Since $H(a, a) = \lim_{r \to \infty} N(r; f^*D)/\pi r^2$, the Riemann form H may be recovered by the counting functions $N(r; f^*D)$ for 1-parameter analytic subgroups.

Yamanoi [Ya01] proved very lately that

Theorem 4.5. *Let the notation be as above. For arbitrary $\epsilon > 0$*

$$N(r, f^*D) - N_1(r, f^*D) < \epsilon T_f(r, c_1(D)) \; \|_\epsilon.$$

Hence we have

$$(1 - \epsilon)T_f(r, c_1(D)) < N_1(r, f^*D) \; \|_\epsilon.$$

Remark 4.6. In the above estimate the term $\epsilon T_f(r, c_1(D))$ cannot be replaced by

$$O(\log(rT_f(r; c_1(\bar{D})))) \; \|_E.$$

In fact, let $E = \mathbf{C}/(\mathbf{Z} + i\mathbf{Z})$ be an elliptic curve, and let D be an irreducible divisor on E^2 with cusp of order l at $0 \in E^2$. Let $f : z \in \mathbf{C} \to (z, z^2) \in E^2$. Then $f(\mathbf{C})$ is Zariski-dense in E^2, and

$$T_f(r; L(D)) \sim r^4(1 + o(1)).$$

Note that $f^{-1}(0) = \mathbf{Z} + i\mathbf{Z}$ and $f^*D \geq l(\mathbf{Z} + i\mathbf{Z})$. For an arbitrary fixed k_0, we take $l > k_0$, and then have

$$N(r; f^*D) - N_{k_0}(r; f^*D) \geqq (l - k_0)r^2(1 + o(1)).$$

The above left-hand side cannot be bounded by $O(\log r)$. This gives also a counter-example to [Kr91], Lemma 4.

348

References

[Ax72] J. Ax, Some topics in differential algebraic geometry II, *Amer. J. Math.* **94** (1972), 1205–1213.

[Bl26] A. Bloch, Sur les systèmes de fonctions uniformes satisfaisant à l'équation d'une variété algébrique dont l'irrégularité dépasse la dimension, *J. Math. Pures Appl.* **5** (1926), 19–66.

[Ca33] H. Cartan, Sur les zéros des combinaisons linéaires de p fonctions holomorphes
données, *Mathematica* **7** (1933), 5–31.

[Fa83] G. Faltings, Endlichkeitssätze für abelsche Varietäten über Zahlkörpern, *Invent. Math.* **73** (1983), 349–366.

[Fa91] G. Faltings, Diophantine approximation on abelian varieties, *Ann. Math.* **133** (1991), 549–576.

[Fu72] H. Fujimoto, Extension of the big Picard's theorem, *Tohoku Math. J.* **24** (1972), 415–422.

[Fu93] H. Fujimoto, Value Distribution Theory of the Gauss Map of Minimal Surfaces in $\mathbf{R}^m$, *Aspects of Math.* **E21**, 1993.

[Gra65] H. Grauert, Mordells Vermutung über rationale Punkte auf Algebraischen Kurven und Funktionenköper, *Publ. Math. I.H.E.S.* **25** (1965), 131–149.

[Gre72] M. Green, Holomorphic maps into complex projective space, *Trans. Amer. Math. Soc.* **169** (1972), 89–103.

[Gri72] P. Griffiths, Holomorphic mappings: Survey of some results and discussion of open problems, *Bull. Amer. Math. Soc.* **78** (1972), 374–382.

[Kr91] R. Kobayashi, Holomorphic curves into algebraic subvarieties of an abelian variety, *Intern'l J. Math.* **2** (1991), 711–724.

[Kr00] R. Kobayashi, Holomorphic curves in Abelian varieties: The second main theorem and applications, *Japan. J. Math.* **26** (2000), 129–152.

[Ks70] S. Kobayashi, Hyperbolic Manifolds and Holomorphic Mappings, Marcel Dekker, New York, 1970.

[Ks98] S. Kobayashi, Hyperbolic Complex Spaces, Grundlehren der mathematischen Wissenschaften **318**, Springer-Verlag, Berlin-Heidelberg, 1998.

[L60] S. Lang, Integral points on curves, Publ. Math. I.H.E.S., Paris, 1960.

[L66] S. Lang *Introduction to Transcendental Numbers*, Addison-Wesley, Reading, 1966.

[L74] S. Lang, Higher dimensional Diophantine problems, *Bull. Amer. Math. Soc.* **80** (1974), 779–787.

[L83] S. Lang, *Fundamentals of Diophantine Geometry*, Springer-Verlag, New York-Berlin-Heidelberg-Tokyo, 1983

[L87] S. Lang, *Introduction to Complex Hyperbolic Spaces*, Springer-Verlag, New York-Berlin-Heidelberg, 1987.

[L91] S. Lang, *Number Theory III*, Encycl. Math. Sci. vol. **60**, Springer-Verlag, Berlin-Heidelberg-New York-London-Paris-Tokyo-Hong Kong-Barcelona, 1991.

[M63] Ju. Manin, Rational points of algebraic curves over function fields, *Izv. Akad. Nauk. SSSR. Ser. Mat.* **27** (1963), 1395–1440.

[Mc96] McQuillan, M., A dynamical counterpart to Faltings' "Diophantine approximation on Abelian varieties", I.H.E.S. preprint, 1996.

[No77] J. Noguchi, Holomorphic curves in algebraic varieties, Hiroshima Math. J. **7** (1977), 833-853.

[No81a] J. Noguchi, A higher dimensional analogue of Mordell's conjecture over function fields, Math. Ann. **258** (1981), 207-212.

[No81b] J. Noguchi, Lemma on logarithmic derivatives and holomorphic curves in algebraic varieties, Nagoya Math. J. **83** (1981), 213-233.

[No85] J. Noguchi, Hyperbolic fibre spaces and Mordell's conjecture over function fields, Publ. RIMS, Kyoto University **21** (1985), 27-46.

[No88] J. Noguchi, Moduli spaces of holomorphic mappings into hyperbolically imbedded complex spaces and locally symmetric spaces, Invent. Math. **93** (1988), 15-34.

[No91] J. Noguchi, Hyperbolic Manifolds and Diophantine Geometry, Sugaku Exposition Vol. **4** pp. 63-81, Amer. Math. Soc., Rhode Island, 1991.

[No92] J. Noguchi, Meromorphic mappings into compact hyperbolic complex spaces and geometric Diophantine problems, International J. Math. **3** (1992), 277-289.

[No96] J. Noguchi, On Nevanlinna's second main theorem, Geometric Complex Analysis, Proc. the Third International Research Institute, Math. Soc. Japan, Hayama, 1995, pp. 489-503, World Scientific, Singapore, 1996.

[No97] J. Noguchi, Nevanlinna-Cartan theory over function fields and a Diophantine equation, J. reine angew. Math. **487** (1997), 61-83.

[No98] J. Noguchi, On holomorphic curves in semi-Abelian varieties, Math. Z. **228** (1998), 713-721.

[NO$\frac{84}{90}$] J. Noguchi and T. Ochiai, Geometric Function Theory in Several Complex Variables, Japanese edition, Iwanami, Tokyo, 1984; English Translation, Transl. Math. Mono. **80**, Amer. Math. Soc., Providence, Rhode Island, 1990.

350

[NW02] J. Noguchi and J. Winkelmann, Holomorphic curves and integral points off divisors, Math. Z. **239** (2002), 593-610.

[NWY00] J. Noguchi, J. Winkelmann and K. Yamanoi, The value distribution of holomorphic curves into semi-Abelian varieties, C.R. Acad. Sci. Paris t. **331** (2000), Serié I, 235-240.

[NWY02] J. Noguchi, J. Winkelmann and K. Yamanoi, The second main theorem for holomorphic curves into semi-Abelian varieties, Acta Math. **188** no. 1 (2002), 129-161.

[O77] T. Ochiai, On holomorphic curves in algebraic varieties with ample irregularity, Invent. Math. **43** (1977), 83-96.

[R93] M. Ru, Integral points and the hyperbolicity of the complement of hypersurfaces, J. reine angew. Math. **442** (1993), 163-176.

[RW91] M. Ru and P.-M. Wong, Integral points of $\mathbf{P}^n - \{2n+1$ hyperplanes in general position$\}$, Invent. Math. **106** (1991), 195-216.

[Sh91] M. Shirosaki, Another proof of the defect relation for moving targets, Tôhoku Math. J. **43** (1991), 355-360.

[SY96] Y.-T. Siu and S.-K. Yeung, A generalized Bloch's theorem and the hyperbolicity of the complement of an ample divisor in an Abelian variety, Math. Ann. **306** (1996), 743-758.

[SiY97] Siu, Y.-T. and Yeung, S.-K., Defects for ample divisors of Abelian varieties, Schwarz lemma, and hyperbolic hypersurfaces of low degrees, Amer. J. Math. **119** (1997), 1139-1172.

[St85] N. Steinmetz, Eine Verallgemeinerung des zweiten Nevanlinnaschen Hauptsatzes, J. für Math. **368** (1985), 134-141.

[Vo87] P. Vojta, Diophantine Approximations and Value Distribution Theory, Lecture Notes in Math. vol. **1239**, Springer, Berlin-Heidelberg-New York, 1987.

[Vo96] P. Vojta, Integral points on subvarieties of semiabelian varieties, I, Invent. Math. **126** (1996), 133-181.

[W96] J. T.-Y. Wang, An effective Roth's theorem of function fields, Rocky Mt. J. Math. **26**, 1225-1234.

[W96] J. T.-Y. Wang, S-integral points of $\mathbf{P}^n - \{2n + 1$ hyperplanes in general position$\}$ over number fields and function fields, Trans. Amer. Math. Soc. **348** (1996), 3379-3389.

[W00] J. T.-Y. Wang, Cartan's conjecture with moving targets of same growth and effective Wirsing's theorem over function fields, Math. Z. **234** (2000), 739-754.

[Ya01] K. Yamanoi, Holomorphic curves in Abelian varieties and intersections with higher codimensional subvarieties, preprint, 2001.

A HISTORICAL COMMENT ABOUT THE GVT IN SHORT INTERVAL

Pan Chengbiao

Peking University and China Agricultural University

pancb@pku.edu.cn

Abstract   In this article, the author introduces the history, progress and method in the Goldbach-Vinogradov Theorem in short interval by which every sufficiently large odd integer could be expressed as the sum of three almost equal prime numbers.

Keywords: Goldbach problem, circle method, estimate for exponential sums.

It is well known that in 1937, I. M. Vinogradov proved the ternary Goldbach conjecture for large odd integer, that is, a large odd integer N is a sum of three primes. This result is usually called the **Goldbach-Vinogradov Theorem (GVT)** or **Three Primes Theorem (TPT)**.

Goldbach-Vinogradov Theorem (GVT). *Let N be an odd integer, $T(N)$ denote the number of solutions of the Diophantine equation with prime variables*

$$N = p_1 + p_2 + p_3.$$

Then we have

$$T(N) = \frac{1}{2}\mathcal{S}_3(N) \cdot \frac{N^2}{\log^3 N} + O(\frac{N^2}{\log^4 N}),$$

where

$$\mathcal{S}_3(N) = \prod_{p|N}\left(1 - \frac{1}{(p-1)^2}\right) \prod_{p \nmid N}\left(1 + \frac{1}{(p-1)^3}\right) > \frac{1}{2}, \qquad 2 \nmid N. \qquad (1)$$

A natural and important generalization of **GVT** is to discuss the following problem: Is every large odd integer N can be represented as **a sum of three almost equal primes?** It is known as **the ternary**

Goldbach conjecture in short interval, that is, we expect to prove the following theorem.

Theorem 1. *Let N be a large odd integer. There exists $1 \leq U = U(N) = o(N)$ such that the Diophantine equation with prime variables*

$$N = p_1 + p_2 + p_3, \qquad \frac{N}{3} - U < p_1, p_2, p_3 \leq \frac{N}{3} + U, \qquad (2)$$

is solvable for every large odd integer N.

The aim is **to find U as small as possible.** This theorem is usually called **the Goldbach-Vinogradov Theorem in short interval** or **Three Primes Theorem in short interval.**

A.   The approach by circle method

In 1951, Haselgrove[1] claimed that the equation (2) is solvable if $U = N^{63/64+\varepsilon}$, but no proof has been published. In 1959, using circle method, Vinogradov's method for estimating the linear exponential sum with prime variables in short intervals and the theory of L-functions, my elder brother Pan Chengdong[1] proved that if there is

$$\zeta(\frac{1}{2} + it) \ll |t|^{c+\varepsilon}, \qquad (3)$$

then the equation (2) is solvable for $U = N^{(5+12c)/(6+12c)+\varepsilon}$. Moreover, the number of solutions of the equation (2)

$$T(N, U) = 3S_3(N) \cdot \frac{U^2}{\log^3 N} + O(\frac{U^2}{\log^4 N}). \qquad (4)$$

In 1949, Min Sihe proved that the constant c in (3) is not greater than $\frac{15}{92}$, and so

$$U = N^{\frac{160}{183}+\varepsilon}. \qquad (5)$$

And then in 1965, following Pan's method, Chen Jingrun[1] improved it to

$$U = N^{\frac{2}{3}+\varepsilon} \qquad (6)$$

and it is the possible best result obtained by Vinogradov's method.

In 1978, my brother and I proceeded to writing the book **Goldbach Conjecture**, and we intended to include this theorem in the book. However, when I read my brother's proof in detail, I found a trivial but serious mistake in his proof, and in Chen's proof there was the same mistake too. Although the method proposed in Pan[1] was often successfully used to discuss other problems in short interval, but no one clearly pointed out the mistake by that time. Now let me explain it in detail.

Obviously, we have

$$T(N, U) = \int_{-\frac{1}{Y}}^{1-\frac{1}{Y}} S_0^3(\alpha; \frac{N}{3}, 2U)e(-N\alpha)d\alpha, \tag{7}$$

where Y is any real number, and

$$S_0(\alpha; x, A) = \sum_{x-A<p\leq x} e(p\alpha). \tag{8}$$

This leads us to study **the linear exponential sum with prime variable in short interval $S_0(\alpha; x, A)$. The idea of circle method** can be described as follows: $S_0(\alpha; x, A)$ **is conjectured to be comparatively 'small' unless α is 'near' to an irreducible fraction with a 'small' denominator.** This suggests that we should **divide the interval $[-1/Y, 1 - 1/Y]$ into two parts E_1 and E_2: E_1 being the set of numbers 'near' to irreducible fractions with 'small' denominators, and E_2 being rest of the interval.** And then the integral in (7) will be divided into two parts correspondingly. Furthermore, the integral on E_1 is conjectured to be the main term of it, and the integral on E_2 its remainder term. In order to divide the interval $[-1/Y, 1 - 1/Y]$ into such two parts E_1 and E_2, we have to use the so-called **Farey dissection.** Let $1 \leq Q \leq Y$, $1 \leq \tau$ be parameters chosen suitably. Let

$$\alpha = \frac{a}{q} + \lambda, \quad (a, q) = 1. \tag{9}$$

If $\tau \geq 2Q^2$, the small intervals

$$I(\frac{a}{q}, \frac{1}{\tau}) = [\frac{a}{q} - \frac{1}{\tau}, \frac{a}{q} + \frac{1}{\tau}], \ (a, q) = 1, \ 0 \leq a < q, \ 1 \leq q \leq Q, \tag{10}$$

are non-overlapping. Let

$$E_1 = E(Q, \tau) = \bigcup_{1\leq q\leq Q} \bigcup_{0\leq a<q, (a,q)=1} [\frac{a}{q} - \frac{1}{\tau}, \frac{a}{q} + \frac{1}{\tau}], \tag{11}$$

and

$$E_2 = [-\frac{1}{Y}, 1 - \frac{1}{Y}]\backslash E_1. \tag{12}$$

We usually call E_1 major arcs and E_2 minor arcs. By Dirichlet's lemma (Lemma 1), if $\alpha \in E_2$, there must be a and q, $(a, q) = 1$, such that λ in (9) satisfies

$$E_{21} : 1 \leq q \leq Q, \ \frac{1}{\tau} < |\lambda| \leq \frac{1}{qY}, \tag{13}$$

or

$$E_{22} : Q < q \leq Y, \ |\lambda| \leq \frac{1}{qY}. \tag{14}$$

If $Y \leq \tau$, E_{21} **disappears. Ordinarily, we take** $Y = \tau$. **And if** $Y = Q$, E_{22} **disappears, but as far as we known this kind of choice is never used in circle method before 1987 (see Pan[2, 3]). Thus we have**

$$T(N,U) = \int_{E_1} S_0^3(\alpha; \frac{N}{3}, 2U)e(-N\alpha)d\alpha + \int_{E_2} S_0^3(\alpha; \frac{N}{3}, 2U)e(-N\alpha)d\alpha$$

$$= T_1(N,U) + T_2(N,U). \tag{15}$$

The circle method is intended to calculate the integral $T_1(N,U)$, and to prove it to be the main term of $T(N,U)$, that is, $T_2(N,U)$ to be its remainder term.

It is more convenient to treat the linear exponential sum with prime variable in short intervals

$$S(\alpha; x, A) = \sum_{x-A<n\leq x} \Lambda(n)e(n\alpha) \tag{16}$$

instead of $S_0(\alpha; x, A)$, where $\Lambda(n)$ is Mangoldt function, since

$$S_0(\alpha; x, A) = (\log x)^{-1}S(\alpha; x, A) + O(x^{-1}A^2(\log x)^{-1}), \tag{17}$$

and $S(\alpha; x, A)$ can be treated easily by using analytic method, that is, the theory of L-functions. It is easy to see that

$$S(\frac{a}{q} + \lambda; x, A) = \phi^{-1}(q) \sum_{\chi \bmod q} G(a, \overline{\chi})S(\lambda, \chi) + O(\log^2 x), \tag{18}$$

where $\phi(q)$ is Euler function, Gauss sum

$$G(a, \chi) = \sum_{l=1}^{q} \chi(l)e(\frac{al}{q}), \qquad \chi \bmod q, \tag{19}$$

$$S(\lambda, \chi) = \sum_{x-A<n\leq x} \chi(n)\Lambda(n)e(n\alpha) = \int_{x-A}^{x} e(\lambda u)d\psi(u, \chi), \tag{20}$$

and

$$\psi(u, \chi) = \sum_{n\leq u} \chi(n)\Lambda(n). \tag{21}$$

From these formulas above, by Lemma 3 we have

$$S(\lambda, \chi) = E(\chi) \int_{x-A}^{x} e(\lambda u) du - \sum_{|\gamma| \leq T} \int_{x-A}^{x} e(\lambda u) u^{\rho-1} du$$

$$+ \int_{x-A}^{x} e(\lambda u) dR_1(u, T, \chi), \tag{22}$$

and by Lemma 9

$$S(\lambda, \chi) = \frac{1}{2\pi i} \int_{x-A}^{x} e(\lambda u) du \int_{b-iT}^{b+iT} -\frac{L'}{L}(s, \chi) u^{s-1} ds$$

$$+ \frac{1}{2\pi i} \int_{x-A}^{x} e(\lambda u) R_2(u, T, \chi) du. \tag{23}$$

This shows that we can get **explicit formulas** of $S(a/q + \lambda; x, A)$, which give us close relations between Dirichlet L-functions and $S(a/q + \lambda; x, A)$.

And so, there are **three approaches** to treat the linear exponential sums with prime variable in short intervals $S(\alpha; x, A)$ and $S_0(\alpha; x, A)$:

(i) the Vinogradov method (Lemma 2);
(ii) the zero-density method (when (22) is used);
(iii) the complex integral method (when (23) is used).

The lemmas which are needed in these methods are listed in Section C. Obviously, if **zero-density method** is applied, we have to estimate

$$I(\lambda, q) = q^{\frac{1}{2}} \phi^{-1}(q) \sum_{\chi \bmod q} \left| \sum_{|\gamma| \leq T} \int_{x-A}^{x} u^{\rho-1} e(\lambda u) du \right|, \tag{24}$$

and if complex integral method is applied, we have to estimate

$$J(\lambda, q) = q^{\frac{1}{2}} \phi^{-1}(q) \sum_{\chi \bmod q} \left| \int_{b-iT}^{b+iT} -\frac{L'}{L}(s, \chi) ds \int_{x-A}^{x} e(\lambda u) u^{s-1} du \right|. \tag{25}$$

Now let me point out what is the mistake in Pan[1]. In his paper, he taken

$$A = x^{\frac{5+12c}{6+12c} + \varepsilon}, \tag{26}$$

and

$$Q = \log^{15} x, \quad \tau = A^{1 - \frac{(\nu - (1-\nu)\varepsilon)}{(3-\nu)}}, \quad Y = \tau, \tag{27}$$

where c is the same constant in (3). By Farey dissection according to (27), we get the major arcs E_1 and minor arcs E_2 defined in (11) and (12) respectively, and E_{21} disappear.

To estimate $S(\alpha; x, A)$ **on** E_{22}, taking $Q_2 = \exp((\log \log x)^3)$, E_{22} is divided into two parts:

$$E_{221} : Q < q \leq Q_2, \ |\lambda| \leq \frac{1}{q\tau}, \quad E_{222} : Q_2 < q \leq \tau, \ |\lambda| \leq \frac{1}{q\tau}. \quad (28)$$

Since

$$A > x^{\frac{2}{3}+h} \quad \text{and} \quad A^2\tau^{-1} > x^{1+h} \quad (29)$$

for some positive h, and $\exp((\log \log x)^3) < q \leq \tau$ if $\alpha \in E_{222}$, he got

$$S(\alpha; x, A) = O(A \log^{-4} x), \quad \alpha \in E_{222}, \quad (30\text{-}1)$$

by Vinogradov method (Lemma 2) and then, using Lemma 5 and a simple estimation of zero-density similar to Lemma 6 he got

$$S(\alpha; x, A) = O(A \log^{-4} x), \quad \alpha \in E_{221}, \quad (30\text{-}2)$$

also. At last he want to evaluate $T_1(N, U)$, that is, to discuss $S(\alpha; x, A)$ on E_1. Using (18) and (22) there is $\alpha \in E_1$

$$S(\frac{a}{q} + \lambda; x, A) = \mu(q)\phi^{-1}(q) \int_{x-A}^{x} e(\lambda u)du + O(I(\lambda, q))$$
$$+ O(xAT^{-1}|\lambda| \log^{17} x) + O(xT^{-1} \log^{17} x), \quad \alpha \in E_1, \quad (31)$$

where $I(\lambda, q)$ is given by (24). **When he estimated** $I(\lambda, q)$ **the mistake appears.** Using Lemmas 4, 5 and a simple estimation of zero-density similar to Lemma 6 he correctly derived

$$I(\lambda, q) = q^{\frac{1}{2}}\phi^{-1}(q) \sum_{\chi \bmod q} \left| \sum_{|\gamma| \leq T} \int_{x-A}^{x} u^{\rho-1} e(\lambda u)du \right|$$
$$\ll A \exp(-c(3) \log^{\frac{1}{5}} x), \quad \alpha \in E_1. \quad (32)$$

But by using these Lemmas he wrote down **a wrong bound:**

$$I(\lambda, q) = q^{\frac{1}{2}}\phi^{-1}(q) \sum_{\chi \bmod q} \left| \sum_{|\gamma| \leq T} \int_{x-A}^{x} u^{\rho-1} e(\lambda u)du \right|$$
$$\ll |\lambda| \exp(-c(3) \log^{\frac{1}{5}} x), \quad \alpha \in E_1, \quad (33)$$

because **there is no Integral Mean Value Theorem in complex analysis.** From (31) and (32), we **only** can get

$$S(\frac{a}{q} + \lambda; x, A) = \mu(q)\phi^{-1}(q) \int_{x-A}^{x} e(\lambda u)du$$
$$+ O(A \exp(-c(3) \log^{\frac{1}{5}} x)), \quad \alpha \in E_1, \quad (34)$$

and then we have $(x = N/3 + U, A = 2U)$

$$T_1(N, U) = 3U^2(\log N)^{-3}S_3(N) + O(\frac{U^3}{\tau} \cdot \exp(-c(4)\log^{\frac{1}{5}}N)). \quad (35)$$

Since $\tau = (2U)^{1-(\nu-(1-\nu)\varepsilon)/(3-\nu)} < (2U)^{1-\mu}$, $\mu > 0$, (35) is useless. **If τ were taken to be $A\log^{-b}x$, we could obtain**

$$T_1(N, U) = 3U^2(\log N)^{-3}S_3(N) + O(U^2(\log N)^{-4}). \quad (36)$$

But there would be $A^2\tau^{-1} = A\log^b x < x^{1-h}$, thus **we could not use Vinogradov method (Lemma 2) to estimate $S(\alpha; x, A)$ on E_{222} successfully.** It is easy to get

$$\int_{E_{11}} S_0^3(\alpha; \frac{N}{3}, 2U)e(-N\alpha)d\alpha$$
$$= 3U^2(\log N)^{-3}S_3(N) + O(U^2(\log N)^{-4}), \quad (37\text{-}1)$$

$$\int_{E_2} S_0^3(\alpha; \frac{N}{3}, 2U)e(-N\alpha)d\alpha = O(U^2(\log N)^{-4}), \quad (37\text{-}2)$$

where $E_{11} = E(\log^{15} x, A\log^{-b} x)$. Therefore, by the method of Pan[1], we can only obtain that

$$T(N, U) = 3U^2(\log N)^{-3}S_3(N)$$
$$+ \int_{E_{12}} S_0^3(\alpha; \frac{N}{3}, 2U)e(-N\alpha)d\alpha$$
$$+ O(U^2(\log N)^{-4}), \quad (37\text{-}3)$$

where

$$E_{12} = E_1/E_{11} : [\frac{a}{q} - \frac{1}{\tau}, \frac{a}{q} - \frac{\log^b x}{A}] \bigcup [\frac{a}{q} + \frac{\log^b x}{A}, \frac{a}{q} + \frac{1}{\tau}],$$
$$1 \leq q \leq \log^{15} x, \ 1 \leq a \leq q, \ (a, q) = 1. \quad (38)$$

It means that E_{12} **is the set of numbers 'far' from irreducible fractions with 'small' denominators.** And so, for correcting this mistake it is ought to prove that $S_0(\alpha; x, A)$ **is comparatively 'small' if α is 'far' from an irreducible fraction with a 'small' denominator,** such that one could prove

$$\int_{E_{12}} S_0^3(\alpha; \frac{N}{3}, 2U)e(-N\alpha)d\alpha = O(U^2(\log N)^{-4}). \quad (39)$$

358

This view of point is **a slight different from the original idea of circle method**. Of course, by Dirichlet's lemma if $\alpha \in E_{12}$ there must be a and q, $(a, q) = 1$, such that λ in (9) satisfies $\log^{15} x < q \leq A \log^{-b} x$, $|\lambda| \leq 1/(qA \log^{-b} x)$.

I discussed this gap with Chengdong and Jingrun several times, and we tried to correct it by all possible ways which we could found at that time, but all of the efforts were failed. And so, there was only a very weak result of the GVT in short interval written in our book. It was very sad to us. This fact was a secret, and no one known it until 1988. From then on, my brother was determined to find a correct proof for his theorem. As I known that from 1978 to 1987 seldom had a week passed for nine years that he did not try in vain to correct his mistake. In the middle of 1987, he was extraordinary happy and gay to tell me that by means of **the pure analytic method** he had found an elegant new proof! I checked his proof step by step very carefully, and finally, with the same spirit I believed that it was fully correct. When I told it to Professor Chen, he was very happy and said that it was a very meaningful work.

Now let me introduce the new idea and new method proposed by my brother himself. Precisely he (see Pan[2, 3]) proved that **Theorem 1 is true for $U = N^{91/96+\varepsilon}$, and the asymptotic formula (4) holds.**

His proof can be outlined as follows. Taking

$$A = x^{\frac{91}{96}+\varepsilon}, \quad Q = Y = \log^{c(1)} x, \quad \tau = A \log^{-c(2)} x, \tag{40}$$

the Farey dissection is

$$E_1 = E(\log^{c(1)} x, A \log^{-c(2)} x), \quad \text{and} \quad E_2 = [-\frac{1}{Y}, 1 - \frac{1}{Y}]\backslash E_1. \tag{41}$$

By Dirichlet's lemma, if $\alpha \in E_2$, there must be a and q, $(a, q) = 1$, such that λ in (9) satisfies

$$1 \leq q \leq \log^{c(1)} x, \quad \frac{\log^{c(2)} x}{A} < |\lambda| \leq \frac{1}{q \log^{c(1)} x}. \tag{42}$$

It seems to be the first time to choose $Q = Y$ in circle method. Now the asymptotic formula (36) holds. To estimate $S(\alpha; x, A)$, that is $S_0(\alpha; x, A)$, $\alpha \in E_2$, taking $\tau_1 = A^8 x^{-7-\varepsilon}$ we have $E_2 = E_{21} \bigcup E_{22}$:

$$E_{21} : \frac{1}{\tau} < |\lambda| \leq \frac{1}{\tau_1}, \quad E_{22} : \frac{1}{\tau_1} < |\lambda| \leq \frac{1}{q \log^{c(1)} x}. \tag{43}$$

Using the same method ($I(\lambda, q)$ and zero-density method) estimating E_{221} in Pan[1], he got

$$S(\alpha; x, A) = O(A \log^{-4} x), \quad \alpha \in E_{21}. \tag{44}$$

But it is more difficult to bound $S(\alpha; x, A)$, $\alpha \in E_{22}$, and it just is the key point of the proof. Applying $J(\lambda, q)$ and several kinds of integral mean value theorems of Dirichlet L-functions (including short interval case), he successfully proved that

$$S(\alpha; x, A) = O(A \log^{-4} x), \quad \alpha \in E_{22}. \tag{45}$$

Consequently, from (7), (15), (36), (44) and (45), the proof was completed.

As we known that **it is the first time to estimate the linear exponential sum with prime variable in short intervals by pure analytic method without using Vinogradov method.**

In 1989, using his **pure analytic method** my brother (see Pan[4]) further proved that **Theorem 1 is true for** $U = N^{2/3} \log^c N$, **and the asymptotic formula (4) holds.** In this case, he took

$$A = x^{\frac{2}{3}} \log^c x, \quad Q = \log^{c(1)} x, \quad Y = x^{\frac{1}{6}}, \quad \tau = A \log^{-c(2)} x, \tag{46}$$

and the Farey dissection is

$$E_1 = E(\log^{c(1)} x, A \log^{-c(2)} x), \quad \text{and} \quad E_2 = [-x^{-\frac{1}{6}}, 1 - x^{-\frac{1}{6}}] \backslash E_1. \tag{47}$$

By Dirichlet's lemma, if $\alpha \in E_2$, there must be a and q, $(a, q) = 1$, such that λ in (9) satisfies

$$E_{21}: 1 \le q \le \log^{c(1)} x, \quad \frac{\log^{c(2)} x}{A} < |\lambda| \le \frac{1}{qx^{\frac{1}{6}}}, \tag{48-1}$$

or

$$E_{22}: \log^{c(1)} x < q \le x^{\frac{1}{6}}, \quad |\lambda| \le \frac{1}{qx^{\frac{1}{6}}}. \tag{48-2}$$

Now the asymptotic formula (36) holds too. In this paper he used the zero-density method only, that is, all the estimations of $S(\alpha; x, A)$ ($\alpha \in E_2$) are reduced to estimate $I(\lambda, q)$. Let

$$\tau_1 = 10x^{-1}A^2 < \tau, \tag{49}$$

E_{21} is divided into two parts: $1 \le q \le \log^{c(1)} x$,

$$E_{211}: \frac{\log^{c(2)} x}{A} < |\lambda| \le \frac{1}{\tau_1}, \quad E_{212}: \frac{1}{\tau_1} < |\lambda| \le \frac{1}{qx^{\frac{1}{6}}}; \tag{50-1}$$

and E_{22} is also divided into two parts: $\log^{c(1)} x < q \le x^{1/6}$,

$$E_{221}: \ \log^{c(1)} x < q \le x^{\frac{1}{6}}, \ |\lambda| \le \frac{1}{\tau_1}, \ E_{222}: \ \frac{1}{\tau_1} < |\lambda| \le \frac{1}{qx^{\frac{1}{6}}}. \quad (50\text{-}2)$$

According to different cases using different zero-density theorems, he obtained the same estimation, that is, we always have

$$S(\alpha; x, A) = O(A \log^{-4} x), \quad \alpha \in E_2. \quad (51)$$

Consequently, from (7), (15), (36) and (51), the proof was completed.

It ought to point out that **if using Vinogradov method, the best result we could get is** $U = N^{2/3+\varepsilon}$, **not to be** $U = N^{2/3} \log^c N$.

In 1989, also using pure analytic method Zhan Tao[1] proved that **Theorem 1 is true for** $U = N^{5/8} \log^c N$, **and the asymptotic formula (4) holds.** Zhan took

$$A = x^{\frac{5}{8}} \log^c x, \ Q = \log^{c(1)} x, \ Y = x^{-1} A^2 (\log x)^{c(3)}, \ \tau = A \log^{-c(2)} x, \quad (52)$$

and the Farey dissection is the same type in Pan[4]. The difference between these two papers is that Zhan reduced the estimation of $S(\alpha; x, A)$ on $\alpha \in E_{22}$ to that of $J(\lambda, q)$. In other words, Zhan's approach is the same of Pan[3], but he used more stronger analytic technique and results. It is the best result by means of the pure analytic method.

By the way, Wolke[1] proved that under GRH Theorem 1 is true for $N^{1/2} \log^{7+\varepsilon} N$, and the asymptotic formula (4) holds.

On the other hand, along the original approach in Pan[1], that is, combining the analytic method and Vinogradov method, Jia[1, 2] proved that **Theorem 1 is true for** $U = N^{13/17+\varepsilon}$, $N^{2/3+\varepsilon}$, **and the asymptotic formula (4) holds.** It corrects the mistakes in Pan[1] and Chen[1] directly. In Jia[2], he took

$$A = x^{\frac{2}{3}+\varepsilon}, \ Q = \log^{20} x, \ Y = x^{-1-\frac{2\varepsilon}{3}} A^2 = x^{\frac{1}{3}+\frac{4\varepsilon}{3}}, \ \tau = A \log^{-6} x, \quad (53)$$

and the Farey dissection is

$$E_1 = E(\log^{20} x, A \log^{-6} x), \ \text{and} \ E_2 = [-\frac{1}{Y}, 1 - \frac{1}{Y}] \backslash E_1. \quad (54)$$

By Dirichlet's lemma, if $\alpha \in E_2$, there must be a and $q, (a, q) = 1$, such that λ in (9) satisfies

$$E_{21}: \ 1 \le q \le \log^{20} x, \ \frac{1}{\tau} < |\lambda| \le \frac{1}{qY}, \quad (55\text{-}1)$$

or

$$E_{22}: \ \log^{20} x < q \le Y, \ |\lambda| \le \frac{1}{qY}. \quad (55\text{-}2)$$

Taking

$$Q \leq Q_2 = \exp((\log \log x)^3) \leq Y, \qquad (56)$$

E_{22} is divided into

$$E_{221}: \quad \log^{20} x < q \leq \exp((\log \log x)^3); \quad E_{222}: \quad \exp((\log \log x)^3) < q \leq Y. \qquad (57)$$

And then, applying zero-density method to estimate $S(\alpha; x, A)$ ($\alpha \in E_{21}$ and E_{222}), and applying Vinogradov method to estimate $S(\alpha; x, A)$ ($\alpha \in E_{221}$), the result follows at once.

By the way, if in Pan[1] we take

$$A = x^{\frac{91}{96}+\varepsilon}, \quad Q = \log^c x, \quad Y = x^{-1-\frac{2\varepsilon}{3}} A^2, \quad \tau = A \log^{-6} x,$$

we can get the corresponding Farey division. And then, we can use Lemma 12 in Pan[3] to estimate E_{12} in Pan[1]. Thus, the mistake in Pan[1] is corrected, and GVT in short interval is true for $U = N^{91/96+\varepsilon}$.

B.   The approach by sieve method

A new method to treat the GVT in short interval was proposed by Jia Chaohua in 1990. Let's describe it. By The Prime Number Theorem in short interval, we know that there exists a positive number $r < 1$ such that if $r \leq \nu < 1$ we have

$$\pi(y) - \pi(y - w) > cw \log^{-1} w, \quad y > w > y^\nu > 100. \qquad (58)$$

The best lower bound is $r = 11/20 - 1/384$ (M[1]). Let N be a large odd integer, $N/3 > U \geq N^\nu$, $r \leq \nu < 1$. Then we have

$$\pi(\frac{N}{3} + U) - \pi(\frac{N}{3} - U) > cU \log^{-1} U. \qquad (59)$$

Let set

$$A = A(N, U) = \{n = N - p: \frac{N}{3} - U < p \leq \frac{N}{3} + U\}. \qquad (60)$$

Clearly, $n \in A$ is a large even integer, and by $D(n)$ we denote the number of solutions of the Diophantine equation with prime variables

$$n = p_1 + p_2, \quad \frac{N}{3} - U < p_1 \leq \frac{N}{3} + U, \quad \frac{N}{3} - U < p_2 \leq \frac{N}{3} + U. \qquad (61)$$

Thus, there is

$$T(N, U) = \sum\nolimits^* D(n), \qquad (62)$$

where $\sum^*$ denotes the sum is over $n \in A$.

Jia[3] discovered that Theorem 1 with $U = N^\nu$ can be derived from the following Theorem 2 immediately if $\nu \geq 11/20 - 1/384$.

Theorem 2. *Assume that N is a large odd integer. Then there exists a positive number $0 < \nu < 1$ such that for $U = N^\nu$ and the even numbers n*

$$2|n, \quad \frac{2N}{3} - N^\nu \leq \frac{2N}{3} + N^\nu, \tag{63}$$

except for $O(N^\nu \log^{-2} N)$ exceptional values, we have

$$D(n) \gg S_2(n) N^\nu \log^{-2} N, \tag{64}$$

where

$$S_2(n) = \prod_{p \nmid n}\left(1 - \frac{1}{(p-1)^2}\right)\prod_{p|n}\left(1 + \frac{1}{(p-1)}\right) \geq 1, \quad 2|n. \tag{65}$$

This kind of result can be regarded as **a problem of short-interval-Goldbach number** (see (63)) **in short interval** (see (61)).

Obviously, from Theorem 2, (62), (60) and (59), Theorem 1 follows, and

$$T(N,U) \gg S_3(N) U^2 (\log N)^{-3}, \tag{66}$$

provided $U = N^\nu$, $11/20 - 1/384 \leq \nu$. **By using sieve method and circle method,** Jia Chaohua[3, 4, 5, 6, 7] proved that Theorem 2 is true for

$$\nu = 0.646 + \varepsilon, \quad 0.636 + \varepsilon, \quad 0.6 + \varepsilon, \quad \frac{23}{39} + \varepsilon, \quad \frac{7}{12} + \varepsilon,$$

respectively. H. Mikawa proved $\nu = 7/12 + \varepsilon$ also. Since the zero-density estimation (Lemma 7) is used in the proof of Jia and Mikawa, $\nu = 7/12 + \varepsilon$ is the best possible result. Now the best result $\nu = 4/7$ is due to Baker and Harman[1]. However, **we can't obtain the asymptotic formula (4) by using sieve method.**

$D(n)$ can be treated by sieve method as follows. Let K be a positive integer, $y \geq 2$, and

$$P_K(y) = \prod_{p \leq y, \, (p, K)=1} p,$$

$\mathcal{B}$ a finite sequence of integers, and sifting function

$$S(\mathcal{B}; P_K, y) = \#\{b \in \mathcal{B} : (b, P_K(y)) = 1\}. \tag{67}$$

Thus for n satisfying (62) there is

$$D(n) = S(\mathcal{H}(n); P_n, \left(\frac{2N}{3}\right)^{\frac{1}{2}}) + O(N^\varepsilon), \tag{68}$$

where

$$\mathcal{H}(n) = \{a = n - p; \ \frac{N}{3} - U < p \le \frac{N}{3} + U\}. \tag{69}$$

In order to get a better lower bound of sifting function $S(\mathcal{H}(n); P_n, (2N/3)^{1/2})$, we need to use ingenious and flexible combinatorial methods in the theory of sieves. It is well known that **the remainders appeared in this case are extra complicated and extra difficult to handle.**

Jia successfully overcame these difficulties by applying circle method, large sieve method, some methods of exponential sums and the theory of L-functions together. It is very complicated, and so I'll not talk about it in detail.

In addition, under the influence of my brother's papers,

(i) Liu Jianya and Zhan Tao established **a non-trivial estimate of the non-linear exponential sums with prime variable in short intervals**

$$S_k(\alpha; x, A) = \sum_{x-A < n \le x} \Lambda(n) e(n^k \alpha) \tag{70}$$

by the analytic method, and got a number of results on the Goldbach-Waring Problems in short interval;

(ii) Jia not only proposed a new proof of Theorem 2 by sieve method, but also proposed some new meaningful problems to study, such as, almost all short intervals containing prime numbers, the exceptional set of Goldbach numbers in a short interval.

All of the above are what I want to talk today, the real story is finished. As I looked back on the past, my mind thronged with thoughts. It seems to be that it is not very important that there is some gap in a paper, the most important thing is that: Is there any new ideas, new methods, new problems and new techniques in it? Perhaps, I could say that my brother' paper- **Some new results in additive number theory, Acta Math. Sin. 9(1959), 315-329**-published about forty years ago has a definite influence to the development of analytic number theory in China in the last decade.

C. Some Lemmas

Lemma 1 (Dirichlet). *For any real numbers $Y \ge 1$ and α, there must be integers a and q, $(a, q) = 1, 1 \le q \le Y$, such that $|\alpha - a/q| < 1/qY$.*

Lemma 2 (Vinogradov). *Let $1 \le A \le x$, and $|\alpha - a/q| \le 1/q^2, (a, q) = 1, 1 \le q \le x$. Then for arbitrarily positive number ε, we have*

$$S(\alpha; x, A) \ll A \exp(b(\log \log x)^2)(q^{\frac{1}{2}} x^{\frac{1}{2}} A^{-1} + q^{-\frac{1}{2}} + x^{\frac{1}{3}+\varepsilon} A^{\frac{1}{2}}).$$

A non-trivial bound can be obtained by Lemma 2 only if $A > x^{2/3+h}$, $A^2\tau^{-1} > x^{1+h}$ for some positive h and $\exp((\log\log x)^3) < q \leq \tau$.

Lemma 3. *Suppose that $2 \leq T \ll u$, χ is a character mod q, and $\rho = \rho_\chi = \beta_\chi + i\gamma_\chi = \beta + i\gamma$ is the non-trivial zero of the Dirichlet L-function $L(s, \chi)$. Then we have*

$$\psi(u, \chi) = \sum_{n \leq u} \chi(n)\Lambda(n) = E(\chi)u + \sum_{|\gamma| \leq T} \frac{u^\rho - 1}{\rho} + R_1(u, T, \chi),$$

$$R_1(u, T, \chi) = O(uT^{-1}\log^2(qu)),$$

where $E(\chi) = 1$, if χ is principal; $E(\chi) = 0$, otherwise.

Lemma 4. *For any $\varepsilon > 0$ there exists a positive constant $c = c(\varepsilon)$ such that, if χ is any real chracter mod q, then $L(\sigma, \chi) \neq 0$ for $\sigma \geq 1 - q^{-\varepsilon}$.*

Lemma 5. *Let $q \geq 1$, $s = \sigma + it$. Then there is a positive constant c such that $\prod_{\chi \bmod q} L(s, \chi)$ has no zero in the region*

$$\sigma \geq 1 - \frac{c}{\log q + (\log(|t| + 2))^{\frac{4}{5}}},$$

except for the possible exceptional zero mod q.

Lemma 6. *Let $\frac{1}{2} \leq h \leq 1$, $T \geq 2$, χ be a character mod q, and suppose $N(h, T, \chi)$ is the number of the zeros of $L(s, \chi)$ in the region: $s = \sigma + it$, $h \leq \sigma \leq 1$, $|t| \leq T$. Then we have*

$$N(h, T, q) = \sum_{\chi \bmod q} N(h, T, \chi)$$

$$\ll \min(qT\log(qT), (qT)^{5(1-h)/2}(\log qT)^{13}).$$

Lemma 7. *Let $\frac{1}{2} \leq h \leq 1, T \geq 2$. Then for any $\varepsilon > 0$ we have*

$$N(h, T, q) = \sum_{\chi \bmod q} N(h, T, \chi) \ll \min(qT\log(qT), (qT)^{(\frac{12}{5}+\varepsilon)(1-h)}).$$

Lemma 8. *Let $\frac{1}{2} \leq h \leq 1$, $T \geq 2$ and $T \geq H \geq T^{4/3}$. Then we have*

$$N(h, T+H, q) - N(h, T, q) \ll (qH)^{4(1-h)/(3-2h)}(\log qH)^c, \quad \frac{1}{2} \leq h \leq \frac{3}{4};$$

$$N(h, T+H, q) - N(h, T, q) \ll (qH)^{8(1-h)/3}(\log qH)^c, \quad \frac{3}{4} \leq h \leq 1;$$

$$N(h, T+H, q) - N(h, T, q) \ll (qH)^{12(1-h)/5}(\log qH)^c, \quad \frac{5}{6} \leq h \leq 1.$$

Lemma 9. *When $2 \leq T \ll u$, χ is a character mod q, and $b = 1 + 1/\log u$, we have*

$$\psi(u, \chi) = \frac{1}{2\pi i} \int_{b-iT}^{b+iT} -\frac{L'}{L}(s, \chi)\frac{u^s}{s}ds + R_2(u, T, \chi),$$

$$R_2(u, T, \chi) = O(uT^{-1}\log^2 u).$$

Lemma 10. (i) *Let*

$$M(s, \chi) = \sum_{n \leq X} \frac{\mu(n)\chi(n)}{n^s},$$

$\mu(n)$ *Möbius function, we have*

$$-\frac{L'}{L}(s, \chi) = -\frac{L'}{L}(s, \chi)(1 - L(s, \chi)M(s, \chi)) - L'(s, \chi)M(s, \chi).$$

(ii) *Let*

$$M(s) = \sum_{n \leq X} \frac{\mu(n)}{n^s},$$

we have

$$-\frac{\zeta'}{\zeta}(s) = \sum_{j=1}^{10} \binom{10}{j}(-1)^{j-1}\zeta'(s)\zeta^{j-1}(s)M(s) - \frac{\zeta'}{\zeta}(s)(1 - \zeta(s)M(s))^{10}.$$

Lemma 11. *Assume that $|t| \geq 2$ and χ is a primitive character mod q. Then we have*

$$L(\sigma + it, \chi) \ll (q|t|)^{(1-\sigma)/2}\log(q|t|), \qquad 0 \leq \sigma \leq 1.$$

Lemma 12. *If $T \geq 2$ and q is a positive integer, then we have*

$$\sum_{\chi \bmod q} \int_{-T}^{T} |L(\tfrac{1}{2} + it, \chi)|^4 dt \ll qT(\log qT)^5.$$

Lemma 13. *Let ε be arbitrarily positive number, $T \geq 2$ and $T \geq H \geq T^{7/8+\varepsilon}$. Then for any $b > 0$,*

$$\sum_{\chi \bmod q} \int_{T}^{T+H} |L(\tfrac{1}{2} + it, \chi)|^4 dt \ll qH(\log qH)^c$$

holds uniformly for $q \leq (\log T)^b$, c being an absolute constant.

Lemma 14. *Let ε be arbitrarily positive number, $T \geq 2$ and $T \geq H \geq T^{1/2+\varepsilon}$. Then for any $q > 1$ we have*

$$\sum_{\chi \bmod q} \int_T^{T+H} |L^{(k)}(\frac{1}{2} + it, \chi)|^2 dt \ll qH(\log qH)^{2k+1}, \qquad k = 0, 1.$$

Lemma 15. *Let ε be arbitrarily positive number, $T \geq H \geq 1$, then we have*

$$\sum_{\chi \bmod q} \int_T^{T+H} |L^{(k)}(\frac{1}{2} + it, \chi)|^4 dt \ll (qH + qT^{\frac{2}{3}})(qT)^\varepsilon, \qquad k = 0, 1.$$

References

Baker, R. C. and Harman, G.
[1] The three primes theorem with almost equal summands, Phil. Trans. Soc. London A, 356(1998), 763-780.

Chen Jingrun
[1] On large odd numbers as sums of three almost equal primes, Sci. Sin. 14(1965), 1113-1117.

Haselgrove, C. B.
[1] Some theorems in the analytic theory of numbers, J. London Math. Soc. 26(1951), 273-277.

Jia Chaohua
[1] Three primes theorem in a short interval, Acta Math. Sin. 32(1989), 464-473.
[2] Three primes theorem in a short interval (II), International Symposium in Memory of Hua Loo Keng Vol.1, 103-116, Science Press and Springer-Verlag, 1991.
[3] Three primes theorem in a short interval (III), Sci. China Ser. A, 34(1991), 1039-1056.
[4] Three primes theorem in a short interval (IV), Advances in Math. (China), 20(1991), 109-126.
[5] Three primes theorem in a short interval (V), Acta Math. Sin. New Ser. 7(1991), 135-170.
[6] Three primes theorem in a short interval (VI), Acta Math. Sin. 34(1991), 832-850.

[7] Three primes theorem in a short interval (VII), Acta Math. Sin. New Ser. 10(1994), 369-387.

Mozzochi
[1] J. Number Theory, 24(1986), 181-187.

Pan Chengdong
[1] Some new results in additive number theory, Acta Math. Sin. 9(1959), 315-329.
[2] On estimation of trigonometric sums over primes in short intervals (I), Sci. Sinica, 32(1989), 408-416. (with Pan Chengbiao).
[3] On estimation of trigonometric sums over primes in short intervals (II), Sci. Sinica, 32(1989), 641-653. (with Pan Chengbiao).
[4] On estimation of trigonometric sums over primes in short intervals (III), Chin. Ann. of Math. 11B(1990), 138-147. (with Pan Chengbiao).
[5] Representation of large odd numbers as sums of three almost equal primes, Acta Sci. Nat. Univ. Sichuan, Special Issue, 1990, 172-183. (with Pan Chengbiao).

Wolke, D.
[1] Uber Goldbach-Zerlegungen mit nahezu gleichen Summanden, J. Number Theory 39(1991), 237-244.

Zhan Tao
[1] On the representation of large odd integers as sums of three almost equal primes, Acta Math. Sin. New Ser. 7(1991), 259-272.

CONVEXITY AND INTERSECTION OF RANDOM SPACES

Mariya Shcherbina
Institute for Low Temperatures, Ukr. Ak. Sci, 47 Lenin Av. Kharkov, Ukraine
shcherbi@ilt.kharkov.ua

Brunello Tirozzi
Department of Physics, University of Rome "La Sapienza", P.A. Moro 2, 00185, Roma, Italy
tirozzi@krishna.phys.uniroma1.it

Abstract   The problem of finding the volume of the intersection of the N dimensional sphere with $p = \alpha N$ random half spaces when α is less than a critical value α_c and when $N, p \to \infty$ is solved rigorously. The asymptotic expression coincides with the one found by E. Gardner ([4]), using non rigorous replica calculations in neural network theory. When α is larger than α_c the volume of the intersection goes to 0 more rapidly than $\exp(-N \text{ const})$. We use the cavity method. The convexity of the volume and the Brunn Minkowski theorem ([3]) have a central role in the proof.

Keywords: Random systems, Convexity.

1.   Introduction

For very large integer N consider the N-dimensional sphere S_N of radius $N^{1/2}$ centered in the origin and a set of $p = \alpha N$ independent random half spaces. $\Pi_\mu = \{ J \in \mathbf{R}^N : N^{-1/2}(\xi^{(\mu)}, J) \geq k \}$ $(\mu = 1, \ldots, p)$, where $\xi^{(\mu)}$ are i.i.d. random vectors with i.i.d. Bernoulli components $\xi_j^{(\mu)}$ and k is the distance of Π_μ from the origin. The problem is to find the maximum value of α such that the volume of the intersection of S_N with $\cap \Pi_\mu$ behaves like $e^{-N \text{ const}} \sigma_N$, where σ_N is the volume of S_N.

This geometrical question is motivated by the problem of the retrieval of patterns in neural networks. The retrieval of patterns depends on the neural dynamic. The neural dynamic is defined as an evolution of the space of neural activities $\sigma \equiv (\sigma_1, \ldots, \sigma_N)$, $\sigma_i = \pm 1$, generated by the

369

370

threshold dynamics

$$\sigma_i(t+1) = \text{sign}\{\sum_{j=1,j\neq i}^{N} J_{ij}\sigma_j(t)\} \quad (i=1,\ldots,N), \tag{1}$$

where $\sigma(t)$, $\sigma(t+1)$ are the vectors of neural activities at time t and $t+1$ respectively. The value $+1$ for σ_i represents an "active neuron" while -1 is a quiescent neuron. In this language the retrieval means that the activities $\sigma(t)$ converge, for suitable initial conditions, to some pattern $\xi^{(\mu)}$.

The interaction (or synaptic) matrix $\{J_{ij}\}$ (not necessarily symmetric) depends on the concrete model, but usually it satisfies the conditions

$$\sum_{j=1,j\neq i}^{N} J_{ij}^2 = NR \quad (i=1,\ldots,N), \tag{2}$$

where R is some fixed number which could be taken equal to 1.

The problem is to find an interaction matrix $\{J_{ij}\}$ such that some set ξ of chosen vectors $\xi \equiv \{\xi^{(\mu)}\}_{\mu=1}^{p}$ (patterns) are the fixed points of the dynamics (1). This implies the conditions:

$$\xi_i^{(\mu)} \sum_{j=1,j\neq i}^{N} J_{ij}\xi_j^{(\mu)} > 0 \quad (i=1,\ldots,N). \tag{3}$$

Sometimes condition (3) is not sufficient to have $\xi^{(\mu)}$ as the end points of the dynamics. To have some "basin of attraction" (that is some neighbourhood of $\xi^{(\mu)}$, starting from which we for sure arrive in $\xi^{(\mu)}$) one should introduce some positive parameter k and impose the conditions:

$$\xi_i^{(\mu)} \sum_{j=1,j\neq i}^{N} J_{ij}\xi_j^{(\mu)} > k \quad (i=1,\ldots,N). \tag{4}$$

So in this paper we consider only $k > 0$.

Gardner [4] was the first who solved this kind of inverse problem. She asked the question: for which $\alpha = \frac{p}{N}$ the interaction $\{J_{ij}\}$, satisfying (2) and (4) exists? What is the typical fractional volume of these interactions? Since all the conditions (2) and (4) are factorised with respect to i, this problem, after a simple transformation, can be replaced by the following. For the system of $p \sim \alpha N$ i.i.d. random patterns $\xi \equiv \{\xi^{(\mu)}\}_{\mu=1}^{p}$ with i.i.d. $\xi_i^{(\mu)}$ $(i=1,\ldots,N)$ assuming values ± 1 with probability $\frac{1}{2}$, consider

$$\Theta_{N,p}(\xi, k) = \sigma_N^{-1} \int_{(\boldsymbol{J},\boldsymbol{J})=N} d\boldsymbol{J} \prod_{\mu=1}^{p} \theta(N^{-1/2}(\xi^{(\mu)}, \boldsymbol{J}) - k), \qquad (5)$$

where the function $\theta(x)$, as usually, is zero in the negative semi-axis and 1 in the positive and σ_N is the Lebesgue measure of N-dimensional sphere of radius $N^{1/2}$. Then, the question of interest is the behaviour of $\frac{1}{N} \log \Theta_{N,p}(\xi, k)$ in the limit $N, p \to \infty$, $\frac{p}{N} \to \alpha$. Gardner [4] had solved this problem by using the so-called replica trick, which is completely non-rigorous from the mathematical point of view but sometimes very useful in the physics of spin glasses (see [7] and references therein). She obtained that for any $\alpha < \alpha_c(k)$, where

$$\alpha_c(k) \equiv (\frac{1}{\sqrt{2\pi}} \int_{-k}^{\infty} (u + k)^2 e^{-u^2/2} du)^{-1},$$

the volume of the space of synaptic couplings which have the patterns $\xi^{(\mu)}$ as fixed points decays exponentially with N while for $\alpha \geq \alpha_c$ the volume of the intersection decays as $N \to \infty$ faster than e^{-LN} with any positive L. Our main goal is to prove rigorously the results of [4]. The methods used by us is typical of statistical mechanics of the disordered systems which are systems of N variables (in our case J_i) with some random function of these variables, which define the "interaction" between them. Since the randomness of the interaction induces the randomness of the integrals, which appear in the problem, a natural question arises about the dependence of the integrals from the choice of the random patterns. For this reason we introduce the operation of expectation E with respect to the random patterns. We define, according to statistical mechanics, the free energy of the system $F_{N,p}(\xi, k)$

$$F_{N,p}(\xi, k) = \frac{1}{N} \log \Theta_{N,p}(\xi, k)$$

and the self-averaging of the free-energy as

$$\lim_{N \to \infty} E(F_{N,p}(\xi, k) - EF_{N,p}(\xi, k))^2 = 0.$$

In other words the free energy is self averaging if, in the limit of large N, it tends in probability to its average E with respect to the patterns.

To formulate our main theorem we should remark that since $\Theta_{N,p}(\xi, k)$ can be zero with nonzero probability (e.g., if for some $\mu \neq \nu$ $\xi^{(\mu)} = -\xi^{(\nu)}$), we cannot, as usually in statistical mechanics, just study

$$\log \Theta_{N,p}(\xi, k).$$

To avoid this difficulty, we take some large enough M and replace the log- function by the function $\log_{(MN)}$, defined as $\log_{(MN)} X = \log \max \{X, e^{-MN}\}$.

Theorem 1. *For any $\alpha \leq \alpha_c(k)$ $N^{-1} \log_{(MN)} \Theta_{N,p}(\xi, k)$ is self-averaging in the limit $N, p \to \infty$, $p/N \to \alpha$ and for M large enough there exists*

$$\lim_{N,p \to \infty} E\{N^{-1} \log_{(MN)} \Theta_{N,p}(\xi, k)\}$$

$$= \min_{0 \leq q \leq 1} \left[\alpha E \left\{ \log H \left(\frac{u\sqrt{q} + k}{\sqrt{1-q}} \right) \right\} + \frac{1}{2} \frac{q}{1-q} + \frac{1}{2} \log(1-q) \right], \quad (6)$$

where $H(x) \equiv \frac{1}{\sqrt{2\pi}} \int_x^\infty e^{-t^2/2} dt$, u is a Gaussian random variable with zero mean variance 1 and $E\{\cdots\}$ is the averaging with respect to u.

For $\alpha > \alpha_c(k)$ $E\{N^{-1} \log_{(MN)} \Theta_{N,p}(\xi, k)\} \to -\infty$, as $N \to \infty$ and then $M \to \infty$.

We remark here that the self-averaging of $N^{-1} \log \Theta_{N,p}(\xi, k)$ was proven also in [17]. In the next two sections we explain the main ideas underlying the proof.

A complete analysis will be presented elsewhere ([13]).

2. Convexity and Decay of Correlations

It can be easily seen that the Gardner problem (5) is very similar to problems of statistical mechanics, where the integrals with respect to N variables in the limit $N \to \infty$ are studied. But due to technical reasons it is not convenient to study directly the model (5) with θ functions. That is why we use a common trick: substitute the θ-functions appearing in the expression of the partition function (5) by some smooth functions which depend on a small parameter ε and tend, as $\varepsilon \to 0$, to the θ-functions. We choose for this purpose $H(-x\varepsilon^{-1/2})$ with $H(x)$ defined in Theorem 1 but the particular form of this function is not important for us. The most important fact is that its logarithm should be a convex function. To substitute in (5) the integration over S_N by the integration over the whole $\mathbf{R}^N$ we use another well known trick in statistical mechanics. We add to the Hamiltonian a term depending on the additional free parameter z. At the end of our considerations we can choose this parameter in order to provide the condition that for large N only a small neighborhood of S_N gives the main contribution to our integral. Thus, we consider the Hamiltonian of the form

$$\mathcal{H}_{N,p}(\boldsymbol{J}, \xi, k, h, z, \varepsilon) \equiv - \sum_{\mu=1}^{p} \log \mathrm{H}\left(\frac{k - (\xi^{(\mu)}, \boldsymbol{J})N^{-1/2}}{\sqrt{\varepsilon}}\right)$$

$$+ \frac{z}{2}(\boldsymbol{J}, \boldsymbol{J}) + h(\boldsymbol{h}, \boldsymbol{J}). \tag{7}$$

Here the last term $h(\boldsymbol{h}, \boldsymbol{J})$ is the scalar product of the variables $\boldsymbol{J}$ with some vector $\boldsymbol{h}$ with independent random components introduced for getting the self averaging of the order parameters of the theory (see below) [6], [8], [9]. The free energy and the Gibbs average for this Hamiltonian are

$$Z_{N,p}(\xi, k, h, z, \varepsilon) = \sigma_N^{-1} \int d\boldsymbol{J} e^{-\mathcal{H}_{N,p}(\boldsymbol{J}, \xi, k, h, z, \varepsilon)},$$

$$\langle \cdots \rangle = \int (\cdots) d\boldsymbol{J} \frac{e^{-\mathcal{H}_{N,p}(\boldsymbol{J}, \xi, k, h, z, \varepsilon)}}{Z_{N,p}(\xi, k, h, z, \varepsilon)},$$

$$F_{N,p}(\xi, k, h, z, \varepsilon) \equiv \frac{1}{N} \log Z_{N,p}(\xi, k, h, z, \varepsilon).$$

Now we have the typical problem of statistical mechanics which we solve by a method usually called the cavity method. The idea of the cavity method is to choose one variable, e.g. J_N and to try to express $\langle J_N \rangle$ through the Gibbs average of the others J_i, and then, using the symmetry of the Hamiltonian, write the self-consistent equations for the so-called order parameters of the problem $q \equiv \frac{1}{N}\sum\langle J_i\rangle^2$ and $R \equiv \frac{1}{N}\sum\langle J_i^2\rangle$. This procedure allows us to reduce the problem to a finite number of nonlinear equations. The rigorous version of the cavity method was proposed in [8] and developed in [9], [10],[11],[12]. The key problem of the application of the cavity method is the proof of the vanishing of the correlation functions $\langle J_i J_j \rangle - \langle J_i\rangle\langle J_j\rangle$ as $N \to \infty$. We derived this property from a geometrical statement contained in theorem 2. This result is the analogous of the result of [2], which allows to prove the vanishing of the correlation functions for a large class of models of statistical mechanics which are generated by convex hamiltonians and so also for the model with the hamiltonian $\mathcal{H}_{N,p}(\boldsymbol{J}, \xi, k, h, z, \varepsilon)$. Thus we use general definitions in order to show the generality of this theorem.

Let $\{\Phi_N(\boldsymbol{J})\}_{N=1}^{\infty}$ $(\boldsymbol{J} \in \mathbf{R}^N)$ be a system of convex functions which have third derivatives bounded in any compact set. Consider also a system of convex domains $\{\Gamma_N\}_{N=1}^{\infty}$ $(\Gamma_N \subset \mathbf{R}^N)$ whose boundaries consist of a finite number (may be depending on N) of smooth pieces. Define the Gibbs measure and the free energy, corresponding to $\Phi_N(\boldsymbol{J})$ in Γ_N:

$$\langle \cdots \rangle_{\Phi_N} \equiv \Sigma_N^{-1} \int_{\Gamma_N} d\boldsymbol{J}(\cdots) e^{-\Phi_N(\boldsymbol{J})}, \quad \Sigma_N(\Phi_N) \equiv \int_{\Gamma_N} d\boldsymbol{J} e^{-\Phi_N(\boldsymbol{J})}$$

$$f_N(\Phi_N) \equiv \frac{1}{N} \log \Sigma_N(\Phi_N).$$

Denote $\quad \tilde{\Omega}_N(U) \equiv \{J : \Phi_N(J) \le U\},$

$$\Omega_N(U) \equiv \tilde{\Omega}_N(U) \cap \Gamma_N, \tag{8}$$

$\mathcal{D}_N(U) \equiv \tilde{\mathcal{D}}_N(U) \cap \Gamma_N$, where $\tilde{\mathcal{D}}_N(U)$ is the boundary of $\tilde{\Omega}_N(U)$.
Define also

$$f_N^*(U) = \frac{1}{N} \log \int_{J \in \mathcal{D}_N(U)} dJ e^{-NU}.$$

Theorem 2. *Let the functions $\Phi_N(J)$ satisfy the conditions:*

$$\frac{d^2}{dt^2} \Phi_N(J + te)|_{t=0} \ge C_0 > 0,$$

$$\Phi_N(J) \ge C_1(J, J) - NC_2, \tag{9}$$

$$|\nabla \Phi_N(J)| \le N^{1/2} C_3(U) \quad (J \in \tilde{\Omega}_N(U)), \tag{10}$$

where e is an arbitrarily direction ($|e| = 1$), $C_0, C_1, C_2, C_3(U)$ are some positive N-independent constants and $C_3(U)$ is continuous in U, ($U > U_{min} \equiv \min_{J \in \Gamma_N} N^{-1} \Phi_N(J) \equiv N^{-1} \Phi_N(J^*)$).
Assume also that there exists some finite N-independent C_4 such that $f_N(\Phi_N) \ge -C_4$.

Then for any $U > U_{min}$

$$f_N^*(U) = \min_{z>0} \{ f_N(z\Phi_N) + zU \} + O(N^{-1} \log N), \tag{11}$$

and for any $e \in \mathbf{R}^N$ ($|e| = 1$) and any natural p

$$\langle (\dot{J}, e)^p \rangle_{\Phi_N} \le C(p), \quad \frac{1}{N^2} \sum_{i,j} \langle \dot{J}_i \dot{J}_j \rangle_{\Phi_N}^2 \le \frac{C(2)}{N} \quad (\dot{J}_i \equiv J_i - \langle J_i \rangle_{\Phi_N}) \tag{12}$$

with some positive N-independent $C(p)$.

Let us explain the role of convexity in the proof of theorem (2). The Gibbs average of any function of the linear combination (J, e) ($|e| = 1$) can be expressed in terms of a two-dimensional integral with respect to the energy U (the value of the hamiltonian) and $c = (J, e)N^{-1/2}$. The additional function, which appears under this change of variables is the "partial entropy", given by the logarithm of the volume of the intersections of the level surfaces of the Hamiltonian with the hyperplanes $(J, e) = cN^{1/2}$. We study these intersections using a theorem of classical geometry known since the nineteenth century as the Brunn-Minkowski theorem [3]. From this theorem we obtain that the "partial entropy" is a concave function of (U, c). Thus we can apply the Laplace method to evaluate the Gibbs averages. So we obtain the vanishing of the correlation functions, which allows us to find the expression for the free energy. A similar idea was used in [1] where the results of [2] (also based on the Brunn-Minkowski theorem) have been used. We would like to remark that, differently from [1], we cannot just use the results of [2] because they are true for $\mathbf{R}^N$ while the most nontrivial part of our proof (i.e. the limiting transition $\varepsilon \to 0$) is based on similar results for the intersections of p random half spaces.

We show now more explicitly how to realize the above ideas. For any $U > 0$ consider the set $\Omega_N(U)$ defined in (8). Since $\Phi_N(\boldsymbol{J})$ is a convex function, the set $\Omega_N(U)$ is also convex and $\Omega_N(U) \subset \Omega_N(U')$, if $U < U'$. Let

$$V_N(U) \equiv \mathrm{mes}(\Omega_N(U)), \quad S_N(U) \equiv \mathrm{mes}(\mathcal{D}_N(U)),$$
$$F_N(U) \equiv \int_{\boldsymbol{J} \in \mathcal{D}_N(U)} |\nabla \Phi_N(\boldsymbol{J})|^{-1} dS_{\boldsymbol{J}}. \tag{13}$$

Here and below the symbol mes(...) means the Lebesgue measure of the correspondent dimension.

Then it is easy to see that the partition function Σ_N can be represented in the form

$$\Sigma_N = N \int_{U > U_{min}} e^{-NU} F_N(U) dU = \int_{U > U_{min}} e^{-NU} \frac{d}{dU} V_N(U) dU$$
$$= N \int_{U > U_{min}} e^{-NU} V_N(U) dU. \tag{14}$$

Here we have used the relation $F_N(U) = N^{-1} \frac{d}{dU} V_N(U)$ and the integration by parts.

Besides, for a chosen direction $\mathbf{e} \in \mathbf{R}^N$ ($|\mathbf{e}| = 1$), and any real c consider the hyper-plane

$$\mathcal{A}(c, \mathbf{e}) = \left\{ \boldsymbol{J} \in \mathbf{R}^N : (\boldsymbol{J}, \mathbf{e}) = N^{1/2} c \right\}$$

and denote

$$\Omega_N(U, c) \equiv \Omega_N(U) \cap \mathcal{A}(c, \mathbf{e}), \quad V_N(U, c) \equiv \mathrm{mes}(\Omega_N(U, c)),$$
$$\mathcal{D}_N(U, c) \equiv \mathcal{D}_N(U) \cap \mathcal{A}(c, \mathbf{e}), \quad F_N(U, c) \equiv \int_{\boldsymbol{J} \in \mathcal{D}_N(U, c)} |\nabla \Phi_N(\boldsymbol{J})|^{-1} dS_{\boldsymbol{J}}. \tag{15}$$

Then, since $F_N(U, c) = N^{-1} \frac{\partial}{\partial U} V_N(U, c)$, we obtain

$$\Sigma_N = N \int dc dU e^{-NU} F_N(U, c) = N \int dc dU e^{-NU} V_N(U, c),$$
$$\langle (\boldsymbol{J}, \mathbf{e})^p \rangle_{\Phi_N} = \frac{N^{p/2} \int dc dU c^p e^{-NU} V_N(U, c)}{\int dc dU e^{-NU} V_N(U, c)}. \tag{16}$$

Denote

$$s_N(U) \equiv \frac{1}{N} \log V_N(U), \quad s_N(U,c) \equiv \frac{1}{N} \log V_N(U,c). \quad (17)$$

The functions $s_N(U)$ and $s_N(U,c)$ are the complete and the "partial" entropies mentioned before and formula (16) is the two dimensional integral on the energy levels U and on the hyper planes $\mathcal{A}(c,\mathbf{e})$. Then the relations (14), (16) give us

$$\Sigma_N = N \int \exp\{N(s_N(U) - U)\}dU,$$

$$\left\langle (\dot{\mathbf{J}},\mathbf{e})^p \right\rangle_{\Phi_N} = N^{p/2} \left\langle (c - \langle c \rangle_{(U,c)})^p \right\rangle_{(U,c)}, \quad (18)$$

where

$$\langle \cdots \rangle_{(U,c)} \equiv \frac{\int dU\, dc (\ldots) \exp\{N(s_N(U,c) - U)\}}{\int dU\, dc \exp\{N(s_N(U,c) - U)\}}. \quad (19)$$

The equations (11) and (12) can be obtained by the standard Laplace method, if we prove that $s_N(U)$ and $s_N(U,c)$ are concave functions and they are strictly concave in the neighbourhood of the points U^* and (U^*,c^*) of maximum of the functions $(s_N(U) - U)$ and $(s_N(U,c) - U)$. To prove this we apply the theorem of Brunn-Minkowski from classical geometry (see e.g. [3]) to the functions $s_N(U)$ and $s_N(U,c)$. To formulate this theorem we need some extra definitions.

Consider two bounded sets in $\mathcal{A}, \mathcal{B} \subset \mathbf{R}^N$. For any positive α and β

$$\alpha \mathcal{A} \times \beta \mathcal{B} \equiv \{\mathbf{s} : \mathbf{s} = \alpha\mathbf{a} + \beta\mathbf{b}, \mathbf{a} \in \mathcal{A}, \mathbf{b} \in \mathcal{B}\}.$$

$\alpha \mathcal{A} \times \beta \mathcal{B}$ is the Minkowski sum of $\alpha \mathcal{A} =$ and $\beta \mathcal{B}$.

The one-parameter family of bounded sets $\{\mathcal{A}(t)\}_{t_1^* \leq t \leq t_2^*}$ is a convex one-parameter family, if for any positive $\alpha < 1$ and $t_{1,2} \in [t_1^*, t_2^*]$ they satisfy the condition

$$\mathcal{A}(\alpha t_1 + (1-\alpha)t_2) \supset \alpha \mathcal{A}(t_1) \times (1-\alpha)\mathcal{A}(t_2).$$

Theorem of Brunn-Minkowski. *Let $\{\mathcal{A}(t)\}_{t_1^* \leq t \leq t_2^*}$ be some convex one-parameter family. Consider $R(t) \equiv (\text{mes}\mathcal{A}(t))^{1/N}$. Then $\frac{d^2 R(t)}{dt^2} \leq 0$ and $\frac{d^2 R(t)}{dt^2} \equiv 0$ for $t \in [t_1', t_2']$ if and only if all the sets $\mathcal{A}(t)$ for $t \in [t_1', t_2']$ are homothetic to each other.*

For the proof of this theorem see, e.g., [3].

To use this theorem for the proof of (11) let us observe that the family $\{\Omega_N(U)\}_{U > U_{min}}$ is a convex one-parameter family and then, according to the Brunn-Minkowski theorem, the function $R(U) = (V_N(U))^{1/N}$ is a concave function. We get that $s_N(U)$ is a concave function:

$$\frac{d^2}{dU^2} s_N(U) = \frac{d^2}{dU^2} \log R(U) = \frac{R''(U)}{R(U)} - \left(\frac{R'(U)}{R(U)}\right)^2 \leq -\left(\frac{R'(U)}{R(U)}\right)^2.$$

$\frac{R'(U)}{R(U)} = \frac{d}{dU} s_N(U) > 1$ for $U < U^*$, and even if $\frac{d}{dU} s_N(U) = 0$ for $U > U^*$, we obtain that $\frac{d}{dU}(s_N(U) - U) = -1$. Thus, using the standard Laplace method, we get

$$f_N(\Phi_N) = s_N(U^*) - U^* + O\left(\frac{\log N}{N}\right)$$

$$= \frac{1}{N} \log V_N(U^*) - U^* + O\left(\frac{\log N}{N}\right), \qquad (20)$$

$$U_* \equiv \frac{1}{N}\langle \Phi_N \rangle_{\Phi_N} = U^* + o(1).$$

Using the condition (10), and taking J^*, which is the minimum point of $\Phi_N(J)$, we get

$$V_N(U^*) \geq N^{-1} \int_{J \in \mathcal{D}_N(U^*)} |(J - J^*, \nabla\Phi_N(J))| \cdot |\nabla\Phi_N(J)|^{-1} dS_J$$

$$\geq S_N(U^*) \frac{U^* - U_{min}}{\max_{J \in \mathcal{D}_N(U^*)} |\nabla\Phi_N(J)|} = N^{-1/2} S_N(U^*) C(U^*). \qquad (21)$$

On the other hand, for any $U < U^*$

$$\frac{S_N(U)}{N^{1/2} V_N(U)} \geq \min_{J \in \mathcal{D}_N(U)} |\nabla\Phi_N(J)| \frac{F_N(U)}{N^{1/2} V_N(U)}$$

$$\geq N^{1/2} \min_{J \in \mathcal{D}_N(U)} \frac{U - U_{min}}{|J - J^*|} \frac{d}{dU} s_N(U) \geq \tilde{C} \frac{d}{dU} s_N(U) > \tilde{C}. \qquad (22)$$

Here we have used (15) and (9). Thus the same inequality is valid also for $U = U^*$. Inequalities (22) and (21) imply that

$$\frac{1}{N} \log S_N(U^*) = \frac{1}{N} \log V_N(U^*) + O(\frac{\log N}{N}).$$

Combining this relation with (20) we get (11).

Let us observe also that for any (U_0, c_0) and (δ_U, δ_c) the family $\{\Omega_N(U_0 + t\delta_U, c_0 + t\delta_c)\}_{t \in [0,1]}$ is a convex one-parameter family and then, according to the Brunn-Minkowski theorem the function $R_N(t) \equiv V^{1/N}(U_0 + t\delta_U, c_0 + t\delta_c)$ is concave. But since in our consideration $N \to \infty$, to obtain that this function is strictly concave in some neighbourhood of the point (U^*, c^*) of maximum of $s_N(U, c) - U$, we shall use some lemma, which is the corollary from the theorem of Brunn-Minkowski.

Lemma *Consider the convex set $\mathcal{M} \subset \mathbf{R}^N$ whose boundary consists of a finite number of smooth pieces. Let the convex one-parameter family $\{\mathcal{A}(t)\}_{t_1^* \leq t \leq t_2^*}$ be given by the intersections of $\mathcal{M}$ with the parallel the hyper-planes $\mathcal{B}(t) \equiv \{J : (J, e) = tN^{1/2}\}$. Suppose that there is some smooth piece $\mathcal{D}$ of the boundary of $\mathcal{M}$, such that for any $J \in \mathcal{D}$ the minimal normal curvature satisfies the inequality $N^{1/2}\kappa_{min}(J) > K_0$, and the Lebesgue measure $S(t)$ of the intersection $\mathcal{D} \cap \mathcal{A}(t)$ satisfies the bound*

$$S(t) \geq N^{1/2}V(t)C(t), \tag{23}$$

where $V(t)$ is the volume of $\mathcal{A}(t)$. Then $\frac{d^2}{dt^2}V^{1/N}(t) \leq -K_0C(t)V^{1/N}(t)$.

As far as we know, the Gardner problem is one of the first problems of spin glass theory completely solved (i.e. for all values of α and k) in a rigorous way. The explanation is that the problem (5) can be reduced to the problem with the convex Hamiltonian (7) in a convex configuration space. It is just this convexity that allows us to prove the vanishing of all correlation functions for all values of α and k, while e.g. in the Hopfield and Sherrington-Kirkpatrick models the vanishing is valid only for small enough α or for high temperatures (see [7] for the physical theory and [11], [12], [14], [15] for the respective rigorous results). Also for the Gardner-Derrida [5] model there is only a justification of the Replica Simmetry solution in a certain region of parameters (see [16]).

3.   The Cavity Method and the Limit $\varepsilon \to 0$

As it was mentioned in Section 2, the vanishing of correlation functions is the key problem of applying of the cavity method to the model (7). It allows us to derive the selfconsistent equations for the order parameters of the model (7) and then find the expression for the free energy, which we use for deriving the result of Theorem 1 when $\varepsilon \to 0$.

Theorem 3. *For any $\alpha, k \geq 0$ and $z > 0$ the functions $F_{N,p}(\xi, k, h, z, \varepsilon)$ are self-averaging in the limit $N, p \to \infty$, $\alpha_N \equiv \frac{p}{N} \to \alpha$:*

$$E\left\{(F_{N,p}(\xi, k, h, z, \varepsilon) - EF_{N,p}(\xi, k, h, z, \varepsilon)\})^2\right\} \to 0 \tag{24}$$

and, if ε is small enough, $\alpha < 2$ and $z \leq \varepsilon^{-1/3}$, then there exists

$$\lim\nolimits_{N,p \to \infty, \alpha_N \to \alpha} E\{F_{N,p}(\xi, k, h, z, \varepsilon)\} = F(\alpha, k, h, z, \varepsilon),$$

$$F(\alpha, k, h, z, \varepsilon) \equiv \max\nolimits_{R>0} \min\nolimits_{0 \leq q \leq R} \left[\alpha E\left\{\log \mathrm{H}\left(\frac{u\sqrt{q}+k}{\sqrt{\varepsilon+R-q}}\right)\right\}\right. \tag{25}$$
$$\left. +\tfrac{1}{2}\frac{q}{R-q} + \tfrac{1}{2}\log(R-q) - \tfrac{z}{2}R + \tfrac{h^2}{2}(R-q)\right],$$

where u is a Gaussian random variable with zero mean and variance 1.

Let us note that the bound $\alpha < 2$ is not important for us, because for any $\alpha > \alpha_c(k)$ ($\alpha_c(k) < 2$ for any k) the free energy of the partition function $\Theta_{N,p}(\xi,k)$ tends to $-\infty$, as $N \to \infty$ (see Theorem 1 for the exact statement). The bound $z < \varepsilon^{-1/3}$ also is not a restriction for us. We could need to consider $z > \varepsilon^{-1/3}$ only if, applying the result on $f_N^*(U)$ of theorem 2 to the Hamiltonian $\mathcal{H}_{N,p}(\boldsymbol{J},\xi,k,h,z,\varepsilon)$, we obtain that the point of minimum $z_{min}(\varepsilon)$ does not satisfy this bound. But it is shown in theorem 1, that for any $\alpha < \alpha_c(k)$ $z_{min}(\varepsilon) < \bar{z}$ with some finite $\bar{z}$ depending only on k and α.

Our last step is the limiting transition $\varepsilon \to 0$, i.e. the proof that θ-functions in (5) can be replaced by $\mathrm{H}(\frac{x}{\sqrt{\varepsilon}})$ with a small difference when ε is small enough. It is the most difficult step from the technical point of view. It is rather straightforward to obtain that the free energy of (7) is an upper bound of $\frac{1}{N}\log\Theta_N,p(\xi,k)$. But the estimate from below is much more complicated. The problem is that to estimate the difference between the free energies corresponding to the two Hamiltonians we, as a rule, need to have them defined in a common configuration space, or at least, we need to know some a priori bounds for some Gibbs averages. In the case of the Gardner problem we do not possess this information. That is why we need to apply our geometrical theorem not only to the model (7) (for these purposes it would be enough to apply the results of [2]) but also to some models, interpolating between (7) and (5), with a complicated random (but convex) configuration space.

Acknowledgments

The authors would like to thank Prof. A.D. Milka for a fruitful discussion of the geometrical aspects of the problem and Prof. M. Talagrand for useful discussions.

References

[1] Bovier A., Gayrard V., Hopfield Models as a Generalized Random Mean Field Models, Mathematical Aspects of Spin Glasses and Neuronal Networks, A. Bovier, P. Picco Eds., Progress in Probability, Birkhauser, **41** (1998), 3–89.

[2] Brascamp H.J., Lieb E.H., On the Extension of the Brunn-Minkowsky and Pekoda-Leindler Theorems, Includings Inequalities for Log Concave functions, and with an application to the diffusion equation. J. Func. Anal. 22 (1976) 366–389.

[3] Hadwiger H., Vorlesungen uber Inhalt, Oberlache und Isoperimetrie. Springer-Verlag, 1957.

380

[4] Gardner G., The Space of Interactions in Neural Network Models, J.Phys.A: Math.Gen. 21 (1988) 271–284.

[5] Gardner E., Derrida B., Optimal Stage Properties of Neural Network Models, J.Phys.A: Math.Gen. 21 (1988) 257–270.

[6] Ghirlanda S., Guerra F., General Properties of Overlap Probability Distributions in Disordered Spin System, J.Phys.A: Math.Gen. 31(1988)9149-9155.

[7] Mezard, M., Parisi, G., Virasoro, M.A., Spin Glass Theory and Beyond, Singapore, World Scientific, 1987.

[8] Pastur L., Shcherbina M., Absence of Self-Averaging of the Order Parameter in the Sherrington-Kirkpatrick Model. J.Stat.Phys. 62 (1991) 1–26.

[9] Pastur L., Shcherbina M., Tirozzi B., The Replica-Symmetric Solution Without Replica Trick for the Hopfield Model. J. Stat. Phys. 74 5/6 (1994) 1161–1183.

[10] Pastur L., Shcherbina M., Tirozzi B., On the Replica Symmetric Equations for the Hopfield Model, J.Math.Phys. 40 8 (1999) 3930–3947.

[11] Shcherbina M., On the Replica Symmetric Solution for the Sherrington-Kirkpatrick Model, Helvetica Physica Acta 70 (1997) 772–797.

[12] Shcherbina M., Some Estimates for the Critical Temperature of The Sherrington-Kirkpatrick Model with Magnetic Field, Mathematical Results in Statistical Mechanics World Scientific, Singapore (1999) 455–474.

[13] Shcherbina M., Tirozzi, B. , Rigorous Solution of the Gardner Problem, to appear.

[14] Talagrand M., Rigorous Results for the Hopfield Model with Many Patterns. Prob. Theor. Rel. Fields 110 (1998) 176–277.

[15] Talagrand M., Exponential Inequalities and Replica Symmetry Breaking for the Sherrington-Kirkpatrick Model, Ann.Probab. 28 (2000) 1018–1068.

[16] Talagrand M., Intersecting Random Half-Spaces: Toward the Gardner-Derrida Problem, Ann.Probab. 28 (2000) 725–758.

[17] Talagrand M., Self Averaging and the Space of Interactions in Neural Networks, Random Structures and Algorithms 14 (1988) 199–213.

GENERALIZED HYPERGEOMETRIC SERIES AND THE SYMMETRIES OF 3-j AND 6-j COEFFICIENTS

K. Srinivasa Rao

The Institute of Mathematical Sciences, CIT Campus, Chennai - 600113, India

rao@imsc.ernet.in

H.D. Doebner

Arnold Sommerfeld Institut fur Mathematische Physik der Technische, Universität Clausthal, Liebnizstrasse 10, D-38678 Clausthal, Germany, F.R.

asibb@ibm.rz.tu-clausthal.de

P. Nattermann

Arnold Sommerfeld Institut fur Mathematische Physik der Technische, Universität Clausthal, Liebnizstrasse 10, D-38678 Clausthal, Germany, F.R.

aspn@pt.tu-clausthal.de

Abstract   The invariance groups for a set of transformations of the non-terminating $_3F_2(1)$ series, and for the set of Bailey transformations for terminating $_4F_3(1)$ series are shown to be S_5 and S_6, respectively. Transformations which relate different basis states are used to discuss the symmetries of the 3-j and 6-j coefficients.

Keywords: angular momentum, basis states, generalized hypergeometric series, group theory of transformations, Saalschützian series, symmetric groups, symmetries of 3-j and 6-j coefficients, transformation formulas.

1.   Introduction

It is shown that from a two-term relation for the non-terminating $_3F_2$ series of unit argument, due to Thomae [1], a set of ten transformations can be derived [2] and that from the Bailey [3] transformation for the terminating Saalschützian $_4F_3$ series of unit argument, a set of twenty transformations can be derived. Though Whipple [4] obtained a set of 120 $_3F_2(1)$ series and tabulated them, he did not recognize the group structure behind them. The group structure arises when a transformation formula between two generalized hypergeometric series is related to a linear transformation between the set of parameters of the two series. An application of two transformation formulas can then be looked upon as the composition of the linear transformations, thereby giving rise to a group structure [4,5,6]. For transformations of the non-terminating $_3F_2(1)$ series and for the terminating Saalschützian $_4F_3(1)$ Bailey transformations, these invariance groups have been shown by Beyer, Louck and Stein [4] to be, respectively, the symmetric groups S_5 and S_6. Here these results are established simply ,á la Hardy [7] as succinct, quintessential one-line statements.

The symmetries of the 3-j and the 6-j coefficients are described in terms of different basis states. The *classical* symmetries of these coefficients are described in the angular momentum basis states. This description implies the existence of sets of series representations for the coefficients. Once the sets of series representations for the 3-j and the 6-j coefficients are rearranged into sets of hypergeometric series, then the symmetries of the latter give rise to the Regge symmetries [8,9] on which are superposed the *classical* symmetries of these coefficients. Here, we obtain the transformations between the basis states which clearly show how one can relate the *classical* symmetries to the *mixed* symmetries – i.e. Regge symmetries on which are superposed the *classical* symmetries.

In section 2, the set of 10 transformations of the non-terminating $_3F_2(1)$ series are derived from a member of the set. A function in five variables is constructed and shown that its symmetric nature in all the five variables is a consequence of the transformations and a simple proof that the symmetry group is S_5 for the set of transformations is established. In section 3, the set of 20 transformations of the terminating Saalschützian $_4F_3(1)$ series are derived from the Bailey transformation. In this case a function of six variables is constructed and it reveals the invariance group structure of this set of transformations to be the symmetric group S_6. Basis states and the transformations of the same leading to a comprehensive understanding of the symmetries of the 3-j and the 6-j coefficients are presented in the sections 4 and 5, respectively. The article ends with a discussion of the results in section 6.

2. Set of ten Thomae transformations

Weber and Erdelyi [10] considered a member of a set of transformations for the terminating $_3F_2$ series of unit argument obtained by

Whipple [11],viz.:

$$
{}_3F_2\left(\begin{array}{c} a,b,-N \\ d,e \end{array}\right) = \Gamma\left(\begin{array}{c} d,d+N-a \\ d+N,d-a \end{array}\right) \ {}_3F_2\left(\begin{array}{c} a,e-b,-N \\ 1+a-d-N,e \end{array}\right)
$$
(1)

where we have used the notation :

$$
\Gamma\left(\begin{array}{c} p, \ q,\cdots \\ r, \ s,\cdots \end{array}\right) = \frac{\Gamma(p)\Gamma(q)\cdots}{\Gamma(r)\Gamma(s)\cdots}
$$
(2)

Weber and Erdelyi used (1) again with the roles of d and e interchanged to transform the rhs of (1) and obtained a second transformation. In ref. [5], we showed explicitly that such a recursive procedure can be continued and that it results in 18 terminating ${}_3F_2$ series and that when the manifestly evident $S_2 \times S_2$ symmetries due to the permutation of the numerator (a,b) and denominator (c,d) parameters are superimposed on these 18 transformations, a set of 72 transformations result and the group theory of these 72 transformations has been studied.

The two-term relation for the non-terminating ${}_3F_2$ series of unit argument given by Thomae [1] is :

$$
{}_3F_2\left(\begin{array}{c} a,b,c \\ d,e \end{array}\right) = \Gamma\left(\begin{array}{c} d,e,s \\ a,s+b,s+c \end{array}\right) \ {}_3F_2\left(\begin{array}{c} d-a,e-a,s \\ s+b,s+c \end{array}\right)
$$
(3)

where $s = d+e-a-b-c$ is the parameter excess. As in the case of the terminating ${}_3F_2$ series, a recursive use of (3) results [2] in the set of **ten** non-terminating Thomae transformations, given in Appendix 1. The non-terminating Thomae transformations which leave a numerator parameter unchanged – viz. (II), (III), (VI), (VII), (VIII) and (X), given in Appendix 1 – can be identified with the terminating ${}_3F_2(1)$ series (II) or (XV) given in the Appendix of ref. [5] with the unchanged parameter set equal to $-N$ and suitable relabeling and permutations of the numerator and/or denominator parameters.

In 1940, in a note, Hardy [7, p.111] stated that the theorem that

$$
\frac{1}{\Gamma(s)\Gamma(\beta_1)\Gamma(\beta_2)} F\left(\begin{array}{c} \alpha_1,\alpha_2,\alpha_3 \\ \beta_1,\beta_2 \end{array}\right)
$$
(4)

is a symmetric function of five arguments $\beta_1,\beta_2,s+\alpha_1,s+\alpha_2,s+\alpha_3$, where $s = \beta_1+\beta_2-\alpha_1-\alpha_2-\alpha_3$, is a consequence of the transformation (3).

Accordingly, we construct the function

$$f(x_1, x_2, x_3, x_4, x_5) = \frac{1}{\Gamma(s, 2x_4, 2x_5)} \; {}_3F_2 \left(\begin{array}{c} 2x_1 - s, 2x_2 - s, 2x_3 - s \\ 2x_4, \quad 2x_5 \end{array} \right)$$

(5)

where $s = x_1 + x_2 + x_3 - x_4 - x_5$. That this function is symmetric in all the five variables is a consequence of (3) can be proved as follows: the function $f(\vec{x}) \equiv f(x_1, x_2, x_3, x_4, x_5)$ is manifestly invariant for permutations of (x_1, x_2, x_3) and (x_4, x_5). Consider the permutation $p : x_1 \to x_2 \to x_3 \to x_4 \to x_5 \to x_1$, which is a permutation of order 5. Upon relabeling the parameters of the ${}_3F_2(1)$ in $f(\vec{x})$ as ${}_3F_2 \left(\begin{array}{c} a, b, c \\ d, e \end{array} \right)$, it is straight forward to see that corresponding to

$$f(\vec{x}) = f(p.\vec{x})$$

(6)

we get (II) of Appendix 1, on which are superposed permutations of the numerator, denominator parameters. Since $f(\vec{x})$ is invariant under this permutation p of order 5, and under the transposition, $x_4 \to x_5$, manifestly, the group generated by these two generators (viz. p and the transposition) is the complete group of permutations on 5 elements, $i.e.$ the symmetric group S_5 (see, for instance Budden [12]).

In section 4, we will show how this set of transformations when applied to the ${}_3F_2(1)$ representation for the 3-j coefficient will result in the non-terminating ${}_3F_2(1)$ forms for the 3-j coefficient obtained by Raynal [13,14].

3.   Set of twenty Bailey transformations

The two-term relation for the terminating Sasalschützian ${}_4F_3$ of unit argument is given by Bailey [3] as:

$$\begin{aligned} {}_4F_3 &\left(\begin{array}{c} A, B, C, -n \\ E, F, G \end{array} \right) \\ &= \frac{(F-C)_n (G-C)_n}{(F)_n (G)_n} \; {}_4F_3 \left(\begin{array}{c} E-A, \quad E-B, \quad C \quad W, -n \\ E, E+F-A-B, E+G-A-B \end{array} \right) \end{aligned}$$

(7)

where the Pochhammer symbol is defined for $n \geq 1$ as:

$$(a)_n = \frac{\Gamma(a+n)}{\Gamma(a)} = a(a+1)(a+2) \cdots (a+n-1), \quad \text{with} \quad (a)_0 = 1.$$

(8)

As in the cases of the terminating and non-terminating ${}_3F_2$ transformations of unit argument, a recursive use of the terminating Saalschützian

$_4F_3$ transformation of unit argument results in a set of **twenty** transformations, given in Appendix 2. In this case, we construct the function:

$$f(x_1, x_2, x_3, x_4, x_5, x_6)$$
$$= (x_1 + x_2 + x_3 + x_4, x_1 + x_2 + x_3 + x_5, x_1 + x_2 + x_3 + x_6)_n$$
$$\times {}_4F_3\left(\begin{matrix} x_1 + x_2, x_2 + x_3, x_3 + x_1, -n \\ x_1 + x_2 + x_3 + x_4, x_1 + x_2 + x_3 + x_5, x_1 + x_2 + x_3 + x_6 \end{matrix}\right) \quad (9)$$

with $(x, y, \cdots)_n = (x)_n (y)_n \cdots$ and $x_1 + x_2 + x_3 + x_4 + x_5 + x_6 = 1 - n$, for some non-negative integer n, which being symmetric in all the six variables $x_1, x_2, x_3, x_4, x_5, x_6$ is a consequence of (7). For, note that the given function is obviously invariant under permutations x_1, x_2, x_3 and x_4, x_5, x_6. Consider the cyclic permutation of order 6,

$$p: \quad x_1 \to x_6 \to x_4 \to x_2 \to x_5 \to x_3 \to x_1.$$

Relabel the parameters of the $_4F_3(1)$ in $f(\vec{x})$ as: $_4F_3\left(\begin{matrix} A, B, C, -n \\ E, F, G \end{matrix}\right)$. As in section 2, it is straight forward to show that corresponding to $f(\vec{x}) = f(p.\vec{x})$, we get (XII) belonging to the set of twenty terminating Bailey transformations (Appendix 2), after some algebra. Hence, the group generated by the two generators (viz. p and the transposition) is the complete symmetric group of 6 elements, *i.e.* the symmetric group S_6.

4. Basis states and symmetries of 3-j coefficient

The 3-j coefficient is related to a $_3F_2(1)$ as:

$$\begin{pmatrix} j_1 & j_2 & j_3 \\ m_1 & m_2 & m_3 \end{pmatrix} = \delta_{m_1+m_2+m_3,0} \prod_{i,k=1}^{3} [R_{ik}!/(J+1)!]^{1/2}(-1)^{\sigma(pqr)}$$

$$\times \sum_s [s!(R_{2p} - s)!(R_{3q} - s)!(R_{1r} - s)!$$

$$\times \quad (s + R_{3r} - R_{2p})!(s + R_{2r} - R_{3q})!]^{-1} \quad (10)$$

$$\begin{pmatrix} j_1 & j_2 & j_3 \\ m_1 & m_2 & m_3 \end{pmatrix} = \delta_{m_1+m_2+m_3,0} \prod_{i,k=1}^{3} [R_{ik}!/(J+1)!]^{1/2}$$

$$\times (-1)^{\sigma(pqr)}[\Gamma(1 - A, 1 - B, 1 - C, D, E)]^{-1}$$

$$\times {}_3F_2(A, B, C; D, E; 1) \quad (11)$$

where

$$A = -R_{2p}, B = -R_{3q}, C = -R_{1r}, D = 1 + R_{3r} - R_{2p}, E = 1 + R_{2r} - R_{3q},$$

(12)

$$J = j_1 + j_2 + j_3, \quad \Gamma(x, y, \cdots) = \Gamma(x)\Gamma(y)\cdots$$

(13)

for all six permutations of $(pqr) = (123)$ with

$$\sigma(pqr) = \begin{cases} R_{3p} - R_{2q} & \text{for even permutations,} \\ R_{3p} - R_{2q} + J & \text{for odd permutaions.} \end{cases}$$

(14)

The R_{ik} are elements of a 3×3 square symbol used by Regge (1958) to represent the 3-j coefficient as:

$$\begin{pmatrix} j_1 & j_2 & j_3 \\ m_1 & m_2 & m_3 \end{pmatrix} = \left\| \begin{matrix} -j_1 + j_2 + j_3 & j_1 - j_2 + j_3 & j_1 + j_2 - j_3 \\ j_1 - m_1 & j_2 - m_2 & j_3 - m_3 \\ j_1 + m_1 & j_2 + m_2 & j_3 + m_3 \end{matrix} \right\|$$

$$= \| R_{ik} \|$$

(15)

The 3×3 square symbol is a *magic* square, in that all column and row sums add to $J = j_1 + j_2 + j_3$. Further, the nine elements of $\| R_{ik} \|$ satisfy the nine relations:

$$R_{lp} + R_{mp} = R_{nq} + R_{nr}$$

(16)

for cyclic permutations of both (lmn) and $(pqr)=(123)$. We wish to draw attention to the fact that the *classical* symmetries of the 3-j coefficient are those which arise due to permutations of the columns of the 3-j symbol $\begin{pmatrix} j_1 & j_2 & j_3 \\ m_1 & m_2 & m_3 \end{pmatrix}$ on which the $m_i \rightarrow -m_i$ (spatial reflections) are superposed. These 12 symmetries immediately imply the existence of the six series representations. However, each of these six series representations exhibits 12 symmetries and they are the *mixed* symmetries – *i.e.* Regge symmetries of the 3-j coefficient on which the *classical* symmetries are superposed. That this set of six series representations, or equivalently the set of six $_3F_2(1)$s is *necessary and sufficient* to account for the 72 symmetries of the 3-j coefficient has been shown by Srinivasa Rao [15].

The twelve *classical* symmetries can be generated in the angular momentum or jm-basis denoted by the column vector whose elements are

$$|j, m\rangle \equiv |j_1, j_2, j_3, m_1, m_2, m_3\rangle \quad \varepsilon \quad \mathbb{R}^6_{j,m}.$$

(17)

The *classical* symmetry generators $g_i^{(j,m)}, i = 1, 2, 3$ for the 12-element group are:

$$
\begin{pmatrix}
0 & 1 & 0 & 0 & 0 & 0 \\
1 & 0 & 0 & 0 & 0 & 0 \\
0 & 0 & 1 & 0 & 0 & 0 \\
0 & 0 & 0 & 0 & 1 & 0 \\
0 & 0 & 0 & 1 & 0 & 0 \\
0 & 0 & 0 & 0 & 0 & 1
\end{pmatrix},
\begin{pmatrix}
1 & 0 & 0 & 0 & 0 & 0 \\
0 & 0 & 1 & 0 & 0 & 0 \\
0 & 1 & 0 & 0 & 0 & 0 \\
0 & 0 & 0 & 1 & 0 & 0 \\
0 & 0 & 0 & 0 & 0 & 1 \\
0 & 0 & 0 & 0 & 1 & 0
\end{pmatrix},
\begin{pmatrix}
1 & 0 & 0 & 0 & 0 & 0 \\
0 & 1 & 0 & 0 & 0 & 0 \\
0 & 0 & 1 & 0 & 0 & 0 \\
0 & 0 & 0 & -1 & 0 & 0 \\
0 & 0 & 0 & 0 & -1 & 0 \\
0 & 0 & 0 & 0 & 0 & -1
\end{pmatrix}.
\tag{18}
$$

The $_3F_2(1)$ has 3 numerator and 2 denominator parameters and has 12 symmetries due to its invariance under the permutations of its numerator and denominator parameters. Let $\mathbb{R}_{\beta,\alpha}^5$ be a 5-dimensional (β, α) space and let the column (ket) vector in this basis be:

$$
|\beta, \alpha\rangle \equiv |\beta_1, \beta_2, \beta_3, \alpha_1, \alpha_2\rangle \quad \varepsilon \quad \mathbb{R}_{\beta,\alpha}^5. \tag{19}
$$

It is straight forward to write down the generators $\gamma_i^{(\beta,\alpha)}$, $i = 1, 2, 3$, of this 12 element group as:

$$
\begin{pmatrix}
0 & 1 & 0 & 0 & 0 \\
1 & 0 & 0 & 0 & 0 \\
0 & 0 & 1 & 0 & 0 \\
0 & 0 & 0 & 1 & 0 \\
0 & 0 & 0 & 0 & 1
\end{pmatrix},
\begin{pmatrix}
1 & 0 & 0 & 0 & 0 \\
0 & 0 & 1 & 0 & 0 \\
0 & 1 & 0 & 0 & 0 \\
0 & 0 & 0 & 1 & 0 \\
0 & 0 & 0 & 0 & 1
\end{pmatrix},
\begin{pmatrix}
1 & 0 & 0 & 0 & 0 \\
0 & 1 & 0 & 0 & 0 \\
0 & 0 & 1 & 0 & 0 \\
0 & 0 & 0 & 0 & 1 \\
0 & 0 & 0 & 1 & 0
\end{pmatrix}.
\tag{20}
$$

Let T_1 be a transformation from the $\mathbb{R}_{j,m}^6$ space into the 5-dimensional $\mathbb{R}_{\beta,\alpha}^5$ space:

$$
T_1 : \mathbb{R}_{j,m}^6 \rightarrow \mathbb{R}_{\beta,\alpha}^5 \tag{21}
$$

The kernel of T_1 is:

$$
Ker(T_1) = \{(j_1, j_2, j_3, m_1, m_2, m_3) | m_1 + m_2 + m_3 = 0\} \equiv \mathbb{H}, \tag{22}
$$

the normal vector space of the physical hyperplance $\mathbb{H}$. Let T_2 be the reverse of the above transformation ($\mathbb{R}_{\beta,\alpha}^{\hat{5}}$ is a normal vector space of $\mathbb{H}$):

$$
i.e. \quad T_2 : \mathbb{R}_{\beta,\alpha}^5 \rightarrow \mathbb{R}_{\beta,\alpha}^{\hat{5}} \oplus \mathbb{H} \approx \mathbb{R}_{j,m}^6, \tag{23}
$$

with

$$
\text{in } \mathbb{R}_{\beta,\alpha}^5 \quad T_1.T_2 = \text{identity}, \tag{24}
$$

$$
\text{in } \mathbb{R}_{j,m}^6 \quad T_2.T_1 = \text{identity}. \tag{25}
$$

However, there are six pairs of transformations $T_1^{(pqr)}$ and $T_2^{(pqr)}$ that can be written down corresponding to the permutations of (pqr) in (26) – (28). To be explicit, for the case $(pqr) = (123)$, these transformations are given by the (5×6) and (6×5) matrices:

$$
T_1^{(123)} = \begin{pmatrix} 1 & 0 & 0 & -\frac{2}{3} & \frac{1}{3} & \frac{1}{3} \\ 0 & 1 & 0 & -\frac{1}{3} & \frac{2}{3} & -\frac{1}{3} \\ 1 & 1 & -1 & 0 & 0 & 0 \\ 1 & 0 & -1 & -\frac{1}{3} & \frac{2}{3} & -\frac{1}{3} \\ 0 & 1 & -1 & -\frac{2}{3} & \frac{1}{3} & \frac{1}{3} \end{pmatrix} \quad \text{and} \quad T_2^{(123)} = \frac{1}{2} \begin{pmatrix} 1 & 0 & 1 & 0 & -1 \\ 0 & 1 & 1 & -1 & 0 \\ 1 & 1 & 0 & -1 & -1 \\ -1 & 0 & 1 & 0 & -1 \\ 0 & 1 & -1 & 1 & 0 \\ 1 & -1 & 0 & -1 & 1 \end{pmatrix}
$$

$$(26)$$

These transformation matrices, $T_1^{(123)}$ and $T_2^{(123)}$, transform the basis vectors as:

$$|\beta, \alpha\rangle = T_1^{(123)} |j, m\rangle \quad \text{and} \quad |j, m\rangle = T_2^{(123)} |\beta, \alpha\rangle \qquad (27)$$

so that we have

$$\beta_1 = R_{21}, \; \beta_2 = R_{32}, \; \beta_3 = R_{13}, \; \alpha_1 = R_{21} - R_{33}, \; \alpha_2 = R_{32} - R_{23}, \quad (28)$$

after taking (22) into account and

$$j_1 = \frac{1}{2}(\beta_1 + \beta_3 - \alpha_2), j_2 = \frac{1}{2}(\beta_2 + \beta_3 - \alpha_1), j_3 = \frac{1}{2}(\beta_1 + \beta_2 - \alpha_1 - \alpha_2),$$

$$m_1 = \frac{1}{2}(-\beta_1 + \beta_3 - \alpha_2), m_2 = \frac{1}{2}(\beta_2 - \beta_3 + \alpha_1), m_3 = -m_1 - m_2.$$

$$(29)$$

From the generators $g_i^{(j,m)}$ in $\mathbb{R}_{j,m}^6$ we obtain the generators $g_i^{(\beta,\alpha)}$ in $\mathbb{R}_{\beta,\alpha}^5$ through:

$$g_i^{(\beta,\alpha)} = T_1 \cdot g_i^{(j,m)} \cdot T_2. \qquad (30)$$

The union of the generators $\gamma_i^{\beta,\alpha}$ and $g_i^{\beta,\alpha}$, viz. $\gamma_i^{\beta,\alpha} \cup g_i^{\beta,\alpha}$, generate the 72 symmetries for the 3-j coefficient.

Conversely, we could start with the three generators $\gamma_i^{(\beta,\alpha)}$ (36) in $\mathbb{R}_{\beta,\alpha}^5$ and obtain the generators $\gamma_i^{(j,m)}$ in $\mathbb{R}_{j,m}^6$ through:

$$\gamma_i^{(j,m)} = T_2 \cdot \gamma_i^{(\beta,\alpha)} \cdot T_1 \qquad (31)$$

and together, $\gamma_i^{(\beta,\alpha)} \cup \gamma_i^{(j,m)}$, these six generators generate the 72 symmetries of the 3-j coefficient. It is to be noted, that one could start with any pair of the transformation matrices $T_1^{(pqr)}, T_2^{(pqr)}$ in this procedure detailed above. In Appendix 3, we give the sets of generators $g_i^{(\beta,\alpha)}$ and $\gamma_i^{(j,m)}$ for $(pqr) = (123)$.

The $_3F_2(1)$ on the lhs of (11), for the 3-j coefficient can be transformed using any one of the set of Thomae transformations given in the Appendix 1. If we do so, we get the ten formulas for the 3-j coefficient, given explicitly by Raynal [13], using the parameters of Whipple [11] who studied the symmetry properties of the $_3F_2(1)$. The $_3F_2(1)$ in (11) is a terminating one, since its numerator parameters are all negative integers. Therefore, when the Thomae transformations corresponding to (I), (IV) and (IX) are used to transform the $_3F_2(1)$, we get the non-terminating forms of Raynal [13] – *i.e.* Raynal's eq.(14) with a, b interchanged and $m_i \rightarrow -m_i$, Raynal's eq.(13) and eq.(14), respectively. For the other Thomae transformations we get Raynal's eq.(15) – eq.(17) and the same with a, b interchanged and $m_i \rightarrow -m_i$. Thus, the nine different forms of Raynal [13] are directly obtained from the explicit transformations of the $_3F_2(1)$ given in the Appendix 1 (without resorting to Whipple parameters).

5. Basis states and symmetries of 6-j coefficient

The highly symmetric form of the 6-j coefficient was first noted by Regge [8], even though Racah [16] derived the single sum series for this coefficient. The independence of the 6-j coefficient from the 3-j coefficient elevates it to a fundamental status in the theory of angular momentum. Regge [8] observed:

$$\begin{Bmatrix} a & b & e \\ d & c & f \end{Bmatrix} = \Delta(abe)\,\Delta(cde)\,\Delta(acf)\,\Delta(bdf)$$

$$\times \sum_p (-1)^p\,(p+1)! \left[\prod_{i=1}^{4}(p-\alpha_i)!\prod_{j=1}^{3}(\beta_j-p)!\right]^{-1}, \quad (32)$$

to be invariant to the permutation of the four α and three β parameters. In (32),

$$\alpha_1 = a+b+e, \qquad \beta_1 = a+b+c+d,$$
$$\alpha_2 = c+d+e, \qquad \beta_2 = b+c+e+f,$$
$$\alpha_3 = a+c+f, \qquad \beta_3 = a+d+e+f,$$
$$\alpha_4 = b+d+f, \qquad p_{max} \leq p \leq p_{min},$$
$$p_{max} = max(\alpha_1, \alpha_2, \alpha_3, \alpha_4), \qquad p_{min} = min(\beta_1, \beta_2, \beta_3), \quad (33)$$

$$\Delta(xyz) = \left[\frac{(-x+y+z)!(x-y+z)!(x+y-z)!}{(x+y+z+1)!}\right]^{1/2},$$

and the α, β parameters satisfy the condition:

$$\sum_{i=1}^{4} \alpha_i = \sum_{j=1}^{3} \beta_j. \tag{34}$$

The *classical* symmetries of the 6-j coefficient are the ones which arise due to the permutations of the columns of the symbol on the rhs of (32) and the interchange of any pair of elements in one row of the symbol with the corresponding elements in the other row. These are called the *tetrahedral* symmetries. On the other hand, the lhs expression for the 6-j coefficient clearly exhibits 144 symmetries that arise due to the permutations of the four α and three β parameters. These 144 *mixed* symmetries are made of the six Regge [8] symmetries on which are superposed the 24 *classical* symmetries. Let $\mathbb{R}_x^6$ be the six dimensional angular momentum space and let the column (ket) vector in this space be:

$$|\vec{x}\rangle \equiv |a, b, c, d, e, f\rangle \quad \varepsilon \quad \mathbb{R}_x^6. \tag{35}$$

The *classical* symmetry generators g_i^x, $i = 1, 2, 3$, in $\mathbb{R}_x^6$ are:

$$\begin{pmatrix} 0 & 1 & 0 & 0 & 0 & 0 \\ 1 & 0 & 0 & 0 & 0 & 0 \\ 0 & 0 & 0 & 1 & 0 & 0 \\ 0 & 0 & 1 & 0 & 0 & 0 \\ 0 & 0 & 0 & 0 & 1 & 0 \\ 0 & 0 & 0 & 0 & 0 & 1 \end{pmatrix}, \begin{pmatrix} 1 & 0 & 0 & 0 & 0 & 0 \\ 0 & 0 & 0 & 0 & 1 & 0 \\ 0 & 0 & 0 & 0 & 0 & 1 \\ 0 & 0 & 0 & 1 & 0 & 0 \\ 0 & 1 & 0 & 0 & 0 & 0 \\ 0 & 0 & 1 & 0 & 0 & 0 \end{pmatrix}, \begin{pmatrix} 0 & 0 & 0 & 1 & 0 & 0 \\ 0 & 0 & 1 & 0 & 0 & 0 \\ 0 & 1 & 0 & 0 & 0 & 0 \\ 1 & 0 & 0 & 0 & 0 & 0 \\ 0 & 0 & 0 & 0 & 1 & 0 \\ 0 & 0 & 0 & 0 & 0 & 1 \end{pmatrix}. \tag{36}$$

Let $\mathbb{R}_y^7$ be a 7-dimensional space and let the column (ket) vector in this basis be:

$$|\vec{y}\rangle \equiv |\alpha_1, \alpha_2, \alpha_3, \alpha_4, \beta_1, \beta_2, \beta_3\rangle \quad \varepsilon \quad \mathbb{R}_y^7. \tag{37}$$

It is straight forward to write down the generators γ_i^y, $i = 1, \cdots 4$, for the $S_4 \times S_3$ group as:

$$
\begin{pmatrix}
0 & 1 & 0 & 0 & 0 & 0 & 0 \\
0 & 0 & 1 & 0 & 0 & 0 & 0 \\
0 & 0 & 0 & 1 & 0 & 0 & 0 \\
1 & 0 & 0 & 0 & 0 & 0 & 0 \\
0 & 0 & 0 & 0 & 1 & 0 & 0 \\
0 & 0 & 0 & 0 & 0 & 1 & 0 \\
0 & 0 & 0 & 0 & 0 & 0 & 1
\end{pmatrix}, \quad
\begin{pmatrix}
0 & 1 & 0 & 0 & 0 & 0 & 0 \\
1 & 0 & 0 & 0 & 0 & 0 & 0 \\
0 & 0 & 1 & 0 & 0 & 0 & 0 \\
0 & 0 & 0 & 1 & 0 & 0 & 0 \\
0 & 0 & 0 & 0 & 1 & 0 & 0 \\
0 & 0 & 0 & 0 & 0 & 1 & 0 \\
0 & 0 & 0 & 0 & 0 & 0 & 1
\end{pmatrix},
$$

$$
\begin{pmatrix}
1 & 0 & 0 & 0 & 0 & 0 & 0 \\
0 & 1 & 0 & 0 & 0 & 0 & 0 \\
0 & 0 & 1 & 0 & 0 & 0 & 0 \\
0 & 0 & 0 & 1 & 0 & 0 & 0 \\
0 & 0 & 0 & 0 & 0 & 1 & 0 \\
0 & 0 & 0 & 0 & 0 & 0 & 1 \\
0 & 0 & 0 & 0 & 1 & 0 & 0
\end{pmatrix}, \quad
\begin{pmatrix}
1 & 0 & 0 & 0 & 0 & 0 & 0 \\
0 & 1 & 0 & 0 & 0 & 0 & 0 \\
0 & 0 & 1 & 0 & 0 & 0 & 0 \\
0 & 0 & 0 & 1 & 0 & 0 & 0 \\
0 & 0 & 0 & 0 & 0 & 1 & 0 \\
0 & 0 & 0 & 0 & 1 & 0 & 0 \\
0 & 0 & 0 & 0 & 0 & 0 & 1
\end{pmatrix}. \quad (38)
$$

These are projected onto the invariant subspace (34), using projection matrix:

$$
\mathbb{P} = I_{7\times 7} - \frac{1}{7}\hat{P} \tag{39}
$$

where $I_{7\times 7}$ is the 7-dimensional identity matrix and

$$
\hat{P} = \begin{pmatrix}
1 & 1 & 1 & 1 & -1 & -1 & -1 \\
1 & 1 & 1 & 1 & -1 & -1 & -1 \\
1 & 1 & 1 & 1 & -1 & -1 & -1 \\
1 & 1 & 1 & 1 & -1 & -1 & -1 \\
-1 & -1 & -1 & -1 & 1 & 1 & 1 \\
-1 & -1 & -1 & -1 & 1 & 1 & 1 \\
-1 & -1 & -1 & -1 & 1 & 1 & 1
\end{pmatrix}. \tag{40}
$$

Let T_3 be the transformation from $\mathbb{R}_x^6$ to $\mathbb{R}_y^7$:

$$
T_3 \quad : \quad \mathbb{R}_x^6 \quad \rightarrow \quad \mathbb{R}_y^7 \tag{41}
$$

and let T_4 be the reverse of the above transformation from $\mathbb{R}_y^7$ to $\mathbb{R}_x^6$:

$$
T_4 \quad : \quad \mathbb{R}_y^7 \quad \rightarrow \quad \mathbb{R}_x^6. \tag{42}
$$

Explicitly, T_3 and T_4 are given by:

$$T_3 = \begin{pmatrix} 1 & 1 & 0 & 0 & 1 & 0 \\ 0 & 0 & 1 & 1 & 1 & 0 \\ 1 & 0 & 1 & 0 & 0 & 1 \\ 0 & 1 & 0 & 1 & 0 & 1 \\ 1 & 1 & 1 & 1 & 0 & 0 \\ 1 & 0 & 0 & 1 & 1 & 1 \\ 0 & 1 & 1 & 0 & 1 & 1 \end{pmatrix}, \quad T_4 = \frac{1}{2} \begin{pmatrix} 1 & 0 & 1 & 0 & 0 & 0 & -1 \\ 1 & 0 & 0 & 1 & 0 & -1 & 0 \\ 0 & 1 & 1 & 0 & 0 & -1 & 0 \\ 0 & 1 & 0 & 1 & 0 & 0 & -1 \\ 1 & 1 & 0 & 0 & -1 & 0 & 0 \\ 0 & 1 & 1 & 1 & -1 & 0 & 0 \end{pmatrix} \cdot \mathbb{P} \quad (43)$$

such that

$$\text{in} \quad \mathbb{R}_x^6 \quad T_3.T_4 = identity,$$
$$\text{in} \quad \mathbb{R}_y^7 \quad T_4.T_3 = identity \quad (44)$$

and note that the former is an identity only after the condition (34) is taken into account.

From the generators g_i^x (36) in $\mathbb{R}_x^6$ we obtain the corresponding generators g_i^y in $\mathbb{R}_y^7$ through:

$$g_i^y = T_3 \cdot g_i^x \cdot T_4. \quad (45)$$

Conversely, starting with the five generators γ_i^y (38) in $\mathbb{R}_y^7$ the corresponding generators γ_i^x in $\mathbb{R}_x^6$ can be obtained through:

$$\gamma_i^x = T_4 \cdot \gamma_i^y \cdot T_3. \quad (46)$$

These 6×6 matrices are directly related to the mixed symmetries [8] of the 6-j coefficient – i.e. Regge symmetries on which the *classical* symmetries are superposed – when they operate on the basis vector $|\vec{x}\rangle$. These four generators γ_i^x also generate the 144-element group of symmetries of the 6-j coefficient. The generators g_i^y and γ_i^x are given in Appendix 3.

In passing, it should be mentioned that the highly symmetric form (32) which exhibits the 144 symmetries of the 6-j coefficient cannot be written down as a $_4F_3(1)$. To do so, the summation index has to be changed from p to $n = \beta_1 - p$ (say). In which case, the resultant form can be cast into a $_4F_3(1)$:

$$\begin{Bmatrix} a & b & e \\ d & c & f \end{Bmatrix}$$

$$=\Delta(abe)\,\Delta(cde)\,\Delta(acf)\,\Delta(bdf)\,(-1)^{\beta_1}$$

$$\times \sum_n (-1)^n (\beta_1 - n + 1)! \left[\prod_{i=1}^{4} (\beta_1 - \alpha_i - n)! \; \prod_{j=1}^{3} (\beta_j - \beta_1 + n)! \right]^{-1}$$

$$=\Delta(abe)\,\Delta(cde)\,\Delta(acf)\,\Delta(bdf)(-1)^{E+1}\Gamma(1-E)$$

$$\times \left[\Gamma(1-A, 1-B, 1-C, D, F, G)\right]^{-1} {}_4F_3(A,B,C,D;E,F,G;1),$$

$$(47)$$

where

$$A = e - a - b, \quad B = e - c - d, \quad C = f - a - c, \quad D = f - b - d,$$
$$E = a + b + c + d, \quad F = e + f - a - d, \quad G = e + f - b - c, \qquad (48)$$
$$\Gamma(x, y, \ldots) = \Gamma(x)\Gamma(y)\ldots.$$

However, this procedure has destroyed the symmetry in (32), in that one of the β parameters is shifted to the numerator of the series and therefore, the series is invariant to the permutation of the four αs and of β_2 and β_3 only. Equivalently, in the corresponding $_4F_3(1)$, the 4! numerator parameter permutations (of A,B,C,D) and 2 of the denominator parameter permutations (of F and G) will give rise to the 48 symmetries. This is attributable to the nature of the parameters: all four of the numerator parameters A,B,C,D of the $_4F_3(1)$ are negative integers, since angular momenta a, b, c, d, e, f can only take integer or half-integer values in quantum physics, while F and G have to be positive integers for the $_4F_3(1)$ to be defined. E being a negative integer, if it is permuted with the positive integer parameter F or G, it will result in an *unphysical*[†] symmetry for the 6-j coefficient like that found by Minton [17] when he used the Bailey transformation (7). Thus, this $_4F_3(1)$ exhibits only 48 of the 144 symmetries and hence it was shown by Srinivasa Rao *et.al.* [18] that there exist a set of three $_4F_3(1)$s which is *necessary and sufficient* to account for the 144 symmetries of the 6-j coefficient.

Equivalently, it has also been shown by Srinivasa Rao and Venkatesh [19] that there exist a set of four $_4F_3(1)$s, obtained from (32) by replacing

---

[†]In the sense that 'unphysical' (negative) angular momenta occur as arguments of the 6-j coefficient.

$p - \alpha_i$, (i=1,2,3,4) by n, successively and that each one of those series will account for only 36 of the 144 symmetries. The two sets of $_4F_3(1)$s have been shown by Srinivasa Rao and Rajeswari [20] to be intimately related to each other by the reversal of series of any given $_4F_3(1)$.

6.   Discussion

The set of ten non-terminating $_3F_2(1)$ transformations are obtained by a simple recursive procedure from the given Thomae transformation. A succinct, one-line statement (6) for the same is also given in terms of a function which has been constructed to show its manifest invariance to the permutations of five parameters, there by exhibiting that the invariance group for the Thomae transformations is S_5, as shown by Beyer *et.al.* [4]. The use of these transformations in the expression for the 3-j coefficient directly results in the nine formulas for it, obtained by Raynal [13] using Whipple [11] parameterization.

The set of twenty transformations when the recursive procedure is used on the given terminating Salschützian hypergeoemtric $_4F_3(1)$ Bailey transformation [3, ch.7] are given explicitly. A function $f(\vec{x})$ constructed with the $_4F_3(1)$ exhibits manifest invariance under the permutation of six parameters. The Bailey transformations are then simply written as $f(\vec{x}) = f(p.\vec{x})$, where p is a permutation belonging to S_6, the invariance group of the transformations. Though the 6-j coefficient is expressible as a member of any one of two equivalent sets of $_4F_3(1)$s, the use of the Bailey transformation on it does not result in any meaningful symmetries for the 6-j coefficient and in ref. [18] it was shown that it *at best* results in a $j \to -j - 1$ substitution.

Transformations from the angular momentum basis states to certain parameter spaces for the 3-j and 6-j coefficients have been obtained and these relate the *classical* symmetries of these coefficients to their 72-element and 144-element symmetry groups, respectively, which are manifestations of their Regge symmetries on which the classical symmetries are superimposed. This procedure is natural and transparent as opposed to the methodology adopted to describe the symmetries in terms of canonical parameters for these angular momentum coefficients used by Lockwood [21] and elaborated by Venkatesh [22], in which the parameter which determines the number of terms in the series is kept on a different and unnatural footing.

Askey [23] has stressed the need to develop other methods to study hypergeometric functions, since the very powerful differential equation point of view does not work well for $p \geq 3$ or $q \geq 2$ in $_pF_q(z)$. Here, we have shown how group theoretical methods can be used to describe transformations of hypergeometric series and how the use of transformations of the basis states enables us to understand the symmetries of the 3-j and 6-j coefficients.

Acknowledgments

The authors acknowledge the use of Mathematica to verify the statements made in this article. One of us (KSR) wishes to thank the Alexander von Humboldt Stiftung for partial financial support. He also thanks Dr. J. Noffke for help with the computer management and the Arnold Sommerfeld Institute for hospitality.

Appendix 1.

Let us denote the parameters of the $_3F_2$ on the rhs and the lhs of the Thomae transformation (3) by column vectors:

$$\vec{x} = (a,b,c,d,e) \quad \text{and} \quad \vec{x'} = (a',b',c',d',e'). \tag{49}$$

They are then related by the linear transformation:

$$\vec{x'} = g_1 \vec{x} \tag{50}$$

where the transformation matrix g_1 is :

$$g_1 = \begin{pmatrix} -1 & 0 & 0 & 1 & 0 \\ -1 & 0 & 0 & 0 & 1 \\ -1 & -1 & -1 & 1 & 1 \\ -1 & 0 & -1 & 1 & 1 \\ -1 & -1 & 0 & 1 & 1 \end{pmatrix} \tag{51}$$

A recursive use of (3) results in the ten transformations of the nonterminating $_3F_2(1)$ given below:

$$_3F_2\left(\begin{array}{c} a,b,c \\ d,e \end{array} \right) = \Gamma\left(\begin{array}{c} d,e,s \\ a,s+b,s+c \end{array} \right) {}_3F_2\left(\begin{array}{c} d-a,e-a,s \\ s+b,s+c \end{array} \right), \tag{I}$$

$$= \Gamma\left(\begin{array}{c} d,s \\ d-a,s+a \end{array} \right) {}_3F_2\left(\begin{array}{c} e-c,e-b,a \\ e,s+a \end{array} \right), \tag{II}$$

$$= \Gamma \left(\begin{matrix} e, s \\ e - c, s + c \end{matrix} \right) \; {}_3F_2 \left(\begin{matrix} c, d - b, d - a \\ s + c, d \end{matrix} \right), \tag{III}$$

$$= \Gamma \left(\begin{matrix} d, e, s \\ c, s + a, s + b \end{matrix} \right) \; {}_3F_2 \left(\begin{matrix} s, d - c, e - c \\ s + a, s + b \end{matrix} \right), \tag{IV}$$

$$= {}_3F_2 \left(\begin{matrix} a, b, c \\ d, e \end{matrix} \right) \qquad \text{(identity)}, \tag{V}$$

$$= \Gamma \left(\begin{matrix} e, s \\ e - a, s + a \end{matrix} \right) \; {}_3F_2 \left(\begin{matrix} d - c, d - b, a \\ d, s + a \end{matrix} \right), \tag{VI}$$

$$= \Gamma \left(\begin{matrix} d, s \\ d - c, s + c \end{matrix} \right) \; {}_3F_2 \left(\begin{matrix} c, e - b, e - a \\ s + c, e \end{matrix} \right), \tag{VII}$$

$$= \Gamma \left(\begin{matrix} e, s \\ e - b, s + b \end{matrix} \right) \; {}_3F_2 \left(\begin{matrix} b, d - c, d - a \\ d, s + b \end{matrix} \right), \tag{VIII}$$

$$= \Gamma \left(\begin{matrix} d, e, s \\ b, s + a, s + c \end{matrix} \right) \; {}_3F_2 \left(\begin{matrix} d - b, s, e - b \\ s + a, s + c \end{matrix} \right), \tag{IX}$$

$$= \Gamma \left(\begin{matrix} d, s \\ d - b, s + b \end{matrix} \right) \; {}_3F_2 \left(\begin{matrix} e - c, e - a, b \\ s + b, e \end{matrix} \right). \tag{X}$$

If on these ten transformations we superpose the manifest permutational symmetries of the numerator and denominator parameters, then we get a set of 120 elements. The 5×5 matrix representation for this set can be obtained by taking into account in addition to the generator g_1, the generators of the S_3 and S_2 groups:

$$g_2 = \begin{pmatrix} 0\,1\,0\,0\,0 \\ 0\,0\,1\,0\,0 \\ 1\,0\,0\,0\,0 \\ 0\,0\,0\,1\,0 \\ 0\,0\,0\,0\,1 \end{pmatrix}, \quad g_3 = \begin{pmatrix} 0\,1\,0\,0\,0 \\ 1\,0\,0\,0\,0 \\ 0\,0\,1\,0\,0 \\ 0\,0\,0\,1\,0 \\ 0\,0\,0\,0\,1 \end{pmatrix}, \quad g_4 = \begin{pmatrix} 1\,0\,0\,0\,0 \\ 0\,1\,0\,0\,0 \\ 0\,0\,1\,0\,0 \\ 0\,0\,0\,0\,1 \\ 0\,0\,0\,1\,0 \end{pmatrix} \tag{52}$$

A scaling on the ${}_3F_2(\vec{x})$ through the transformation:

$$ {}_3\tilde{F}_2(\vec{x}) = \frac{1}{\Gamma(d, e, s)} {}_3F_2(\vec{x}) \tag{53}$$

enables us to write the Thomae transformation (3) as:

$$ {}_3\tilde{F}_2(\vec{x}) = {}_3\tilde{F}_2(g_1\vec{x}). \tag{54}$$

The other three generating transformations become:

$$ {}_3\tilde{F}_2(\vec{x}) = {}_3\tilde{F}_2(g_2\vec{x}) = {}_3\tilde{F}_2(g_3\vec{x}) = {}_3\tilde{F}_2(g_4\vec{x}). \tag{55}$$

In general, this implies that the scaled non-terminating $_3\tilde{F}_2$ with unit argument satisfies:

$$_3\tilde{F}_2(\vec{x}) = {}_3\tilde{F}_2(g\vec{x}), \qquad \forall \ \ g \ \epsilon \ G_T \qquad (56)$$

where G_T is the 120 element invariance group of the non-terminating $_3F_2$ transformations. That this invariance group is the symmetric group S_5 has been proved by Beyer, Louck and Stein [4].

Appendix 2.

The twenty Bailey transformations of the terminating $_4F_3(1)$ are given below. The explicit set of transformations are being presented here for the first time. Though the one-line statement

$$f(\vec{x}) \ = \ f(p.\vec{x}), \quad p \ \epsilon \ S_6 \quad \text{and} f(x) \ \text{asin}(9),$$

is useful to summarize these, explicit forms are essential for actual applications.

$$_4F_3\begin{pmatrix} A, B, C, -n \\ E, F, G \end{pmatrix} =$$

$$= \frac{(F-C)_n(G-C)_n}{(F)_n(G)_n} \ {}_4F_3\begin{pmatrix} E-A, E-B, C, -n \\ E, E+F-A-B, E+G-A-B \end{pmatrix} \qquad (I)$$

$$= {}_4F_3\begin{pmatrix} A, B, C, -n \\ E, F, G \end{pmatrix} \qquad \text{(identity)} \qquad (II)$$

$$= \frac{(E-C)_n(G-C)_n}{(E)_n(G)_n} \ {}_4F_3\begin{pmatrix} F-A, F-B, C, -n \\ E+F-A-B, F, E+G-A-B \end{pmatrix} \qquad (III)$$

$$= \frac{(E-C)_n(F-C)_n}{(E)_n(F)_n} \ {}_4F_3\begin{pmatrix} G-A, G-B, C, -n \\ E+G-A-B, F+G-A-B, G \end{pmatrix} \qquad (IV)$$

$$= \frac{(F-B)_n(G-B)_n}{(F)_n(G)_n} \ {}_4F_3\begin{pmatrix} E-A, B, E-C, -n \\ E, E+F-A-C, E+G-A-C \end{pmatrix} \qquad (V)$$

$$= (-1)^n \frac{(A)_n(G-B)_n(G-C)_n}{(E)_n(F)_n(G)_n}$$
$$\times {}_4F_3\begin{pmatrix} E-A, F-A, E+F-A-B-C, -n \\ E+F-A-B, E+F-A-C, E+F+G-2A-B-C \end{pmatrix} \ (VI)$$

$$= (-1)^n \frac{(A)_n(F-B)_n(F-C)_n}{(E)_n(F)_n(G)_n}$$
$$\times {}_4F_3\begin{pmatrix} E-A, G-A, E+G-A-B-C, -n \\ E+G-A-B, E+G-A-C, E+F+G-2A-B-C \end{pmatrix} (VII)$$

$$= \frac{(F-A)_n(G-A)_n}{(F)_n(G)_n} \ {}_4F_3\begin{pmatrix} A, E-B, E-C, -n \\ E, E+F-B-C, E+G-B-C \end{pmatrix} \qquad (VIII)$$

$$= (-1)^n \frac{(G-A)_n (B)_n (G-C)_n}{(E)_n (F)_n (G)_n}$$
$$\times {}_4F_3 \left(\begin{matrix} E-B, F-B, E+F-A-B-C, -n \\ E+F-A-B, E+F-A-C, E+F+G-A-2B-C \end{matrix} \right) \quad (IX)$$

$$\times {}_4F_3 \left(\begin{matrix} E-B, G-B, E+G-A-B-C, -n \\ E+G-A-B, E+G-B-C, E+F+G-A-2B-C \end{matrix} \right) \quad (X)$$

$$= \frac{(E-A)_n (G-A)_n}{(E)_n (G)_n} \; {}_4F_3 \left(\begin{matrix} A, F-B, F-C, -n \\ E+F-B-C, F, F+G-B-C \end{matrix} \right) \quad (XI)$$

$$= (-1)^n \frac{(A)_n (E-B)_n (E-C)_n}{(E)_n (F)_n (G)_n}$$
$$\times {}_4F_3 \left(\begin{matrix} F-A, G-A, F+G-A-B-C, -n \\ F+G-A-B, F+G-A-C, E+F+G-2A-B-C \end{matrix} \right) \quad (XII)$$

$$= \frac{(E-B)_n (G-B)_n}{(E)_n (G)_n} \; {}_4F_3 \left(\begin{matrix} F-A, B, F-C, -n \\ E+F-A-C, F, F+G-A-C \end{matrix} \right) \quad (XIII)$$

$$= \frac{(E-A)_n (F-A)_n}{(E)_n (F)_n} \; {}_4F_3 \left(\begin{matrix} A, G-B, G-C, -n \\ E+G-B-C, F+G-B-C, G \end{matrix} \right) \quad (XIV)$$

$$= \frac{(E-B)_n (F-B)_n}{(E)_n (F)_n} \; {}_4F_3 \left(\begin{matrix} G-A, B, G-C, -n \\ E+G-A-C, F+G-A-C, G \end{matrix} \right) \quad (XV)$$

$$= (-1)^n \frac{(G-A)_n (G-B)_n (C)_n}{(E)_n (F)_n (G)_n}$$
$$\times {}_4F_3 \left(\begin{matrix} E-C, F-C, E+F-A-B-C, -n \\ E+F-A-C, E+F-B-C, E+F+G-A-B-2C \end{matrix} \right)$$
$$\quad (XVI)$$

$$= (-1)^n \frac{(E-A)_n (B)_n (E-C)_n}{(E)_n (F)_n (G)_n}$$
$$\times {}_4F_3 \left(\begin{matrix} F-B, G-B, F+G-A-B-C, -n \\ F+G-A-B, F+G-B-C, E+F+G-A-2B-C \end{matrix} \right)$$
$$\quad (XVII)$$

$$= (-1)^n \frac{(F-A)_n (F-B)_n (C)_n}{(E)_n (F)_n (G)_n}$$
$$\times {}_4F_3 \left(\begin{matrix} E-C, G-C, E+G-A-B-C, -n \\ E+G-A-C, E+G-B-C, E+F+G-A-B-2C \end{matrix} \right)$$
$$\quad (XVIII)$$

$$= (-1)^n \frac{(E-A)_n (E-B)_n (C)_n}{(E)_n (F)_n (G)_n}$$
$$\times {}_4F_3 \left(\begin{matrix} F-C, G-C, F+G-A-B-C, -n \\ F+G-A-C, F+G-B-C, E+F+G-A-B-2C \end{matrix} \right)$$
$$\quad (XIX)$$

$$= (-1)^n \frac{(A)_n (B)_n (C)_n}{(E)_n (F)_n (G)_n}$$
$$\times {}_4F_3 \left(\begin{matrix} E+F-A-B-C, F+G-A-B-C, E+G-A-B-C, -n \\ E+F+G-2A-B-C, E+F+G-A-2B-C, E+F+G-A-B-2C \end{matrix} \right)$$
$$\quad (XX)$$

Let us denote the parameters of the $_4F_3$ on the rhs and the lhs of the Bailey transformation (7) by column vectors:

$$\vec{x} = (A, B, C, E, F, G) \quad \text{and} \quad \vec{x'} = (A', B', C', E', F', G'). \tag{57}$$

They are then related by the linear transformation:

$$\vec{x'} = g_1 \vec{x} \tag{58}$$

where the transformation matrix g_1 is:

$$g_1 = \begin{pmatrix} -1 & 0 & 0 & 1 & 0 & 0 \\ 0 & -1 & 0 & 1 & 0 & 0 \\ 0 & 0 & 1 & 0 & 0 & 0 \\ 0 & 0 & 0 & 1 & 0 & 0 \\ -1 & -1 & 0 & 1 & 1 & 0 \\ -1 & -1 & 0 & 1 & 0 & 1 \end{pmatrix} \tag{59}$$

If on the twenty transformations given above we superpose the manifest permutational symmetries of the numerator and denominator parameters, then we get a set of 720 elements. The 6×6 matrix representation for this set can be obtained by taking into account in addition to the generator g_1, the generators of the two S_3 groups :

$$g_2 = \begin{pmatrix} 0 & 1 & 0 & 0 & 0 & 0 \\ 0 & 0 & 1 & 0 & 0 & 0 \\ 1 & 0 & 0 & 0 & 0 & 0 \\ 0 & 0 & 0 & 1 & 0 & 0 \\ 0 & 0 & 0 & 0 & 1 & 0 \\ 0 & 0 & 0 & 0 & 0 & 1 \end{pmatrix} \quad \text{and} \quad g_3 = \begin{pmatrix} 0 & 1 & 0 & 0 & 0 & 0 \\ 1 & 0 & 0 & 0 & 0 & 0 \\ 0 & 0 & 1 & 0 & 0 & 0 \\ 0 & 0 & 0 & 1 & 0 & 0 \\ 0 & 0 & 0 & 0 & 1 & 0 \\ 0 & 0 & 0 & 0 & 0 & 1 \end{pmatrix}, \tag{60}$$

$$g_4 = \begin{pmatrix} 1 & 0 & 0 & 0 & 0 & 0 \\ 0 & 1 & 0 & 0 & 0 & 0 \\ 0 & 0 & 1 & 0 & 0 & 0 \\ 0 & 0 & 0 & 0 & 1 & 0 \\ 0 & 0 & 0 & 0 & 0 & 1 \\ 0 & 0 & 0 & 1 & 0 & 0 \end{pmatrix} \quad \text{and} \quad g_5 = \begin{pmatrix} 1 & 0 & 0 & 0 & 0 & 0 \\ 0 & 1 & 0 & 0 & 0 & 0 \\ 0 & 0 & 1 & 0 & 0 & 0 \\ 0 & 0 & 0 & 0 & 1 & 0 \\ 0 & 0 & 0 & 1 & 0 & 0 \\ 0 & 0 & 0 & 0 & 0 & 1 \end{pmatrix}. \tag{61}$$

A scaling on the $_4F_3(\vec{x})$ through the transformation :

$$_4\tilde{F}_3(\vec{x}) = (E)_n (F)_n (G)_n {}_4F_3(\vec{x}) \tag{62}$$

enables us to write the Bailey transformation as :

$$_4\tilde{F}_3(\vec{x}) = {}_4\tilde{F}_3(g_1\vec{x}). \tag{63}$$

The other four generating transformations become :

$$_4\tilde{F}_3(\vec{x}) = {_4}\tilde{F}_3(g_2\vec{x}) = {_4}\tilde{F}_3(g_3\vec{x}) = {_4}\tilde{F}_3(g_4\vec{x}) = {_4}\tilde{F}_3(g_5\vec{x}). \qquad (64)$$

In general, this implies that the scaled terminating $_4\tilde{F}_3$ with unit argument satisfies :

$$_4\tilde{F}_3(\vec{x}) = {_4}\tilde{F}_3(g\vec{x}), \qquad \forall \quad g \in G_T \qquad (65)$$

where G_T is the 720 element invariance group of the terminating $_4F_3$ transformations. That this invariance group is the symmetric group S_6 has been proved by Beyer, Louck and Stein [4].

Appendix 3.

Explicitly the generators $g_i^{(\beta,\alpha)}$ in $\mathbb{R}_{\beta,\alpha}^5$, for $(pqr) = (123)$, are:

$$\begin{pmatrix} 0 & 0 & 1 & -1 & 0 \\ 0 & 0 & 1 & 0 & -1 \\ 0 & 0 & 1 & 0 & 0 \\ -1 & 0 & 1 & 0 & 0 \\ 0 & -1 & 1 & 0 & 0 \end{pmatrix}, \quad \begin{pmatrix} 1 & 0 & 0 & 0 & 0 \\ 1 & 0 & 0 & -1 & 0 \\ 1 & 0 & 0 & 0 & -1 \\ 1 & -1 & 0 & 0 & 0 \\ 1 & 0 & -1 & 0 & 0 \end{pmatrix}, \quad \begin{pmatrix} 0 & 0 & 1 & 0 & -1 \\ 0 & 0 & 1 & -1 & 0 \\ 0 & 0 & 1 & 0 & 0 \\ 0 & -1 & 1 & 0 & 0 \\ -1 & 0 & 1 & 0 & 0 \end{pmatrix}.$$
$$(66)$$

The generators $\gamma_i^{(j,m)}$ in $\mathbb{R}_{j,m}^6$, for $(pqr) = (123)$ are:

$$\frac{1}{2}\begin{pmatrix} 1 & 1 & 0 & 1/3 & 1/3 & -2/3 \\ 1 & 1 & 0 & -1/3 & -1/3 & 2/3 \\ 0 & 0 & 2 & 0 & 0 & 0 \\ 1 & -1 & 0 & 1 & -1 & 0 \\ 1 & -1 & 0 & -1 & 1 & 0 \\ -2 & 2 & 0 & 0 & 0 & 0 \end{pmatrix}, \quad \frac{1}{2}\begin{pmatrix} 1 & 0 & 1 & -1/3 & 2/3 & -1/3 \\ 0 & 2 & 0 & 0 & 0 & 0 \\ 1 & 0 & 1 & 1/3 & -2/3 & 1/3 \\ -1 & 0 & 1 & 1 & 0 & -1 \\ 2 & 0 & -2 & 0 & 0 & 0 \\ -1 & 0 & 1 & -1 & 0 & 1 \end{pmatrix},$$

$$\frac{1}{2}\begin{pmatrix} 1 & 1 & 0 & -1/3 & -1/3 & 2/3 \\ 1 & 1 & 0 & 1/3 & 1/3 & -2/3 \\ 0 & 0 & 2 & 0 & 0 & 0 \\ -1 & 1 & 0 & 1 & -1 & 0 \\ -1 & 1 & 0 & -1 & 1 & 0 \\ 2 & -2 & 0 & 0 & 0 & 0 \end{pmatrix}. \qquad (67)$$

If we denote the sets of matrix generators $g_i^{(\beta,\alpha)}$ by a, b, c, then they satisfy the properties:

$$a^2 = b^2 = c^2 = I, \quad bc = cb, \quad ac = ca, \quad ba = (ab)^2 \to (ab)^3 = 1. \quad (68)$$

The only independent elements that can be generated by forming all possible products of all possible powers of a, b, c are:

$$I, \ a, \ b, \ c, \ ab, \ bc, \ ca, \ ba, \ abc, \ bac, \ aba, \ babc, \tag{69}$$

where I is the 5-dimensional unit matrix. The multiplication table for these elements reveals that they form a group with the following subgroups:

$$\{1,a\}, \ \{1,b\}, \ \{1,c\}, \ \{1,a,c,ac\}, \ \{1,b,c,bc\}, \{1,a,b,ab,ba,aba\}, \tag{70}$$

with $\{1,c\}$ being the center of the group, since c commutes with every element of the group.

In the case of the 6-j coefficient, the generators g_i^y in $\mathbb{R}_y^7$, after the condition (34) is taken into account, are:

$$
\begin{pmatrix}
1&0&0&0&0&0&0\\
0&1&0&0&0&0&0\\
0&0&0&1&0&0&0\\
0&0&1&0&0&0&0\\
0&0&0&0&1&0&0\\
0&0&0&0&0&0&1\\
0&0&0&0&0&1&0
\end{pmatrix},\
\begin{pmatrix}
1&0&0&0&0&0&0\\
0&0&0&1&0&0&0\\
0&0&1&0&0&0&0\\
0&1&0&0&0&0&0\\
0&0&0&0&0&1&0\\
0&0&0&0&1&0&0\\
0&0&0&0&0&0&1
\end{pmatrix},\
\begin{pmatrix}
0&1&0&0&0&0&0\\
1&0&0&0&0&0&0\\
0&0&0&1&0&0&0\\
0&0&1&0&0&0&0\\
0&0&0&0&1&0&0\\
0&0&0&0&0&1&0\\
0&0&0&0&0&0&1
\end{pmatrix}. \tag{71}
$$

The generators γ_i^x in $\mathbb{R}_x^6$ obtained are:

$$
\frac{1}{2}
\begin{pmatrix}
0&0&0&2&0&0\\
0&1&1&0&1&-1\\
0&1&1&0&-1&1\\
2&0&0&0&0&0\\
0&-1&1&0&1&1\\
1&-1&0&0&1&1
\end{pmatrix},\
\frac{1}{2}
\begin{pmatrix}
1&-1&1&1&0&0\\
-1&1&1&1&0&0\\
1&1&1&-1&0&0\\
1&1&-1&1&0&0\\
0&0&0&0&2&0\\
0&0&0&0&0&2
\end{pmatrix},
$$

$$
\frac{1}{2}
\begin{pmatrix}
1&0&0&-1&1&1\\
1&1&-1&1&0&0\\
1&-1&1&1&0&0\\
-1&0&0&1&1&1\\
0&1&1&0&1&-1\\
0&1&1&0&-1&1
\end{pmatrix},\
\frac{1}{2}
\begin{pmatrix}
2&0&0&0&0&0\\
0&1&-1&0&1&1\\
0&-1&1&0&1&1\\
0&0&0&2&0&0\\
0&1&1&0&1&-1\\
0&1&1&0&-1&1
\end{pmatrix}. \tag{72}
$$

These latter 6×6 matrices are directly related to the Regge symmetries [8] of the 6-j coefficient when they operate on the basis vector $|x\rangle$ and

402

they are:

$$\left\{ \begin{matrix} d & \frac{1}{2}(b+c+e-f) & \frac{1}{2}(-b+c-e+f) \\ a & \frac{1}{2}(b+c-e+f) & \frac{1}{2}(b-c+e+f) \end{matrix} \right\},$$

$$\left\{ \begin{matrix} \frac{1}{2}(a-b+c+d) & \frac{1}{2}(-a+b+c+d) & e \\ \frac{1}{2}(a+b-c+d) & \frac{1}{2}(a+b+c-d) & f \end{matrix} \right\},$$

$$\left\{ \begin{matrix} \frac{1}{2}(a-d+e+f) & \frac{1}{2}(a+b-c+d) & \frac{1}{2}(b+c+e-f) \\ \frac{1}{2}(-a+d+e+f) & \frac{1}{2}(a-b+c+d) & \frac{1}{2}(b+c-e+f) \end{matrix} \right\},$$

$$\left\{ \begin{matrix} a & \frac{1}{2}(b-c+e+f) & \frac{1}{2}(b+c+e-f) \\ d & \frac{1}{2}(-b+c+e+f) & \frac{1}{2}(b+c-e+f) \end{matrix} \right\}, \tag{73}$$

which are mixed symmetries in the sense they are the Regge symmetries of the 6-j coefficient on which the *classical* symmetries are superposed.

References

[1] J. Thomae, J. Reine Angew. Math. **87** (1879) 26.

[2] K. Srinivasa Rao, H.-D. Doebner and P. Nattermann, Proc. of the 5th Wigner Symposium, Ed. by P. Kasperkovitz and D. Grau, World Scientific (1998) p.97.

[3] W.N. Bailey, it Generalized Hypergeometric Series, Cambridge Univ. Press. (1935).

[4] W.A. Beyer, J.D. Louck and P.R. Stein, J. Math. Phys. **28** (1987) 497.

[5] K. Srinivasa Rao, J. Van der Jeugt, J. Raynal, R. Jagannathan and V. Rajeswari, J. Phys. A : Math. Gen. **25** (1992) 861.

[6] J. Van der Jeugt and K. Srinivasa Rao, J. Math. Phys. (to appear).

[7] G.H. Hardy, *Ramanujan: Twelve Lectures on subjects suggested by his life and work*, Chelsea, New York (1940).

[8] T. Regge, Nuo. Cim. **10** (1958) 544.

[9] T. Regge, Nuo. Cim. **11** (1959) 116.

[10] M. Weber and A. Erdelyi, Am. Math. Mon. **59** (1952) 163.

[11] F.J.W. Whipple, Proc. London Math. Soc. **23** (1925) 104.

[12] F.J. Budden, *The Fascination of Groups*, Cambridge Univ. Press (1972).

[13] J. Raynal, J. Math. Phys. **19** (1978) 467.

[14] J. Raynal, J. Math. Phys. **20** (1979) 2398.

[15] K. Srinivasa Rao, J. Phys.**11A** (1978) L69.

[16] G. Racah, Phys. Rev. **62** (1942) 438.

[17] B.M. Minton, J. Math. Phys. **11** (1970) 3061.

[18] K. Srinivasa Rao, T.S. Santhanam and K. Venkatesh, J. Math. Phys. **16** (1975) 1528.

[19] K. Srinivasa Rao and K. Venkatesh, *Group Theoretical methods in Physics*, Ed. by R.T.Sharp and B.Kolman, Academic Press (1977) p.649; K. Srinivasa Rao

and K. Venkatesh, *Group Theoretical methods in Physics*, Ed. by P.Kramer and A.Rieckers, Lecture Notes in Phys., Springer - Verlag (1978) p.501.

[20] K. Srinivasa Rao and V. Rajeswari, Int. J. Theor. Phys.**24** (1985) 983.

[21] L.A. Lockwood, J. Math. Phys. **17** (1976) 1671; ibid **18** (1977) 45.

[22] K. Venkatesh, J. Math. Phys. **19** (1978) 1973, 2060; ibid **21** (1980) 1555.

[23] R. Askey's Foreword in G. Gasper and M. Rahman, *Basic Hypergeometric Series*, Cambridge University Press, Cambridge (1990) p. xiii.

STABILITY AND NEW NON-ABELIAN ZETA FUNCTIONS

Lin Weng

Graduate School of Mathematics, Kyushu University, Fukuoka, Japan

weng@math.kyushu-u.ac.jp

(former address)

Graduate School of Mathematics, Nagoya University, Japan

weng@math.nagoya-u.ac.jp

Abstract   In this paper, we first use classification of uni-modular lattices as a motivation to introduce semi-stable lattices. Then, as an integration over moduli spaces of semi-stable lattices, using a new arithmetic cohomology, we define a new type of non-abelian zeta functions for number fields. This is a natural generalization of what Iwasawa and Tate did for Dedekind functions. Basic facts for these zeta functions such as functional equations, singularities and residues at simple poles are discussed. Finally we introduce and study new non-abelian zeta functions for curves over finite fields, as a natural generalization of Artin's (abelian) zeta functions, using moduli spaces of semi-stable bundles. Some interesting examples are also given here.

Keywords: arithmetic cohomology, adelic moduli spaces, reciprocity law, stability, zeta functions.

The aim of this paper is to describe our new constructions of non-abelian zeta functions.

1.   Motivation

1.1   Classification of Unimodular Lattices

Recall that a full rank lattice $\Lambda \subset \mathbf{R}^r$ is said to be *integral* if for any $x \in \Lambda$, (x, x) is an integer, and that an integral lattice Λ is called *unimodular*, if the volume of its fundamental domain is one. It is a classical yet still very challenging problem to classify all unimodular lattices.

This classification problem has two different aspects, i.e., the local one and the global one. For the local study, we are mainly interested in enumerating all unimodular lattices, which in recent years proves to be very fruitful. However for the global study, besides the earlier works done by Minkowski and Siegel, such as the mass formula and asymptotic upper bounds for the numbers of unimodular lattices in terms of volumes of Siegel domains, less progress has been recorded. (For details, please see [3].)

Our new non-abelian zeta functions for number fields are motivated by such a global study of unimodular lattices. The difference between ours and that of Siegel is that instead of using all lattices of volume one, we consider only a well-behavior part of them, i.e., these which are semi-stable. For this purpose, let me introduce the following:

Definition. A lattice Λ is called *stable* (resp. *semi-stable*) if for any proper sublattice Λ',

$$\mathrm{Vol}(\Lambda')^{\mathrm{rank}(\Lambda)} \quad > \quad (\mathrm{resp.} \geq) \quad \mathrm{Vol}(\Lambda)^{\mathrm{rank}(\Lambda')}.$$

Remark. The stability of Mumford in geometry has been proved to be one of the most popular and important discoveries in modern mathematics. By contrast, stability proves to be relatively less popular in arithmetic, despite the fact that it was first introduced by Stuhler [19, 20] as earlier as in 70's: For example, many mathematicians, including myself, independently introduce the stability in arithmetic later. (See e.g., [15] and [23].) I would like to thank Deninger for introducing me the works of Stuhler and Grayson [7, 8] shortly before this symposium.

Standard properties about Harder-Narasimhan filtrations and Jordan-Hölder filtrationsfiltrations, Jordan-Hölder — hold here as well. That is to say, we have the following:

Proposition. *Let Λ be a lattice. Then*
(1) There exists a unique filtration of proper sublattices,

$$0 = \Lambda_0 \subset \Lambda_1 \subset \cdots\cdots \subset \Lambda_s = \Lambda$$

such that Λ_i/Λ_{i-1} is semi-stable and

$$\mathrm{Vol}(\Lambda_{i+1}/\Lambda_i)^{\mathrm{rank}(\Lambda_i/\Lambda_{i-1})} > \mathrm{Vol}(\Lambda_i/\Lambda_{i-1})^{\mathrm{rank}(\Lambda_{i+1}/\Lambda_i)}.$$

(2) If moreover Λ is semi-stable, then there exists a filtration of proper sublattices,

$$0 = \Lambda^{t+1} \subset \Lambda^t \subset \cdots\cdots \subset \Lambda^0 = \Lambda$$

such that Λ^j/Λ^{j+1} *is stable and*

$$\text{Vol}(\Lambda^j/\Lambda^{j+1})^{\text{rank}(\Lambda^{j-1}/\Lambda^j)} = \text{Vol}(\Lambda^{j-1}/\Lambda^j)^{\text{rank}(\Lambda^j/\Lambda^{j+1})}.$$

Furthermore, the graded lattice $\text{Gr}(\Lambda) := \oplus\Lambda^{j-1}/\Lambda^j$, *the so-called Jordan-Hölder graded lattice of* Λ, *is uniquely determined by* Λ.

In particular for unimodular lattices, we have the following

Corollary. *Unimodular lattices are semi-stable. Moreover, a unimodular lattice is stable if and only if it contains no proper unimodular sublattice.*

Thus it suffices to classify unimodular lattices from the point of view of stability, which then leads to the following consideration.

Denote by $\mathcal{M}_{\mathbf{Q},r}(1)$ the collection, or better, the moduli space, of all rank r semi-stable lattices of volume one. Then one checks that $\mathcal{M}_{\mathbf{Q},r}(1)$ admits a natural metric and is indeed compact ([23]). For example, with reduction theory from geometry of numbers, we know that

$$\mathcal{M}_{\mathbf{Q},2}(1) \simeq \left\{ \begin{pmatrix} a & 0 \\ b & \frac{1}{a} \end{pmatrix} : 1 \le a \le \sqrt{\frac{2}{\sqrt{3}}}, \sqrt{a^2 - a^{-2}} \le b \le a - \sqrt{a^2 - a^{-2}} \right\}$$
$$\cdot \, SO(2).$$

Moreover,

$$\left\{ \begin{pmatrix} a & 0 \\ b & \frac{1}{a} \end{pmatrix} : 1 \le a \le \sqrt{\frac{2}{\sqrt{3}}}, \sqrt{a^2 - a^{-2}} \le b \le a - \sqrt{a^2 - a^{-2}} \right\}$$

may be viewed as a closed bounded domain in the upper half plane, which admits a natural metric, i.e., the Poincaré metric.

In this way, we may view unimodular lattices naturally as certain special points in these geometric moduli spaces. (Recall that unimodular lattices are integral.) So the problem of classifying unimodular lattices looks very much similar to that of finding rational points in algebraic varieties, and hence geometry of the moduli spaces should play a crucial role. Thus, along with the line of Minkowski's geometry of numbers, the first thing we have to do is to evaluate volumes of these moduli spaces with respect to the associated natural metrics. It is for this purpose that we introduce our non-abelian zeta functions for number fields: Theory of Dedekind zeta functions tells us that, regulators of number fields, or better, volumes of the lattices generated by fundamental units, may be read from the residues of Dedekind zeta functions at the simple poles $s = 1$.

1.2   Geometric Arithmetic Riemann-Roch

To introduce a more general zeta function, from Artin's definition of (abelian) zeta function for curves defined over finite fields ([2]), and Iwasawa's interpretation of Dedekind zeta function ([11]), we see that it is better to have a cohomology theory in arithmetic such that the duality and the Riemann-Roch are satisfied. We claim that this can be rigorously developed following Tate's Thesis. Here we use only a simple example to indicate how it works.

Consider only rank 1 lattices over $\mathbf{Q}$: They are parameterized by $\mathbf{R}_{>0}$, say the lattices with the forms $\Lambda_t := \mathbf{Z} \cdot \sqrt{t}$, $t \in \mathbf{R}_{>0}$. Then the Poisson summation formula says that

$$\sum_{n \in \mathbf{Z}} e^{-\pi t n^2} = \frac{1}{\sqrt{t}} \sum_{n \in \mathbf{Z}} e^{-\pi n^2/t}.$$

That is to say,

$$\sum_{\alpha \in \Lambda_t} e^{-\pi |\alpha|^2} = \frac{1}{\mathrm{Vol}(\Lambda_t)} \sum_{\beta \in \Lambda_t^\vee} e^{-\pi |\beta|^2}.$$

Here $\Lambda_t^\vee$ denotes the dual lattice of Λ_t. Thus, if we set

$$h^0(\mathbf{Q}, \Lambda_t) := \log \Big(\sum_{\alpha \in \Lambda_t} e^{-\pi |\alpha|^2} \Big),$$

then

$$h^0(\mathbf{Q}, \Lambda_t) - h^0(\mathbf{Q}, \Lambda_t^\vee) = \deg(\Lambda_t). \qquad (*)$$

This is simply the analog of the Riemann-Roch in geometry. Indeed, as for $\mathbf{Q}$, the metrized dualizing sheaf is simply the standard lattice $\mathbf{Z} \subset \mathbf{R} = \mathbf{Q}_\infty$. (See e.g. [14].) So $(*)$ becomes

$$h^0(\mathbf{Q}, \Lambda_t) - h^0(\mathbf{Q}, K_\mathbf{Q} \otimes \Lambda_t^\vee) = \deg(\Lambda_t) - \frac{1}{2} \deg K_\mathbf{Q},$$

where $\deg K_\mathbf{Q} = \log |\Delta_\mathbf{Q}| = \log 1 = 0$ with $\Delta_\mathbf{Q}$ the discriminant of $\mathbf{Q}$.

The above discussion works for any number fields as well: We could have stability and corresponding moduli spaces, well-defined h^0 and h^1 for metrized vector sheaves which satisfy the duality and Riemann-Roch. Moreover, such a discussion works equally well in adelic language. Say, we have the adelic moduli spaces $\mathcal{M}_{\mathbf{A}_F, r}(t)$ over which natural Tamagawa measures $d\mu$ exist, and for all $g \in \mathcal{M}_{\mathbf{A}_F, r}(t)$,

$$h^0(\mathbf{A}_F, g) - h^1(\mathbf{A}_F, g) = \deg(g) - \frac{r}{2} \log |\Delta_F|.$$

For more details, please see [23, 26, 27]. All in all, we then could introduce our new *non-abelian zeta function* for a number field F by setting it to be

$$\xi_{F,r}(s) := \left(|\Delta_F|^{\frac{r}{2}}\right)^s \int_{g \in \mathcal{M}_{\mathbf{A}_F,r}(t), t \in \mathbf{R}_{>0}} \left(e^{h^0(\mathbf{A}_F,g)} - 1\right)\left(e^{-s}\right)^{\deg(g)} \cdot d\mu(g),$$

$$\mathrm{Re}(s) > 1$$

where Δ_F denotes the discriminant of F, $\mathcal{M}_{\mathbf{A}_F,r}(t)$ denotes the (adelic) moduli space of rank r semi-stable lattices of volume t over F and $d\mu$ denotes the natural Tamagawa measure on $\mathcal{M}_{\mathbf{A}_F,r}(t)$.

Theorem ([23]). *With the same notation as above, we have*
(0) $\xi_{F,1}(s) = w_F \cdot \xi_F(s)$ *where w_F denotes the number of roots of unity in F and $\xi_F(s)$ denotes the completed Dedekind zeta function for F;*
(1)

$$\xi_{F,r}(s) = \left(|\Delta_F|^{\frac{r}{2}}\right)^s \int_{g \in \mathcal{M}_{\mathbf{A}_F,r}(t), t \in \mathbf{R}_{>0}} \left(e^{h^0(\mathbf{A}_F,g)} - 1\right)\left(e^{-s}\right)^{\deg(g)} \cdot d\mu(g)$$

converges absolutely and uniformly when $\mathrm{Re}(s) \geq 1 + \delta$ for any $\delta > 0$;
(2) $\xi_{F,r}(s)$ *admits a unique meromorphic continuation to the whole complex s-plane with only two simple poles at $s = 0, 1$ whose residues are* $\mathrm{Vol}(\mathcal{M}_{\mathbf{A}_F,r}(t))$ *for one and hence for all t;*
(3) (Functional Equation) $\xi_{F,r}(s) = \xi_{F,r}(1 - s)$.

Remarks. (a) The proof of (0) above is essentially the content of Iwasawa's ICM talk at MIT (see e.g. [11]);
(b) A Riemann-Roch theorem for number fields was first proved in Tate's thesis in 1950. In 60's, Lang gave his Riemann-Roch theorem ([13]). Then, in 80's, in an attempt to understand Arakelov's work, Szpiro gave his version of Riemann-Roch theorem using the Arakelov-Euler characteristic ([14]). However in all these studies, neither h^0 nor h^1 is defined. In fact, it was not until the end of 90's, two totally different versions of h^0 for Arakelov divisors had been introduced by Neukirch ([17]) and van der Geer-Schoof ([6]) respectively. Our h^0 follows Tate and Iwasawa, which is quite different from that of Deninger ([4]) who uses infinite dimensional spaces and regularized determinants.
(c) Most suitable definition for non-abelian zeta functions of number fields should be

$$\xi_{F,r}(s) := \left(|\Delta_F|^{\frac{r}{2}}\right)^s \int_{\Lambda \in \mathcal{M}_{\mathbf{A}_F,r}(t), t \in \mathbf{R}_+} \frac{e^{h^0(\mathbf{A}_F,\Lambda)} - 1}{\#\mathrm{Aut}(\Lambda)} \left(e^{-s}\right)^{\deg(\Lambda)} d\mu(\Lambda),$$

$$\mathrm{Re}(s) > 1$$

where Aut denotes the automorphism group.

2. New Local Non-Abelian Zeta Functions for Curves

In this section, we introduce our non-abelian zeta functions for curves defined over finite fields.

2.1 Moduli Spaces of Semi-Stable Bundles

Let C be a regular, reduced and irreducible projective curve defined over an algebraically closed field $\bar{k}$. Then according to Mumford [16], a vector bundle V on C is called semi-stable (resp. stable) if for any proper subbundle V' of V,

$$\mu(V') := \frac{d(V')}{r(V')} \leq (\text{resp.} <)\frac{d(V)}{r(V)} =: \mu(V).$$

Here d denotes the degree and r denotes the rank.

Proposition 1. *Let V be a vector bundle over C. Then*
(a) ([9]) there exists a unique filtration of subbundles of V, the so-called Harder-Narasimhan filtration of V,

$$\{0\} = V_0 \subset V_1 \subset V_2 \subset \cdots \subset V_{s-1} \subset V_s = V$$

such that for $1 \leq i \leq s - 1$, V_i/V_{i-1} is semi-stable and $\mu(V_i/V_{i-1}) > \mu(V_{i+1}/V_i)$;
(b) (see e.g. [18]) if moreover V is semi-stable, there exists a filtration of subbundles of V, a Jordan-Hölder filtration of V,

$$\{0\} = V^{t+1} \subset V^t \subset \cdots \subset V^1 \subset V^0 = V$$

such that for all $0 \leq i \leq t$, V^i/V^{i+1} is stable and $\mu(V^i/V^{i+1}) = \mu(V)$. Moreover, $\mathrm{Gr}(V) := \oplus_{i=0}^t V^i/V^{i+1}$, the associated (Jordan-Hölder) graded bundle of V, is determined uniquely by V.

Following Seshadri, two semi-stable vector bundles V and W are called S-equivalent, if their associated Jordan-Hölder graded bundles are isomorphic, i.e., $\mathrm{Gr}(V) \simeq \mathrm{Gr}(W)$. Applying Mumford's general result on geometric invariant theory ([16]), Seshadri proves the following

Theorem 2 ([18]). *Let C be a regular, reduced, irreducible projective curve defined over an algebraically closed field. Then over the set $\mathcal{M}_{C,r}(d)$ of S-equivalence classes of rank r and degree d semi-stable vector bundles over C, there is a natural normal, projective algebraic variety structure.*

Now assume that C is defined over a finite field k. Naturally we may talk about k-rational bundles over C, i.e., bundles which are defined over k. Moreover, from geometric invariant theory, projective varieties $\mathcal{M}_{C,r}(d)$ are defined over a certain finite extension of k. Thus it makes sense to talk about k-rational points of these moduli spaces too. The relation between these two types of rationality is given by Harder and Narasimhan based on a discussion about Brauer groups:

Proposition 3 ([9]). *There exists a finite field $\mathbf{F}_q$ such that for any d, the subset of $\mathbf{F}_q$-rational points of $\mathcal{M}_{C,r}(d)$ consists exactly of all S-equivalence classes of $\mathbf{F}_q$-rational bundles in $\mathcal{M}_{C,r}(d)$.*

From now on, without loss of generality, we always assume that finite fields $\mathbf{F}_q$ (with q elements) satisfy the property in this Proposition. Also for simplicity, we write $\mathcal{M}_{C,r}(d)$ for $\mathcal{M}_{C,r}(d)(\mathbf{F}_q)$, the subset of $\mathbf{F}_q$-rational points, and call them moduli spaces by an abuse of notations.

2.2 Local Non-Abelian Zeta Functions

Let C be a regular, reduced, irreducible projective curve defined over the finite field $\mathbf{F}_q$ with q elements. Define the *rank r non-abelian zeta function* $\zeta_{C,r,\mathbf{F}_q}(s)$ by setting

$$\zeta_{C,r,\mathbf{F}_q}(s) := \sum_{V \in [V] \in \mathcal{M}_{C,r}(d), d \geq 0} \frac{q^{h^0(C,V)} - 1}{\#\mathrm{Aut}(V)} \cdot (q^{-s})^{d(V)}, \qquad \mathrm{Re}(s) > 1.$$

Proposition 1. *With the same notation as above, $\zeta_{C,1,\mathbf{F}_q}(s)$ is nothing but the classical Artin zeta function for curve C. That is to say,*

$$\zeta_{C,1,\mathbf{F}_q}(s) = \sum_{D \geq 0} \frac{1}{N(D)^s}, \qquad \mathrm{Re}(s) > 1.$$

Here D runs over all effective divisors of C.

Proof. By definition, the classical Artin zeta function ([2]) for C is given by

$$\zeta_C(s) := \sum_{D \geq 0} \frac{1}{N(D)^s}.$$

Here $N(D) = q^{d(D)}$ with $d(\Sigma_P n_P P) = \Sigma_P n_P d(P)$. Thus by first grouping effective divisors according to their rational equivalence classes $\mathcal{D}$, then taking the sum on effective divisors in the same class, we obtain

$$\zeta_C(s) = \sum_{\mathcal{D}} \sum_{D \in \mathcal{D}, D \geq 0} \frac{1}{N(D)^s}.$$

Clearly,

$$\sum_{D \in \mathcal{D}, D \geq 0} \frac{1}{N(D)^s} = \frac{q^{h^0(C,\mathcal{D})} - 1}{q - 1} \cdot (q^{-s})^{d(\mathcal{D})}.$$

Therefore,

$$\zeta_C(s) = \sum_{L \in \mathrm{Pic}^d(C), d \geq 0} \frac{q^{h^0(C,L)} - 1}{\#\mathrm{Aut}(L)} \cdot (q^{-s})^{d(L)},$$

due to the fact that $\mathrm{Aut}(V) \simeq \mathbf{F}_q^*$. Here as usual, $\mathrm{Pic}^d(C)$ denotes the (subset of rational points of) degree d Picard variety of C, which we view as the moduli space of degree d (semi-stable) line bundles. $\quad\square$

Moreover, we have the following

Theorem 2 ([24]). *(1) The non-abelian zeta function $\zeta_{C,r,\mathbf{F}_q}(s)$ is well-defined for $\mathrm{Re}(s) > 1$, and admits a meromorphic extension to the whole complex s-plane.*
(2) (Rationality) If we set $t := q^{-s}$ and introduce the non-abelian Z-function of C by setting

$$\zeta_{C,r,\mathbf{F}_q}(s) =: Z_{C,r,\mathbf{F}_q}(t) := \sum_{V \in [V] \in \mathcal{M}_{C,r}(d), d \geq 0} \frac{q^{h^0(C,V)} - 1}{\#\mathrm{Aut}(V)} \cdot t^{d(V)}, \qquad |t| < 1,$$

then there exists a polynomial $P_{C,r,\mathbf{F}_q}(s) \in \mathbf{Q}[t]$ such that

$$Z_{C,r,\mathbf{F}_q}(t) = \frac{P_{C,r,\mathbf{F}_q}(t)}{(1 - t^r)(1 - q^r t^r)}.$$

(3) (Functional Equation) If we set the rank r non-abelian ξ-function $\xi_{C,r,\mathbf{F}_q}(s)$ by setting

$$\xi_{C,r,\mathbf{F}_q}(s) := \zeta_{C,r,\mathbf{F}_q}(s) \cdot (q^s)^{r(g-1)},$$

then

$$\xi_{C,r,\mathbf{F}_q}(s) = \xi_{C,r,\mathbf{F}_q}(1 - s).$$

One may prove this theorem by using the vanishing theorem, duality, and the Riemann-Roch theorem.

Corollary. *With the same notation as above,*
(1) $P_{C,r,\mathbf{F}_q}(t) \in \mathbf{Q}[t]$ is a degree $2rg$ polynomial;
(2) Denote all reciprocal roots of $P_{C,r,\mathbf{F}_q}(t)$ by $\omega_{C,r,\mathbf{F}_q}(i)$, $i = 1, \ldots, 2rg$. Then after a suitable rearrangement,

$$\omega_{C,r,\mathbf{F}_q}(i) \cdot \omega_{C,r,\mathbf{F}_q}(2rg - i) = q, \qquad i = 1, \ldots, rg;$$

(3) For each $m \in \mathbf{Z}_{\geq 1}$, there exists a rational number $N_{C,r,\mathbf{F}_q}(m)$ such that

$$Z_{r,C,\mathbf{F}_q}(t) = P_{C,r,\mathbf{F}_q}(0) \cdot \exp\left(\sum_{m=1}^{\infty} N_{C,r,\mathbf{F}_q}(m) \frac{t^m}{m} \right).$$

Moreover,

$$N_{C,r,\mathbf{F}_q}(m) = \begin{cases} r(1+q^m) - \displaystyle\sum_{i=1}^{2rg} \omega_{C,r,\mathbf{F}_q}(i)^m, & \text{if } r \mid m; \\ -\displaystyle\sum_{i=1}^{2rg} \omega_{C,r,\mathbf{F}_q}(i)^m, & \text{if } r \nmid m; \end{cases}$$

(4) For any $a \in \mathbf{Z}_{>0}$, denote by ζ_a a primitive a-th root of unity and set $T = t^a$. Then

$$\prod_{i=1}^{a} Z_{C,r}(\zeta_a^i t) = (P_{C,r,\mathbf{F}_q}(0))^a \cdot \exp\left(\sum_{m=1}^{\infty} N_{r,C,\mathbf{F}_q}(ma) \frac{T^m}{m} \right).$$

3. Global Non-Abelian Zeta Functions for Curves

In this section, we introduce new non-abelian zeta functions for curves defined over number fields via the Euler product formalism.

3.1 Preparations

Let C be a regular, reduced, irreducible projective curve of genus g defined over the finite field $\mathbf{F}_q$ with q elements. Then the rationality of $\zeta_{C,r,\mathbf{F}_q}(s)$ says that there exists a degree $2rg$ polynomial $P_{C,r,\mathbf{F}_q}(t) \in \mathbf{Q}[t]$ such that

$$Z_{C,r,\mathbf{F}_q}(t) = \frac{P_{C,r,\mathbf{F}_q}(t)}{(1-t^r)(1-q^r t^r)}.$$

Set

$$P_{C,r,\mathbf{F}_q}(t) = \sum_{i=0}^{2rg} a_{C,r,\mathbf{F}_q}(i) t^i.$$

By the functional equation for $\xi_{C,r,\mathbf{F}_q}(t)(s)$, we have

$$P_{C,r,\mathbf{F}_q}(t) = P_{C,r,\mathbf{F}_q}\left(\frac{1}{qt}\right) \cdot q^{rg} \cdot t^{2rg}.$$

Thus by comparing coefficients on both sides, we get the following

Lemma 1. *With the same notation as above, for* $i = 0, 1, \ldots, rg - 1$,

$$a_{C,r,\mathbf{F}_q}(2rg - i) = a_{C,r,\mathbf{F}_q}(i) \cdot q^{rg-i}.$$

To further determine these coefficients, following Harder and Narasimhan (see e.g. [9] and [5]), who first consider the β-series invariants below, we introduce the following invariants:

$$\alpha_{C,r,\mathbf{F}_q}(d) := \sum_{V \in [V] \in \mathcal{M}_{C,r}(d)(\mathbf{F}_q)} \frac{q^{h^0(C,V)}}{\#\mathrm{Aut}(V)},$$

$$\beta_{C,r,\mathbf{F}_q}(d) := \sum_{V \in [V] \in \mathcal{M}_{C,r}(d)(\mathbf{F}_q)} \frac{1}{\#\mathrm{Aut}(V)},$$

$$\gamma_{C,r,\mathbf{F}_q}(d) := \sum_{V \in [V] \in \mathcal{M}_{C,r}(d)(\mathbf{F}_q)} \frac{q^{h^0(C,V)} - 1}{\#\mathrm{Aut}(V)}.$$

One checks that all $\alpha_{C,r,\mathbf{F}_q}(d), \beta_{C,r,\mathbf{F}_q}(d)$ and $\gamma_{C,r,\mathbf{F}_q}(d)$'s may be calculated from $\alpha_{C,r,\mathbf{F}_q}(i), \beta_{C,r,\mathbf{F}_q}(j)$ with $i = 0, \ldots, r(g-1)$ and $j = 0, \ldots, r-1$.

Proposition 2 (An Ugly Formula). ([24]) *With the same notation as above,*

$$a_{C,r,\mathbf{F}_q}(i)$$
$$= \begin{cases} \alpha_{C,r,\mathbf{F}_q}(d) - \beta_{C,r,\mathbf{F}_q}(d), \\ \qquad if\ 0 \leq i \leq r - 1; \\[2mm] \alpha_{C,r,\mathbf{F}_q}(d) - (q^r + 1)\alpha_{C,r,\mathbf{F}_q}(d - r) + q^r\beta_{C,r,\mathbf{F}_q}(d - r), \\ \qquad if\ r \leq i \leq 2r - 1; \\[2mm] \alpha_{C,r,\mathbf{F}_q}(d) - (q^r + 1)\alpha_{C,r,\mathbf{F}_q}(d - r) + q^r\alpha_{C,r,\mathbf{F}_q}(d - 2r), \\ \qquad if\ 2r \leq i \leq r(g-1) - 1; \\[2mm] -(q^r + 1)\alpha_{C,r,\mathbf{F}_q}(r(g-2)) + q^r\alpha_{C,r,\mathbf{F}_q}(r(g-3)) + \alpha_{C,r,\mathbf{F}_q}(r(g-1)), \\ \qquad if\ i = r(g-1); \\[2mm] \alpha_{C,r,\mathbf{F}_q}(d) - (q^r + 1)\alpha_{C,r,\mathbf{F}_q}(d - r) + \alpha_{C,r,\mathbf{F}_q}(d - 2r)q^r, \\ \qquad if\ r(g-1) + 1 \leq i \leq rg - 1; \\[2mm] 2q^r\alpha_{C,r,\mathbf{F}_q}(r(g-2)) - (q^r + 1)\alpha_{C,r,\mathbf{F}_q}(r(g-1)), \\ \qquad if\ i = rg. \end{cases}$$

3.2   Global Non-Abelian Zeta Functions for Curves

Let C be a regular, reduced, irreducible projective curve of genus g defined over a number field F. Let S_{bad} be the collection of all infinite places and these finite places of F at which C does not have good reductions. As usual, a place v of F is called good if $v \notin S_{\text{bad}}$.

Thus, in particular, for any good place v of F, the v-reduction of C, denoted as C_v, gives a regular, reduced, irreducible projective curve defined over the residue field $F(v)$ of F at v. Denote the cardinal number of $F(v)$ by q_v. Then, by the construction of Section 2, we obtain the associated rank r non-abelian zeta function $\zeta_{C_v, r, \mathbf{F}_{q_v}}(s)$. Moreover, from the rationality of $\zeta_{C_v, r, \mathbf{F}_{q_v}}(s)$, there exists a degree $2rg$ polynomial $P_{C_v, r, \mathbf{F}_{q_v}}(t) \in \mathbf{Q}[t]$ such that

$$Z_{C_v, r, \mathbf{F}_{q_v}}(t) = \frac{P_{C_v, r, \mathbf{F}_{q_v}}(t)}{(1 - t^r)(1 - q^r t^r)}.$$

Clearly,

$$P_{C_v, r, \mathbf{F}_{q_v}}(0) = \gamma_{C_v, r, \mathbf{F}_{q_v}}(0) \neq 0.$$

Thus it makes sense to introduce the polynomial $\tilde{P}_{C_v, r, \mathbf{F}_{q_v}}(t)$ with constant term 1 by setting

$$\tilde{P}_{C_v, r, F(v)}(t) := \frac{P_{C_v, r, F(v)}(t)}{P_{C_v, r, F(v)}(0)}.$$

Now by definition, *the rank r non-abelian zeta function $\zeta_{C, r, F}(s)$ of C over F* is the following Euler product

$$\zeta_{C, r, F}(s) := \prod_{v: \text{good}} \frac{1}{\tilde{P}_{C_v, r, \mathbf{F}_{q_v}}(q_v^{-s})}, \qquad \mathrm{Re}(s) \gg 0.$$

Clearly, when $r = 1$, $\zeta_{C, r, F}(s)$ coincides with the classical Hasse-Weil zeta function for C over F ([10]).

Conjecture. *Let C be a regular, reduced, irreducible projective curve of genus g defined over a number field F. Then its associated rank r global non-abelian zeta function $\zeta_{C, r, F}(s)$ admits a meromorphic continuation to the whole complex s-plane.*

Recall that even when $r = 1$, i.e., for the classical Hasse-Weil zeta functions, this conjecture is still open. However, in general, we have the following

Theorem 3 ([24]). *Let C be a regular, reduced, irreducible projective curve defined over a number field F. When $\mathrm{Re}(s) > 1 + g + (r^2 - r)(g - 1)$, the associated rank r global non-abelian zeta function $\zeta_{C, r, F}(s)$ converges.*

This theorem may be deduced from a result of (Harder-Narasimhan) Siegel on Tamagawa numbers of SL_r, the ugly yet very precise formula for local zeta function above, Clifford Lemma for semi-stable bundles, and Weil's theorem on the Riemann hypothesis for Artin zeta functions ([22]), together with the following Proposition. For details, please consult [24].

Proposition 4. *With the same notation as above, when* $q \to \infty$,
(a) For $0 \le d \le r(g-1)$,

$$\frac{\alpha_{C,r,\mathbf{F}_q}(d)}{q^{d/2+r+r^2(g-1)}} = O(1);$$

(b) For all d,

$$\beta_{C,r,\mathbf{F}_q}(d) = O\left(q^{r^2(g-1)}\right);$$

(c)

$$\frac{q^{(r-1)(g-1)}}{\gamma_{C,r,\mathbf{F}_q}(0)} = O\left(1\right).$$

3.3 Working Hypothesis

Like in the theory for abelian zeta functions, we want to use our non-abelian zeta functions to study non-abelian aspect of arithmetic of curves. Motivated by the classical analytic class number formula for Dedekind zeta functions and its counterpart BSD conjecture for Hasse-Weil zeta functions of elliptic curves, we expect that our non-abelian zeta function could be used to understand the Weil-Petersson volumes of moduli spaces of stable bundles.

For doing so, we then also need to introduce local factors for 'bad' places: For Γ-factors, we take these coming from the functional equation for $\zeta_F(rs) \cdot \zeta_F(r(s-1))$, where $\zeta_F(s)$ denotes the standard Dedekind zeta function for F; while for finite bad places, we do as follows: first, use the semi-stable reduction for curves to find a semi-stable model for $\mathcal{C}$, then use Seshadri's moduli spaces of parabolic bundles to construct polynomials for singular fibers, which usually have degree lower than $2rg$. With all this being done, we then can introduce the so-called completed rank r non-abelian zeta function for $\mathcal{C}$ over F, or better, the completed rank r non-abelian zeta function $\xi_{X,r,\mathcal{O}_F}(s)$ for a semi-stable model $X \to \text{Spec}(\mathcal{O}_F)$ of $\mathcal{C}$. Here $\mathcal{O}_F$ denotes the ring of integers of F. (If necessary, we take a finite extension of F.)

Conjecture. $\xi_{X,r,\mathcal{O}_F}(s)$ *is holomorphic and satisfies the functional equation*

$$\xi_{X,r,\mathcal{O}_F}(s) = \pm \xi_{X,r,\mathcal{O}_F}\left(1 + \frac{1}{r} - s\right).$$

Moreover, we expect that for certain classes of curves, the inverse Mellin transform of our non-abelian zeta functions are naturally associated to certain modular forms of weight $1 + \frac{1}{r}$.

Example. ([25]) From our study for non-abelian zeta functions of elliptic curves, we obtain the following 'absolute Euler product' for rank 2 zeta functions of elliptic curves

$$\zeta_2(s) = \prod_{p \text{ prime}} \frac{1}{1 + (p-1)p^{-s} + (2p-4)p^{-2s} + (p^2-p)p^{-3s} + p^2 p^{-4s}}.$$

On the other hand, in his talk [12], Kohnen introduces us the following result of Andrianov ([1]). The so-called genus two spinor L-function stands in the form

$$\prod_p \frac{1}{1 - \lambda(p)p^{-s} + (\lambda(p)^2 - \lambda(p^2) - p^{2k-4})p^{-2s} - \lambda(p)p^{2k-3}p^{-3s} + p^{4k-6}p^{-4s}}.$$

Clearly, if we set $k = 2$, $\lambda(p) = 1 - p$ and $\lambda(p^2) = p^2 - 4p + 4$, we see that *formally* the above two zeta functions coincide. This suggests that there might be a close relation between them. The following is a speculation I made after the discussion with Deninger and Kohnen during the symposium.

The relation between the above two zeta functions should be in the same style as the Shimura correspondence for half weight and integral weight modular forms.

To convince the reader, let me point out the following facts:
(1) Andrianov's zetas have a Hecke theory, are coming from certain weight 2 modular forms, and have the local factors

$$\frac{1 - \lambda(p)t + \left(\lambda(p)^2 - \lambda(p^2) - p^{2k-4}\right)t^2 - \lambda(p)p^{2k-3}t^3 + p^{4k-6}t^4}{1 - p^{2k-4}t^2};$$

(2) Our working hypothesis concerning weight 3/2 modular forms are made mainly from the fact that our local factor takes the form

$$\frac{1 + (p-1)t + (2p-4)t^2 + (p^2-p)t^3 + p^2 t^4}{(1-t^2)(1-p^2 t^2)},$$

in which an additional factor $1 - p^2 t^2$ appears in the denominator.

418

References

[1] A.N. Andrianov, Euler products that correspond to Siegel's modular forms of genus 2, *Russian Math. Surveys* **29**:3(1974), 45–116.

[2] E. Artin, Quadratische Körper im Gebiete der höheren Kongruenzen, I,II, *Math. Zeit*, **19** 153–246 (1924)

[3] J.H. Conway & N.J.A. Sloane, *Sphere packings, lattices and groups*, Springer-Verlag, 1993.

[4] Ch. Deninger, Motivic *L*-functions and regularized determinants, *Proc. Sympos. Pure Math*, **55** (1), AMS, (1994), 707–743.

[5] U.V. Desale & S. Ramanan, Poincaré polynomials of the variety of stable bundles, *Math. Ann* **216**, 233–244 (1975).

[6] G. van der Geer & R. Schoof, Effectivity of Arakelov Divisors and the Theta Divisor of a Number Field, math.AG/9802121

[7] D.R. Grayson, Reduction theory using semistability. *Comment. Math. Helv.* **59** (1984), no. 4, 600–634.

[8] D.R. Grayson, Reduction theory using semistability. II, *Comment. Math. Helv.* **61** (1986), no. 4, 661–676.

[9] G. Harder & M.S. Narasimhan, On the cohomology groups of moduli spaces of vector bundles over curves, *Math Ann.* **212** (1975), 215–248.

[10] H. Hasse, *Mathematische Abhandlungen*, Walter de Gruyter, 1975.

[11] K. Iwasawa, Letter to Dieudonné, April 8, 1952, in *Advanced Studies in Pure Math.* **21** (1992), 445-450.

[12] W. Kohnen, this proceeding.

[13] S. Lang, *Algebraic Number Theory*, Springer-Verlag, 1986.

[14] S. Lang, *Fundamentals on Diophantine Geometry*, Springer-Verlag, 1983.

[15] A. Moriwaki, Stable sheaves on arithmetic curves, personal note dated in 1992.

[16] D. Mumford, *Geometric Invariant Theory*, Springer-Verlag, 1965.

[17] J. Neukirch, *Algebraic Number Theory*, Springer-Verlag, 1999.

[18] C. S. Seshadri, Fibrés vectoriels sur les courbes algébriques, *Asterisque* **96**, 1982.

[19] U. Stuhler, Eine Bemerkung zur Reduktionstheorie quadratischer Formen, *Arch. Math.* (Basel) **27** (1976), no. 6, 604–610.

[20] U. Stuhler, Zur Reduktionstheorie der positiven quadratischen Formen. II, *Arch. Math.* (Basel) **28** (1977), no. 6, 611–619.

[21] J. Tate, Fourier analysis in number fields and Hecke's zeta functions, Thesis, Princeton University, 1950

[22] A. Weil, *Sur les courbes algébriques et les variétés qui s'en déduisent*, Herman, 1948

[23] L. Weng, Riemann-Roch theorem, stability and new zeta functions for number fields, math.AG/0007146

[24] L. Weng, Constructions of new non-abelian zeta functions for curves, math.AG/0102064

[25] L. Weng, Refined Brill-Noether locus and non-abelian zeta functions for elliptic curves, math.AG/0101183

[26] L. Weng, A Program for Geometric Arithmetic, math.AG/0111241

[27] L. Weng, A note on arithmetic cohomologies for number fields, math.AG/0112164

A HYBRID MEAN VALUE OF L-FUNCTIONS AND GENERAL QUADRATIC GAUSS SUMS*

Zhang Wenpeng

Department of Mathematics, Northwest University
Xi'an, Shaanxi, P.R.China

wpzhang@nwu.edu.cn

Abstract   The main purpose of this paper is using the estimates for character sums and the analytic method to study the $2k$-th power mean of Dirichlet L-functions with the weight of general quadratic Gauss sums, and give an interesting asymptotic formula.

Keywords: General quadratic Gauss sums; L-functions; Asymptotic formula.

1. Introduction

Let $q \geq 2$ be an integer, χ denotes a Dirichlet character modulo q. For any integer n, we define the general quadratic Gauss sums $G(n, \chi; q)$ as follows:

$$G(n, \chi; q) = \sum_{a=1}^{q} \chi(a) e\left(\frac{na^2}{q}\right), \tag{1}$$

where $e(y) = e^{2\pi i y}$. This summation is more important, because it is a generalization of the classical Gauss sums and quadratic Gauss sums. But about the properties of $G(n, \chi; q)$, we know very little at present, even if we do not know how large $|G(n, \chi; q)|$ is. The main purpose of this paper is using the estimates for character sums and the analytic method to study the asymptotic distribution of the $2k$-th power mean

$$\sum_{\chi \neq \chi_0} |G(n, \chi; p)|^4 \cdot |L(1, \chi)|^{2k}, \tag{2}$$

and give a sharper asymptotic formula, where p is an odd prime, $L(s, \chi)$ denotes the Dirichlet L-function corresponding to character $\chi \bmod p$. In fact, we shall prove the following:

*This work is supported by the N.S.F. and the P.N.S.F. of P.R.China

Theorem 1. *Let p be an odd prime. Then for any fixed positive integer k and any integer n with $(p, n) = 1$, we have the asymptotic formula*

$$\sum_{\chi \neq \chi_0} |G(n, \chi; p)|^4 \cdot |L(1, \chi)|^{2k} = \frac{\pi^{4k-2}}{2 \cdot 6^{2k-2}} \cdot p^3 \cdot \prod_{p_1} \left(1 - \frac{1 - C_{2k-2}^{k-1}}{p_1^2}\right) + O\left(p^{\frac{5}{2}+\epsilon}\right),$$

where $\sum\limits_{\chi \neq \chi_0}$ *denotes the summation over all non-principal characters modulo p,* $\prod\limits_{p_1}$ *denotes the product over all prime p_1, ϵ denotes any fixed positive number and* $C_m^n = \frac{m!}{n!(m-n)!}$.

Theorem 2. *Let p be a prime with $p \equiv 3 \mod 4$. Then for any fixed positive integer k and any integer n with $(p, n) = 1$, we have the asymptotic formula*

$$\sum_{\chi \neq \chi_0} |G(n, \chi; p)|^6 \cdot |L(1, \chi)|^{2k} = \frac{5}{3} \cdot \frac{\pi^{4k-2}}{6^{2k-2}} \cdot p^4 \cdot \prod_{p_1} \left(1 - \frac{1 - C_{2k-2}^{k-1}}{p_1^2}\right) + O\left(p^{\frac{7}{2}+\epsilon}\right).$$

From these two theorems we may immediately deduce the following corollaries:

Corollary 1. *For any odd prime p and integer n with $(p, n) = 1$, we have the asymptotic formula*

$$\sum_{\chi \neq \chi_0} |G(n, \chi; p)|^4 \cdot |L(1, \chi)|^4 = \frac{5}{24} \cdot \pi^4 \cdot p^3 + O\left(p^{\frac{5}{2}+\epsilon}\right).$$

Corollary 2. *For any odd prime $p \equiv 3 \mod 4$ and integer n with $(p, n) = 1$, we have the asymptotic formula*

$$\sum_{\chi \neq \chi_0} |G(n, \chi; p)|^6 \cdot |L(1, \chi)|^4 = \frac{25}{36} \cdot \pi^4 \cdot p^4 + O\left(p^{\frac{7}{2}+\epsilon}\right).$$

2. Some lemmas

In order to complete the proof of the theorems, we need following several lemmas.

Lemma 1. *Let $f(x)$ be a k-th degree polynomial, a_0 be the coefficient of k-th term, and Δ denote the difference operator. Then we have*

$$\Delta^k f(x) = k!a_0, \qquad \Delta^l f(x) = 0 \qquad (l \geq k + 1)$$

In particular, for $f(n) = \left(C^n_{k+n-1}\right)^2$, we get

$$\Delta^{2k-2} f(n) = C^{k-1}_{2k-2}.$$

Proof. This result can be easily proved by the properties of the difference operator and mathematical induction.   □

Lemma 2. *Let q be an integer with $q \geq 3$, $d_k(n)$ denote the k-th divisor function. Then for any complex variable s with $\mathrm{Re}\,(s) > 1$, we have*

$$\sideset{}{'}\sum_{n=1}^{\infty} \frac{d_k^2(n)}{n^s} = \zeta^{2k-1}(s) \prod_{p \mid q} \left(1 - \frac{1}{p^s}\right)^{2k-1} \prod_{p \nmid q} \left(1 - \frac{1 - C^{k-1}_{2k-2}}{p^s}\right),$$

where $\sum'$ denotes the summation over all positive integers n with $(n, q) = 1$, $\zeta(s)$ is the Riemann zeta-function.

Proof. It is clear that $d_k^2(n)$ is a multiplicative function and the series $\sideset{}{'}\sum_{n=1}^{\infty} \frac{d_k^2(n)}{n^s}$ is absolutely convergent. Thus from Euler's infinite product representation we have

$$\sideset{}{'}\sum_{n=1}^{\infty} \frac{d_k^2(n)}{n^s} = \prod_{p \nmid q} \left(1 + \frac{d_k^2(p)}{p^s} + \frac{d_k^2(p^2)}{p^{2s}} + \cdots + \frac{d_k^2(p^n)}{p^{ns}} + \cdots\right).$$

For $n > 1$, let $n = p_1^{m_1} \cdots p_r^{m_r}$ denotes the factorization of n into prime powers, then (See reference [5], 6.4.12)

$$d_k(n) = \prod_{j=1}^{r} C^{m_j}_{k+m_j-1},$$

Hence

$$\sideset{}{'}\sum_{n=1}^{\infty} \frac{d_k^2(n)}{n^s} = \prod_{p \nmid q} \left(1 + \frac{(C_k^1)^2}{p^s} + \frac{(C_{k+1}^2)^2}{p^{2s}} + \cdots + \frac{(C_{k+n-1}^n)^2}{p^{ns}} + \cdots\right).$$

Let

$$S = 1 + \frac{(C_k^1)^2}{p^s} + \frac{(C_{k+1}^2)^2}{p^{2s}} + \cdots + \frac{(C_{k+n-1}^n)^2}{p^{ns}} + \cdots$$

then from Lemma 1 we get

$$S \left(1 - \frac{1}{p^s}\right)^{2k-2} = 1 + \frac{C^{k-1}_{2k-2}}{p^s} + \frac{C^{k-1}_{2k-2}}{p^{2s}} + \cdots + \frac{C^{k-1}_{2k-2}}{p^{ns}} + \cdots$$

$$= 1 + \frac{C^{k-1}_{2k-2}}{p^s} \times \frac{1}{1 - \frac{1}{p^s}},$$

so that

$$S = \frac{1 - \frac{1 - C_{2k-1}^{k-1}}{p^s}}{\left(1 - \frac{1}{p^s}\right)^{2k-1}}.$$

Note that $\zeta(s) = \prod_p \left(1 - \frac{1}{p^s}\right)^{-1}$, hence we have

$$\sum_{n=1}^{\infty}{}' \frac{d_k{}^2(n)}{n^s} = \zeta^{2k-1}(s) \prod_{p|q}\left(1 - \frac{1}{p^s}\right)^{2k-1} \prod_{p\nmid q}\left(1 - \frac{1 - C_{2k-2}^{k-1}}{p^s}\right).$$

This proves Lemma 2. $\qquad\square$

Lemma 3. *Let q be an integer with $q \geq 3$, χ denotes a Dirichlet character modulo q, and write $A(y,\chi) = A(y,\chi,k) = \sum_{N<n\leq y} \chi(n)d_k(n)$. Then for any integer $k \geq 1$, we have the estimate*

$$\sum_{\chi\neq\chi_0} |A(y,\chi)|^2 \ll y^{2-\frac{4}{2^k}+\epsilon}\phi^2(q),$$

where $\phi(n)$ is the Euler function, ϵ denotes any fixed positive number.

Proof. For any positive numbers a and b with $ab = X$, note that

$$\sum_{mn\leq X} f(m)g(n) = \sum_{m\leq a} f(m) \sum_{n\leq X/a} g(n) +$$

$$\sum_{n\leq b} g(n) \sum_{m\leq X/b} f(m) - \left(\sum_{m\leq a} f(m)\right) \cdot \left(\sum_{n\leq b} g(n)\right)$$

(See Theorem 3.17 of [1]), from this partition identity, the estimate for character sums

$$\sum_{\chi\neq\chi_0} \left|\sum_{N\leq n\leq y} \chi(n)\right|^2 \leq \sum_{N\leq n\leq y\leq N+q} \sum_{N\leq m\leq y\leq N+q} \sum_{\chi\neq\chi_0} \chi(n)\overline{\chi}(m) \ll \phi^2(q)$$

(This is the case in Lemma 3 with $k = 1$) and mathematical induction (on k) we can easily deduce Lemma 3. $\qquad\square$

Lemma 4. *Let p be an odd prime, χ denote a Dirichlet character modulo p. Then we have the estimate*

$$\sum_{r=1}^{p-1} \left|\sum_{\chi\neq\chi_0} \chi(r)|L(1,\chi)|^{2k}\right| = O\left(p^{1+\epsilon}\right).$$

Proof. For convenience, firstly we let

$$A(\chi, y) = \sum_{\frac{p}{r} < n \le y} \chi(n) d_k(n), \qquad B(\chi, y) = \sum_{p < n \le y} \chi(n) d_k(n).$$

Then applying the Abel's identity we have

$$L^k(1, \chi) = \sum_{1 \le n \le \frac{p}{r}} \frac{\chi(n) d_k(n)}{n} + \int_{\frac{p}{r}}^{+\infty} \frac{A(\chi, y)}{y^2} dy$$

$$= \sum_{n=1}^{p} \frac{\chi(n) d_k(n)}{n} + \int_{p}^{+\infty} \frac{B(\chi, y)}{y^2} dy$$

Hence from the definition of Dirichlet L-function we get

$$\sum_{\chi \ne \chi_0} \chi(r) |L(1, \chi)|^{2k}$$

$$= \sum_{\chi \ne \chi_0} \chi(r) \left| \sum_{n_1=1}^{\infty} \cdots \sum_{n_k=1}^{\infty} \frac{\chi(n_1 \cdots n_k)}{n_1 \cdots n_k} \right|^2$$

$$= \sum_{\chi \ne \chi_0} \chi(r) \left| \sum_{n=1}^{\infty} \frac{\chi(n) d_k(n)}{n} \right|^2$$

$$= \sum_{\chi \ne \chi_0} \chi(r) \left(\sum_{1 \le n \le \frac{p}{r}} \frac{\chi(n) d_k(n)}{n} + \int_{\frac{p}{r}}^{+\infty} \frac{A(\chi, y)}{y^2} dy \right)$$

$$\times \left(\sum_{m=1}^{p} \frac{\overline{\chi}(m) d_k(m)}{m} + \int_{p}^{+\infty} \frac{B(\overline{\chi}, y)}{y^2} dy \right)$$

$$= \sum_{\chi \ne \chi_0} \chi(r) \left(\sum_{1 \le n \le \frac{p}{r}} \frac{\chi(n) d_k(n)}{n} \right) \left(\sum_{m=1}^{p} \frac{\overline{\chi}(m) d_k(m)}{m} \right)$$

$$+ \sum_{\chi \ne \chi_0} \chi(r) \left(\sum_{1 \le n \le \frac{p}{r}} \frac{\chi(n) d_k(n)}{n} \right) \left(\int_{p}^{+\infty} \frac{B(\overline{\chi}, y)}{y^2} dy \right)$$

$$+ \sum_{\chi \ne \chi_0} \chi(r) \left(\sum_{m=1}^{p} \frac{\overline{\chi}(m) d_k(m)}{m} \right) \left(\int_{\frac{p}{r}}^{+\infty} \frac{A(\chi, y)}{y^2} dy \right)$$

$$+ \sum_{\chi \ne \chi_0} \chi(r) \left(\int_{\frac{p}{r}}^{+\infty} \frac{A(\chi, y)}{y^2} dy \right) \left(\int_{p}^{+\infty} \frac{B(\overline{\chi}, y)}{y^2} dy \right)$$

$$\equiv M_1 + M_2 + M_3 + M_4,$$

so that

$$\sum_{r=1}^{p-1}\left|\sum_{\chi\neq\chi_0}\chi(r)\,|L(1,\chi)|^{2k}\right|\leq\sum_{r=1}^{p-1}\left(|M_1|+|M_2|+|M_3|+|M_4|\right).\qquad(3)$$

Now we shall estimate each term in expression (3). From the orthogonality relationship for character sums modulo p we know that for $(p,mn)=1$,

$$\sum_{\chi\ \mathrm{mod}\ p}\chi(n)\overline{\chi}(m)=\begin{cases}p-1,&\text{if }n\equiv m\mod p;\\0,&\text{otherwise.}\end{cases}$$

From this identity we can easily get

$$M_1=\sum_{\chi\neq\chi_0}\chi(r)\left(\sum_{n\leq\frac{p}{r}}\frac{\chi(n)d_k(n)}{n}\right)\left(\sum_{m\leq p}\frac{\overline{\chi}(m)d_k(m)}{m}\right)$$

$$=\sum_{\chi\bmod p}\chi(r)\left(\sum_{n\leq\frac{p}{r}}\frac{\chi(n)d_k(n)}{n}\right)\left(\sum_{m\leq p}\frac{\overline{\chi}(m)d_k(m)}{m}\right)$$

$$-\sum_{n\leq\frac{p}{r}}\frac{d_k(n)}{n}\sum_{m\leq p-1}\frac{d_k(m)}{m}$$

$$=(p-1)\sum_{\substack{n\leq\frac{p}{r}\ m\leq p-1\\rn\equiv m(p)}}\frac{d_k(n)d_k(m)}{nm}+O\left(p^\epsilon\right)$$

$$=(p-1)\sum_{n\leq\frac{p}{r}}\frac{d_k(n)d_k(rn)}{rn^2}+O\left(p^\epsilon\right)$$

$$\leq(p-1)\frac{d_k(r)}{r}\sum_{n\leq\frac{p}{r}}\frac{|d_k(n)|^2}{n^2}+O\left(p^\epsilon\right)$$

$$\leq(p-1)\frac{d_k(r)}{r}\sum_{\substack{n=1\\(n,p)=1}}^{\infty}\frac{|d_k(n)|^2}{n^2}+O\left(p^\epsilon\right).\qquad(4)$$

So from (4) we immediately obtain

$$\sum_{r=1}^{p-1}|M_1|=O\left(p\sum_{r=1}^{p-1}\frac{d_k(r)}{r}\right)+O\left(p^\epsilon\sum_{r=1}^{p-1}1\right)=O(p^{1+\epsilon}),\qquad(5)$$

where we have used the estimate $d_k(n) \ll n^{\epsilon_1}$.

$$\sum_{r=1}^{p-1} |M_2| = \sum_{r=1}^{p-1} \left| \sum_{\chi \neq \chi_0} \chi(r) \left(\sum_{n \leq \frac{p}{r}} \frac{\chi(n)d_k(n)}{n} \right) \left(\int_p^{+\infty} \frac{B(\overline{\chi}, y)}{y^2} dy \right) \right|$$

$$\leq \sum_{r=1}^{p-1} \left| \sum_{\chi \neq \chi_0} \chi(r) \left(\sum_{n \leq \frac{p}{r}} \frac{\chi(n)d_k(n)}{n} \right) \left(\int_p^{p \cdot 2^{k-2}} \frac{B(\overline{\chi}, y)}{y^2} dy \right) \right|$$

$$+ \sum_{r=1}^{p-1} \left| \sum_{\chi \neq \chi_0} \chi(r) \left(\sum_{n \leq \frac{p}{r}} \frac{\chi(n)d_k(n)}{n} \right) \left(\int_{p^{3 \cdot 2^{k-2}}}^{+\infty} \frac{B(\overline{\chi}, y)}{y^2} dy \right) \right|$$

$$\leq \sum_{r=1}^{p-1} \left| \sum_{\chi \bmod p} \chi(r) \left(\sum_{n \leq \frac{p}{r}} \frac{\chi(n)d_k(n)}{n} \right) \left(\int_p^{p^{3 \cdot 2^{k-2}}} \frac{\sum_{p<m\leq y} \overline{\chi}(m)d_k(m)}{y^2} dy \right) \right|$$

$$+ \sum_{r=1}^{p-1} \left(\sum_{n \leq \frac{p}{r}} \frac{d_k(n)}{n} \right) \left(\int_p^{p^{3 \cdot 2^{k-2}}} \frac{\sum_{p<m\leq y} d_k(m)}{y^2} dy \right)$$

$$+ p^\epsilon \sum_{r=1}^{p-1} \int_{p^{3 \cdot 2^{k-1}}}^{+\infty} \frac{1}{y^2} \sum_{\chi \neq \chi_0} |B(\overline{\chi}, y)| dy. \qquad (6)$$

Using the Cauchy inequality and Lemma 3 we immediately get the estimate

$$\sum_{\chi \neq \chi_0} |B(\overline{\chi}, y)| \leq \phi^{\frac{1}{2}}(p) \left(\sum_{\chi \neq \chi_0} |B(\overline{\chi}, y)|^2 \right)^{\frac{1}{2}} \ll p^{\frac{3}{2}} y^{1 - \frac{2}{2^k} + \epsilon}. \qquad (7)$$

So from (6) and (7) we have

$$\sum_{r=1}^{p-1} |M_2| \leq (p-1) \int_p^{p^{3 \cdot 2^{k-2}}} \frac{1}{y^2} \left[\sum_{r=1}^{p-1} \sum_{n \leq \frac{p}{r}} \sum_{\substack{p<m\leq y \\ rn \equiv m(p)}} \frac{d_k(n)d_k(m)}{n} \right] dy$$

$$+ O \left(p^{\frac{5}{2}+\epsilon} \int_{p^{3 \cdot 2^{k-2}}}^{+\infty} y^{-1 - \frac{2}{2^k} + \epsilon} dy \right) + O \left(p^{1+\epsilon} \right)$$

$$= O \left(p \sum_{r=1}^{p-1} \sum_{n \leq \frac{p}{r}} \frac{d_k(n)}{n} \int_p^{p^{3 \cdot 2^{k-2}}} \frac{y \cdot \frac{1}{q} \cdot p^\epsilon}{y^2} dy \right) + O \left(p^{1+\epsilon} \right)$$

$$= O \left(p^{1+\epsilon} \right). \qquad (8)$$

Similarly, we also have

$$
\sum_{r=1}^{p-1} |M_3| = \sum_{r=1}^{p-1} \left| \sum_{\chi \neq \chi_0} \chi(r) \left(\sum_{m \leq p} \frac{\overline{\chi}(m) d_k(m)}{m} \right) \left(\int_{\frac{p}{r}}^{+\infty} \frac{A(\chi, y)}{y^2} dy \right) \right|
$$

$$
\leq \sum_{r=1}^{p-1} \left| \sum_{\chi \neq \chi_0} \chi(r) \left(\sum_{m \leq p} \frac{\overline{\chi}(m) d_k(m)}{m} \right) \left(\int_{\frac{p}{r}}^{p^{3 \cdot 2^{k-2}}} \frac{A(\chi, y)}{y^2} dy \right) \right|
$$

$$
+ \sum_{r=1}^{p-1} \left| \sum_{\chi \neq \chi_0} \chi(r) \left(\sum_{m \leq p} \frac{\chi(m) d_k(m)}{m} \right) \left(\int_{p^{3 \cdot 2^{k-2}}}^{+\infty} \frac{A(\chi, y)}{y^2} dy \right) \right|
$$

$$
\leq \sum_{r=1}^{p-1} \left| \sum_{\chi \bmod p} \chi(r) \left(\sum_{m \leq p} \frac{\chi(m) d_k(m)}{m} \right) \left(\int_{\frac{p}{r}}^{p^{3 \cdot 2^{k-2}}} \frac{\sum_{\frac{p}{r} < n \leq y} \chi(n) d_k(n)}{y^2} dy \right) \right|
$$

$$
+ \sum_{r=1}^{p-1} \left(\sum_{m \leq p} \frac{d_k(m)}{m} \right) \left(\int_{\frac{p}{r}}^{p^{3 \cdot 2^{k-2}}} \frac{\sum_{\frac{p}{r} < n \leq y} d_k(n)}{y^2} dy \right)
$$

$$
+ p^\epsilon \sum_{r=1}^{p-1} \int_{p^{3 \cdot 2^{k-2}}}^{+\infty} \frac{1}{y^2} \sum_{\chi \neq \chi_0} |A(\chi, y)| dy
$$

$$
\ll p \sum_{r=1}^{p-1} \int_{\frac{p}{r}}^{p^{3 \cdot 2^{k-2}}} \frac{1}{y^2} \left[\sum_{\substack{m \leq p \\ rn \equiv m(p)}} \sum_{\frac{p}{r} < n \leq y} \frac{d_k(m) d_k(n)}{m} \right] dy + p^{1+\epsilon}
$$

$$
\ll p \sum_{r=1}^{p-1} \int_p^{p^{3 \cdot 2^{k-1}}} \frac{1}{y^2} \left[\sum_{\substack{m \leq p \\ rn \equiv m(q)}} \sum_{\frac{p}{r} < n \leq y} \frac{d_k(m) d_k(n)}{m} \right] dy + p^{1+\epsilon}
$$

$$
+ p \sum_{r=1}^{p-1} \int_1^p \frac{1}{y^2} \left[\sum_{\substack{m \leq p \\ rn \equiv m(p)}} \sum_{\frac{p}{r} \leq n \leq y} \frac{d_k(m) d_k(n)}{m} \right] dy
$$

$$
\ll p \sum_{r=1}^{p-1} \sum_{m \leq p} \frac{d_k(m)}{m} \int_p^{p^{3 \cdot 2^{k-2}}} \frac{y \cdot \frac{1}{p} \cdot p^\epsilon}{y^2} dy + p^{1+\epsilon} + p \sum_{m \leq p} \frac{d_k(m)}{m} \int_1^p \frac{y \cdot p^\epsilon}{y^2} dy
$$

$$
\ll p^{1+\epsilon}. \tag{9}
$$

where we have used the estimate $d_k(n) \ll n^{\epsilon_1}$, and for any fixed positive integers l and m, the number of the solutions of equation $rn = lp + m$ (for r and n) is $d_k(lp + m) \ll p^\epsilon$.

Using the same method we can also get the estimate

$$\sum_{r=1}^{p-1} \left| \sum_{\chi \neq \chi_0} \chi(r) \left(\int_{\frac{p}{r}}^{+\infty} \frac{A(\chi, y)}{y^2} dy \right) \left(\int_{p}^{+\infty} \frac{B(\overline{\chi}, y)}{y^2} dy \right) \right| = O(p^{1+\epsilon}).$$

(10)

Combining (3), (5), (8), (9) and (10) we obtain

$$\sum_{r=1}^{p-1} \left| \sum_{\chi \neq \chi_0} \chi(r) |L(1, \chi)|^{2k} \right| = O(p^{1+\epsilon}).$$

This proves Lemma 4. $\qquad\qquad\qquad\qquad\qquad\qquad\qquad\qquad\qquad\qquad\qquad\square$

Lemma 5. *Let p be an odd prime. Then we have the asymptotic formula*

$$\sum_{\substack{\chi \neq \chi_0 \\ \chi(-1)=1}} |L(1, \chi)|^{2k} = \frac{1}{2} \cdot \zeta^{2k-1}(2) \cdot \prod_{p_1} \left(1 - \frac{1 - C_{2k-2}^{k-1}}{p_1^2} \right) + O(p^\epsilon).$$

Proof. From Lemma 2 and the result of [4] we can easily deduce this asymptotic formula. $\qquad\qquad\qquad\qquad\qquad\qquad\qquad\qquad\qquad\qquad\square$

Lemma 6. *For any integer $q \geq 1$, we have the calculation formula*

$$G(1; q) = \frac{1}{2}\sqrt{q}(1 + i)\left(1 + e^{\frac{-\pi i q}{2}} \right) = \begin{cases} \sqrt{q} & \text{if } q \equiv 1 \ (\mod 4); \\ 0 & \text{if } q \equiv 2 \ (\mod 4); \\ i\sqrt{q} & \text{if } q \equiv 3 \ (\mod 4); \\ (1 + i)\sqrt{q} & \text{if } q \equiv 0 \ (\mod 4). \end{cases}$$

where $G(1; q) = \sum_{a=1}^{q} e\left(\frac{a^2}{q} \right)$.

Proof. This is a remarkable formula of Gauss. See Theorem 9.16 of [1]. $\qquad\qquad\qquad\qquad\qquad\qquad\qquad\qquad\qquad\qquad\qquad\qquad\qquad\square$

Lemma 7. *Let p be an odd prime, χ be any nonprincipal even character (i.e. $\chi(-1) = 1$ and $\chi \neq \chi_0$) mod p. Then for any integer n with $(n, p) = 1$, we have the identity*

$$|G(n, \chi; p)|^2 = 2p + \left(\frac{n}{p} \right) G(1; p) \sum_{a=1}^{p-1} \chi(a) \left(\frac{a^2 - 1}{p} \right),$$

where $\left(\frac{n}{p}\right)$ is the Legendre symbol.

Proof. We first proof that if $p \nmid n$, then

$$G(n;p) = \left(\frac{n}{p}\right) G(1;p). \tag{11}$$

In fact if n is a quadratic residue modulo p, then $\left(\frac{n}{p}\right) = 1$, and there exists at least one integer c with $p \nmid c$ such that $c^2 \equiv n \pmod{p}$. It is clear that if $\{a_1, a_2, \cdots, a_p\}$ is a complete residue system modulo p, then $\{ca_1, ca_2, \cdots, ca_p\}$ is also a complete residue system modulo p. So for $p \nmid n$ and $c^2 \equiv n \pmod{p}$, we have

$$G(n;p) = \sum_{a=1}^{p} e\left(\frac{c^2 a^2}{p}\right) = \sum_{a=1}^{p} e\left(\frac{(ca)^2}{p}\right) = \sum_{a=1}^{p} e\left(\frac{a^2}{p}\right) = \left(\frac{n}{p}\right) G(1;p).$$

If n is a quadratic non-residue modulo p, then $\left(\frac{n}{p}\right) = -1$, and $n \cdot 1^2$, $n \cdot 2^2, \ldots, n \cdot \left(\frac{p-1}{2}\right)^2$ are all quadratic non-residues modulo p. So we have

$$G(n;p) = 1 + 2 \sum_{a=1}^{\frac{p-1}{2}} e\left(\frac{na^2}{p}\right)$$

$$= 1 + 2 \left[\sum_{a=1}^{\frac{p-1}{2}} e\left(\frac{na^2}{p}\right) + \sum_{a=1}^{\frac{p-1}{2}} e\left(\frac{a^2}{p}\right)\right] - 2 \sum_{a=1}^{\frac{p-1}{2}} e\left(\frac{a^2}{p}\right)$$

$$= 1 + 2 \sum_{a=1}^{p-1} e\left(\frac{a}{p}\right) - \sum_{a=1}^{p-1} e\left(\frac{a^2}{p}\right)$$

$$= 2 \sum_{a=1}^{p} e\left(\frac{a}{p}\right) - \sum_{a=1}^{p} e\left(\frac{a^2}{p}\right) = -G(1;p) = \left(\frac{n}{p}\right) G(1;p).$$

So for any $p \nmid n$, (11) is correct.

From (11) we know that if χ is a non-principal even character mod p, then

$$|G(n,\chi;p)|^2 = \sum_{a=1}^{p-1}\sum_{b=1}^{p-1}\chi(a)\overline{\chi}(b)e\left(\frac{na^2-nb^2}{p}\right) = \sum_{a=1}^{p-1}\sum_{b=1}^{p-1}\chi(a\overline{b})e\left(\frac{n(a^2-b^2)}{p}\right)$$

$$= \sum_{a=1}^{p-1}\sum_{b=1}^{p-1}\chi(a)e\left(\frac{nb^2(a^2-1)}{p}\right) = \sum_{a=1}^{p-1}\chi(a)\left[\sum_{b=1}^{p}e\left(\frac{nb^2(a^2-1)}{p}\right)-1\right]$$

$$= 2p + \sum_{a=2}^{p-2}\chi(a)G(n(a^2-1);p) - \sum_{a=1}^{p-1}\chi(a)$$

$$= 2p + \left(\frac{n}{p}\right)G(1;p)\sum_{a=1}^{p-1}\chi(a)\left(\frac{a^2-1}{p}\right).$$

This proves Lemma 7. $\qquad\qquad\qquad\qquad\qquad\qquad\qquad\qquad\qquad\qquad\square$

3.   Proof of the theorems

In this section, we shall complete the proof of the theorems. First note that if χ is an odd character modulo p, then

$$G(n,\chi;p) = \sum_{a=1}^{p}\chi(a)e\left(\frac{a^2n}{p}\right) = 0.$$

So for any integer n with $(p,n) = 1$, from Lemma 7 we have

$$\sum_{\chi\neq\chi_0}|G(n,\chi;p)|^4\cdot|L(1,\chi)|^{2k} = \sum_{\substack{\chi\neq\chi_0\\\chi(-1)=1}}|G(n,\chi;p)|^4\cdot|L(1,\chi)|^{2k}$$

$$= \sum_{\substack{\chi\neq\chi_0\\\chi(-1)=1}}\left[2p+\left(\frac{n}{p}\right)G(1;p)\sum_{a=1}^{p-1}\chi(a)\left(\frac{a^2-1}{p}\right)\right]^2\cdot|L(1,\chi)|^{2k}$$

$$= \sum_{\substack{\chi\neq\chi_0\\\chi(-1)=1}}\left[4p^2+4p\left(\frac{n}{p}\right)G(1;p)\sum_{a=1}^{p-1}\chi(a)\left(\frac{a^2-1}{p}\right)\right]\cdot|L(1,\chi)|^{2k}$$

$$+ G^2(1;p)\sum_{\substack{\chi\neq\chi_0\\\chi(-1)=1}}\sum_{a=1}^{p-1}\sum_{b=1}^{p-1}\chi(ab)\left(\frac{a^2-1}{p}\right)\left(\frac{b^2-1}{p}\right)\cdot|L(1,\chi)|^{2k}.$$

$$(12)$$

Note that the identity

$$\sum_{a=1}^{p-1}\sum_{b=1}^{p-1}\chi(ab)\left(\frac{a^2-1}{p}\right)\left(\frac{b^2-1}{p}\right)$$

$$=\sum_{a=1}^{p-1}\sum_{b=1}^{p-1}\chi(a)\left(\frac{a^2\overline{b}^2-1}{p}\right)\left(\frac{b^2-1}{p}\right)$$

$$=\sum_{a=1}^{p-1}\sum_{b=1}^{p-1}\chi(a)\left(\frac{a^2-b^2}{p}\right)\left(\frac{b^2-1}{p}\right)$$

$$=2\left(\frac{-1}{p}\right)(p-3)+\sum_{a=2}^{p-2}\sum_{b=1}^{p-1}\chi(a)\left(\frac{a^2-b^2}{p}\right)\left(\frac{b^2-1}{p}\right)\qquad(13)$$

and the estimate

$$\sum_{b=1}^{p-1}\left(\frac{b^2-a^2}{p}\right)\left(\frac{b^2-1}{p}\right)\ll\sqrt{p},\quad a^2\not\equiv1\mod p\qquad(14)$$

(This estimate can be deduced by the lemma 1 of [6]), from (12), (13), (14), Lemma 4, 5 and 6 we have

$$\sum_{\chi\neq\chi_0}|G(n,\chi;p)|^4\cdot|L(1,\chi)|^{2k}=\sum_{\substack{\chi\neq\chi_0\\\chi(-1)=1}}|G(n,\chi;p)|^4\cdot|L(1,\chi)|^{2k}$$

$$=\left(4p^2+G^2(1;p)\cdot2\left(\frac{-1}{p}\right)(p-3)\right)\cdot\sum_{\substack{\chi\neq\chi_0\\\chi(-1)=1}}|L(1,\chi)|^{2k}$$

$$+4p\left(\frac{n}{p}\right)G(1;p)\sum_{a=1}^{p-1}\left(\frac{a^2-1}{p}\right)\sum_{\substack{\chi\neq\chi_0\\\chi(-1)=1}}\chi(a)|L(1,\chi)|^{2k}$$

$$+G^2(1;p)\sum_{a=2}^{p-2}\sum_{b=1}^{p-1}\left(\frac{a^2-b^2}{p}\right)\left(\frac{b^2-1}{p}\right)\sum_{\substack{\chi\neq\chi_0\\\chi(-1)=1}}\chi(a)|L(1,\chi)|^{2k}$$

$$=(6p^2-6p)\sum_{\substack{\chi\neq\chi_0\\\chi(-1)=1}}|L(1,\chi)|^{2k}+O\left(p^{\frac{3}{2}}\sum_{a=1}^{p-1}\left|\sum_{\chi\neq\chi_0}\chi(a)|L(1,\chi)|^{2k}\right|\right)$$

$$=\frac{\pi^{4k-2}}{2\cdot6^{2k-2}}\cdot p^3\cdot\prod_{p_1}\left(1-\frac{1-C_{2k-2}^{k-1}}{p^2}\right)+O(p^{\frac{5}{2}+\epsilon}).$$

This proves Theorem 1.

Now we prove Theorem 2. From Lemma 7 we have

$$\sum_{\substack{\chi \neq \chi_0}} |G(n,\chi;p)|^6 \cdot |L(1,\chi)|^{2k} = \sum_{\substack{\chi \neq \chi_0 \\ \chi(-1)=1}} |G(n,\chi;p)|^6 \cdot |L(1,\chi)|^{2k}$$

$$= \sum_{\substack{\chi \neq \chi_0 \\ \chi(-1)=1}} \left[2p + \left(\frac{n}{p}\right) G(1;p) \sum_{a=1}^{p-1} \chi(a) \left(\frac{a^2-1}{p}\right) \right]^3 \cdot |L(1,\chi)|^{2k} \quad (15)$$

and

$$\sum_{\substack{\chi \neq \chi_0}} |G(n,\chi;p)|^6 \cdot |L(1,\chi)|^{2k} = \sum_{\substack{\chi \neq \chi_0}} |\overline{G(n,\chi;p)}|^6 \cdot |\overline{L(1,\chi)}|^{2k}$$

$$= \sum_{\substack{\chi \neq \chi_0}} \left| \sum_{a=1}^{p-1} \overline{\chi}(a) e\left(\frac{-na^2}{p}\right) \right|^6 \cdot |L(1,\overline{\chi})|^{2k}$$

$$= \sum_{\substack{\chi \neq \chi_0}} \left| \sum_{a=1}^{p-1} \chi(a) e\left(\frac{-na^2}{p}\right) \right|^6 \cdot |L(1,\chi)|^{2k}$$

$$= \sum_{\substack{\chi \neq \chi_0 \\ \chi(-1)=1}} \left[2p + \left(\frac{-n}{p}\right) G(1;p) \sum_{a=1}^{p-1} \chi(a) \left(\frac{a^2-1}{p}\right) \right]^3 \cdot |L(1,\chi)|^{2k}. \quad (16)$$

Adding (15) and (16), and note that $\left(\frac{-1}{p}\right) = -1$, from (13), (14), Lemma 4, 5 and 6 we obtain

$$\sum_{\substack{\chi \neq \chi_0}} |G(n,\chi;p)|^6 \cdot |L(1,\chi)|^{2k}$$

$$= \frac{1}{2} \left[\sum_{\substack{\chi \neq \chi_0}} |G(n,\chi;p)|^6 \cdot |L(1,\chi)|^{2k} + \sum_{\substack{\chi \neq \chi_0}} |\overline{G(n,\chi;p)}|^6 \cdot |\overline{L(1,\chi)}|^{2k} \right]$$

$$= \sum_{\substack{\chi \neq \chi_0 \\ \chi(-1)=1}} \left[(2p)^3 + 3 \cdot 2p \cdot G^2(1;p) \left(\sum_{a=1}^{p-1} \chi(a) \left(\frac{a^2-1}{p}\right) \right)^2 \right] \cdot |L(1,\chi)|^{2k}$$

$$= \frac{5}{3} \cdot \frac{\pi^{4k-2}}{6^{2k-2}} \cdot p^4 \cdot \prod_{p_1} \left(1 - \frac{1 - C_{2k-2}^{k-1}}{p_1^2} \right) + O\left(p^{\frac{7}{2}+\epsilon} \right).$$

This completes the proof of Theorem 2.

434

Acknowledgments

The author express his gratitude to the referee for his very helpful and detailed comments.

References

[1] Tom M. Apostol, *Introduction to Analytic Number Theory*, Springer-Verlag, New York, 1976.

[2] Tom M. Apostol, *Modular Functions and Dirichlet Series in Number Theory*, Springer-Verlag, New York, 1976.

[3] H. Walum, An exact formula for an average of L-series, *Illinois J. Math.* **26** (1982), 1–3.

[4] Zhang Wenpeng, On the fourth power mean of Dirichlet L-functions, *Chinese Science Bulletin* **34** (1990), 647–650.

[5] Pan Chengdong and Pan Chengbiao, *Elements of the Analytic Number Theory*, Science Press, Beijing, 1991.

[6] D.A.Burgess, On character sums and L-series, *Proc. London Math. Soc.* **12** (1962), 193–206.

Index

436

Developments in Mathematics

1. Alladi et al. (eds.): *Analytic and Elementary Number Theory*. 1998
   ISBN 0-7923-8273-0
2. S. Kanemitsu and K. Györy (eds.): *Number Theory and Its Applications*. 1999
   ISBN 0-7923-5952-6
3. A. Blokhuis, J.W.P. Hirschfeld, D. Jungnickel and J.A. Thas (eds.): *Finite Geometries*. Proceedings of the Fourth Isle of Thorns Conference. 2001   ISBN 0-7923-6994-7
4. F.G. Garvan and M.E.H. Ismail (eds.): *Symbolic Computation, Number Theory, Special Functions, Physics and Combinatorics*. 2001   ISBN 1-4020-0101-0
5. S.C. Milne: *Infinite Families of Exact Sums of Squares Formulas, Jacobi Elliptic F. Continued Fractions, and Schur Functions*. 2002   ISBN 1-4020-0491-5
6. C. Jia and K. Matsumoto (eds.): *Analytic Number Theory*. 2002
   ISBN 1-4020-0545-8
7. J. Martínez (ed.): *Ordered Algebraic Structures*. 2002   ISBN 1-4020-0752-3

KLUWER ACADEMIC PUBLISHERS – DORDRECHT / BOSTON / LONDON